INTRODUCTION TO CDMA WIRELESS COMMUNICATIONS

Introduction to CDMA Wireless Communications

Mosa Ali Abu-Rgheff

AMSTERDAM • BOSTON • HEIDELBERG • LONDON • NEW YORK • OXFORD
PARIS • SAN DIEGO • SAN FRANCISCO • SINGAPORE • SYDNEY • TOKYO
Academic Press is an imprint of Elsevier

Academic Press is an imprint of Elsevier
Linacre House, Jordan Hill, Oxford, OX2 8DP
84 Theobald's Road, London WC1X 8RR, UK
30 Corporate Drive, Burlington, MA 01803
525 B Street, Suite 1900, San Diego, California 92101-4495, USA

First edition 2007

British Library Cataloguing in Publication Data
A catalogue record for this book is available from the British Library

Library of Congress Cataloging-in-Publication Data
A catalog record for this book is available from the Library of Congress

ISBN 978-0-75-065252-0

For information on all Academic Press publications
visit our web site at books.elsevier.com

Typeset by Charon Tec Ltd (A Macmillan Company), Chennai, India
www.charontec.com

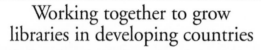

Contents

Preface

This book is based on experience in teaching digital and wireless communications to undergraduate and postgraduate students at the University of Plymouth for many years and on a series of industrial short courses on advanced digital communications. It seems to the author that there are few books available on the market that are suitable for teaching the subject with an in-depth knowledge of the up-to-date development in the CDMA wireless systems. Most of the currently available books are mainly aimed at engineers with prior knowledge of the subject through their work in industry or in research centres.

Therefore, this book provides the principles, standards and applications for the rapidly developing technology and is further enhanced by the author's own research into the topic over more than two decades. Such experience supports the presentation of complex theoretical concepts in a clear and easy to understand rendition. In addition, worked examples, drill exercises and laboratory sessions are included with the text and are used to consolidate the text's theoretical concepts. The book commences by presenting solid foundations in digital communications that are essential to an unequivocal understanding of the CDMA technology. In Chapters 3 through to 7 it guides the reader through the fundamentals and characteristics of cellular CDMA communications. Finally, Chapter 8 introduces the international standards adopted for the design of contemporary CDMA wireless systems.

Intended readers
The book is a highly detailed and thorough documentation of the present theories, system design, international standards, and cutting edge research in the CDMA wireless technology field, and should be of invaluable assistance to university students throughout their course as well as during their industrial employment.

The book serves as a text for teaching digital and wireless communications courses at senior undergraduate and Master levels. Research students studying for a PhD, and graduate engineers working in industry who wish to improve their expertise in the area, will also find the book a useful reference text.

Main features of the book
The book has the following important key features:

- Thorough description of the principle concepts of spread spectrum techniques and their applications in cellular wireless systems.
- Concise, accurate, clear and user-friendly readability style to provide an enjoyable experience to the reader.
- Provides worked examples, exercises and practical sessions based on industry-recognized software platform.

- Easy-to-follow descriptions of the air interface standards used in cdma2000 and 3G WCDMA systems.
- Clear description of the evolutions of the cdma2000.
- Clear description of the development of the 3G WCDMA for High Speed Downlink Packet Access (HSDPA).
- Clear description of the development of the 3G WCDMA for High Speed Uplink Packet Access (HSUPA).

Contents organization

The objective of the book is to provide comprehensive treatment of the direct-sequence spread spectrum techniques and its applications in cellular wireless communications. Although a strong emphasis is placed on the CDMA wireless communications, a review of the important digital communications foundations is used as a platform to explain CDMA principles. The main topics covered in the book are:

- Digital transmission and reception, review of the frequency domain analysis of signals and the characterization of noise random processes, transmission schemes and up-to-date techniques for forward error control coding such as Turbo coding (Chapters 1 and 2).
- Spread spectrum technique principles and spreading sequences (Chapters 3 and 4).
- A comprehensive treatment of the CDMA wireless systems (Chapters 5, 6 and 7).
- Up-to-date international standards of CDMA wireless systems (Chapter 8).

The book begins with a review of the development of wireless communications and guides the reader to the wireless communications that use CDMA technology. This review is followed by an introduction to the analysis and processing of communications signals corrupted by additive white Gaussian noise (Chapter 1). This chapter also reviews the Rayleigh statistical distributions commonly used to model wireless channels, and the chi-square distributions widely used in many signal processing schemes. It goes on to present signal processing tools that are commonly used in evaluating wireless systems performance such as Fast Fourier Transform (FFT), convolution, correlation and spectral density analysis.

The basic principles of digital communications used to explain the foundation of the spread spectrum techniques are introduced in Chapter 2. This chapter reviews the transmission of the N-dimensional signals and discusses the optimum receivers including raised cosine pulse shaping and matched filtering techniques. Channel equalization, such as zero forcing, minimum mean square error, and adaptive equalization, represent powerful techniques to reduce signal distortion. These schemes are introduced in the chapter together with a reassessment of the rake receivers. The commonly used channel coding schemes: convolutional codes and its Viterbi decoding algorithms, the MAP decoding algorithm for recursive systematic convolutional codes, and the powerful turbo coding and its decoding algorithms the max-Log MAP and the Log Map algorithms are also presented in this chapter. Two laboratory sessions based on Matlab software platform, namely: 'matched filtering' and 'signal equalization', are given. It is expected that these laboratory sessions, together with the worked examples, will considerably support the understanding of the theoretical presentations in this chapter.

Chapter 3 presents the most common types of spread spectrum systems and focuses on the direct sequence spread spectrum technique since this type of approach is widely used in contemporary CDMA wireless systems. The basic method of the spreading and despreading to generate the spreading gain that combats jamming is explained in this chapter, and the correlation functions between users' signature waveforms which impact the level of interference are described. The significance of the correlation functions is demonstrated by the fact that they not only have considerable influence on the generation of the multiple access interference but also affect the system time synchronization. These issues are discussed in more detail in this and following chapters. This chapter includes a laboratory session on the principles of the direct sequence spread spectrum systems to further enhance the reader's understanding of the systems considered in the book.

The generation, analysis and characterization of pseudo-random code sequences are considered in Chapter 4. This chapter embarks on a review of the basic algebra concepts when applied to the binary polynomials. This knowledge is then used in the generation and decimation of the maximum-length sequences (m-sequences). A new brand of spreading code sequences that have recently received great interest, namely the complex spreading codes, is presented in detail.

Chapter 5 considers the problem of time synchronization of spread spectrum systems and presents the analysis of the acquisition system in additive white Gaussian noise and the metrics that are used in the evaluation of the system. These metrics include mean acquisition time and probabilities of detection and false alarm. Once the acquisition is achieved, tracking loop is initiated to keep the system locked in synchronous harmony. Loops that are considered include the early-late tracking loops and the τ-dither loops operating in additive white Gaussian noise. The time synchronization of spread spectrum systems operating in mobile fading environments is also discussed and the contemporary research on the topic is summarized.

The cellular CDMA wireless systems normally operate through wideband radio channels and Chapter 6 considers the metrics that characterize such transmission operation. In addition, circumstances that degrade the performance of such systems, such as the near-far transmissions and multiple access interference, are considered in detail. The conventional single-user receiver operating in a multi-user channel is presented with discussion on its development. The system capacity for both the single cell system and multiple cells system is derived and the affect of the erroneous power control on the capacity is determined.

Chapter 7 deals with the problem of multi-user detection and begins with consideration of the optimum multi-user detection followed by a discussion of the sub-optimum detection systems such as decorrelator detection and minimum mean square error detection. The smart antennas and space diversity algorithms that improve the quality of the detection are also examined. Furthermore, the advantages of the beam forming methods and the system layered space–time architecture used in multi-user detection are demonstrated. An important issue that has a considerable affect on the quality of the detector is the multiple access interference, and various interference cancellation techniques are presented.

The international standards related to the CDMA wireless communications are appraised in Chapter 8 with emphasis being placed on the design of the air interface for systems, such as IS-95 A/B and cdma2000 including the later revisions. In addition, the standards of the Wideband CDMA used in the Universal Mobile Telecommunication Services (UMTS) and their revision for High Speed Downlink Packet Access (HSDPA) and the High Speed Uplink Packet Access (HSUPA) are explained.

The author would like to thank his colleagues in the Faculty of Technology. In particular, thanks are due to Professor Martin Tomlinson, Dr Mohammed Zaki Ahmed, Dr Adrian Ambroze from the fixed and mobile communication research group and to Dr David Walton for their contributions that have improved the text presentation. Thanks to my students who over many years have contributed to the clarity of the present text. Thanks are due to the many researchers whose contributions are being sited in the book. Without their contributions the completion of the book would not have been possible.

The author would also like to thank the editor Tim Pitts and the editorial assistant Kate Dennis for their help, patience and encouragement during the development of the book. Finally, the author is especially grateful to his family for their support throughout the preparation and completion of the book.

Mosa Abu-Rgheff

1

Introduction

1.1   Development of CDMA Wireless Communications

Wireless communication has made a huge leap since its first commercial service in the late 1970s and early 1980s. In the UK, the 1G service was provided by Total Access Communications Systems (TACS) in 1985. TACS standard is based upon an earlier Bell Labs system which was developed in the late 1970s and has been deployed in North America under the name Advanced Mobile Phone System (AMPS).

The first move toward digital wireless communications in Europe began in the early 1980s when the Conference of European Post and Telecommunications (CEPT) initiated the work for a new digital cellular standard which would provide the capacity for an ever-increasing demand on the European mobile networks. The 2G wireless system is called the 'Global System for Mobile Communications' and is denoted as GSM. The GSM air interface is based on the Time Division Multiple Access (TDMA) technique, which separates voice calls by time and transports parts of the conversations on the same carrier. The first GSM system was first in service in Finland in 1992. Indeed, GSM is now adopted by world-wide service providers and the GSM standards have gone through different phases of evolution to increase the spectral efficiency, throughput and data speed. The development of GSM is shown in Figure 1.1.

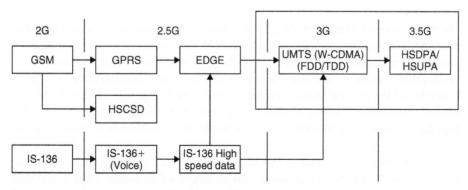

Figure 1.1   *Evolution of 2G networks based on TDMA technology.*

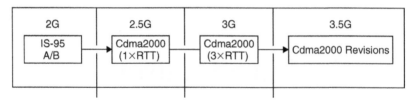

Figure 1.2   *Evolution of 2G networks based on CDMA technology.*

The development of digital wireless communications in North America was initiated in 1990. New digital transmission systems, such as IS-54 and IS-136, are deployed and known as Digital AMPS (D-AMPS) and the transmission algorithm used in these systems is also based on TDMA techniques. D-AMPS shared spectrum with the existing analogue system (AMPS).

The spread spectrum technique was originally used in military (secure) communications. However, in the early 1990s, Qualcom Inc pioneered the technology for commercial cellular communications. In 1993, digital cellular systems based on spread spectrum technology known as Code Division Multiple Access (CDMA) were deployed in the US and implemented according to interim standard 95 (IS-95). Since then, the CDMA technology has undergone various phases of development and revisions to increase data speed and improve system capacity and throughput. These developments and revisions of CDMA technology are captured in Figure 1.2.

1.2   Basic digital communication system

Communication systems consist of three main units: the transmitting unit, the communication transmission media, and the corresponding receiving unit. The communication signals transported through the system carry the information from the source where generated to one or more distance users. Transmission of the information through the non-ideal media causes the information signal to be attenuated and distorted due to channel loss and bandwidth limitation and to pick up extraneous signals such as random noise and interference. A schematic diagram is shown in Figure 1.3.

The source encoder eliminates, or at least reduces, redundancy in the source output by encoding (compressing) techniques using either lossless or lossy coding schemes. In the lossless compression, no information is lost in the compression–decompression process. In other words, the de-compressor results in exactly the same data as the compressor input. Examples of the popular lossless compression are the Lempel-Ziv-Welch (LZW) and Huffman algorithms. The lossy compression permits slight loss of information that is less important to the quality of the data in exchange for flexibility and higher compression ratios.

Applications that tolerate modest loss of information are videoconferences and video phones, where transform compression such as Adaptive Differential Pulse Code Modulation (APCM) and Discrete Cosine Transform (DCT) algorithm are used.

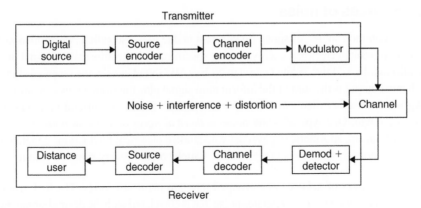

Figure 1.3 *Block diagram of basic digital communication system.*

A channel encoder works on the reverse principles of a source encoder by introducing redundant digits which can be used by the channel decoder to detect and/or correct errors occurring during transmission. Consequently, the combined task of the channel encoder and decoder is to maintain reliable communications over a non-ideal channel. Examples of forward error correcting codec's are the convolutional codes and the most powerful turbo codes. Both of these coding schemes are discussed and their error correction capabilities are assessed in detail in the next chapter.

A modulator maps the compressed and channel encoded baseband digital signal into RF waveforms, the energy of which is located in a frequency band defined by the RF carrier and the data filtering appropriate for transmission over the available channel. In the systems covered by the book, information is converted into phase difference for the transmitted RF waveforms using Quadrature Phase Shift Keying (QPSK) modulation. The demodulator performs the inverse of the modulator producing a distorted, attenuated and noise corrupted copy of the transmitted baseband signal. Although the modulator and the demodulator are shown in Figure 1.3 as separate units, they are combined in a single Modem unit (for example, in most mobile communication systems). The detector acts on the demodulator output using optimum detection schemes such as Viterbi algorithm.

Traditionally, signal processing carried out by the individual units in Figure 1.3 is performed separately. However, some of these signal processes are currently combined to provide spectral efficient systems such as combination of source and channel encoding or channel coding and modulation.

The communication process in the practical systems usually involves other devices, the most important of these are pulse filtering or spectral shaping devices to limit the spectrum occupied by the transmitted information in a band-limited system, the channel equalizers to compensate for the signal distortion during transmission, and the interleavers to act as an effective method to combat burst errors due to signal fading in mobile channels. All of these devices are presented in detail in the next chapter.

1.3 Sources of noise

The most common source of signal impairment is due to the so called *background noise*. This is an unwanted electrical energy sharing spectrum with the information signal during transmission. In Additive White Gaussian Noise (AWGN) channels, the received signal at the channel output is the sum of the information signal plus the white Gaussian noise. The background noise is called *white* because its spectral density is constant in a very wide frequency range. One type of white noise is *thermal noise* or *Johnson noise* after John Bertrand Johnson, an American physicist who studied electrical fluctuations generated by heat. White noise is analogous to white light, which contains every possible colour.

In general, sources of noise can be grouped into two main categories. The first group of sources, the *correlated noise*, generates noise that is correlated with the desired signal. Such noise is normally only generated in the presence of the information signal, from non-linear processing such as non-linear amplification which causes the generation of harmonics. Signals mixing also generate noise due to inter-modulation products that create a lot of distortion. For periodic signals, the correlated noise is also periodic.

The second group of noise sources generate noise that bears no relation to the information signal and the noise presence is independent of the presence of the signal. This noise is called *uncorrelated noise* and can be generated by either external or internal process. The external process could be due to atmospheric effects such as static electricity or lightning. However, the atmospheric noise is insignificant at frequencies above 30 MHz. Another external source of noise is due to extraterrestrial influences due to deep space effects, such as sun's heat noise and black body noise. The extraterrestrial noise contribution is significant up to about 1.5 GHz. Finally, a possible external source of noise could be man-made, i.e. generated by motors, ignition systems and switching equipment.

Internal uncorrelated noise is due to the electron's random motion in the atomic level of metal molecules causing collision with each other. Such random motion generates noise voltage that increases with heat but does not depend on the value of the metal electrical resistance. This noise is commonly encountered in most communication systems and is appropriately called *thermal noise*. The thermal noise process is defined by Gaussian distribution and it has power spectral density (N_0) which is almost constant over a very large frequency spectrum and hence is called white Gaussian noise. The one-sided spectral density (positive frequencies only) of the thermal noise, denoted by (N_0), in watts/Hz, and the thermal Noise Power (N) in dBm at room temperature (290 degrees Kelvin) are

$$N_0 = KT$$

$$N = -174 + 10\log_{10}(B)\, \text{dBm} \tag{1.1}$$

where B is the bandwidth in Hz, T is the absolute temperature in degrees Kelvin and K is Boltzmann's constant $= 1.38 \times 10^{-23}$ joules/Kelvin. The two-sided (positive and negative frequencies) noise spectral density is given by $\frac{N_0}{2}$. A schematic plot of (N_0) vs frequency is shown in Figure 1.4.

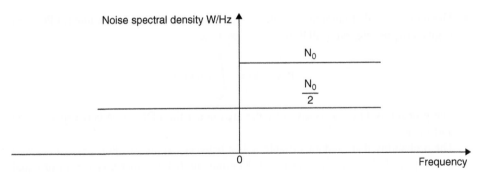

Figure 1.4   *Power spectral density of thermal noise.*

In the *Rayleigh fading* channel, which is encountered in most multipath signal propagations in mobile communications, the noise multiplies rather than adds the information signal, causing the signal to fade during reception with the loss of the information. Rayleigh fading channels are described in more detail in the next section.

1.4   Properties of the probability density functions

A random process where all of its statistical properties do not vary with time is defined as a *stationary process*; otherwise the processes are referred to as *non-stationary*.

Consider a random process described by a random variable (X) which takes on values described by x. The random distribution that is describing the random process is defined by a Probability Density Function (PDF), $f_X(x)$. The PDF has the following properties:

- $f_X(x)$ is always positive, i.e. $f_X(x) \geq 0$ for $-\infty < x < \infty$
- A process is *Wide Sense Stationary* (WSS) random process if its *expected power is finite* $E(x^2(t)) < \infty$. The process *mean value is constant*, and its *autocorrelation depends only on the time difference of the samples*.
- A random process is called *Strict Sense Stationary* (SSS) when all of its distribution parameters are unchanged regardless of the time shift applied to them. An SS stationary process will always be WSS but the reverse will not always hold true.
- When the random process is described by two random events X and Y, then the probability that event X and Y occur at the same time is given by the joint PDF $f_{XY}(x, y) = P(X \leq x, Y \leq y)$
- Integration of PDF with respect to random variable x over the region $-\infty < x < \infty$ is equal to 1 i.e.

$$\int_{-\infty}^{\infty} f_X(x)dx = 1 \quad \text{for } -\infty < x < \infty$$

- The probability that random variable x is greater or equal to certain value (a) $P(x \geq a)$ is given by the integral of PDF from $-\infty$ to a. i.e.

$$P(x \geq a) = \int\limits_{-\infty}^{a} f_X(x)dx$$

The probability $P(x \geq a)$ is given by the area under the PDF curve between $x = -\infty$ and $x = a$.

- The probability that random variable X lies between (a) and (b) $P(a \leq x \leq b)$ is given by the probability that x less or equal (b) minus probability that x is greater or equal (a), i.e.

$$P(a \leq x \leq b) = \int\limits_{a}^{b} f_X(x)dx$$

Clearly $P(a \leq x \leq b)$ is given by the area under $f_X(x)$ between $x = a$ and $x = b$ which is the probability of observing x in this region.

- The random variable X is described by statistical averages or expected values. The mean value of X, m_X, sometime written as $E\{X\}$ or $\overline{X}$ is defined as:

$$m_X = E\{X\} = \int\limits_{-\infty}^{\infty} x f_X(x)dx$$

- The variance of X, σ_X^2 is given by:

$$\sigma_X^2 = E\{(x - m_X)^2\} = \int\limits_{-\infty}^{\infty} (x - m_X)^2 f_X(x)dx$$

which can be simplified to

$$\sigma_X^2 = \left[\int\limits_{-\infty}^{\infty} x^2 f_X(x)dx \right] - m_X^2$$

- The standard deviation of the random variable X is σ_X.

1.5   Examples of probability distributions

1.5.1   Uniform distribution

A random variable taking on values between $x = a$ and $x = b$ where $b \geq a$ is said to have a uniform PDF expressed as:

$$f_X(x) = \frac{1}{b-a} \quad a \leq x \leq b$$
$$= 0 \quad\quad\ \text{elsewhere}$$

(1.2)

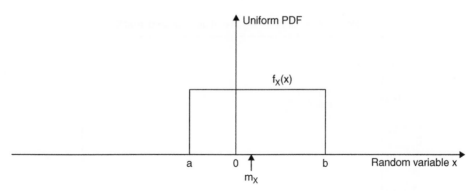

Figure 1.5 *Uniform random distribution.*

It can be shown that the mean value m_X and the variance σ_X^2 of random variable X are given by:

$$m_X = \frac{a+b}{2} \tag{1.3}$$

$$\sigma_X^2 = \frac{(b-a)^2}{12} \tag{1.4}$$

A typical uniform distribution (Figure 1.5) is obtained from a random sequence of equal likely symbols.

1.5.2 Gaussian (normal) distribution

A Gaussian random process X with mean value m_X and variance σ_x^2 can be described by the following probability density function expression

$$f_X(x) = \frac{1}{\sigma_x \sqrt{2\pi}} \exp\left(-\frac{(x-m_x)^2}{2\sigma_x^2}\right) \tag{1.5}$$

The Gaussian white noise PDF is symmetrical about $x = 0$ with mean value of the variable $m_X = 0$. The area under the PDF being 0.5 for $-\infty \le x \le 0$ and 0.5 for $0 \le x \le \infty$. The peak value of the PDF is at $\frac{1}{\sigma\sqrt{2\pi}}$. The tail of the Gaussian PDF extends to $\pm\infty$.

The area under the PDF tail determines the probability of error of digital transmission as we will explain in the next chapter. The Q-function denoted, as $Q(x)$, is a convenient way to compute the area under the tail and hence the probability of error. For a given normal random variable x with mean $m_X = 0$ and $\sigma = 1$, the area under the tail is $Q(x)$, shown shaded in Figure 1.6 covering the area between $2 \le x \le \infty$. Using appendix A, for $x = 2$, $Q(2) = 2.2750 \times 10^{-2}$. The mathematical expression representing $Q(x)$ is:

$$Q(x) = \text{prob} \cdot (x > a) = \int_a^\infty f_X(x) \cdot dx \tag{1.6}$$

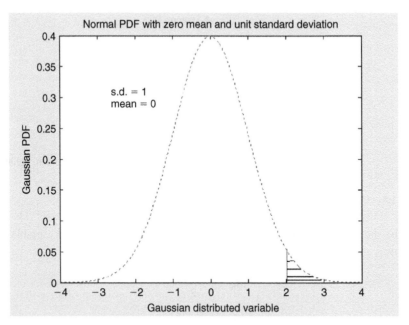

Figure 1.6 *Gaussian PDF with zero mean and unit standard deviation.*

When σ_X decreases, the PDF becomes narrower and taller but when σ_X increases it becomes broader and shallower as shown in Figure 1.7.

When a Gaussian random variable has $m_X \neq 0$ and $\sigma_X \neq 1$, then the Q(.) table in Appendix 1.A can still be used; not for Q(x), but for $Q\left(\frac{x - m_X}{\sigma_x}\right)$. For example $x = 4$, $m_X = 2$ and $\sigma_X = 4$ then $Q\left(\frac{x - m_X}{\sigma_x}\right) = Q(2/4) = Q(0.5) = 3.0854 \times 10^{-1}$, while $Q(4) = 3.1671 \times 10^{-5}$ from Appendix 1.A ($m_X = 0$ and $\sigma_X = 1$).

The function Q(x) is plotted verses x in Figure 1.8 and a tabulation of Q(x) for values of x is given in Appendix 1.A.

Consider a random process Y that is the sum of n statistically independent Gaussian events, each denoted by x_i with mean m_i and variance σ_i^2 where $i = 1, 2, \ldots, n$. It can be shown that the process Y is Gaussian distributed with mean m_Y and variance σ_Y^2 such that

$$m_Y = \sum_{i=1}^{n} m_i \tag{1.7}$$

$$\sigma_Y^2 = \sum_{i=1}^{n} \sigma_i^2 \tag{1.8}$$

On the other hand, when the n random variables are statistically independent and have different random distributions, the central limit theorem predicts the distribution of the random process Y to be Gaussian as $n \to \infty$.

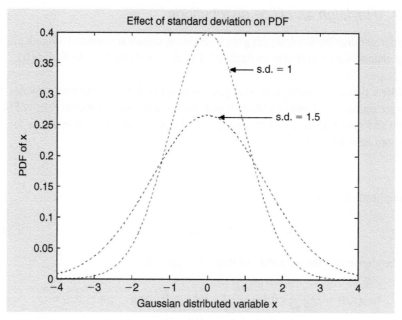

Figure 1.7 *Effect of standard deviation on Gaussian (normal) distribution.*

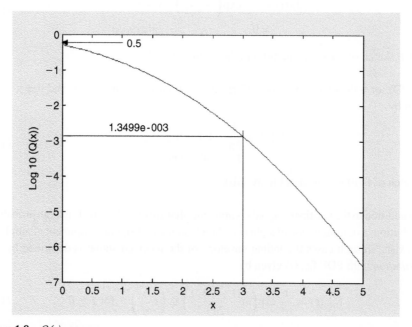

Figure 1.8 *Q(x) verses x.*

1.5.3 Rayleigh and Rice distributions

An important distribution describing the statistics of signals transmitted through multipath fading channels are referred to as *Rayleigh distribution* defined by the PDF $f_R(r)$.

Consider a two-dimensional Gaussian process described by two random variables X and Y that are transformed into amplitude R and phase θ with power spectral density PSD that is finite and symmetrical about frequency $\pm f_c$. Let us assume the mean values for X and Y are both zero and their variances are

$$\sigma_X^2 = \sigma_Y^2 = \sigma^2 \tag{1.9}$$

The amplitude R is given by

$$R^2 = X^2 + Y^2 \quad \text{and} \quad \theta = \tan^{-1}\frac{X}{Y} \tag{1.10}$$

It can be shown that the joint PDF of R and θ is given by

$$\begin{aligned} f_{R,\theta}(r, \theta) &= \frac{r}{2\pi\,\sigma^2} \exp\left(-\frac{r^2}{2\sigma^2}\right) \quad \text{for } r \geq 0 \quad \text{and} \quad -\pi \leq \theta \leq \pi \\ &= 0 \qquad\qquad\qquad\qquad\quad \text{otherwise} \end{aligned} \tag{1.11}$$

The PDF of the envelope of the Gaussian noise $f_R(r)$ is

$$\begin{aligned} f_R(r) &= \frac{r}{\sigma^2} \exp\left(-\frac{r^2}{2\sigma^2}\right) \quad r \geq 0 \\ &= 0 \qquad\qquad\qquad\quad \text{otherwise} \end{aligned} \tag{1.12}$$

A PDF sketch of the envelope $f_R(r)$ is shown in Figure 1.9.

The PDF of θ denoted $f_\theta(\theta)$ has uniform distribution between $\theta = -\pi$ and $\theta = \pi$ and is given by

$$\begin{aligned} f_\theta(\theta) &= \frac{1}{2\pi} \quad \text{for } -\pi \leq \theta \leq \pi \\ &= 0 \quad\;\; \text{otherwise} \end{aligned} \tag{1.13}$$

A sketch of $f_\theta(\theta)$ is shown in Figure 1.10.

If, in addition to the multipath signal components that are described by Rayleigh probability distribution, a dominant line of sight (i.e. direct) sinusoidal signal component of amplitude (A) exists, in such cases the fading envelope of the received signal is described by *Rice distribution* with PDF $f_{rice}(r)$ given by

$$f_{rice}(r) = \frac{r}{\sigma^2} \exp\left(-\frac{(r^2 + A^2)}{2\sigma^2}\right) I_0\left(r\frac{A}{\sigma^2}\right) \quad \text{for } r \geq 0 \tag{1.14}$$

where A = amplitude of the direct sinusoidal signal, $I_0(\ )$ is the zero order modified Bessel function of the first kind.

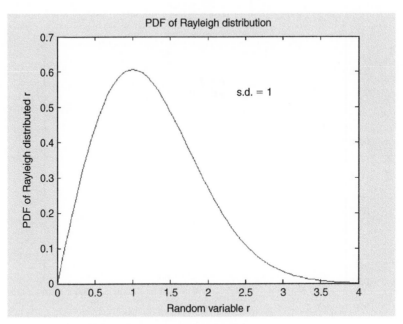

Figure 1.9   *PDF of Rayleigh distribution standard deviation = 1.*

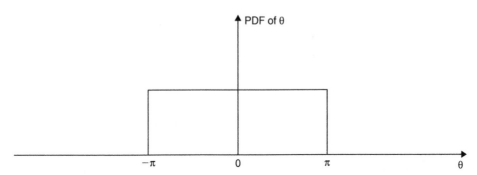

Figure 1.10 $f_\theta(\theta)$ *verses* θ.

The phase distribution is

$$f_{\text{rice}}(\theta) = \frac{1}{2\pi\sigma^2} \exp\left(-\frac{A^2}{2\sigma^2}\right)$$
$$\left[\sigma^2 + \frac{A}{2} \cdot \sqrt{2\pi\sigma^2} \cdot \cos\theta \cdot \exp\left(\frac{A^2 \cos^2\theta}{2\sigma^2}\right) \left[1 + \text{erf}\left(\frac{A\cos\theta}{2\sigma^2}\right)\right]\right] \quad (1.15)$$

The envelope and phase of Rice distributions are plotted in Figures 1.11 and 1.12, respectively.

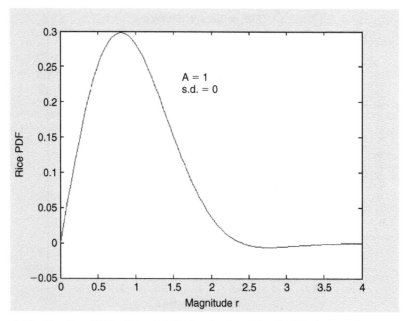

Figure 1.11   *Envelope PDF of Rice distribution A = 1 and Standard deviation = 1.*

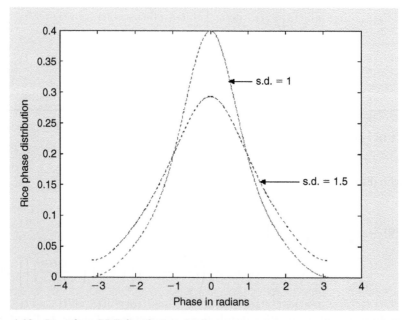

Figure 1.12   *Rice phase PDF distribution, A = 1.*

1.5.4 Binomial distribution

Consider a binary random variable X with two possible values 1 and 0 with probabilities p and $(1 - p)$, respectively. Let the random variable Y be the sum of n statistically independent and identically distributed variables X_i where $I = 1, 2, \ldots, n$ such that:

$$Y = \sum_{i=1}^{n} X_i \tag{1.16}$$

It can be shown that the binominal PDF of Y is given by:

$$f_Y(y) = \sum_{k=0}^{n} \binom{n}{k} p^k (1 - p)^{n-k} \delta(y - k) \tag{1.17}$$

The mean m_y and the variance σ_y^2 are given by

$$m_y = np \tag{1.18}$$

$$\sigma_y^2 = np(1 - p) \tag{1.19}$$

1.5.5 Chi-square distribution

Let X be Gaussian distributed random variable and $Y = X^2$, then the Y is chi-square distributed random variable. When the mean value of X is zero, the distribution of Y is called central chi-square distribution otherwise it is called non-central chi-square distribution. The PDF of Y is given by

$$f_Y(y) = \frac{1}{\sqrt{2\pi y}\,\sigma_x} \exp\left(\frac{y}{2\sigma_x^2}\right) \tag{1.20}$$

where σ_x^2 is variance of the Gaussian zero mean variable X. Now define n Gaussian variables X_i with zero mean and variance σ^2 where $i = 1, 2, \ldots, n$ such that

$$Y' = \sum_{i=1}^{n} X_i^2 \tag{1.21}$$

The distribution of Y is called central chi-square distribution with a PDF given by

$$\begin{aligned} f_{Y'}(y') &= \frac{1}{\sigma^n 2^{\frac{n}{2}} \Gamma\left(\frac{n}{2}\right)} \cdot y'^{\frac{n}{2}-1} \cdot \exp\left(-\frac{y'}{2\sigma^2}\right) \quad \text{for } y' \geq 0 \\ &= 0 \qquad\qquad\qquad\qquad\qquad\qquad\qquad\quad \text{otherwise} \end{aligned} \tag{1.22}$$

where $\Gamma(\ )$ defines the gamma function. Expression (1.22) defines the central chi-square PDF with n degrees of freedom (df). It can easily be seen that (1.22) reduces to (1.20) for $n = 1$. Therefore, (1.20) is the chi-square PDF with one degree of freedom. Expression (1.22) is plotted in Figure 1.13 for a number of dfs.

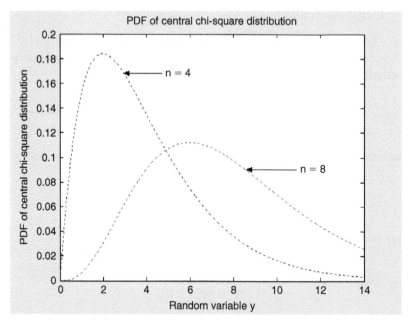

Figure 1.13 *PDF of central chi-square distribution with 4 and 8 dfs.*

Let us now consider the variable Y'' which is the sum of the statistically independent variables X_i, $i = 1, 2, \ldots, n$ with means m_i and identical variance such that

$$Y'' = \sum_{i=1}^{n} X_i^2 \tag{1.23}$$

$$m^2 = \sum_{i=1}^{n} m_i^2 \tag{1.24}$$

where m^2 is the *noncentrality parameter*. The non-central PDF of variable Y'' with n degrees of freedom is

$$\begin{aligned} f_{Y''}(y'') &= \frac{1}{2\sigma^2} \cdot \left(\frac{y''}{m^2}\right)^{\frac{n-2}{4}} \cdot \exp\left(-\frac{m^2 + y''}{2\sigma^2}\right) \cdot I_{\frac{n}{2}-1}\left(\sqrt{y''}\frac{m}{2}\right) &&\text{for } y'' \geq 0 \\ &= 0 &&\text{otherwise} \end{aligned} \tag{1.25}$$

where $I_\alpha(x)$ is the αth-order modified Bessel function of the first kind.

1.6 Equivalent noise bandwidth

Evaluation of the performance of systems carrying signal plus noise requires the knowledge of system bandwidth, as seen by the noise which is not necessarily the same bandwidth seen

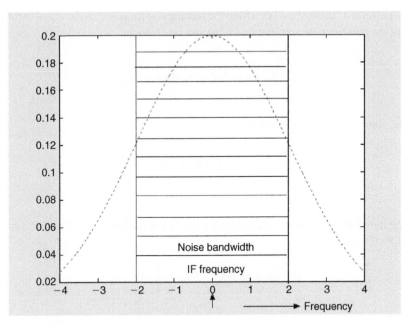

Figure 1.14 *Sketch of the noise bandwidth of a filter.*

by the signals. In this section we consider this issue in more detail because of its importance in the system. Commonly, the receiver front-end stage comprises of a bandpass filter at the Intermediate Frequency (IF) stage to limit the noise power input to the detector stage. Let the one-sided noise power spectral density at the input to the filter be N_0 W/Hz, the frequency transfer response of the filter be $H(f)$ centred at f_o, and the bandwidth (3 db cut-off frequency) of the filter be B, respectively.

Since the received signal and the noise at the input of the IF filter are uncorrelated, we can analyse them separately. We start by considering the noise power at the output of the IF filter by noting that the Noise Power Spectral Density (PSD) at the out of the filter is $N_0|H(f)|^2$ and the output noise power (N) is given by

$$N = \int_0^\infty N_0|H(f)|^2 df \qquad (1.26)$$

Now consider the filter bandwidth that is used in estimating the output noise power, referred to as *noise bandwidth* B_N. We will define the *noise bandwidth* by the width of a rectangular frequency response (*ideal*) filter that passes equal noise power as the *real* IF filter as shown in Figure 1.14. Thus the noise power at the output of the *ideal* filter is

$$N = N_0|H(f)|^2 B_N \qquad (1.27)$$

Since we assumed the noise power within the rectangular frequency to be equal to the noise power of the real filter, we can equate (1.26) and (1.27) to get

$$\int_0^\infty N_0 |H(f)|^2 df = N_0 |H(f_0)|^2 B_N$$

Thus the noise bandwidth is given by:

$$B_N = \frac{\int_0^\infty |H(f)|^2 \cdot df}{|H(f_0)|^2} \tag{1.28}$$

It can be shown that the noise bandwidths for the following lowpass filters that have 3 dB bandwidth (B_{3-dB}) are

$$\text{RC lowpass filter} \quad B_N = \frac{\pi}{2} \cdot B_{3-dB} \tag{1.29}$$

$$\text{Butterworth LP filter of order n} \quad B_N = \frac{\frac{\pi}{2n}}{\sin \frac{\pi}{2n}} \cdot B_{3-dB} \tag{1.30}$$

1.7 Linear filtering of white noise

In many applications, it is required to filter the white noise with the accompanying signal. Linear filters, such as differentiators and integrators, are commonly used as filters in electronic systems. We now consider applying the white noise at the input of such filters and evaluate the output noise power.

1.7.1 White noise differentiation

Consider the differentiator in Figure 1.15 where we applied an input signal $e^{j\omega t}$. The output signal is $j\omega\tau\, e^{j\omega t}$ where τ is a differentiation constant of maximum value one. Therefore, the transfer function $H(f)$ of the differentiator is $j\omega\tau$. The two-sided noise PSD at the

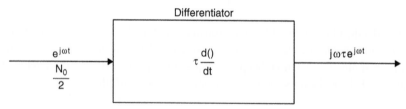

Figure 1.15 *Sketch of a differentiator.*

output of the differentiator is $\frac{N_0}{2}\omega^2\tau^2$. The noise power at the output of the differentiator is given by:

$$N_{\text{diff}} = \frac{N_0}{2}\tau^2 \int\limits_{-B}^{B} (2\pi f)^2 \cdot df = \frac{4}{3}N_0\pi^2\tau^2 B^3 \qquad (1.31)$$

1.7.2 White noise integration

Consider the integrator in Figure 1.16 with the input signal $e^{j\omega t}$. The output signal is $\frac{e^{j\omega t}}{j\omega\tau}$ where $\frac{1}{\tau}$ is the integration constant of maximum value one. Therefore, the transfer function of the integrator is $\frac{1}{j\omega\tau}$. Furthermore, the integrator incurred a delay equal to the integration interval T. The delay can be represented by a factor $e^{-j\omega T}$ so that the integrator output due to this factor is $\frac{e^{-j\omega T}}{j\omega\tau}$. Consequently, the integrator transfer function H(f) is

$$H(f) = \frac{1}{j\omega\tau} - \frac{e^{-j\omega T}}{j\omega\tau}$$

$$= 2j \cdot e^{-j\frac{\omega}{2}T} \cdot \frac{\sin\left(\dfrac{\omega T}{2}\right)}{j\omega\tau} \qquad (1.32)$$

$$\text{And} \quad |H(f)| = \frac{T}{\tau}\frac{\sin(\pi f T)}{\pi f T} \qquad (1.33)$$

Therefore, two-sided noise PSD at integrator output is

$$\frac{N_0}{2}|H(f)|^2 = \frac{N_0}{2}\left(\frac{T}{\tau}\right)^2\left(\frac{\sin(\pi f T)}{\pi f T}\right)^2$$

Thus noise power at integrator output, N_{int}, is given by

$$N_{\text{int}} = \frac{N_0}{2}\left(\frac{T}{\tau}\right)^2 \int\limits_{-\infty}^{\infty} \left(\frac{\sin(\pi f T)}{\pi f T}\right)^2 df$$

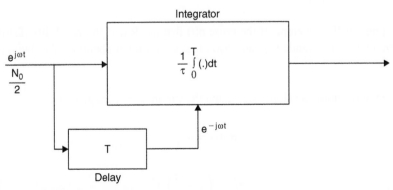

Figure 1.16 *Sketch of an integrator.*

Since $\displaystyle\int_{-\infty}^{\infty} \left(\frac{\sin(\pi fT)}{\pi fT}\right)^2 df = \frac{\pi}{\pi T} = \frac{1}{T}$

Therefore

$$N_{int} = \frac{N_0}{2}\left(\frac{T}{\tau}\right)^2 \frac{1}{T} = \frac{N_0}{2}\frac{T}{\tau^2} \tag{1.34}$$

1.8 Narrowband Gaussian noise

Contemporary communication systems are bandwidth limited. Consequently most of the thermal noise, n(t), with zero mean and variance σ^2 is confined within a relatively narrowband around the operating frequency, f_o, of the communication system. A convenient way to represent this noise is using the *quadrature components*. This representation considers the narrowband noise as an *in phase* component, $n_C(t)$ and a *quadrature* component, $n_S(t)$, each with zero mean values and both have variance equal σ^2. Mathematically we can express the noise n(t) as

$$n(t) = n_c(t)\cos\omega_0 t - n_s(t)\sin\omega_0 t \tag{1.35}$$

where $n_C(t)$ and $n_S(t)$ are uncorrelated Gaussian noise processes with variances σ_c^2 and σ_s^2 of the quadrature noise components $n_C(t)$ and $n_S(t)$ are given by:

$$\sigma_c^2 = \sigma_s^2 = \sigma^2 \tag{1.36}$$

We can write n(t) in a polar form as

$$n(t) = R_n(t) \cdot \cos(\omega_0 t + \phi_n(t))$$

where $R_n(t) = \sqrt{n_c^2(t) + n_s^2(t)}$ for $0 \le R_n(t) \le \infty$

and $\phi_n(t) = \tan^{-1}\dfrac{n_s(t)}{n_{c(t)}}$ $0 \le \phi \le 2\pi$

$R_n(t)$ represents the envelope of the noise n(t) that has Rayleigh probability distribution given by (1.12). The random phase $\phi_n(t)$ has a uniform probability distribution given by (1.13).

The mean of the noise envelope, $\overline{R}_n(t)$, and the variance, σ_R^2, are given by

$$\overline{R}_n(t) = \sigma^2\sqrt{\frac{\pi}{2}} \tag{1.37}$$

$$\sigma_R^2 = \left(2 - \frac{\pi}{2}\right)\sigma^2 \tag{1.38}$$

Let us define the instantaneous power of the narrowband noise as $P_n = R_n^2$ and the PDF of P_n to be $f_P(P_n)$ then:

$$f_P(P_n) = f_P(P_n = R_n^2) \cdot \left| \frac{dR_n}{dP_n} \right|$$

where

$$\frac{dR_n}{dP_n} = \frac{1}{2R_n}$$

so that

$$f_P(P_n) = f_P(P_n = R_n^2) \cdot \frac{1}{2R_n}$$

$$= \frac{1}{2R_n} \frac{R_n}{\sigma^2} \exp\left(-\frac{P_n}{2\sigma^2}\right)$$

Thus

$$f_P(P_n) = \frac{1}{2\sigma^2} \exp\left(-\frac{P_n}{2\sigma^2}\right) \quad \text{for} \quad 0 \leq P_n \leq \infty \tag{1.39}$$

The average power of the narrowband noise $\overline{P}_n$ is equal $2\sigma^2$ and the variance of the power $\sigma_P^2 = 4\sigma^4$. The PDF of the instantaneous noise power is shown in Figure 1.17.

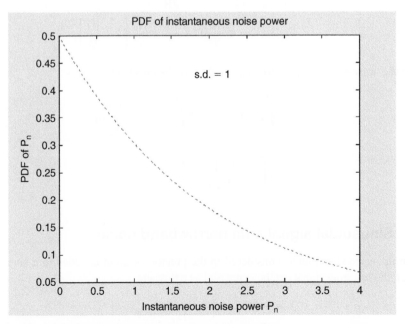

Figure 1.17 *PDF of instantaneous noise power, variance = 1.*

Example 1.1

The power spectral density of noise n(t) is $G_N(f)$ shown below. Given $N_0 = 1.5\mu$ W/Hz and $B = 9.6$ kHz, calculate total noise power.

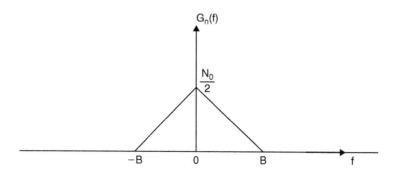

Solution

Expressing $G_n(f) = a|f| + b$

For
$$f = 0 \quad \frac{N_0}{2} = b$$

$$|f| = B \quad 0 = aB + \frac{N_0}{2}$$

Thus
$$a = -\frac{N_0}{2B}$$

Hence
$$G_n(f) = -\frac{N_0}{2B}|f| + \frac{N_0}{2} = \frac{N_0}{2}\left[1 - \frac{|f|}{B}\right]$$

Since the noise PSD is symmetrical around $f = 0$, the noise power, N, is

$$N = 2\int_0^B G_n(f) \cdot df = 2\int_0^B \frac{N_0}{2}\left[1 - \frac{f}{B}\right]df$$

$$= N_0\left[f - \frac{f^2}{2B}\right]_0^B = \frac{N_0}{2}B$$

1.9   Sinusoidal signal plus narrowband noise

When the narrowband noise considered in the previous section is added to a sinusoidal carrier, the resulting signal can be expressed as r(t) where

$$r(t) = A\cos\omega_0 t + n(t)$$

$$= (A + n_c(t))\cos\omega_0 t - n_s(t)\sin\omega_0 t \qquad (1.40)$$

The distribution of the envelope of r(t) follows Rice distribution given by (1.14) and the phase of r(t) follows the uniform distribution given by (1.13).

The SNR of the received signal is $\frac{A^2}{2\sigma^2}$. For large SNR, the distribution of the envelope of r(t) approaches Gaussian distribution with mean value A and variance σ^2.

1.10 Fourier analysis

It is important at this stage to appreciate that signal processing is an essential component of modern communications. Communication signals can be processed in either the real time domain or in the frequency domain. Signal processing aspects of communications such as modulation and forward error correcting coding are presented in detail in the next chapter, while aspects like multiple access interference and multi-user detection are dealt with in Chapters 6 and 7, and the time synchronization techniques in Chapter 5.

In this section we present the processing elements such as the Fourier processing, convolution and signals correlations.

1.10.1 Fourier series

Consider a complex periodic signal, f(t), that possesses a finite number of maximums and minimums and integrable conditions (i.e. satisfied Dirichlet conditions) and the periodic duration T_0. Such a signal can be represented in terms of a finite number of sinusoidal, cosinusoidal signals and a constant term. This representation comprises a series known as *Fourier series* given as

$$f(t) = \sum_{n=-\infty}^{\infty} \alpha_n e^{jn\omega_0 t} \quad \text{for} \quad \varepsilon_0 \le t \le \varepsilon_0 + T_0 \tag{1.41}$$

where

$$\alpha_n = \frac{1}{T_0} \int_{\varepsilon_0}^{\varepsilon_0 + T_0} f(t) e^{-jn\omega_0 t} dt = |\alpha_n| e^{j\phi_n} \tag{1.42}$$

Clearly

$$\alpha_{-n} = \alpha_n^* \tag{1.43}$$

And

$$\phi_n = \tan^{-1} \frac{\text{Imaginary}(\alpha_n)}{\text{Real}(\alpha_n)} \tag{1.44}$$

Thus, using (1.41) to (1.44), we can express the periodic signal f(t) as

$$f(t) = \alpha_0 + \sum_{n=1}^{\infty} [(\alpha_n + \alpha_n^*) \cos(n\omega_0 t) + j(\alpha_n - \alpha_n^*) \sin(n\omega_0 t)] \tag{1.45}$$

where α_0 represents the dc in the periodic signal. We can express (1.45) in trigonometric form as

$$f(t) = \frac{a_0}{2} + \sum_{n=1}^{\infty} a_n \cos(n\omega_0 t) + b_n \sin(n\omega_0 t)$$

where

$$a_0 = 2\alpha_0 = \frac{2}{T_0} \int_{\varepsilon_0}^{\varepsilon_0 + T_0} f(t) \cdot dt \qquad (1.46)$$

$$a_n = \alpha_n + \alpha_n^* = \frac{2}{T_0} \int_{\varepsilon_0}^{\varepsilon_0 + T_0} f(t) \cdot \cos(n\omega_0 t) dt \qquad (1.47)$$

$$b_n = \alpha_n - \alpha_n^* = \frac{2}{T_0} \int_{\varepsilon_0}^{\varepsilon_0 + T_0} f(t) \cdot \sin(n\omega_0 t) dt \qquad (1.48)$$

and

$$\phi_n = \tan^{-1} \frac{b_n}{a_n} \qquad (1.49)$$

1.10.2 Fourier transform

It is clear from the previous section that if $f(t)$ is non-periodic, i.e. $T_0 \to \infty$, the coefficient α_n (and hence a_n and b_n) $\to 0$ and the signal analysis using the Fourier series breaks down. It can be shown that as T_0 approaches infinity, Fourier series summation translates to an integral deriving the frequency domain function $F(\omega)$ for a known time domain function $f(t)$ and conversely the knowledge of $F(\omega)$ permits the determination of $f(t)$. The *Fourier transform* of a continuous function $f(t)$ and its inverse are defined by the integral functions:

$$F(\omega) = \int_{-\infty}^{\infty} f(t) e^{-j\omega t} \cdot dt \qquad (1.50)$$

The *inverse Fourier transform* is given by

$$f(t) = \frac{1}{2\pi} \int_{-\infty}^{\infty} F(\omega) \cdot e^{j\omega t} d\omega \qquad (1.51)$$

It is worth noting that for $f(t)$ to be Fourier transformable, it must satisfy Dirichlet conditions. Now in most communication signal analysis, $f(t)$ is considered complex so that

$$f(t) = f_r(t) + j f_i(t) \qquad (1.52)$$

Substituting (1.52) in (1.50), we get

$$F(\omega) = \int\limits_{-\infty}^{\infty} [f_r(t) \cos \omega t + f_i(t) \sin \omega t] dt + j \int\limits_{-\infty}^{\infty} [f_r(t) \cos \omega t - f_i(t) \sin \omega t] dt$$

Therefore, the real and imaginary parts of $F(\omega)$ are

$$F_r(\omega) = \int\limits_{-\infty}^{\infty} [f_r(t) \cos \omega t + f_i(t) \sin \omega t] dt \qquad (1.53)$$

$$F_i(\omega) = - \int\limits_{-\infty}^{\infty} [f_r(t) \cos \omega t - f_i(t) \sin \omega t] dt \qquad (1.54)$$

The inverse terms are

$$f_r(t) = \frac{1}{2\pi} \int\limits_{-\infty}^{\infty} [F_r(\omega) \cos \omega t - F_i(\omega) \sin \omega t] d\omega \qquad (1.55)$$

$$f_i(t) = \frac{1}{2\pi} \int\limits_{-\infty}^{\infty} [F_r(\omega) \sin \omega t + F_i(\omega) \cos \omega t] d\omega \qquad (1.56)$$

1.10.3 Fast Fourier transform

In digital communications, the signal $f(t)$ consists of a large number of discrete samples taken at regular short interval Δt. Thus the continuous $f(t)$ corresponds to the discrete $f(n\Delta t) \equiv f(n)$ where $n = 0, 1, \ldots, N - 1$. Similarly the continuous function (ωt) can be represented be a discrete quantity $2\pi k \, \Delta f n \, \Delta t$. Substituting for $\Delta f = \frac{1}{N\Delta t}$, we get $\omega t \approx \frac{2\pi}{N} nk$. Now $F(\omega) = F(k\Delta\omega) \equiv F(k)$. Since the integration in the continuous Fourier transform translates to a summation, the Discrete Fourier Transform (DFT) form of (1.50) becomes:

$$F(k) = \sum_{n=0}^{N-1} f(n) e^{-jk\frac{2\pi}{N}n} \quad \text{for } k = 0, 1, \ldots\ldots\ldots, N - 1 \qquad (1.57)$$

The inverse DFT (IDFT) is

$$f(n) = \frac{1}{N} \sum_{k=0}^{N-1} F(k) e^{jk\frac{2\pi}{N}n} \quad \text{for } n = 0, 1, \ldots\ldots\ldots, N - 1 \qquad (1.58)$$

Since DFT and IDFT basically involve the same type of computations, the efficient computational algorithms for the DFT apply as well to the computation of the IDFT.

The computation of a single DFT sample using (1.57) requires N^2 complex multiplications, $N(N-1)$ complex additions making the total computation $(2N^2 - N)$ complex operations.

The computation of N-point DFT using (1.57) involves terms that re-appear in the computation causing computation redundancy and, therefore, even with a reasonable number of samples, such computation tends to include millions of complex multiplication and addition operations. For example, a 1024-point DFT requires 1,048,576 complex multiplications and 1,047,552 complex additions. The DFT computation can be reduced considerably by removing such redundancies. Let us pause for a moment here to examine the last statement further.

Denote the complex exponential in (1.57) by writing:

$$W_N = e^{-j\frac{2\pi}{N}}$$
$$W_N^k = e^{-jk\frac{2\pi}{N}} \tag{1.59}$$

Thus
$$W_N^1 = e^{-j\frac{2\pi}{N}} \tag{1.60}$$

$$W_N^2 = (e^{-j\frac{2\pi}{N}})^2 = e^{-j2\frac{2\pi}{N}} = e^{-j\frac{2\pi}{\frac{N}{2}}} = W_{\frac{N}{2}}^1 \tag{1.61}$$

Furthermore, W_N^k have two important properties: the symmetry property such that

$$W_N^{k+\frac{N}{2}} = W_N^k W_N^{\frac{N}{2}} = W_N^k \left(e^{-j\frac{2\pi}{N}\frac{N}{2}} \right) = W_N^k e^{-j\pi} = -W_N^k \tag{1.62}$$

The periodicity property shows:

$$W_N^{k+N} = W_N^k \tag{1.63}$$

Substituting (1.59) in (1.57), we get:

$$F(k) = \sum_{n=0}^{N-1} f(n) W_N^{nk} \quad \text{for } k = 0, 1, \ldots\ldots, N-1 \tag{1.64}$$

where $W_N^{nk} = e^{-j\frac{2\pi}{N}nk}$ is known as the *twiddle factor*. The inverse DFT (IDFT), f(n), is given by:

$$f(n) = \frac{1}{N} \sum_{k=0}^{N-1} F(k) W_N^{-nk} \quad k = 0, 1, \ldots\ldots, N-1 \tag{1.65}$$

The twiddle factor is a complex function with unit magnitude and phase $-\frac{2\pi}{N}nk$ (also known as *phase factor*), can be represented as a unit radius circle shown in Figure 1.18.

The vectors in Figure 1.18 are spaced by radian frequency $\Delta\omega = \frac{2\pi}{N}$. The periodicity property of the twiddle factor means we only need to compute $-\frac{2\pi}{N}nk \mod 2\pi$. The symmetry property means only half of the twiddle factors need calculation since the other half can

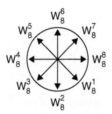

Figure 1.18 *Polar representation of the complex twiddle factor for N = 8.*

be obtained from the inverse of the first half, i.e. $W_8^7 = -W_8^3$. It also means we only need to compute the first half of the DFT, i.e. up to half the sampling frequency since the other half can be obtained from the first half as stated before.

The efficient algorithm for computing the DFT is known as the *Fast Fourier Transform* (FFT), originally developed by Cooley and Tukey in 1965. The basic idea behind the FFT is the process of continually *decomposing and recombining* the N-point transform into a pair of $\frac{N}{2}$-point transforms. For example, consider 1024-point transform and split it into two 512-point DFTs. The computations requirement would be 262,144 complex multiplications and 261,632 complex additions compared to 1,048,576 complex multiplications and 1,047,552 complex additions for a solely DFT. Therefore, carrying out the DFT on smaller input sequences would reduce the computations. The process of decomposing a long input sequence into many small sequences is known as *time decimation*.

We start by expanding (1.64) to give:

$$F(k) = f(0)(W_N^k)^0 + f(1)(W_N^k)^1 + f(2)(W_N^k)^2 + \cdots\cdots\cdots + f(N-1)(W_N^k)^{N-1}$$

$$(1.66)$$

Generally, N is selected to be the even number. However, if N is odd we insert augmenting zeros to the end of the sequence to make N divisible by 2. Let us now break down (1.66) into two equal parts containing the even numbered samples f(2n), i.e. $f(0), f(2), \ldots, f\left(2\left(\frac{N}{2}-1\right)\right)$ and the odd numbered samples f(2n + 1), i.e. $f(1), f(3), \ldots,$ $f\left(2\left(\frac{N}{2}-1\right)+1\right)$. The first divide-by-2 process yields the following expression:

$$F(k) = \left[f(0)(W_N^k)^0 + f(2)(W_N^k)^2 + \cdots\cdots\cdots\cdots + f\left(2\left(\frac{N}{2}-1\right)\right)(W_N^k)^{2\left(\frac{N}{2}-1\right)} \right]$$

$$+ \left[f(1)(W_N^k)^1 + f(3)(W_N^k)^3 + \cdots\cdots\cdots + f\left(2\left(\frac{N}{2}-1\right)+1\right)(W_N^k)^{2\left(\frac{N}{2}-1\right)+1} \right]$$

After the first divide-by-2 process, (1.64) becomes

$$F(k) = \sum_{n=0}^{\frac{N}{2}-1} f(2n)(W_N^k)^{2n} + \sum_{n=0}^{\frac{N}{2}-1} f(2n+1)(W_N^k)^{2n+1} \qquad (1.67)$$

Simplifying the above to give:

$$F(k) = \sum_{n=0}^{\frac{N}{2}-1} f(2n) \left(W_{\frac{N}{2}}^{k}\right)^{n} + W_N^k \sum_{n=0}^{\frac{N}{2}-1} f(2n+1) \left(W_{\frac{N}{2}}^{k}\right)^{n} \tag{1.68}$$

The DFT expression given in (1.68) is the sum of two $\frac{N}{2}$ points DFTs with common twiddle factor $W_{\frac{N}{2}}^{nk}$ that needs to be computed once only. Consequently, it takes less time to compute (1.68) than solely N points DFT. Denote the DFT for the even numbered samples as $F_e(k)$ and the odd numbered samples $F_o(k)$ so that:

$$F(k) = F_e(k) + W_N^k F_o(k) \tag{1.69}$$

where

$$F_e(k) = \sum_{n=0}^{\frac{N}{2}-1} f(2n) \left(W_{\frac{N}{2}}^{k}\right)^{n} \tag{1.70}$$

and

$$F_o(k) = \sum_{n=0}^{\frac{N}{2}-1} f(2n+1) \left(W_{\frac{N}{2}}^{k}\right)^{n} \tag{1.71}$$

Using (1.68) and putting $(k + \frac{N}{2})$ for k, we get

$$F\left(k + \frac{N}{2}\right) = \sum_{n=0}^{\frac{N}{2}-1} f(2n) \left(W_{\frac{N}{2}}^{k+\frac{N}{2}}\right)^{n} + W_N^{k+\frac{N}{2}} \sum_{n=0}^{\frac{N}{2}-1} f(2n+1) \left(W_{\frac{N}{2}}^{k+\frac{N}{2}}\right)^{n} \tag{1.72}$$

Now

$$W_{\frac{N}{2}}^{k+\frac{N}{2}} = W_{\frac{N}{2}}^{k} e^{-j\frac{2\pi}{N}\frac{N}{2}} = W_{\frac{N}{2}}^{k} e^{-j2\pi} = W_{\frac{N}{2}}^{k} \tag{1.73}$$

Substituting (1.73) in (1.72), we get

$$F\left(k + \frac{N}{2}\right) = \sum_{n=0}^{\frac{N}{2}-1} f(2n) \left(W_{\frac{N}{2}}^{k}\right)^{n} + W_N^{k+\frac{N}{2}} \sum_{n=0}^{\frac{N}{2}-1} f(2n+1) \left(W_{\frac{N}{2}}^{k}\right)^{n}$$

$$F\left(k + \frac{N}{2}\right) = F_e(k) + W_N^{k+\frac{N}{2}} F_o(k) \tag{1.74}$$

Combining (1.69) and (1.74) gives a pair of equations that can be used for $k = 0, 1, \ldots, \left(\frac{N}{2} - 1\right)$.

$$F(k) = F_e(k) + W_N^k F_o(k) \tag{1.75}$$

$$F\left(k + \frac{N}{2}\right) = F_e(k) + W_N^{k+\frac{N}{2}} F_o(k) \tag{1.76}$$

Now let us carry out a second 'divide-by-2' process on (1.64) to give

$$F(k) = \sum_{n=0}^{\frac{N}{4}-1} f(4n)\,(W_N^k)^{4n} + \sum_{n=0}^{\frac{N}{4}-1} f(4n+1)\,(W_N^k)^{4n+1}$$

$$+ \sum_{n=0}^{\frac{N}{2}-1} f(4n+2)\,(W_N^k)^{4n+2} + \sum_{n=0}^{\frac{N}{4}-1} f(4n+3)\,(W_N^k)^{4n+3}$$

Therefore after two 'divide-by-2' processes, (1.64) becomes:

$$F(k) = \sum_{n=0}^{\frac{N}{4}-1} f(4n)\,(W_N^k)^{4n} + W_N^k \sum_{n=0}^{\frac{N}{4}-1} f(4n+1)\,(W_N^k)^{4n}$$

$$+ W_N^{2k} \sum_{n=0}^{\frac{N}{2}-1} f(4n+2)\,(W_N^k)^{4n} + W_N^{3k} \sum_{n=0}^{\frac{N}{4}-1} f(4n+3)\,(W_N^k)^{4n} \qquad (1.77)$$

We can re-write (1.77) in a simple form

$$F(k) = \sum_{n=0}^{\frac{N}{4}-1} f(4n)\left(W_{\frac{N}{4}}^k\right)^n + W_{\frac{N}{2}}^k \sum_{n=0}^{\frac{N}{4}-1} f(4n+2)\left(W_{\frac{N}{4}}^k\right)^n$$

$$+ W_N^k \sum_{n=0}^{\frac{N}{4}-1} f(4n+1)\left(W_{\frac{N}{4}}^k\right)^n + W_N^k W_{\frac{N}{2}}^k \sum_{n=0}^{\frac{N}{4}-1} f(4n+3)\left(W_{\frac{N}{4}}^k\right)^n \qquad (1.78)$$

Equation (1.78) can be written in terms of the even numbered DFT $F'_e(k)$ and the DFT of the odd numbered DFT $F'_o(k)$ where:

$$F'_e(k) = \sum_{n=0}^{\frac{N}{4}-1} f(4n)\left(W_{\frac{N}{4}}^k\right)^n + W_{\frac{N}{2}}^k \sum_{n=0}^{\frac{N}{4}-1} f(4n+2)\left(W_{\frac{N}{4}}^k\right)^n \quad \text{for } k = 0, 1, \ldots, \frac{N}{4}-1 \ \ (1.79)$$

Write

$$F'_e(k) = F_e^{21}(k) + W_{\frac{N}{2}}^k F_e^{22}(k) \qquad (1.80)$$

Where

$$F_e^{21}(k) = \sum_{n=0}^{\frac{N}{4}-1} f(4n)\left(W_{\frac{N}{4}}^k\right)^n \qquad (1.81)$$

$$F_e^{22}(k) = \sum_{n=0}^{\frac{N}{4}-1} f(4n+2)\left(W_{\frac{N}{4}}^k\right)^n \qquad (1.82)$$

$$F'_o(k) = \sum_{n=0}^{\frac{N}{4}-1} f(4n+1)\left(W_{\frac{N}{4}}^k\right)^n + W_{\frac{N}{2}}^k \sum_{n=0}^{\frac{N}{4}-1} f(4n+3)\left(W_{\frac{N}{4}}^k\right)^n \qquad (1.83)$$

$$F'_o(k) = F_o^{23}(k) + W_{\frac{N}{2}}^k F_o^{24}(k) \qquad (1.84)$$

$$F_o^{23}(k) = \sum_{n=0}^{\frac{N}{4}-1} f(4n+1)\left(W_{\frac{N}{4}}^k\right)^n \qquad (1.85)$$

$$F_o^{24}(k) = \sum_{n=0}^{\frac{N}{4}-1} f(4n+3)\left(W_{\frac{N}{4}}^k\right)^n \qquad (1.86)$$

We use lengthy derivation processes to reach (1.80) to (1.86). However, these expressions can be checked using (1.75) and (1.76), which are the most important expressions in FFT processing. This decimation process is repeated until computing 2-point DFT functions. These decimation processes are called '*passes*' and there are n passes in N-point FFT where $N = 2^n$.

It is worth noting that the original order of $N = 8$ samples are $f(0), f(1), f(2), \ldots, f(7)$ and that time decimation changes the sequence order of the $F(k)$ samples. However, to keep the $F(k)$ samples in the correct order, i.e. $F(0), F(1), \ldots, F(7)$, the order of the sequence entering the FFT should be $f(0), f(4), f(2), f(6), f(1), f(5), f(3),$ and $f(7)$. The original samples are assumed to be stored in a binary memory address, say sample $f(1)$ is stored in the address 001. However, the second sample entering the FFT is sample at the *bit-reversal address*, that is sample in memory address 100, i.e. $f(4)$ instead of the original sample at 001 as shown in the Table below.

Original sequence order		Re-ordered sequence input	
Decimal	Binary	Binary	Decimal
0	000	000	0
1	001	100	4
2	010	010	2
3	011	110	6
4	100	001	1
5	101	101	5
6	110	011	3
7	111	111	7

Thus decimation in time of the input samples brings about the re-ordering of the samples using *bit-reversal*. The block diagram for computing 8-point FFT is depicted in Figure 1.19.

An essential method used in the FFT computation is known as the *butterfly* which combines a 2-point input and the twiddle factor multiplication. A butterfly represented in flow graph is shown in Figure 1.20.

$$F_{out}(a) = F_i(a) + W_N^k F_i(b) \tag{1.87}$$

$$F_{out}(b) = F_i(a) - W_N^k F_i(b) \tag{1.88}$$

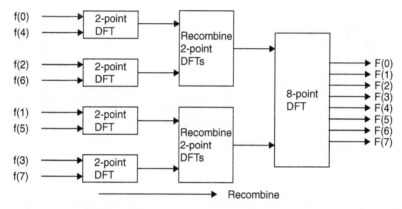

Figure 1.19   *Computing 8-point FFT.*

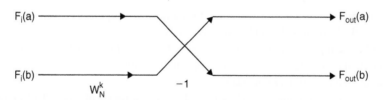

Figure 1.20   *A butterfly for radix-2 FFT algorithm.*

There are $\frac{N}{2}$ butterflies in each decimation stage in an N-point FFT, and there are $\log_2 N$ stages. Furthermore, each butterfly computation involves one complex multiplication and two complex additions. Therefore, the computation of the N-point FFT involves $\frac{N}{2} \log_2 N$ complex multiplications and $N \log_2 N$ complex additions making the total computation $\frac{3}{2} N \log_2 N$ complex operations. The computation requirement for N-point DFT and FFT are shown in Figure 1.21.

Usually when one speeds up an algorithm, this speed-up results in an increased cost of overheads in the algorithm execution. In FFT, the computational savings, however, do not come at the expense of accuracy.

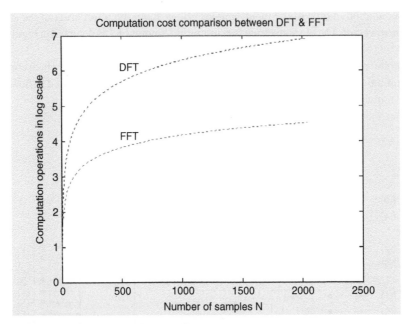

Figure 1.21 *Computation cost of DFT and FFT.*

In conclusion, we list the computation procedure for the FFT algorithm in a step by step order:

- Pad the input sequence with zeros until the number of samples N is nearest to the power of 2, i.e. 125 samples should be padded to $128 = 2^7$.
- Re-order the input sequence using bit reversal, i.e. $6 = 110$ goes to $011 = 3$.
- Compute the twiddle factors W_N^k involved in the stage.
- Compute the butterflies in the stage.
- Compute and recombine the DFTs in each stage starting with stage 1 (2-point DFT) and proceed to determine N-point DFT.

The Inverse FFT (IFFT), expressed by (1.58), transforms the signal spectra into their corresponding waveforms in time domain. The IFFT computation is carried out using the FFT algorithm described above with two alterations: the first is twiddle factor phase is changed to W_N^{-k} and the summation is scaled by $\frac{1}{N}$.

So far we have focused our attention on the *Decimation In Time* (DIT) algorithm but we can decimate in the frequency domain. The *Decimation In Frequency* (DIF) algorithm was derived in 1966 (commonly known as the Sande-Tukey FFT algorithm). It decomposes the N-point DFT into two $\frac{N}{2}$-point transforms, one transform containing the first $\frac{N}{2}$, i.e. samples of input sequence $\left[f(0), f(1), f(2), \ldots, f\left(\frac{N}{2} - 1\right) \right]$ and the other transform

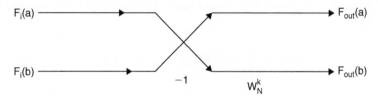

Figure 1.22 *Butterfly of radix-2 DIF N-point FFT.*

containing the second $\frac{N}{2}$ samples $\left[f\left(\frac{N}{2}\right), f\left(\frac{N}{2}+1\right), f\left(\frac{N}{2}+2\right), \ldots, f(N-1) \right]$. This process of decomposition continues till the last 2-point transforms. The DIF butterfly is shown in Figure 1.22 where

$$F_{out}(a) = F_i(a) + F_i(b)$$

$$F_{out}(b) = [F_i(a) - F_i(b)]W_N^k \qquad (1.89)$$

The computation of the DIF FFT algorithm, similar to the DIT FFT algorithm, recombines the DFTs starting in the 2-point DFTs and proceeds to the N-point FFT. Both algorithms are based on the same concept of *decimations* and *recombinations*, one in time and the other in frequency and the costs of computing both algorithms are equal. However, the choice between the two algorithms depends on which of the two is most appropriate to the application in hand.

Example 1.2

Consider the 4-point DIT FFT with samples f(0), f(1), f(2), and f(3). Compute the 4-point DFT using step by step method to show in depth the algorithm computation.

Solution

Re-order the samples using bit reversal.

Original samples	Binary addresses	Bit reverses addresses	Required samples for computation
f(0)	00	00	f(0)
f(1)	01	10	f(2)
f(2)	10	01	f(1)
f(3)	11	11	f(3)

The DFT for the even numbered samples [f(0), f(2)] is denoted $F_e(k)$ and the DFT for the odd numbered samples [f(1), f(3)] is denoted $F_o(k)$. Since $N=4$ so $n=2$, i.e. there are two passes (0, 1).

The FFT butterfly for 4-point DFT is shown below. First we calculate the even and the odd numbered FFT.

$$F_e(k) = \sum_{n=0}^{1} f(2n)(W_2^k)^n \quad \text{for } k = 0, 1$$

$$F_e(k) = \sum_{n=0}^{1} f(2n)(W_2^k)^n = f(0) + f(2)W_2^k$$

$$F_e(0) = f(0) + f(2)W_2^0$$

$$F_e(1) = f(0) + f(2)W_2^1 = f(0) - f(2)W_2^0$$

Similarly

$$F_o(k) = \sum_{n=0}^{1} f(2n+1)\left(W_{\frac{N}{2}}^k\right)^n = f(0) + f(3)W_2^k \quad \text{for } k = 0, 1$$

$$F_o(0) = f(1) + f(3)W_2^0$$

$$F_o(1) = f(1) - f(3)W_2^0$$

These are the results of Pass-0

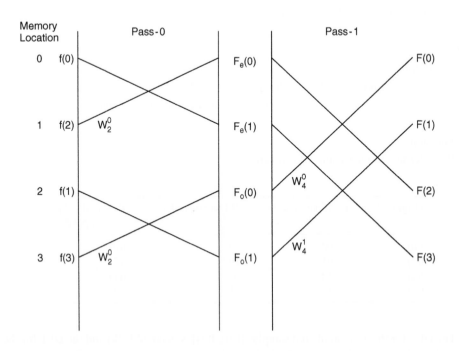

From (1.75) and (1.76) we have

$$F(k) = F_e(k) + W_4^k F_o(k)$$

$$F(0) = F_e(0) + W_4^0 F_o(0) = f(0) + f(2)W_2^0 + W_4^0[f(1) + f(3)W_2^0]$$

$$F(0) = f(0) + f(2) + f(1) + f(3)$$

$$F(1) = F_e(1) + W_4^1 F_o(1) = f(0) - f(2)W_2^0 + W_4^1[f(1) - f(3)W_2^0]$$

$$F(1) = f(0) - f(2) - j[f(1) - f(3)]$$

$$F(k+2) = F_e(k) + W_4^{k+2} F_o(k)$$

$$F(2) = F_e(0) + W_4^2 F_o(0)$$

$$F(2) = F_e(0) - F_o(0) = f(0) + f(2)W_2^0 - [f(1) + f(3)W_2^0]$$

$$F(2) = f(0) + f(2) - [f(1) + f(3)]$$

$$F(3) = F_e(1) + W_4^3 F_o(1)$$

$$F(3) = F_e(1) + jF_o(1) = f(0) - f(2) + J[f(1) - f(3)]$$

Example 1.3

Consider the samples f(0), f(1), ..., f(7). Calculate the DIT 8-point FFT.

Solution
Let us now consider re-ordering the samples using bit reversal.

Memory location	Binary addresses	Bit reverses addresses	Sample location
x(0)	000	000	x(0)
x(1)	001	100	x(4)
x(2)	010	010	x(2)
x(3)	011	110	x(6)
x(4)	100	001	x(1)
x(5)	101	101	x(5)
x(6)	110	011	x(3)
x(7)	111	111	x(7)

The 8-point radix-2 butterfly for the DIT FFT is

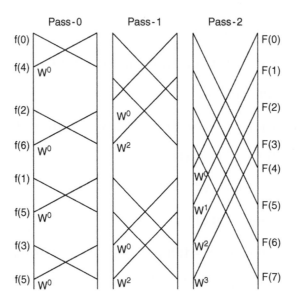

computations at the end of pass-0

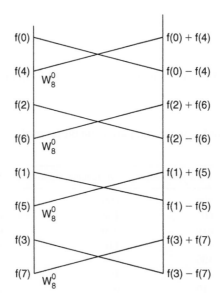

computations at the end of pass-1

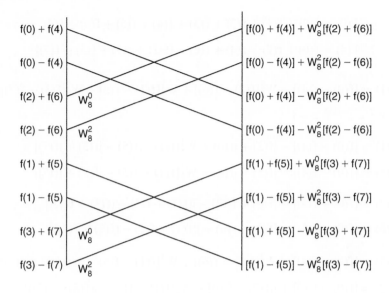

Pass-2 computations

$[f(0) + f(4)] + W_8^0[f(2) + f(6)]$ $F(0)$

$[f(0) - f(4)] + W_8^2[f(2) - f(6)]$ $F(1)$

$[f(0) + f(4)] - W_8^0[f(2) + f(6)]$ $F(2)$

$[f(0) - f(4)] - W_8^2[f(2) - f(6)]$ $F(3)$

$[f(1) + f(5)] + W_8^0[f(3) + f(7)]$ W_8^0 $F(4)$

$[f(1) - f(5)] + W_8^2[f(3) - f(7)]$ W_8^1 $F(5)$

$[f(1) + f(5)] - W_8^0[f(3) + f(7)]$ W_8^2 $F(6)$

$[f(1) - f(5)] - W_8^2[f(3) - f(7)]$ W_8^3 $F(7)$

Using the above butterfly, we get

$$F(0) = f(0) + f(1) + f(2) + f(3) + f(4) + f(5) + f(6) + f(7)$$

$$F(4) = f(0) + f(2) + f(4) + f(6) - [f(1) + f(3) + f(5) + f(7)]$$

$$F(1) = [f(0) - f(4)] + W_8^2[f(2) - f(6)] + W_8^1[f(1) - f(5) + W_8^2[f(3) - f(7)]]$$

Now $W_8^2 = e^{-j2\frac{2\pi}{8}} = -j$

$$F(1) = [f(0) - f(4)] - jf(2) + jf(6) + W_8^1[f(1) - f(5) - jf(3) + jf(7)]$$

$$F(5) = [f(0) - f(4)] - jf(2) + jf(6) - W_8^1[f(1) - f(5) - jf(3) + jf(7)]$$

$$F(2) = [f(0) + f(4)] - [f(2) + f(6)] - j[f(1) + f(5) - (f(3) + f(7))]$$

$$F(6) = [f(0) + f(4)] - [f(2) + f(6)] + j[f(1) + f(5) - (f(3) + f(7))]$$

$$F(3) = [f(0) - f(4)] - W_8^2[f(2) - f(6)] + W_8^3[f(1) - f(5) - W_8^2(f(3) - f(7))]$$

$$F(3) = [f(0) - f(4)] + j[f(2) - f(6)] + W_8^3[f(1) - f(5) + j(f(3) - f(7))]$$

$$F(7) = [f(0) - f(4)] + j[f(2) - f(6)] - W_8^3[f(1) - f(5) + j(f(3) - f(7))]$$

Summary of the results:

$$F(0) = f(0) + f(1) + f(2) + f(3) + f(4) + f(5) + f(6) + f(7)$$

$$F(1) = [f(0) - f(4)] - jf(2) + jf(6) + W_8^1[f(1) - f(5) - jf(3) + jf(7)]$$

$$F(2) = [f(0) + f(4)] - [f(2) + f(6)] - j[f(1) + f(5) - (f(3) + f(7))]$$

$$F(3) = [f(0) - f(4)] - W_8^2[f(2) - f(6)] + W_8^3[f(1) - f(5) - W_8^2(f(3) - f(7))]$$

$$F(4) = f(0) + f(2) + f(4) + f(6) - [f(1) + f(3) + f(5) + f(7)]$$

$$F(5) = [f(0) - f(4)] - jf(2) + jf(6) - W_8^1[f(1) - f(5) - jf(3) + jf(7)]$$

$$F(6) = [f(0) + f(4)] - [f(2) + f(6)] + j[f(1) + f(5) - (f(3) + f(7))]$$

$$F(7) = [f(0) - f(4)] + j[f(2) - f(6)] - W_8^3[f(1) - f(5) + j(f(3) - f(7))]$$

1.11   Signals convolution

The interaction of an input signal with a linear time invariant (LTI) system is commonly described by the convolution integral. This interaction is measured by the response of the LTI system to a single impulse applied to its input and is referred to as the *impulse response* of the system h(t). An impulse applied at instant 0 to the LTI system yield h(t) shown in Figure 1.23.

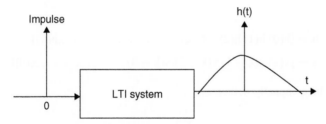

Figure 1.23   *Impulse response of LTI system.*

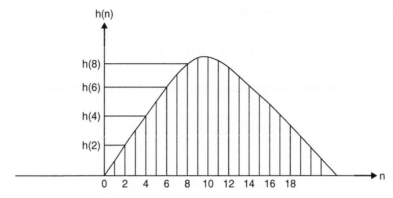

Figure 1.24   *Sampled impulse response.*

Consider the LTI system in Figure 1.23 with an input x(t) and impulse response h(t). The output y(t) is given by convolving x(t) with h(t) as expressed in the following equation

$$y(t) = x(t)\,{}^*\,h(t) \tag{1.90}$$

The asterisk (*) in (1.90) denotes the convolution operation, expressing (1.90) in the following convolution integral

$$y(t) = \int_{-\infty}^{\infty} x(\lambda) \cdot h(t - \lambda) \cdot d\lambda \tag{1.91}$$

$$= \int_{-\infty}^{\infty} x(t - \lambda) \cdot h(\lambda) \cdot d\lambda \tag{1.92}$$

Thus, y(t) is given by the area under the product function $x(\lambda) \cdot h(t - \lambda)$ plotted verses λ for given time t. Consider sampling the impulse response h(t) at $t = nT_S$ at the sampling rate $\frac{1}{T_s}$. The sampled impulse response h(n) is shown in Figure 1.24.

Consider a finite casual sequence of sampled signal x(n) and sampled impulse response such that n = 0, 1, 2, …, k. The sampled output y(n) is given here:

$$y(k) = \mathbf{h} \cdot \mathbf{x}^{T} \tag{1.93}$$

where

$$\mathbf{h} = [h(0)\, h(1)\, h(2) \dots\dots\dots\dots\dots\dots\dots\dots h(k)] \qquad (1.94)$$

$$\mathbf{x} = [x(k)\, x(k-1)\, x(k-2)\, x(k-3) \dots\dots\dots\dots\dots x(0)] \qquad (1.95)$$

y(k) can also be compactly written as:

$$y(k) = \sum_{n=0}^{k} h(k-n) \cdot x(n) \qquad (1.96)$$

Equivalently, (1.96) can be written as:

$$y(k) = \sum_{n=0}^{k} h(n) \cdot x(k-n) \qquad (1.97)$$

In general, the input samples could extend from $-\infty$ to ∞ so that (1.96) and (1.97) become

$$y(k) = \sum_{n=-\infty}^{\infty} h(k-n) \cdot x(n) \qquad (1.98)$$

$$y(k) = \sum_{n=-\infty}^{\infty} h(n) \cdot x(k-n) \qquad (1.99)$$

Example 1.4

Two signals, $x_1(t)$ and $x_2(t)$ are sampled at the rate of one sample per second and eight samples are taken from each signal. Signal $x_1(t)$ is a single rectangular pulse of amplitude 1 volt and duration 4 seconds and the second signal is $x_2(t) = e^{-t}$.

Calculate the convolution y(n) given by the 8-sample sequences taken from $x_1(t)$ and $x_2(t)$.

Solution
The sequences of the samples signals are

$x_1(t) = 1, 1, 1, 1, 0, 0, 0, 0$

$x_2(t) = 1.0000, 3.6788e\text{-}001, 1.3534e\text{-}001, 4.9787e\text{-}002, 1.8316e\text{-}002, 6.7379e\text{-}003,$

$\qquad 2.4788e\text{-}003, 9.1188e\text{-}004$

Using the appropriate MATLAB commands, the convolution y(n) is

$y(k) = [1.0000, 1.3679, 1.3862, 1.3863, 3.8632e\text{-}1, 1.8439e\text{-}2, 1.2352e\text{-}4, 1.1255e\text{-}7,$

$\qquad 1.3888e\text{-}11, 2.3195e\text{-}16, 5.2429e\text{-}22, 0, 0, 0, 0]$

Note the length of vector $\mathbf{y} = $ length of vector $\mathbf{x1} + $ length of vector $\mathbf{x2} - 1 = 15$.

A fast algorithm for computing the discrete convolution when both h(n) and x(n) are periodic sequences of length N is to use DFT. Let Y(f), H(f) and X(f) be the DFTs of y(k), h(n) and x(n) respectively so that

$$Y(f) = [X(f) \cdot H(f)] \qquad (1.100)$$

$$y(k) = IFFT(X(f) \cdot H(f)) \qquad (1.101)$$

1.12 Signals deconvolution

In practical LTI systems the measured output y(n) is often obtained when the input sequence is known so that the LTI system is identified by its impulse response h(n). This process is called *deconvolution* which is the undoing of the convolution; also known as the *inverse filtering*. Deconvolution is used for *system identification*. In such a process the required term is h(n) given the output y(k) and input x(n) (Figure 1.25).

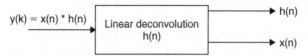

Figure 1.25 *Schematic of deconvolution.*

Expanding (1.96), we get

$$y(k) = h(k)x(0) + \sum_{n=1}^{k} h(k - n) \cdot x(n)$$

When k = 0 and assuming x(0) ≠ 0

$$h(0) = \frac{y(0)}{x(0)} \qquad (1.102)$$

Therefore, the system identifier produces h(k) where

$$h(k) = \frac{1}{x(0)} \left[y(k) - \sum_{n=1}^{k} h(k - n) \cdot x(n) \right] \quad \text{for } k \geq 1 \qquad (1.103)$$

In defining (1.103), we have assumed that the system is *noiseless*. The estimate of h(k) given by (1.103) still holds for small levels of noise but as the system noise increases the estimates of h(k) become unreliable.

Another application of the deconvolution arises when y(k) and system h(k) are both known and required to estimate the *most likely input*. Expanding (1.97) we get:

$$y(k) = h(0).x(k) + \sum_{n=1}^{k} h(n) \cdot x(k - n)$$

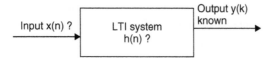

Figure 1.26   *Blind deconvolution.*

For $k = 0$ and assuming $h(0) \neq 0$

$$x(0) = \frac{y(0)}{h(0)}$$

Therefore, the input $x(k)$ is

$$x(k) = \frac{1}{h(0)}\left[y(k) - \sum_{n=1}^{k} h(n) \cdot x(k-n) \right] \quad \text{for } k \geq 1 \tag{1.104}$$

Again the terms in (1.104) do not contain system noise term.

In some applications, the system impulse response $h(n)$ is not known and the input $x(n)$ is determined from the measured output $y(k)$ in a process known as *blind deconvolution*, the concept of which is shown in Figure 1.26.

Blind deconvolution is widely used in image signal processing to remove the blurring that degrades the quality of the original image. We will not develop this topic any further as it is beyond the scope of this book.

1.13   Signals correlation

Correlation is a measure of the similarity between two signals as one is shifted with respect to the other. The correlation is maxima at the time when the two signals match best. If the two signals are identical, this maximum is when the two copies are synchronous (no delay). Correlation is widely used in applications such as the detection of signals corrupted by channel noise, the estimation of time delay, the time synchronization, pattern matching, and cross spectral analysis.

Correlation is equivalent to the time reversed convolution of the two signals. The correlation of two signals is called the *cross-correlation* and the correlation of the signal with a copy of itself is called the *autocorrelation*. The average cross correlation function $R_{12}(\tau)$ of two periodic signals $s_1(t)$ and $s_2(t)$, period T_0 is defined as

$$R_{12}(\tau) = \frac{1}{T_0} \cdot \int_{-\frac{T_0}{2}}^{\frac{T_0}{2}} s_1(t) \cdot s_2(t + \tau) \cdot dt \tag{1.105}$$

If the two signals $s_1(t)$ and $s_2(t)$ are non-periodic, then the cross correlation is given by

$$R_{12}(\tau) = \int\limits_{-\infty}^{\infty} s_1(t) \cdot s_2(t + \tau) \cdot dt \qquad (1.106)$$

In general

$$R_{12}(\tau) \neq R_{21}(\tau) \qquad (1.107)$$

But

$$R_{12}(\tau) = R_{21}(-\tau) \qquad (1.108)$$

The autocorrelation function is an even function of time shift τ so that

$$R_{xx}(\tau) = R_{xx}(-\tau) \qquad (1.109)$$

The *discrete time cross-correlation* function $R_{xy}(\tau)$ between two discrete signals $x(n)$ and $y(n)$ of length N is defined by

$$R_{xy}(\tau) = \frac{1}{N} \sum_{n=0}^{N-1} x(n + \tau) \cdot y(n) \qquad (1.110)$$

where $\tau = 0, \pm 1, \pm 2, \ldots\ldots\ldots\ldots\ldots\ldots\ldots\ldots$

The *discrete time autocorrelation function* $R_{xx}(\tau)$ between discrete signal $x(n)$ of length N and a copy of itself is defined by

$$R_{xx}(\tau) = \frac{1}{N} \sum_{n=0}^{N-1} x(n + \tau) \cdot x(n) \qquad (1.111)$$

The function $R_{xx}(\tau)$ acquires a maximum value $[R_{xx}(0)]$ at $\tau = 0$.

$$[R_{xx}(0)] = E_x \qquad (1.112)$$

where E_x is the energy in the discrete signal $x(n)$. The above equation has an important application in data detection. It suggests that at zero time shift, the level of the auto-correlation is proportional to the signal energy which helps in optimizing the detection process.

The subject of energy spectral density of discrete signals and power spectral density of periodic signals expressed in terms of the correlation functions are taken up again in Section 1.12, and the correlation functions between users' signature waveforms in multiple access spread spectrum system including periodic and aperiodic correlation functions and their contribution to multiple access interference are considered in Chapter 3 (Section 3.6).

Example 1.5

Find the autocorrelation function of the sinusoidal signal

$$s(t) = A \cos \omega_0 t$$

where T_0 is the period of the signal.

Solution

$$R_{ss}(\tau) = \frac{1}{T_0} \int_{-\frac{T_0}{2}}^{\frac{T_0}{2}} A \cos \omega_0 t \cdot A \cos \omega_0(t+\tau) \cdot dt$$

$$= \frac{A^2}{T_0} \int_{-\frac{T_0}{2}}^{\frac{T_0}{2}} \cos \omega_0 t \cdot \cos \omega_0(t+\tau) \cdot dt$$

$$= \frac{A^2}{2T_0} \int_{-\frac{T_0}{2}}^{\frac{T_0}{2}} [\cos \omega_0(2t+\tau) + \cos \omega_0 \tau] \cdot dt$$

$$= \frac{A^2}{2T_0} \left[\frac{\sin \omega_0(2t+\tau)}{2\omega_0} + t \cos \omega_0 \tau \right]_{-\frac{T_0}{2}}^{\frac{T_0}{2}}$$

Now

$$\left[\frac{\sin \omega_0(2t+\tau) = \sin 2\omega_0 t \cos \omega_0 \tau + \cos 2\omega_0 t \sin \omega_0 \tau}{2\omega_0} \right]_{-\frac{T_0}{2}}^{\frac{T_0}{2}} = 0$$

Thus

$$R_{ss}(\tau) = \frac{A^2}{2} \cos \omega_0 \tau$$

Example 1.6

A received signal $r(t)$ consisting of the signal $s(t)$ and the noise $n(t)$ such that:

$$r(t) = s(t) + n(t)$$

The received signal is sampled and N samples are taken. Calculate the autocorrelation $R_{rr}(\tau)$ of the received signal.

Solution

$$r(k) = s(k) + n(k) \quad k = 0, 1, 2, \ldots\ldots\ldots N-1$$

$R_{rr}(\tau) = \frac{1}{N} \sum_{k=0}^{N-1} r(k)r(k + \tau)$ where we assumed both s(k) and n(k) are real functions. Substitute for r(k) in the above equation:

$$R_{rr}(\tau) = \frac{1}{N} \sum_{k=0}^{N-1} [s(k) + n(k)][s(k + \tau) + n(k + \tau)]$$

$$= \frac{1}{N} \sum_{k=0}^{N-1} [s(k)\, s(k + \tau) + n(k)n(k + \tau) + n(k)\, s(k + \tau) + s(k)n(k + \tau)]$$

$$= R_{ss}(\tau) + R_{nn}(\tau) + 2\, R_{sn}(\tau)$$

where $R_{ss}(\tau)$ and $R_{nn}(\tau)$ are the signal and noise autocorrelation functions respectively and $R_{sn}(\tau)$ is the cross-correlation between s(k) and n(k). If we assume s(t) and n(t) to be uncorrelated and n(t) is thermal white noise of zero mean then:

$$R_{sn}(\tau) = 0 \text{ and } R_{nn}(\tau) = \frac{N_0}{2}\delta(\tau) \text{ (see next section) so that}$$

$$R_{rr}(\tau) = R_{ss}(\tau) + \frac{N_0}{2}\delta(\tau)$$

Example 1.7

Calculate the cross-correlation function $R_{12}(\tau)$ between the discrete sequence $x_1(k)$ and $x_2(k)$ and their autocorrelation functions $R_{11}(\tau)$ and $R_{22}(\tau)$ given that

$x_1(k)$	2	−1	0	3	5	−1	−2	1	7	8
$x_2(k)$	0	−2	3	5	2	−1	3	4	−2	1

Solution

k	X1	X2	R12	R11	R22	k	X1	X2	R12	R11	R22
0	2	0	2	16	0	10			49	71	16
1	−1	−2	−5	6	−2	11			52	−23	−9
2	0	3	10	−5	7	12			−4	−23	21
3	3	5	5	1	−9	13			7	41	26
4	5	2	−6	61	−2	14			58	61	−2
5	−1	−1	6	41	26	15			59	19	−9
6	−2	3	37	−23	21	16			10	−5	7
7	1	4	14	−23	−9	17			−16	6	−2
8	7	−2	−12	71	16	18			0	1.6	0
9	8	1	20	158	73						

1.14   Spectral density of discrete signals

For wide-sense stationary signals, the Wiener-Khinchine theorem stated that the power spectral density PSD, $P_x(f)$, can be computed from the Fourier transform of the finite autocorrelation function $R_{xx}(\tau)$:

$$P_x(f) = \Im[R_{xx}(\tau)] \tag{1.113}$$

We have assumed that $R_{xx}(\tau)$ tends to zero for large τ and $\Im(\cdot)$ denotes Fourier transform of the function $(\cdot)$.

Consider a baseband digital data $x(k)$ of zero mean and unit variance with pulse shape $f(t)$ and transmitted at the rate R_b b/s. Let the DFT of the pulse shape be denoted by $F(f)$ so that for the rectangular pulse shape:

$$F(f) = T_b \left(\frac{\sin \pi f T_b}{\pi f T_b} \right) \text{ where } T_b \text{ is sample interval.} \tag{1.114}$$

The PSD of $x(t)$ is given by

$$P_x(f) = R_b |F(f)|^2 \Im[R_{xx}(\tau)] \tag{1.115}$$

Now consider a linear system of impulse response and frequency transfer function are $h(t)$ and $H(f)$ as shown in Figure 1.27 where the PSD at the input and output are denoted by $P_x(f)$ and $P_y(f)$:

Figure 1.27   *PSD of LTI system.*

The PSD at the output of the system is

$$P_{y(f)} = |H(f)|^2 P_x(f) \tag{1.116}$$

Now consider a white noise process $n(t)$ with two-sided PSD $P_n(f)$:

$$P_n(f) = \frac{N_0}{2} \tag{1.117}$$

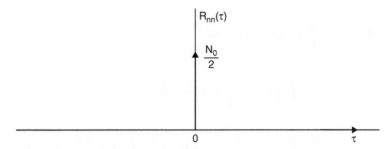

Figure 1.28 *Autocorrelation function of white noise.*

The autocorrelation function $R_{nn}(\tau)$ of the white noise is obtained by taking the inverse Fourier transform

$$R_{nn}(\tau) = \frac{N_0}{2}\delta(\tau) \tag{1.118}$$

$R_{nn}(\tau)$ as function τ is shown in Figure 1.28.

Example 1.8

Determine and plot the autocorrelation of the noise considered in Example 1.1 (plotted below).

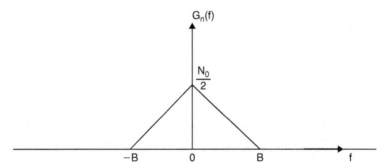

Solution

The triangular-shaped PSD suggests that it is produced by the convolution of a rectangular-shaped spectrum with a copy of itself. The Inverse Fourier Transform (IFT) of the convolution is a multiplication in the time domain. The IFT of a rectangular spectra is sinc($\cdot$) function in the time domain. Consequently, the expected autocorrelation function is sinc$^2(\cdot)$. In the following analysis, we derive the expression for the autocorrelation given by the IFT of the noise PSD $G_n(f)$ so that

$$R_{nn}(\tau) = \Im^{-1}[G_n(f)]$$

$$G_n(f) = \frac{N_0}{2}\left[1 - \frac{|f|}{B}\right]$$

$$R_{nn}(\tau) = \int_{-\infty}^{\infty} G_n(f) \exp(j\omega\tau) df$$

$$= \frac{N_0}{2} \int_{-\infty}^{\infty} \left[1 - \frac{|f|}{B} \right] \cdot \exp(j\omega\tau) df$$

$$= \frac{N_0}{2} \int_{-B}^{B} \left[1 - \frac{f}{B} \right] \cdot \cos\omega\tau df + j\frac{N_0}{2} \int_{-B}^{B} \left[1 - \frac{f}{B} \right] \cdot \sin\omega\tau df$$

Let
$$x = \omega\tau = 2\pi f\tau$$

Thus
$$dx = 2\pi\tau \, df$$

$$R_{nn}(\tau) = \frac{N_0}{2} \int_{-2\pi\tau B}^{0} \left[1 + \frac{x}{2\pi\tau B} \right] \cdot \cos\omega\tau \, df + j\frac{N_0}{2} \int_{-2\pi\tau B}^{0} \left[1 + \frac{x}{2\pi\tau B} \right] \cdot \sin\omega\tau \, df$$

$$+ \frac{N_0}{2} \int_{0}^{2\pi\tau B} \left[1 - \frac{x}{2\pi\tau B} \right] \cdot \cos\omega\tau \, df + j\frac{N_0}{2} \int_{0}^{2\pi\tau B} \left[1 - \frac{x}{2\pi\tau B} \right] \cdot \sin\omega\tau \, df$$

$$R_{nn}(\tau) = \frac{N_0}{2} \left(\frac{B}{2} \right)^2 \sin c^2 \pi\tau B$$

Example 1.9

A Gaussian white noise of two-sided noise power density $\frac{N_0}{2}$ W/Hz is applied to the input of an RC lowpass filter. Derive an expression for the noise power density and the noise power at the output of the filter.

Solution

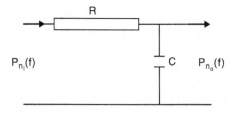

$P_{n_i}(f) = \frac{N_0}{2}$ W/Hz = noise power spectral density at filter input. The noise power spectral density at filter output is

$$P_{n_0}(f) = |H(f)|^2 P_{n_i}(f) \, \text{W/Hz}$$

where $|H(f)|$ is the magnitude of transfer function of the RC filter which is given by:

$$H(\omega) = \frac{1}{R + \frac{1}{j\omega C}} \frac{1}{j\omega C} = \frac{1}{1 + j\omega RC}$$

Therefore

$$|H(f)|^2 = \frac{1}{1 + (\omega RC)^2}$$

$$P_{n_0}(f) = \frac{1}{1 + (\omega RC)^2} \frac{N_0}{2} W/Hz$$

Noise power at filter output, N, is

$$N = \int_{-\infty}^{\infty} P_{n_0}(f) df = \int_{-\infty}^{\infty} \frac{1}{1 + (\omega RC)^2} \frac{N_0}{2} df$$

Let $x = \omega RC = 2\pi f RC$

Thus $dx = 2\pi RC\, df$

So that

$$N = \frac{N_0}{2} \frac{1}{2\pi RC} \int_{-\infty}^{\infty} \frac{dx}{1 + x^2}$$

Now $\int_{-\infty}^{\infty} \frac{dx}{1+x^2}$ is a standard integral equal π. Substituting in the above expression for N we get:

$$N = \frac{N_0}{2} \frac{1}{2\pi RC} \pi = \frac{N_0}{4RC}$$

Example 1.10

Find the power spectral density of a sinusoidal waveform

$$y(t) = A \cos(\omega_0 t + \theta_0)$$

Solution

The autocorrelation function of waveform y(t) is derived in Example 1.5

$$R_{yy}(\tau) = \frac{A^2}{2} \cos \omega_0 \tau$$

The power spectral density of y(t) is given by the FFT of $R_{yy}(\tau)$

$$P_y(f) = \Im[R_{yy}(\tau)] = \frac{A^2}{4} \delta(f - f_0) + \frac{A^2}{4} \delta(f + f_0)$$

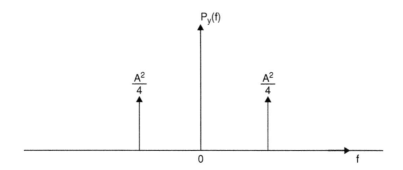

1.15 Summary

This chapter begins with a review of the development of wireless communication systems that use CDMA technology. This review introduces the road map for the topics that are dealt with in the rest of the book and which are shown in bordered bold lines in Figures 1.1 and 1.2. The chapter prepares the reader for coping with this advanced technology by gaining knowledge and understanding of the building blocks of the digital communication systems operating in an additive white Gaussian noise (AWGN) environment and in Rayleigh fading channels.

This chapter shows how to analyse and process the information signals and evaluate the noise corruption of the communication signals. Presented in detail are the statistical distributions of signals and noise such as: uniform distribution describing the phase distribution and distribution of equal probable binary data; Gaussian noise found in almost all systems and the Rayleigh and Rice distributions that are most important to wireless channels; Binomial distributions for describing digital and statistical problems; multi-freedom chi-square distribution derived from Gaussian distribution and widely used in many signal processing schemes in the communication fields.

In most systems, the input white noise is filtered so we present the analysis of the narrowband white noise and distribution of its power. Furthermore, signal processing such as FFT, convolution and correlation are essential tools for the system performance evaluation.

Problems

1.1 Determine the autocorrelation function and the PSD of signal s(t) where

$$s(t) = Ae^{-2\pi \left(\frac{t}{d}\right)^2}$$

1.2 Determine the order of a lowpass Butterworth filter if it has an equal noise bandwidth to a lowpass RC filter when both filters have equal 3-dB bandwidth.
1.3 Consider a narrowband Gaussian noise with zero mean and variance $1.65V^2$. Determine the mean and variance of the noise envelope and power.

1.4 Given the frequency response of a lowpass filter $H(\omega)$ is

$$H(\omega) = \frac{\alpha}{\beta + j\omega}$$

Determine the filter impulse response.

1.5 Calculate the 8-point DFT of the sequence [1,0,0,1,1,0,1,1]. Check your answer by calculating the DFT to compare with the given sequence.

1.6 Compute, using the appropriate MATLAB commands, the 16-point and 32-point FFT of the sequence x(n) and compare the two FFT, given

$$X(n) = 1 \quad \text{for } n = 0, 1, \dots, 7$$
$$= 0 \quad \text{for } n = 8, 9, \dots, 15$$

1.7 Determine the discrete sequence x(n) given the DFT coefficients are

$$7 + j0$$
$$-3.232 + j4.273$$
$$3 + j0$$
$$0.423 - j3.723$$

1.8 Use the DIT (Cooley-Tukey) algorithm to compute the FFT of the sequence in problem (1.5). Compare your result with the coefficients from DFT.

1.9 Use the appropriate MATLAB commands to calculate the autocorrelation functions and the cross-correlation function of the data sequences $x_1(t)$ and $x_2(t)$ given that

$$x_1(t) = [100110110]$$
$$X_2(t) = [01011011]$$

1.10 A Gaussian noise of two-sided density $\frac{N_0}{2}$ is applied at input of an RC lowpass filter. Determine the autocorrelation of the noise at the filter output. Plot the autocorrelation function verses the shift τ.

Bibliography and further reading

Bracewell, R.N. (1986) *The Fourier Transform and its Applications*, McGraw-Hill.

Brigham, E.O. (1974) *The Fast Fourier Transform*, Prentice-Hall.

Carlson, A.B. (1986) *Communication Systems*, McGraw-Hill.

Cooley, J. W. and Tukey, J. W. (1965) An algorithm for the machine calculation of complex Fouriers series, Math. Comput. 19, 297–301.

Couch, L.W. (1993) *Digital and Analog Communication Systems*, Macmillan.

Ifeachor, E.C. and Jervis, B.W. (2002) *Digital Signal Processing: A practical approach*, Prentice Hall.

Poularikas, A.D. and Seely, S. (1991) *Signals and Systems*, PWS Kent.

Taub, H. and Schilling, D.L. (1986) *Principles of Communication Systems*, McGraw-Hill.

Appendix 1.A

x	Q(x)	x	Q(x)	x	Q(x)
0.00	5.0000e-001	1.70	4.4565	3.40	3.3693
0.05	4.8006	1.75	4.0059	3.45	2.8029
0.10	4.6017	1.80	3.5930	3.50	2.3263
0.15	4.4038	1.85	3.2157	3.55	1.9262
0.20	4.2074	1.90	2.8717	3.60	1.5911
0.25	4.0129	1.95	2.5588	3.65	1.3112
0.30	3.8209	2.00	2.2750	3.70	1.0780
0.35	3.6317	2.05	2.0182	3.75	8.8417e-005
0.40	3.4458	2.10	1.7864	3.80	7.2348
0.45	3.2636e	2.15	1.5778	3.85	5.9059
0.50	3.0854	2.20	1.3903	3.90	4.8096
0.55	2.9116	2.25	1.2224	3.95	3.9076
0.60	2.7425	2.30	1.0724	4.00	3.1671
0.65	2.5785	2.35	9.3867e-003	4.05	2.5609
0.70	2.4196	2.40	8.1975	4.10	2.0658
0.75	2.2663	2.45	7.1428	4.15	1.6624
0.80	2.1186	2.50	6.2097	4.20	1.3346
0.85	1.9766	2.55	5.3861	4.25	1.0689
0.90	1.8406	2.60	4.6612	4.30	8.5399e-006
0.95	1.7106	2.65	4.0246	4.35	6.8069
1.00	1.5866	2.70	3.4670	4.40	5.4125
1.05	1.4686	2.75	2.9798	4.45	4.2935
1.10	1.3567	2.80	2.5551	4.50	3.3977
1.15	1.2507	2.85	2.1860	4.55	2.6823e
1.20	1.1507	2.90	1.8658	4.60	2.1125
1.25	1.0565	2.95	1.5889	4.65	1.6597
1.30	9.6800e-002	3.00	1.3499	4.70	1.3008
1.35	8.8508	3.05	1.1442	4.75	1.0171
1.40	8.0757	3.10	9.6760e-004	4.80	7.9333e-007
1.45	7.3529	3.15	8.1635	4.85	6.1731
1.50	6.6807	3.20	6.8714	4.90	4.7918
1.55	6.0571	3.25	5.7703	4.95	3.7107
1.60	5.4799	3.30	4.8342	5.00	2.8665
1.65	4.9471	3.35	4.0406		

2

Introduction to Digital Communications

2.1 Introduction

Digital communications refer to the exchange of digital information (electronically) from one location (source) to another (destination). It requires an electronic system at the source to convert the information from its natural form in analogue – such as a human voice – video picture or music to digits – such as binary-coded numbers, graphic symbols, microprocessor codes or data-base information. These digits are then processed (compressed, filtered, pulse shaped) to a form most suitable for sending (transmitting) over the media (channel). Consequently, the communication system carrying the digital information comprises of three main parts: processing and transmission at source; a communication channel over which the digital information is transmitted; and receiving and processing at the destination. Each application has its own communication system and the elements in the system are unique to the particular application. Modern communication systems are developed for application in mobile communications and ubiquitous networks for data communications.

The purpose of this chapter is to review the essential elements used in communication systems and based on spread spectrum techniques described in IS-95 and UMTS standards. The chapter starts with a reminder of the basics for the digital transmission theory in Section 2.2 and gradually introduces the reader to the principles of matched filtering in Section 2.2.6, and raised cosine pulse shaping in Section 2.2.9. Channel equalization techniques are dealt with in Section 2.3 and the modulation in Section 2.4. The digital modulation and demodulation used in IS-95 and UMTS standard-based system, such as the Quadrature Phase Shift Keying (QPSK) and its offset variant O-QPSK, are also explored in Section 2.4. The Rake receivers normally used in mobile communication systems to combat multipath fading are described in Section 2.5.

The channel coding, introduced in Section 2.6, presents the convolutional codes and its Viterbi decoding algorithm. The Map decoding algorithm for recursive systematic convolutional codes and the turbo coding and its decoding algorithms the Max-Log MAP and the Log Map algorithms are also presented in this section. The Shannon channel capacity and the Shannon limit are presented in Section 2.7 while the performance of an ideal digital communication system, as defined by Shannon theory, is discussed in Section 2.10 and the chapter is summarized in Section 2.9.

2.2 Review of digital transmission theory

The transmission theory deals with the physical layer of the communication system and defines the channel to transport the information from source to destination. The system comprises of three main components: the transmitter, the channel and the receiver (see Figure 2.1).

We will assume that the source outputs a sequence of binary digits, processed by the source encoder to remove any redundancy in the data so that the information is represented by the minimum number of binary digits. The channel encoder adds a certain redundancy to the information that can be used by the receiver to detect or correct errors which occur due to signal corruption by the channel additive noise and interference. The digital modulator maps the discrete signal into appropriate signal waveforms that are sent to the destination through wire or wireless channel.

The transmitted waveforms are frequently corrupted by a Gaussian distributed additive noise, and interference from other transmitters operating within the frequency spectrum. In mobile communication systems, there are two other sources which contribute to the waveform corruption. The first source is due to the fact that there is no line of sight between transmitter and receiver antennas so that the transmitted waveforms would be reflected and diffracted by surrounding obstacles, creating multiple transmission paths for

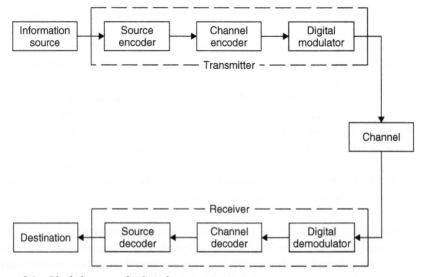

Figure 2.1   *Block diagram of a digital communication system.*

the waveforms to arrive at the receiver. These multiple copies of the original signal arrive at different times and with different phases generating interference between consecutive symbols, called Inter Symbol Interference (ISI) causing received signal fading. The other possible source of signal corruption is due to signals from multiple users transmitted asynchronously generating an interference called Multiple Access Interference (MAI).

The digital demodulator converts the corrupted received waveform back into a sequence of digits, and the channel decoder uses the available redundancy to estimate the original digitals with as small a number of errors as possible. The source decoder constructs an approximation to the original source signal.

Clearly, the communication system requires certain information storage such as buffers together with fast computing processors to achieve an efficient and accepted service quality. The information transmission rate is limited by the channel Shannon capacity. In a multiple users' system, reducing the individual rate would increase the number of users. Furthermore, increasing the channel signal-to-noise power ratio would improve the system performance, as we will see later in the chapter.

2.2.1 Data transmission codes

Transmission codes are data formats which are optimally compatible with the characteristics of the transmission channel. Key issues that guide the choice of any specific format are the code bandwidth requirement and the available channel bandwidth, since a match between the two bandwidths reduces the likelihood of signal distortion. The ease of clock recovery from the received data code is another key issue since availability of a synchronous clock at the receiver is important for symbol synchronization in the detection process, otherwise a fraction of the available transmitted power has to be allocated to the transmission of the clock within the data signal. The DC energy contents of the data code are also a major issue since it represents a reduction in the transmission efficiency.

Common types of transmission codes are: unipolar, polar, and Manchester codes. Unipolar signalling alternates between two states: the low state representing digit '0' of 0 volts and the high state representing digit '1' of +5 volts in the positive logic as shown in Figure 2.2. The negative logic reverse the voltages of the logic states which in logic '0' is +5 volts and logic '1' is 0 volts. The bandwidth (B) of a unipolar signal is inversely proportional to the duration of the data pulse (T_b), that is $B = \frac{1}{T_b}$ and unipolar symbols contain DC component.

In the polar (known also as bipolar or antipodal) signalling, the logic '0' is represented by −5 volts and the logic '1' is represented by +5 volts as shown in Figure 2.3. Again, the bandwidth of polar signal $B = \frac{1}{T_b}$ and polar symbols contain no DC component.

The Manchester code symbols possess one transition in mid-symbol duration and, therefore, such a code is self-clocking, meaning an accurate clock can be reconstructed from the received Manchester symbols. The DC component of the Manchester symbols is zero.

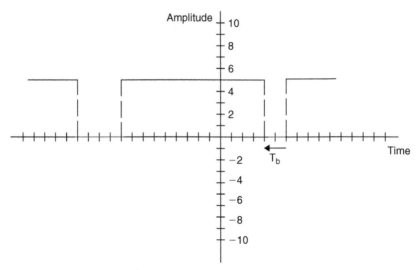

Figure 2.2   *Unipolar line encoding.*

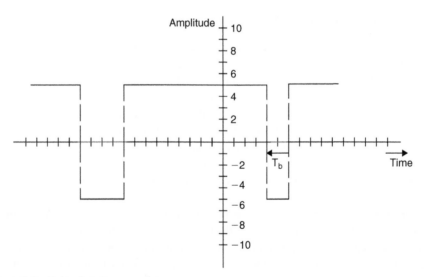

Figure 2.3   *Polar data time waveform.*

Logic '0' is encoded as low-high transition and logic '1' is encoded as high-low transition as shown in Figure 2.4. The bandwidth requirement of Manchester encoded symbols is doubled compared with unipolar and polar encoded digits.

The encoded symbols shown in Figures 2.2 and 2.3 are also known as non-return to zero (NRZ) codes since the amplitude of the encoded symbol is constant during encoded symbol interval. There is another type of encoding called *Return to Zero* (RZ) where the encoded

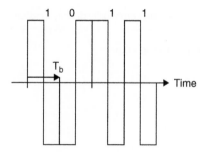

Figure 2.4   *Manchester encoded symbols.*

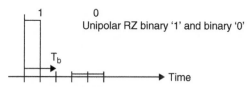

Figure 2.5   *Return to zero encoded unipolar symbols.*

symbol waveform returns to a zero volt level during the second half of the encoded symbol duration as shown in Figure 2.5. The bandwidth requirement of the return to zero unipolar encoded digits is doubled compared with non-return to zero unipolar and polar encoded digits.

The power spectral density of unipolar, polar and Manchester codes are shown in Figure 2.6. The pronounced disadvantage of unipolar and polar codes is the waste of power due to

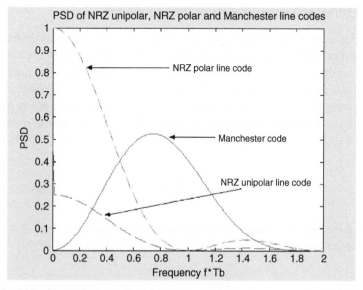

Figure 2.6   *PSD of transmission codes.*

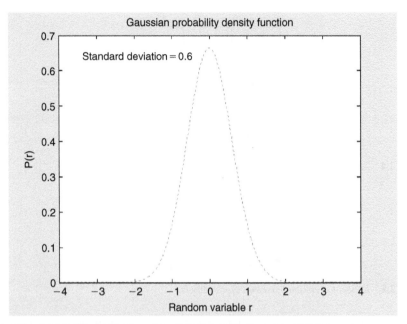

Figure 2.7   *Probability density function of Gaussian distributed random process.*

spectrum approaching DC. The Manchester code has a zero DC level on a bit-by-bit basis and its PSD approaches zero near DC.

2.2.2   General theory of digital transmission

In the development of the digital transmission theory, we will assume that the transmission media is a linear Additive White Gaussian Noise (AWGN) channel with a bandwidth wide enough to accommodate the information without causing any distortion, and the background noise is an additive zero mean white Gaussian process. Since the noise is a random signal, meaning that it is impossible to precisely determine its value at any given time, only statistical estimations are used to predict its behaviour. The additive white noise can be defined using the Gaussian probability distribution expressed in (2.1) and plotted in Figure 2.7:

$$P(r) = \frac{1}{\sigma\sqrt{2\pi}} e^{-\frac{1}{2}\left(\frac{r}{\sigma}\right)^2} \tag{2.1}$$

The white noise has a constant two-sided power spectral density (PSD) $G_n(f)$ over the whole spectrum given by:

$$G_n(f) = \frac{N_0}{2} \text{ w/Hz} \tag{2.2}$$

It is worth noting that samples of the channel noise are uncorrelated and interact with the transmitted symbols independently.

The transmission of N-dimensional signal can be represented by a vector s_i consisting of N elements drawn from a discrete alphabet of size M such that we are dealing with M-ary symbol m_i where $N \leq M$. Each symbol has a priori probability of being transmitted $P(m_i)$ for $I = 1, 2, 3, \ldots, M$. In most practical cases, the symbols are likely to be equally probable such that:

$$P(m_i) = \frac{1}{M} \tag{2.3}$$

For example, an 8-ary symbols is represented by a discrete numbers $[0, 1, 2, \ldots, 7]$ and the transmitted signals are 3-dimensional since $N = 3$ and the binary values s_i are:

$$000, 001, 010, 011, 100, 101, 110, 111$$

We can represent the M-ary N-dimensional signal $s_i(t)$ of duration T seconds generated from vector s_i using a set of N-expansion coefficients (s_{ij}) and N orthogonal basis functions (ϕ_j) (Haykin, 1988):

$$s_i(t) = \sum_{j=1}^{N} s_{ij}\phi_j(t) \quad 0 \leq t \leq T; \; i = 1, 2, \ldots, M \tag{2.4}$$

where the expansion coefficients are defined by:

$$s_{ij} = \int_0^T s_i(t)\phi_j(t)dt \tag{2.5}$$

The real-valued basis functions are normalized to have unit energy and they are orthogonal with respect to each other so that:

$$\int_0^T \phi_i(t)\phi_j(t)dt = 1 \quad \text{if } i = j$$
$$= 0 \quad \text{if } i \neq j \tag{2.6}$$

A block diagram showing the generation of the N-dimensional signal $s_i(t)$ is given in Figure 2.8.

The AWGN channel can be modelled as shown in Figure 2.9.

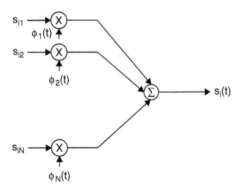

Figure 2.8   *Generation of N-dimensional signal.*

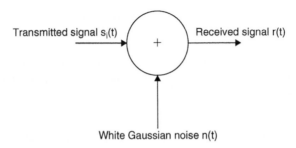

Figure 2.9   *Model of a linear AWGN channel.*

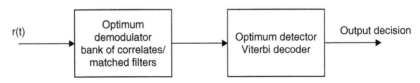

Figure 2.10   *Optimum receiver.*

The received signal at the output of the AWGN channel in the interval $0 \leq t \leq T$ is given by:

$$r(t) = s_i(t) + n(t) \quad i = 1, 2, \ldots, M \tag{2.7}$$

The receiver can be subdivided into two parts (Proakis, 1995): signal demodulator and the detector as shown in Figure 2.10. The function of the demodulator is to convert the received analogue signal r(t) into N-dimensional vector $\mathbf{r} = [r_1, r_2, \ldots, r_N]$. The detector is to decide which of the transmitted signal waveforms was transmitted based on the statistics of $\mathbf{r}$. In the remainder of this section we deal with the optimum demodulator using a bank of correlators or matched filters. The optimum detector uses Viterbi algorithm to minimize the probability of error.

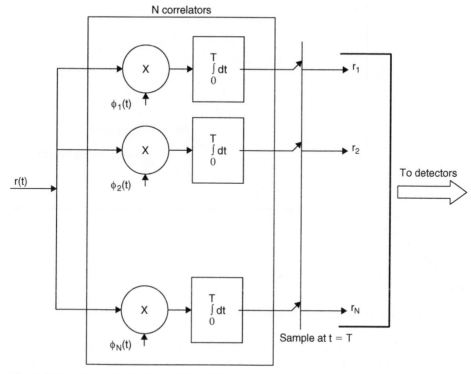

Figure 2.11   *Generation of the set of coefficients {r_i}.*

Using a bank of N correlators shown in Figure 2.11, we can compute the expansion coefficients at the receiver {r_j} for $j = 1, 2, \ldots, N$ so that the jth output of a correlator is given as:

$$\int_0^T r(t)\phi_j(t)dt = \int_0^T [s_i(t) + n(t)]\phi_j(t)dt$$

$$= \int_0^T s_i(t) \cdot \phi_j(t) \cdot dt + \int_0^T n(t) \cdot \phi_j(t) \cdot dt \qquad (2.8)$$

Therefore, $r_i = s_{ij} + n_j \quad j = 1, 2, \ldots\ldots\ldots, N$ \qquad\qquad (2.9)

Where s_{ij} is given by (2.5) and:

$$n_j = \int_0^T n(t) \cdot \phi_j(t) \cdot dt \quad j = 1, 2, \ldots\ldots\ldots\ldots, N \qquad (2.10)$$

The relationship between transmitted signal vector, s_i, the received signal vector, **r**, and noise vector **n** is shown in Figure 2.12.

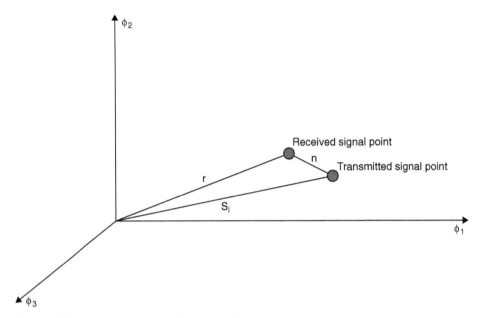

Figure 2.12   *Noise perturbation of transmitted signal.*

The detector then uses these samples to decide which symbol among M possibilities was transmitted. Elements of **r** are independent Gaussian random variables with mean values equal to s_{ij} and variance equal $\frac{N_0}{2}$. Since the elements of the vector **r** are statistically independent, we can express the conditional probability **r** denoted as f_r (**r**|m_i) as the product of the conditional PDF of its individual elements (Haykin, 1988):

$$f_r(\mathbf{r}|m_i) = \prod_{j=1}^{N} f_{r_j}(r_j|m_i) \quad i = 1, 2, \ldots\ldots\ldots\ldots, M \tag{2.11}$$

The conditional PDFs f_r (**r**|m_i) for each symbol M_i, i = 1, 2, ..., M are called *likelihood functions* which are the characterization of the memoryless Gaussian channel. The conditional PDF of each element is defined by Gaussian distribution with mean s_{ij} and variance $\sigma^2 = \frac{N_0}{2}$. Thus, using (2.1) we have:

$$f_{r_j}(r_j|m_i) = \frac{1}{\sqrt{\pi N_0}} \exp\left[-\frac{1}{N_0}(r_j - s_{ij})^2\right] \quad \begin{array}{l} j = 1, 2, \ldots\ldots, N \\[4pt] i = 1, 2, \ldots\ldots, M \end{array} \tag{2.12}$$

Substituting (2.12) in (2.11), we define the likelihood functions of the AWGN channel as:

$$f_r(\mathbf{r}|m_i) = \frac{1}{\sqrt{(\pi N_0)^N}} \exp\left[-\frac{1}{N_0}\sum_{j=1}^{N}(r_j - s_{ij})^2\right] \tag{2.13}$$

We now consider the maximum-likelihood detection of equal probable symbols where we use the observations **r** of the transmitted symbol m_i to find its estimate $\hat{m}_i$ which would minimize the average probability of symbol error. The optimum decision rule is defined as:

$$\text{Set} \quad \hat{m}_i = m_i \quad \text{if} \quad P(m_i \text{ sent}|\mathbf{r}) \geq P(m_k \text{ sent}|\mathbf{r}) \quad \text{for all } i \neq k \quad (2.14)$$

Using Bayes's rule, the left-hand side of (2.14) can be simplified as:

$$\text{Set} \quad \hat{m}_i = m_i \quad \text{if} \quad [P(m_i)/f_r(\mathbf{r})][f_r(\mathbf{r}|m_i)] \text{ is maximum for } i = k \quad (2.15)$$

Now since the priori $P(m_i)$ is the same for equal probable symbols and $f_r(\mathbf{r})$ is independent of the transmitted symbols, we can write (2.15) as:

$$\text{Set} \quad \hat{m}_i = m_i \quad \text{if} \quad f_r(\mathbf{r}|m_i) \text{ is maximum for } i = k \quad (2.16)$$

The likelihood function $f_r(\mathbf{r}|m_i)$ is always positive, the log-likelihood function is

$$\text{Set} \quad \hat{m}_i = m_i \quad \text{if} \quad \ln f_r(\mathbf{r}|m_i) \text{ is maximum for } i = k \quad (2.17)$$

The Maximum Likelihood (ML) detector computes the metric $\ln f_r(\mathbf{r}|m_i)$ for each transmitted m_i and compares them to decide in favour of the symbol which maximizes the metric $\ln f_r(\mathbf{r}|m_i)$.

Taking the natural log of both sides of (2.13), we get the log likelihood function as the following metric:

$$\ln [f_r(\mathbf{r}|m_i)] = -\frac{N}{2} \ln (\pi N_0) - \frac{1}{N_0} \left[\sum_{j=1}^{N} (r_j - s_{ij})^2 \right] \quad i = 1, 2, \ldots\ldots\ldots, M \quad (2.18)$$

The 1st term $\left(-\frac{N}{2} \ln (\pi N_0)\right)$ gives a constant value which is the same for all the symbols, so we can ignore this term and restate the decision rule as:

$$\text{Set} \quad \hat{m}_i = m_i \quad \text{if} \quad -\frac{1}{N_0} \left[\sum_{j=1}^{N} (r_j - s_{ij})^2 \right] \text{ is maximum for } i = k \quad (2.19)$$

which is equivalent to:

$$\text{Set} \quad \hat{m}_i = m_i \quad \text{if} \quad \left[\sum_{j=1}^{N} (r_j - s_{ij})^2 \right] \text{ is minimum for } i = k \quad (2.20)$$

Note that $\left[\sum_{j=1}^{N} (r_j - s_{ij})^2 \right] = \|r - s_i\|^2$ where $\|r - s_i\|$ is the distance between the received signal point and the transmitted signal point shown in Figure 2.12. This distance is called the '*Euclidean distance*'. Therefore, the *ML detector* decides in favour of the symbols that minimizes the Euclidean distance and is also called '*minimum distance detector*'.

2.2.3   *Statistical detection theory for binary transmission*

Consider a source of binary data emitting equal probable symbols. Symbol '0' is represented by V_0 volt and binary '1' by V_1. The PDF of the binary data is binary as well such that binary '0' PDF is $\frac{1}{2}\delta(x - V_0)$ and binary '1' PDF is $\frac{1}{2}\delta(x - V_1)$. The PDF of received signal plus AWGN is given by the convolution of their individual PDFs and, since they are independent random variables (information signal and Gaussian noise are independent with respect to each other), the data PDF will bias the bell shaped PDF of the AWG noise at V_0 and V_1 as shown in Figure 2.13.

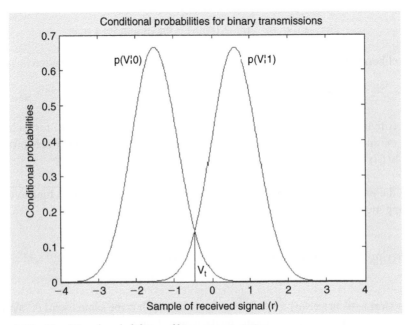

Figure 2.13   *Conditional probabilities of binary transmission.*

Let us now consider two *dependent* random events A and B. Using Bayes's rule, their joint probability P(A,B) can be expressed in terms of their conditional probabilities:

$$P(A, B) = P(A) \cdot P(B|A)$$
$$= P(B) \cdot P(A|B) \tag{2.21}$$

Thus
$$P(A|B) = \frac{P(A) \cdot P(B|A)}{P(B)} \tag{2.22}$$

where P(A) and P(B) are a priori probabilities of random events A and B, respectively; P(A|B) is the probability of event A conditioned event B occurred and P(B|A) is the probability of event B conditioned event A occurred. Let the received sample be V and let event A correspond to sending binary '0' and event B correspond to receiving sample V then substitute in (2.22) and we get:

$$P(0|V) = \frac{P(0) \cdot P(V|0)}{P(V)} \tag{2.23}$$

Similarly when event A corresponds to sending binary '1' we get:

$$P(1|V) = \frac{P(1) \cdot P(V|1)}{P(V)} \tag{2.24}$$

The optimum decision rule that produces minimum number of symbols in error can be defined as follows:

If the probability of receiving binary '0' > probability of receiving binary '1' when in both cases the received sample is V, then decode binary '0' else decode binary '1'.

This rule can be expressed mathematically as follows:

$$P(0|V) \underset{>}{\overset{1}{\underset{0}{<}}} P(1|V) \tag{2.25}$$

Substituting (2.23) and (2.24) in (2.25) we get:

$$\frac{P(0) \cdot P(V|0)}{P(V)} \underset{>}{\overset{1}{\underset{0}{<}}} \frac{P(1) \cdot P(V|1)}{P(V)} \tag{2.26}$$

We can simplify (2.26) as:

$$\frac{P(V|0)}{P(V|1)} \underset{>}{\overset{1}{\underset{0}{<}}} \frac{P(1)}{P(0)} \tag{2.27}$$

Define the likelihood ratio λ as:

$$\lambda = \frac{P(V|0)}{P(V|1)} \tag{2.28}$$

Therefore,

$$\lambda = \frac{P(1)}{P(0)} \tag{2.29}$$

For AWGN channel, the conditional PDFs $P(V|0)$ and $P(V|1)$ can be expressed as:

$$P(V|0) = \frac{1}{\sigma\sqrt{2\pi}} \exp\left[-\frac{(V-V_0)^2}{2\sigma^2}\right] \tag{2.30}$$

$$P(V|1) = \frac{1}{\sigma\sqrt{2\pi}} \exp\left[-\frac{(V - V_1)^2}{2\sigma^2}\right] \tag{2.31}$$

Substituting (2.30) and (2.31) in (2.27) we get:

$$\frac{\exp\left[-\dfrac{(V - V_0)^2}{2\sigma^2}\right]}{\exp\left[-\dfrac{(V - V_1)^2}{2\sigma^2}\right]} \overset{1}{\underset{\underset{0}{>}}{<}} \frac{P(1)}{P(0)} \tag{2.32}$$

Taking $\log_e$ for both sides of (2.32) we get:

$$2V(V_0 - V_1) + (V_1^2 - V_0^2) \overset{1}{\underset{\underset{0}{>}}{<}} 2\sigma^2 \cdot \ln\frac{P(1)}{P(0)} \tag{2.33}$$

Therefore, the threshold voltage V_t for minimum probability of error is given by:

$$2V_t(V_0 - V_1) - (V_0^2 - V_1^2) = 2\sigma^2 \cdot \ln\frac{P(1)}{P(0)}$$

The optimum threshold voltage V_t is:

$$V_t = \frac{(V_0^2 - V_1^2) + 2\sigma^2 \ln\dfrac{P(1)}{P(0)}}{2(V_0 - V_1)} \tag{2.34}$$

Example 2.1

A digital signal, represented by a binary voltage, +5.5 volts and −4.5 volts, is transmitted over an AWGN channel with noise standard deviation 1.12 volt. Compute the voltage of the optimum threshold when the a priori probabilities of the binary source are as follows:

  i. $P(0) = 0.4$
 ii. $P(1) = 0.25$
iii. $P(0) = 0.5$

Solution
Since $P(0) + P(1) = 1.0$

Standard deviation $= \sigma = 1.12$ so $\sigma^2 = 1.25$

  i. $P(0) = 0.4$ and $P(1) = 0.6$
 ii. $P(1) = 0.25$ and $P(0) = 0.75$
iii. $P(0) = 0.5$ and $P(1) = 0.5$

Substituting the above values in (2.32) gives:

i. $V_t = 0.45$ volt
ii. $V_t = 0.64$ volt
iii. $V_t = 0.5$ volt

We are ready to derive the optimum threshold voltage V_t for equal probable data symbol as follows.

2.2.4 Optimum threshold voltage

2.2.4.1 Polar data

For this case, $V_0 = -a$ volts and $V_1 = a$ volts. $P(0) = P(1) = 0.5$. Substituting in (2.34) we get:

$$V_t = -\frac{2\sigma^2 \ln \frac{P(1)}{P(0)}}{2a}$$

Since $P(0) = P(1)$ so that:

$$V_t = 0 \text{ volt} \tag{2.35}$$

2.2.4.2 Unipolar data

The binary symbols $V_0 = 0$ volt and $V_1 = a$ volts so that:

$$V_t = \frac{a^2 + 2\sigma^2 \ln \frac{P(0)}{P(1)}}{2a}$$

But $P(0) = P(1)$ so that:

$$V_t = \frac{a}{2} \text{ volt} \tag{2.36}$$

2.2.5 Minimum probability of error

There are two possibilities for an error to occur in binary transmission through AWGN channel: binary '1' in error or binary '0' in error. Since these two errors are independent, the total probability of error is the sum of these probabilities. Each probability of error is given by the a priori probability of the individual symbol being sent times the area under the PDF tails on either side of the optimum threshold voltage V_t.

Let us denote these probabilities by $P(\text{error}|V_0)$ and $P(\text{error}|V_1)$. Expressed mathematically, when binary '0' is sent the probability of making an error is $[P(0) \times P(\text{error}|V_0)]$ and when binary '1' sent the probability of making error is $[P(1) \times P(error|V_1)]$. Therefore, the total probability of making error P_e is given by:

$$P_e = P(0) \cdot P(error|V_0) + P(1) \cdot P(error|V_1) \tag{2.37}$$

Now we express the tails area of the two PDFs as follows:

$$P(error|V_0) = \int_{V_t}^{\infty} P(V|V_0)dV \tag{2.38}$$

$$P(error|V_1) = \int_{-\infty}^{V_t} P(V|V_1)dV \tag{2.39}$$

Substituting (2.30) and (2.31) in (2.38) and (2.39), respectively, we get:

$$P(error|V_0) = \int_{V_t}^{\infty} \frac{1}{\sigma\sqrt{2\pi}} \exp\left[-\frac{(V-V_0)^2}{2\sigma^2}\right] dV \tag{2.40}$$

To simplify (2.40), we write:

$$x = \frac{V - V_0}{\sigma} \text{ so that } dx = \frac{dV}{\sigma} \text{ and when } V = V_t \text{ then } x = \frac{V_t - V_0}{\sigma}.$$

Therefore, (2.40) now becomes:

$$P(error|V_0) = \int_{x}^{\infty} \frac{1}{\sqrt{2\pi}} \exp\left[-\frac{x^2}{2}\right] dx = Q(x)$$

$$P(error|V_0) = Q\left(\frac{V_t - V_0}{\sigma}\right) \tag{2.41}$$

Similarly, we can show that:

$$P(error|V_1) = Q\left(\frac{V_1 - V_t}{\sigma}\right) \tag{2.42}$$

Note that in deriving (2.42) we use the following symmetry property inherited in Figure 2.13:

$$\int_{-\infty}^{V_t} P(V|V_1)dV = \int_{-V_t}^{\infty} P(V|V_1)dV \tag{2.43}$$

Substituting (2.41) and (2.42) in (2.37) we have:

$$P_e = P(0) \cdot Q\left(\frac{V_t - V_0}{\sigma}\right) + P(1) \cdot Q\left(\frac{V_1 - V_t}{\sigma}\right) \tag{2.44}$$

For equal probable data $P(0) = P(1) = 0.5$ and $V_t = \frac{V_0 + V_1}{2}$ so that (2.44) becomes:

$$P_e = Q\left(\frac{V_1 - V_0}{2\sigma}\right) \tag{2.45}$$

Let $\Delta V = V_1 - V_0$, then (2.45) reduces to (Glover and Grant, 1998):

$$P_e = Q\left(\frac{\Delta V}{2\sigma}\right) \tag{2.46}$$

It is more convenient to express (2.46) in terms of signal-to-noise power ratio rather than ΔV. Consider first the unipolar encoded binary symbols so that the normalized peak power S_{peak} is:

$$S_{peak} = \Delta V^2 \tag{2.47}$$

then the average signal power S is:

$$S = \frac{\Delta V^2}{2} \tag{2.48}$$

The normalized Gaussian noise power N is:

$$N = \sigma^2 \tag{2.49}$$

Thus, for unipolar encoded binary signalling:

$$P_e = Q\left(\sqrt{\frac{1}{2}\frac{S}{N}}\right) \tag{2.50}$$

The peak and average signal power in polar signalling are equal:

$$S_{peak} = S = \left(\frac{\Delta V}{2}\right)^2 \tag{2.51}$$

and the probability of error for polar signalling is:

$$P_e = Q\left(\sqrt{\frac{S}{N}}\right) \tag{2.52}$$

A comparison of (2.50) with (2.52) indicates an advantage in P_e of 3 dB when polar data rather than unipolar data signalling is used assuming signalling level is kept the same in both systems. Both (2.50) and (2.52) are plotted in Figure 2.14.

2.2.6 Principles of matched filtering

The matched filter is a linear filter which maximizes the input Signal-to-Noise power Ratio (SNR) at its output. The impulse response of the matched filter is derived from the

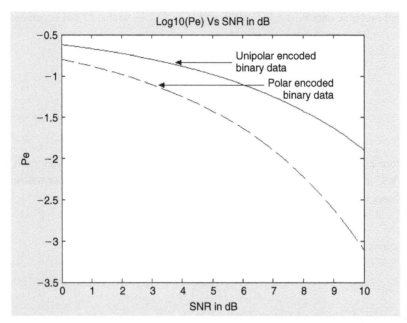

Figure 2.14 *Probability of error for NRZ Unipolar and NRZ Polar data transmitted through Gaussian channel.*

waveform applied to its input. Such characteristics of the matched filter make it unique from any known filter in the sense that each received waveform has its own equivalent matched filter. The filter effectively matches its frequency response to the spectrum of the input waveforms to minimize the noise power input to the detector.

Consider the filter shown in Figure 2.15 where the received signal $s_i(t)$ is corrupted by additive white Gaussian noise $n(t)$. The output signal of the filter is $s_0(t)$ and the output noise is $n_0(t)$. The input signal $s_i(t)$ is considered to be discrete such that:

$$s_i(t) = 0 \quad \text{for } t < 0 \quad \text{and} \quad t > T \tag{2.53}$$

where T is symbol duration of the discrete signal.

The purpose of the matched filtering is to maximize the SNR at its output, i.e. at $t = T$.

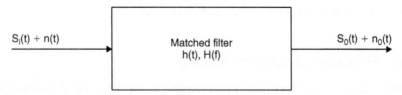

Figure 2.15 *Input–output of matched filter.*

The output signal power at $t = T$ is $S = |s_0(T)|^2$ (2.54)

The average output noise power at $t = T$ is $N = E[n_0^2(T)]$ (2.55)

Thus the SNR at filter output at $t = T$ is $\dfrac{S}{N} = \dfrac{|s_0(T)|^2}{E[n_0(T)]}$ (2.56)

Denote the impulse response and the frequency transfer function of the matched filter h(t) and H(f), respectively, so that:

$$S_0(f) = S_i(f) \cdot H(f)$$ (2.57)

The output signal $s_0(t)$ is given by inverse Fourier transform of the right-hand side of (2.57):

$$s_0(t) = \int_{-\infty}^{\infty} S_0(f) \cdot H(f) \cdot e^{j\omega t} \cdot df$$ (2.58)

The output signal power $S = |s_0(T)|^2$ at $t = T$ is given by:

$$|s_0(T)|^2 = \left| \int_{-\infty}^{\infty} S_i(f) \cdot H(f) \cdot e^{j\omega T} \cdot df \right|^2$$

The output noise power N is given by:

$$N = E[n_0^2(T)] = \int_{-\infty}^{\infty} \frac{1}{2} N_0 |H(f)|^2 \cdot df$$ (2.59)

Therefore,
$$\frac{S}{N} = \frac{\left| \int_{-\infty}^{\infty} S_i(f) \cdot H(f) \cdot e^{j\omega T} \cdot df \right|^2}{\frac{1}{2} N_0 \int_{-\infty}^{\infty} |H(f)|^2 \cdot df}$$ (2.60)

We use the following Schwartz inequality to simplify (2.60):

$$\left| \int_{-\infty}^{\infty} U(f) \cdot V(f) \cdot df \right|^2 \leq \int_{-\infty}^{\infty} |U(f)|^2 \cdot df \cdot \int_{-\infty}^{\infty} |V(f)|^2 \cdot df$$ (2.61)

where U(f) and V(f) can be complex variables. Both sided of (2.61) are equal when

$$U(f) = C \cdot V^*(f)$$ (2.62)

where C is real constant and $V^*(f)$ denotes the complex conjugate of $V(f)$. Now substituting $U(f)$ and $V(f)$ in (2.60) we get:

$$U(f) = H(f) \tag{2.63}$$

$$V(f) = S_i(f) \cdot e^{j\omega T} \tag{2.64}$$

$$\left| \int_{-\infty}^{\infty} S_i(f) \cdot H(f) \cdot e^{j\omega T} \cdot df \right|^2 \leq \int_{-\infty}^{\infty} |H(f)|^2 \cdot df \cdot \int_{-\infty}^{\infty} |S_i(f) \cdot e^{j\omega T}|^2 \cdot df \tag{2.65}$$

Substituting (2.65) in (2.60), we get:

$$\frac{S}{N} \leq \frac{\displaystyle\int_{-\infty}^{\infty} |S_i(f)|^2 \cdot df}{\frac{1}{2}N_0} \tag{2.66}$$

Using Parseval theorem:

$$\int_{-\infty}^{\infty} |S_i(f)|^2 \cdot df = \int_{-\infty}^{\infty} s_i^2(t) \cdot dt = E_b \tag{2.67}$$

E_b = energy per bit of received signal at filter input
N_0 = one-sided PSD of AWGN at filter input

Therefore, (2.66) simplified to (Proakis, 1995):

$$\frac{S}{N} \leq \frac{E_b}{\frac{1}{2}N_0} \tag{2.68}$$

Therefore, maximum SNR at filter output at $t = T$ is given by:

$$\frac{S}{N} = \frac{E_b}{\frac{1}{2}N_0} \tag{2.69}$$

The significance of (2.69) is that the matched filter output SNR at $t = T$ is related to measurable parameters (E_b, N_0) at the input of the filter. Thus maximizing SNR at detector input (matched filter output) requires a matching of the frequency transfer function of the matched filter to the spectra of the received signal. No other lowpass filters, such as Butterworth or Chebyshev, are capable of such spectrum matching and, consequently, their output SNR is always less than $\frac{E_b}{\frac{1}{2}N_0}$.

2.2.7 Matched filter impulse response h(t)

In the derivation of (2.69) we make use of (2.62) which can be re-written as:

$$H(f) = C \cdot S_i^*(f) \cdot e^{-j\omega T} \tag{2.70}$$

Taking the inverse Fourier transform of (2.70) we get:

$$h(t) = \int_{-\infty}^{\infty} H(f) \cdot e^{j\omega t} \cdot df$$

$$= \int_{-\infty}^{\infty} C \cdot S_i^*(f) \cdot e^{j\omega(t-T)} \cdot df$$

$$= C \int_{-\infty}^{\infty} S_i(-f) \cdot e^{j\omega(t-T)} \cdot df$$

$$= C \int_{-\infty}^{\infty} S_i(f) \cdot e^{j\omega(T-t)} \cdot df \tag{2.71}$$

The integral (2.71) is the inverse Fourier transform of $S_i(f)$ but with $s_i(t)$ varying not with time t but with $T - t$. Therefore:

$$h(t) = C \cdot s_i(T - t) \quad \text{for } 0 \le t \le T \tag{2.72}$$

We now show that matched filtering is in fact correlation processing, and that the matched filter itself is nothing but a correlator. We start by using the fundamental theory of filtering (i.e. convolution) relating the output and the input signals. Referring to Figure 2.15, we have:

$$s_0(t) = s_i(t) * h(t) \tag{2.73}$$

where * denoted a convolution process. Using (2.72) to substitute for $h(t)$ in (2.73) we get:

$$s_0(t) = s_i(t) * C \cdot s_i(T - t) \tag{2.74}$$

We express the convolution in (2.74) in terms on an integral:

$$s_0(t) = C \cdot \int_0^t s_i(\tau) \cdot s_i(T - (t - \tau)) \cdot d\tau$$

$$s_0(t) = C \cdot \int_0^t s_i(\tau) \cdot s_i(T - t + \tau) \cdot d\tau$$

Let $\upsilon = T - t$ then

$$s_0(t) = C \cdot \int_0^t s_i(\tau) \cdot s_i(\upsilon + \tau) \cdot d\tau$$

$$= C \cdot R_{s_i}(\upsilon) = R_{si}(T - t) \quad \text{for } 0 \le t \le T \tag{2.75}$$

R_{s_i} is time autocorrelation function of the input signal $s_i(t)$ with time reversed and shifted by T.

Example 2.2

A discrete signal s(t) is transmitted through an AWGN channel where the white noise n(t) has a two-sided noise spectral density equal $\frac{N_0}{2}$. Given that the signal is:

$$s(t) = \begin{array}{ll} A \cdot \cos \omega_0 t & 0 \le t \le T \\ 0 & \text{otherwise} \end{array}$$

  i. find the matched filter impulse response of s(t)
 ii. the SNR at the output of the matched filter
iii. for high frequency signal pulse within T (i.e. $\omega_0 T \gg 1$) what does SNR reduces to?

Solution

  i. The matched filter for an input s(t) has an impulse response h(t) given by s(T − t)

$$h(t) = s(T - t) = \begin{array}{ll} A \cdot \cos \omega_0(T - t) & 0 \le t \le T \\ 0 & T \le t \le \infty \end{array}$$

 ii. The output $SNR = \dfrac{E_b}{\frac{1}{2} N_0}$

$$E_b = \int_0^T [s(t)]^2 dt = A^2 \cdot \int_0^T \cos^2 \omega_0 t \cdot dt$$

$$\text{Now} \qquad = \frac{A^2}{2} \cdot \int_0^T [\cos(2\omega_0 t) + 1] \cdot dt$$

$$= \frac{A^2 \cdot T}{2} \cdot \left[\frac{\sin(2\omega_0 T)}{2\omega_0 T} + 1 \right]$$

Thus the output $SNR = \dfrac{A^2 \cdot T}{N_0} \left[\dfrac{\sin(2\omega_0 T)}{2\omega_0 T} + 1 \right]$

iii. When $\omega_0 T \gg 1$, $\dfrac{\sin(2\omega_0 T)}{2\omega_0 T} \to 0$

Therefore   Output $SNR \to \dfrac{A^2 \cdot T}{N_0}$

Example 2.3

A composite signal, made up of a sinusoidal signal s(t) and an additive white noise n(t) with two-sided spectral density $\frac{N_0}{2}$ W/Hz, is applied to a lowpass RC circuit. Calculate the maximum signal power to noise power ratio $(SNR)_{RC}$ at the circuit output. Compare

(SNR)$_{RC}$ with the maximum signal power to noise power ratio at the output of a matched filter (SNR)$_{MF}$ if the RC circuit is replaced with a matched filter.

Solution

When a pulse of amplitude A and duration T is applied at the input of RC filter, the voltage at the output is:

$$v_0(t) = A\left[1 - \exp\left(-\frac{t}{RC}\right)\right] \quad 0 \le t \le T$$

The signal power at $t = T$ is:

$$S_0 = A^2[1 - \exp(-1)]^2 \quad \text{for } T = RC$$

The average noise power at the RC output is:

$$N = \frac{N_0}{4RC}$$

Thus the SNR at the RC filter output is:

$$(SNR)_{RC} = \frac{A^2 \cdot [1 - \exp(-1)]^2}{N_0} \cdot 4RC$$

The SNR at the output of the matched filter is:

$$(SNR)_{MF} = \frac{E_b}{\frac{1}{2}N_0} = \frac{A^2 \cdot T}{\frac{1}{2}N_0} = \frac{2A^2 \cdot T}{N_0}$$

Now we compare the two SNRs:

$$\frac{(SNR)_{MF}}{(SNR)_{RC}} = \frac{\dfrac{2A^2}{N_0}}{\dfrac{4A^2 \cdot T[1 - \exp(-1)]^2}{N_0}} = \frac{1}{2[1 - \exp(-1)]^2} = 0.97\,\text{dB}$$

2.2.8 Probability of error at the output of matched filter

We now apply the expression for the minimum probability of error optimized with respect to the threshold voltage, given by (2.46), to the output of the matched filter where ΔV resulting in the (difference) energy E_d. Therefore, the SNR at the matched filter output is $\frac{E_d}{\frac{1}{2}N_0}$. The probability of error P_e is:

$$P_e = Q\left(\sqrt{\left(\frac{\Delta V}{2\sigma}\right)^2}\right) = Q\left(\sqrt{\frac{1}{4}\frac{\Delta V^2}{\sigma^2}}\right) \tag{2.76}$$

where

$$E_d = \Delta V^2 \tag{2.77}$$

$$\sigma^2 = \frac{1}{2}N_0 \tag{2.78}$$

The probability of error can be expressed as:

$$P_e = Q\left(\sqrt{\frac{1}{4}\frac{E_d}{\frac{1}{2}N_0}}\right) = Q\left(\sqrt{\frac{1}{2}\frac{E_d}{N_0}}\right) \tag{2.79}$$

When the binary symbols are NRZ unipolar encoded:

$$E_b = \frac{A^2 T_b + 0}{2}$$

$$E_d = A^2 \cdot T_b + 0 = 2E_b \tag{2.80}$$

Therefore,

$$P_e = Q\left(\sqrt{\frac{E_b}{N_0}}\right) \tag{2.81}$$

Similarly for polar binary symbols:

$$E_b = \frac{A^2 T_b + (-A)^2 T_b}{2} = A^2 T_b$$

and

$$E_d = A^2 \cdot T_b + (-A)^2 T_b = 4E_b \tag{2.82}$$

Thus,

$$P_e = Q\left(\sqrt{2\frac{E_b}{N_0}}\right) \tag{2.83}$$

Example 2.4

Consider the transmission of binary symbols through Gaussian channel and matched filtering at the receiver. If $\frac{E_b}{N_0} = 3.4$ dB, calculate the probability of error for NRZ unipolar and NRZ polar signalling.

Solution

$$\frac{E_b}{N_0} = 3.4 \, \text{dB} = 2.19$$

For NRZ unipolar signalling:

$$P_e = Q\left(\sqrt{\frac{E_b}{N_0}}\right) = Q(\sqrt{2.19}) = 0.0694$$

For NRZ polar signalling:

$$P_e = Q\left(\sqrt{2\frac{E_b}{N_0}}\right) = Q(\sqrt{2 \times 2.19}) = 0.0183$$

2.2.9 Binary Nyquist pulse signalling

The ideal pulse shape for baseband signalling over a linear channel such as the AWGN channel is that which generate a flat amplitude response in the frequency range extending from $-B_0$ to B_0 where B_0 is the lowpass bandwidth, equal to half the bit rate $\left(\frac{R_b}{2}\right)$

$$B_0 = \frac{R_b}{2} \tag{2.84}$$

The signalling interval $T_b = \frac{1}{R_b}$. Mathematically, we can express the frequency response $P(f)$ of an ideal pulse shape filter as:

$$P(f) = \frac{1}{R_b}\text{Rect}\left(\frac{f}{R_b}\right) \tag{2.85}$$

The frequency response of the ideal pulse shape filter is plotted in Figure 2.16.

The time domain filtered pulse that produces the ideal brick-shaped response in the frequency domain is:

$$P(t) = \frac{\sin(\pi R_b t)}{\pi R_b t} = \sin c(R_b t) \tag{2.86}$$

The ideal pulse shape in the t-domain (sin c function) is plotted in Figure 2.17 where the correct sampling instants are identified with arrows pointing at them.

Normally synchronous data is signalled through the channel according to a clock set to frequency R_b. For an ideal channel (no distortion or attenuation), the received string of 1s data with sin c pulse waveform are shown in Figure 2.18.

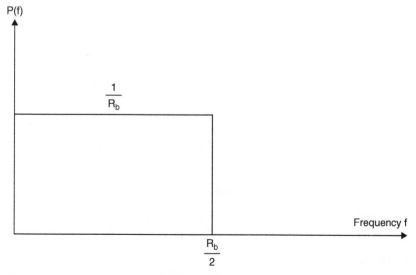

Figure 2.16 *Frequency response of ideal pulse shape filtering.*

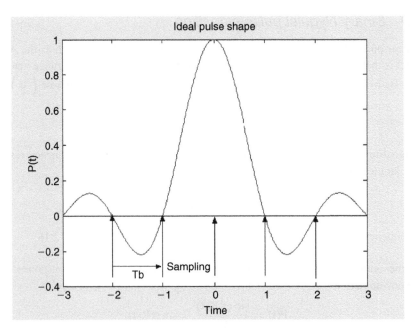

Figure 2.17   *Sin c pulse shape.*

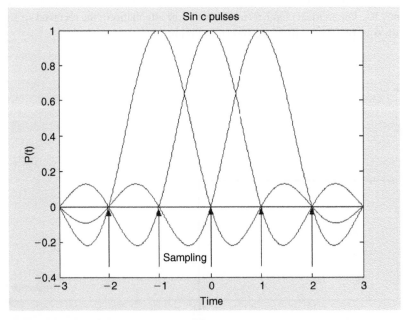

Figure 2.18   *Sequence of sin c pulses.*

Examining Figure 2.18, we can see clearly that sampling at the correct instants will introduce no inter symbol interference from adjacent symbols.

So far we have assumed that the transmission channel is distortion free and with infinite bandwidth to accommodate the spectrum of the rectangular pulses. However, contemporary communication systems operate within band-limited channels. Thus, the rectangular pulses must be filtered before transmission and the effects of the filtering is smearing these pulses into adjacent symbol time slots causing (if this filtering is not of the appropriate type) Inter Symbol Interference (ISI). Therefore, we need a filtering scheme that guarantees the nulling of the ISI at the correct sampling instant.

The time domain signalling scheme of $\frac{\sin x}{x}$ pulse shape eliminates the ISI at the sampling instant constrains and is called the *Nyquist first method*. This method ensures transmission at the maximum rate $2B_0$ where B_0 is the available channel bandwidth. However, such a method encounters practical difficulties since the $\frac{\sin x}{x}$ pulse shape can be physically unrealized. Besides, it requires very accurate synchronization since any timing errors due to inaccurate synchronization result in ISI.

A widely used pulse shape in wireless communications, which is closely approximate to the $\frac{\sin x}{x}$, is the raised-cosine (rc) pulse that exchanges practicability with a slightly wider bandwidth.

The frequency transfer function of the rc filter is given as:

$$H_{rc}(f) = \begin{cases} T_b & 0 \le |f| \le \frac{(1-\alpha)}{2T_b} \\ \frac{T_b}{2}\left[1 + \cos\frac{\pi \cdot T_b}{\alpha}\left(|f| - \frac{1-\alpha}{2T_b}\right)\right] & \frac{(1-\alpha)}{2T_b} \le |f| \le \frac{(1+\alpha)}{2T_b} \\ 0 & |f| > \frac{(1+\alpha)}{2T_b} \end{cases} \quad (2.87)$$

The roll off factor α has values within the range $0 \le \alpha \le 1$. The frequency response of the rc filter for $\alpha = 0, 0.5, 1$ is shown in Figure 2.19. The frequency response drops from unity at $f = 0$ to 0.5 at $f = \frac{R_b}{2}$, that is the 6 dB bandwidth of the filter is equal to $\frac{R_b}{2}$. The impulse response of the rc filter is shown in Figure 2.20 and is given by:

$$h_{rc}(t) = \frac{\sin\frac{\pi \cdot t}{T_b}}{\frac{\pi \cdot t}{T_b}} \cdot \frac{\cos\frac{\alpha \cdot \pi t}{T_b}}{1 - \left(\frac{2 \cdot \alpha \cdot t}{T_b}\right)^2} \quad (2.88)$$

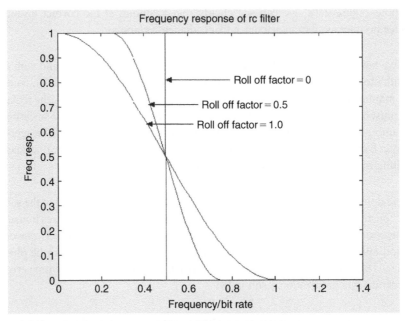

Figure 2.19 *Frequency response of rc filter with α = 0.0, 0.5, 1.0.*

The bandwidth of the binary rc channel, B_{rc} is given by:

$$B_{rc} = (1 + \alpha) \cdot \frac{R_b}{2} \tag{2.89}$$

The ideal raised cosine filter frequency response consists of unity gain at low frequencies, a raised cosine function in the middle, and total attenuation at high frequencies. The width of the middle frequencies is defined by the roll off factor α. The pass band frequency is defined as the 50% signal attenuation (6 dB) point.

It can be seen from Figure 2.20 that the sin x/x (curve with $\alpha = 0.0$) and the raised cosine (rc) functions have the same zero crossings at $t = kT_b$ where $k = \pm1, \pm2, \ldots$, which means that we may transmit data pulses shaped with raised cosine filtering every T_b seconds with zero ISI.

It is more common to utilize the rc filtering as a cascade of two filters: one at the transmitter with frequency transfer function being $H_T(f)$ and the other at the receiver with frequency transfer function being $H_R(f)$ such that:

$$H_T(f) = H_R(f) = \sqrt{H_{rc}(f)} \tag{2.90}$$

Each of the cascaded filters is called square root raised cosine (rrc) filter and collectively achieve the matched rc filtering to eliminate ISI at the sampling instants. Both filters have

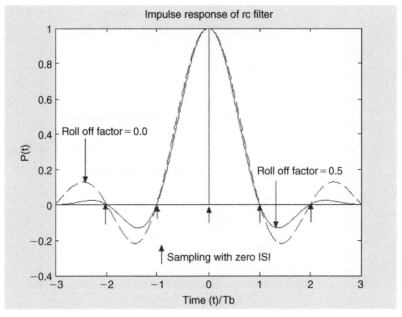

Figure 2.20 *Impulse response of rc filter with α = 0.0, 0.5.*

equal transmission bandwidth as shown in Figure 2.21 (see over). The impulse response of the rrc filter is given by:

$$h_{rrc}(t) = \frac{4\alpha}{\pi\sqrt{T_b}} \cdot \frac{\cos\left(\frac{(1+\alpha)\pi t}{T_b}\right) + \frac{T_b}{4\alpha t}\sin\left(\frac{(1-\alpha)\pi t}{T_b}\right)}{1 - \left(\frac{4\alpha t}{T_b}\right)^2} \qquad (2.91)$$

The impulse response $h_{rrc}(t)$ of a square root raised cosine filter convolved with itself is approximately equal to the impulse response $h_{rc}(t)$ of the raised cosine filter.

2.3 Channel equalizing

The channel bandwidth of the communication systems is limited in order to enable as many users as possible to access the available spectrum. The band-limiting affect of the transmission channel, and the multiple path energy propagation over the wireless channel, cause considerable distortion in the received signal. Severe distortion can result in a tremendous amount of errors in the data recovery. The distortion is mainly caused by the frequency dependent attenuation and stretching of the time domain signal, so that symbols appear beyond their duration causing the ISI.

The most effective technique to encounter signal distortion, in addition to the pulse shaping described in the previous section, is the channel equalization. To activate such a process,

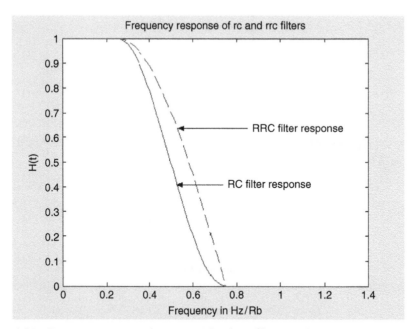

Figure 2.21   *Frequency response of rc compared with rrc filters α = 0.5.*

the channel impulse response has to be known. However, in most cases such channel characterization is unknown at the receiver. Even if it is known at a particular instant of time, it is very likely to be varying with time. The conventional equalization techniques employ a training sequence transmitted in a pre-defined time slot and is known in advance by the receiver. Using the information conveyed to the receiver, fixed or adaptive changes are introduced to the equalizer's coefficients to match the equalizer output closely to the training sequence. The equalized communication system is shown in Figure 2.22.

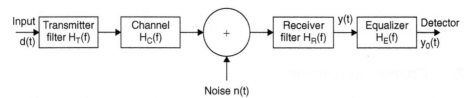

Figure 2.22   *Block diagram of equalized communication system.*

The receiving filter output y(t) is:

$$y(t) = \sum_{k=-\infty}^{\infty} d_k \cdot h'(t - kT_s) + n(t) \qquad (2.92)$$

where d_k is the data at the kth instant of time, T_S is symbol duration, $h'(t)$ is the *equivalent impulse response* of the transmitter filter, and the channel and the receiver filter cascaded, that is:

$$h'(t) = h_T(t) * h_c(t) * h_R(t) \tag{2.93}$$

where $*$ denotes convolution, and $h_T(t)$, $h_C(t)$, $h_R(t)$ are the impulse responses of transmitter filter, channel and receiver filter, respectively. The sampled receiver filter output is (Proakis and Salehi, 2002):

$$y_m = \sum_{k=-\infty}^{\infty} d_k \cdot h'(m-k) + n_m \tag{2.94}$$

$$= d_m h'_{(0)} + \sum_{\substack{k=-\infty \\ k \neq m}}^{\infty} d_k \cdot h'_{(m-k)} + n_m \tag{2.95}$$

where $h'(k) = h'(kT_S)$ and $k = 0, \pm 1, \pm 2, \ldots,$. The 1st term in (2.95) is the desired symbol, the 2nd term represents the ISI and the 3rd term represents the noise sample. It is clear from (2.95) that the ISI term is dependent on the characteristic of the equivalent band-limited transmission channel $h'(t)$. Consequently, the elimination of the ISI necessitates the following characteristics for the (equivalent) channel impulse response:

$$\begin{aligned} h'(m-k) &= 1 \quad m = k \\ &= 0 \quad m \neq k \end{aligned} \tag{2.96}$$

There are two schemes for equalizing (compensating) the signal distortion: *linear schemes* using the transversal filters and the *non-linear schemes* using the decision feedback filters. In the latter, the previously detected symbols are fed back and compared with the input to compensate for the signal in a non-linear method. When the channel has known time invariant frequency response, the parameters of the equalizer can be preset and these equalizers are called *preset equalizers*. However, for time variant channel response, the equalizer parameters need to be updated periodically during transmission. The latter are called *adaptive equalizers*.

2.3.1 Linear equalizers

2.3.1.1 Zero-forcing equalizers

These linear equalizers consist of finite impulse response filters (transversal filters) with $(2N + 1)$ coefficients so that the impulse response of the filter is:

$$h_E(t) = \sum_{k=-N}^{N} f_k \, \delta(t - k\tau) \tag{2.97}$$

where f_k the filter coefficient with k varying between $-N$ to N and τ is time delay between adjacent filter tap. The equivalent pulse input to the equalizer is given by $h'(t)$. So the output of the equalizer $y_0(t)$ is given by:

$$y_0(t) = \sum_{k=-N}^{N} f_k \cdot h'(t - k\tau) + n(t) \qquad (2.98)$$

The samples of the equalizer output taken at $t = mT_S$ where $m = 0, \pm 1, \pm 2, \ldots, \pm N$ are:

$$y_0(mT_s) = \sum_{k=-N}^{N} f_k \cdot h'(mT_s - k\tau) + v_m \qquad (2.99)$$

where v_m represents noise samples at the equalizer output. To reduce the channel distortion due to ISI, we force the condition:

$$\begin{aligned} y_0(mT_s) &= 1 \quad \text{for } m = 0 \\ &= 0 \qquad m \neq 0 \end{aligned} \qquad (2.100)$$

The equalizer described in (2.100) is called *zero-forcing equalizer* and, because it has finite length, eliminates the ISI from a finite number of symbols.

We solve a set of $(2N + 1)$ linear equations to compute the $(2N + 1)$ coefficients that satisfies (2.100). For noise-free transmission, we write (2.99) as:

$$\mathbf{y}_0 = \mathbf{h}'\mathbf{f} \qquad (2.101)$$

where

$$\mathbf{y}_0 = [y_0(-N), y_0(-N + 1), \ldots \ldots \ldots \ldots, y_0(-1), y_0(0), y_0(1), \ldots \ldots \ldots \ldots, y_0(N)]^T$$

$$\mathbf{f} = [f_{-N}, f_{1-N}, \ldots \ldots \ldots \ldots f_{-1}, f_0, f_1, \ldots \ldots \ldots f_N]^T$$

$$\mathbf{h}' = \begin{bmatrix} h'(0)h'(-1) & h'(-2N) \\ h'(1) \ h'(0) & h'(-2N + 1) \\ \\ h'(2N)h'(2N - 1) & h'(0) \end{bmatrix}$$

Therefore,

$$\mathbf{f} = \mathbf{h}'^{-1}\mathbf{y}_0 \qquad (2.102)$$

The variance σ_k^2 of the Gaussian noise component at the receiver filter output is given by:

$$\sigma_k^2 = \int_{-B}^{B} \beta_k(f) \cdot |H_R(f)|^2 \, df \qquad (2.103)$$

where $3_K(f)$ is the spectral density of the noise at the receiver filter input.

$$\beta_k(f) = \frac{N_0}{2} \qquad (2.104)$$

The variance σ_k^2 of the Gaussian noise component v_m and band-limited channel of bandwidth B is given by:

$$\sigma_\nu^2 = \int_{-B}^{B} \beta_k(f) \cdot |H_R(f)|^2 \cdot |H_E(f)|^2 \, df \qquad (2.105)$$

For zero-forcing equalizer, we have:

$$|H_E(f)| = \frac{1}{|H_c(f)|} \qquad (2.106)$$

For root raised cosine pulse filtering:

$$|H_R(f)|^2 = |H_{rc}(f)| \qquad (2.107)$$

Substituting (2.107) in (2.103), we get:

$$\sigma_k^2 = \frac{N_0}{2} \int_{-B}^{B} |H_{rc}(f)| \cdot |H_E(f)|^2 \, df \qquad (2.108)$$

Substituting (2.106) and (2.107) in (2.105) gives:

$$\sigma_\nu^2 = \frac{N_0}{2} \int_{-B}^{B} \frac{|H_{rc}(f)|}{|H_c(f)|^2} \, df \qquad (2.109)$$

When the channel $|H_c(f)|$ is operating in the frequency range $-B \leq f \leq B$. Comparing (2.108) and (2.109) we conclude that, though the zero-forcing equalizer compensates for the channel distortion, it in fact increases the noise power at its output. A block diagram of the linear equalizer is shown in Figure 2.23.

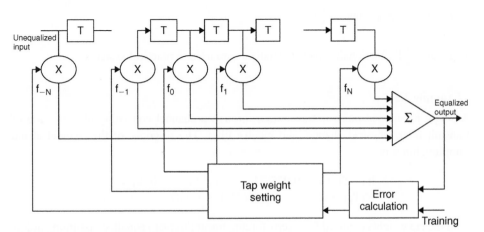

Figure 2.23 *Block diagram on linear equalizer.*

The linear equalizer consists of a tapped delay line which stores the input samples and then outputs a weighted sum of the stored values once per symbol interval. The output is compared with the training sequence to calculate the error. The Wiener algorithm is commonly used to minimize the error and the tap coefficients are updated in preparation for the next symbol interval.

If the time delay τ is equal symbol interval T_S (symbol rate $= R_S$) the linear equalizer is called the *symbol spaced equalizer*. Such equalizers are optimum if preceded by receiving filters matched to the operating characteristics of the channels. However, in such practical cases, the channel characteristics are unknown and consequently the equalizers are sub-optimum. In fact, when $\frac{1}{T_s}$ is less the 2B, the signal above $\frac{1}{T_s}$ (the folding frequency) are aliased into frequencies below $\frac{1}{T_s}$ and the equalizer compensates for the aliased received signal and the channel distortion inherited in the received signal can not be compensated.

On the other hand, when the received signal is sampled at least at twice the symbol rate (i.e. Nyquist rate), so that $\frac{1}{\tau} \geq 2B > \frac{1}{T_s}$, no aliasing occurs and the equalizer compensates for the actual received signal distortion. In such a case when $\tau < T_S$, the equalizer is called a *fractionally spaced equalizer*. Consider raised cosine pulses with roll off factor α. The highest frequency component in the spectrum is:

$$(1+\alpha) \cdot \frac{R_s}{2} = (1+\alpha) \cdot \frac{1}{2T_s} \qquad (2.110)$$

Sampling at Nyquist rate, the highest frequency $= (1+\alpha) \cdot \dfrac{1}{T_s}$ \qquad (2.111)

The samples are passed through the equalizer with tap spacing $\frac{T_s}{1+\alpha}$. Thus when:

$$\alpha = 0 \quad \text{tap spacing} = T_s$$
$$\alpha = 0.5 \quad \text{tap spacing} = \frac{2}{3}T_s$$
$$\alpha = 1 \quad \text{tap spacing} = \frac{1}{2}T_s$$

In general, $\frac{1}{2}T_s$ fractionally spaced equalizers are used in practical systems.

Example 2.5

A single flat-topped pulse is transmitted through a channel consisting of a cascade of transmit filter, physical channel and receive filter. The response of the channel at the sampling times is:

$$h'(-2T_s) = -0.075 \quad h'(-T_s) = 0.2 \quad h'(0) = 1$$
$$h'(T_s) = 0.35 \quad h'(2T_s) = -0.07$$

Find the tap weights required for a zero-forcing linear three-tap equalizer assuming noise-free transmission.

Solution

Use (2.99) with $N = 1$, $\tau = T_s$ and $V_m = 0$ so that (2.99) reduces to:

$$y_0(mT_s) = \sum_{k=-1}^{1} f_k \cdot h'[(m-k)T_s]$$

$$y_0(mT_s) = f_{-1}h'[(m+1)T_s] + f_0h'[(mT_s)] + f_1h'[(m-1)T_s]$$

Now $m = 0, \pm 1$

$$y_0(-T_s) = f_{-1}h'[0] + f_0h'[(-T_s)] + f_1h'[-2T_s]$$

$$y_0(0) = f_{-1}h'[T_s] + f_0h'[0] + f_1h'[-T_s]$$

$$y_0(T_s) = f_{-1}h'[2T_s] + f_0 h'[T_s] + f_1h'[0]$$

Substituting for the channel samples we get

$$\begin{bmatrix} 0 \\ 1 \\ 0 \end{bmatrix} = \begin{bmatrix} 1 & 0.2 & -0.075 \\ 0.35 & 1 & 0.2 \\ -0.07 & 0.35 & 1 \end{bmatrix} \begin{bmatrix} f_{-1} \\ f_0 \\ f_1 \end{bmatrix}$$

Thus $\mathbf{f} = [-0.2685 \quad 1.1803 \quad -0.4319]$

2.3.1.2 *Minimum mean square error (MMSE) equalizers*

The main disadvantage of the zero-forcing equalizer is the enhancement of the noise at the equalizer output. We now relax the condition of zero ISI and aim at minimizing the combined power of the ISI and noise. Consider the output of the linear equalizer expressed in (2.99), but we shall ignore the noise for the moment and set $\tau = T_S$.

$$y_0(mT_s) = \sum_{k=-N}^{N} f_k \cdot y[(mT-k)T_s] \tag{2.112}$$

Let the desired sample at the output of the equalizer at $t = m\,T_S$ be a_m. The mean square error (MSE) between the actual output $y_0(mT_S)$ and the desired output a_m is:

$$MSE = E[y_0(mT_s) - a_m]^2 \tag{2.113}$$

Substituting (2.110) in (2.111) we get:

$$MSE = E\left[\sum_{k=-N}^{N} f_k \cdot y[(m-k)T_s] - a_m \right]^2 \tag{2.114}$$

E in (2.114) represents the ensemble average. Expanding the inside of the brackets of (2.114) we get:

$$MSE = \sum_{k=-N}^{N} \sum_{i=-N}^{N} f_k f_i R_y(k-i) - 2\sum_{i=-N}^{N} R_{ya}(i) + E(a_m^2) \tag{2.115}$$

where R_y and R_{ya} are the autocorrelation of the input samples {y} and the cross-correlation between input samples and the training sequence {a}. Minimization of the MSE in (2.115) is achieved (i.e. MMSE condition) when (Proakis and Salehi, 2002):

$$\sum_{k=-N}^{N} f_k R_y(k-i) = R_{ya}(i) \quad i = 0, \pm 1, \pm 2, \ldots\ldots\ldots, \pm N \qquad (2.116)$$

Expressing (2.116) in Matrix form, the MMSE condition is:

$$\mathbf{R_y f} = \mathbf{R_{ya}} \qquad (2.117)$$

where $\mathbf{f}$ and $\mathbf{R_{ya}}$ are vectors of $(2N+1)$ elements each and $\mathbf{R_y}$ is $(2N+1) \times (2N+1)$ matrix where:

$$\mathbf{R_{ya}} = [R_{ya}(-N)R_{ya}(-N+1)\ldots\ldots\ldots\ldots, R_{ya}(-1), R_{ya}(0),$$
$$R_{ya}(1), \ldots\ldots\ldots\ldots, R_{ya}(N)]^T$$

The correlation functions are given by:

$$R_y(k) = \frac{1}{K}\sum_{i=1}^{K} y(i-k)y(i)$$

$$R_{ya}(k) = \frac{1}{K}\sum_{i=1}^{K} y(i-k)a(i) \qquad (2.118)$$

The optimum tap weights are given by:

$$\mathbf{f} = \mathbf{R_y^{-1} R_{ya}} \qquad (2.119)$$

Example 2.6

The transmission of a signal through AWGN channel results in the following sampled input to the linear equalizer:

$$\mathbf{y} = [0.7 \quad -0.2 \quad 0.5 \quad 0.1 \quad 0.9 \quad -0.3 \quad 0.6 \quad -0.1]$$

The training sequence used is:

$$\mathbf{a} = [1 \quad -1 \quad 1 \quad -1 \quad 1 \quad -1 \quad 1 \quad -1]$$

Determine the tap weights of a three-tap linear equalizer based on the MMSE criterion.

Solution

The discrete correlation functions in (2.116) are given by:

$$R_y(k) = \frac{1}{8} \sum_{i=1}^{8} y(i - k)y(i)$$

$$R_{ya}(k) = \frac{1}{8} \sum_{i=1}^{8} y(i - k)a(i)$$

τ	$R_y(\tau)$	$R_{ya}(\tau)$
−6	0.05	0.1125
−5	−0.0475	−0.1750
−4	0.1225	0.1625
−3	−0.0362	−0.2750
−2	0.1650	0.3125
−1	−0.0762	−0.3875
0	0.2575	0.4000
1	−0.0762	−0.3125
2	0.1650	0.2875
3	−0.0362	−0.2250
4	0.1225	0.2375
5	−0.0475	−0.1250
6	0.0550	0.0875

We use Matlab to calculate tap weights as:

$$f = -0.2337 \quad 0.0954 \quad -0.8756 \quad 1.0512 \quad -0.0749 \quad 0.1663 \quad -0.1447$$

2.3.1.3 Adaptive linear equalizers

So far we have presented algorithms for computing the equalizer coefficients, using the zero forcing and the MMSE criteria, for a channel's characteristics in a given instant of time. If the channel characteristics change with time, the equalizer coefficients need to be updated to combat the change in the interference. Consequently, although our previous algorithms are adequate for minimizing (possibly eliminating) the interference for non-varying channels, they are not suitable for channels which are varying with time and the equalization procedure has to *adapt* to the channel's new characteristic to reduce/eliminate the distortion. These equalizers are called *adaptive equalizers*.

We notice that the equalization algorithms are expressed in a set of linear equations that have the matrix form:

$$\mathbf{A} = \mathbf{B}\,\mathbf{f} \tag{2.120}$$

Where $\mathbf{B}$ is $(2N + 1) \times (2N + 1)$ matrix, $\mathbf{A}$ is $(2N + 1)$ column vector and $\mathbf{f}$ is $(2N + 1)$ column vector representing the equalizer weights. The adaptation of the old weights to the new channel state is generally achieved in an iterative procedure such as the steepest decent scheme. In this scheme an arbitrary weights vector, $\mathbf{f}_0$, is chosen and the gradient vector, $\mathbf{g}_\ell$, is computed from the derivative of the MSE for each of the $(2N + 1)$ weights such that:

$$\mathbf{g}_\ell = \mathbf{B}\mathbf{f}_\ell - \mathbf{A} \quad \ell = 0, 1, 2, \ldots. \tag{2.121}$$

All the iteration procedures for updating the weights have to be completed within the symbol interval to avoid loosing any information. At the end of the ℓth iteration, the new weights are:

$$\hat{\mathbf{f}}_{\ell+1} = \hat{\mathbf{f}}_\ell - \Delta \cdot \hat{\mathbf{g}}_\ell \tag{2.122}$$

Δ is the step-size parameter that greatly affects the convergence of the iteration method. A rough estimate for Δ can be computed using:

$$\Delta = \frac{0.2}{(2N + 1)P_r} \tag{2.123}$$

where P_r is the received signal plus noise power.

It can be shown that an estimate for the gradient vector is (Haykin, 1988):

$$\hat{\mathbf{g}}_\ell = -e_k \cdot \mathbf{y}_k \tag{2.124}$$

where e_k is the error signal between the training sequence symbol a_k and the equalizer output $y_0(kT_S)$, $\mathbf{y}_k$ is the received signal $(2N + 1)$ columns vector.

$$e_k = a_k - y_0(kT_s) \tag{2.125}$$

Substituting (2.124) in (2.122), we get:

$$\mathbf{f}_{\ell+1} = \mathbf{f}_\ell + \Delta \cdot e_k \cdot \mathbf{y}_k \tag{2.126}$$

The adaptation scheme follows (2.126) algorithm is known as Least Mean Square (*LMS*) algorithm.

Example 2.7

We now examine the adaptive equalizer by considering an example. Assume a single pulse of amplitude $(+1)$ transmitted through a band-limited noise-free channel. The sampled output of the receive filter is $\mathbf{y}$

$$\mathbf{y} = [0.32 \quad 0.98 \quad 0.25]$$

Determine a three-tap adaptive linear equalizer when the received power is 0.15 watt. We may assume the initial tap weights to be $[-1 \quad 1 \quad -1]$.

Solution

To give insight to the iteration process of the LMS algorithm which used MMSE criterion, we will show the result at the end of each iteration, and then plot the difference between the equalized output and the desired symbol $a_m = 1$ to show the improvement caused by updating the tap weights. Using (2.123), the best value for Δ which produces minimum mean square error is:

$$\Delta = \frac{0.2}{(2 \times 3 + 1) \times 0.15} = 0.19$$

The output of the equalizer is given by:

$$y_0(mT_s) = \sum_{n=-1}^{1} f_n y[(m - n)T_s]$$

$$= f_{-1}y[(m + 1)T_s] + f_{-2}y(mT_s) + f_1 y[(m - 1)T_s]$$

Therefore, $y_0^k(mT_s) = -y[(m + 1)T_s] + y(mT_s)y[(m - 1)T_s]$

Where k represents the iteration index and:

$$e_m^k = 1 - y_0^k(mT_s)$$

For 1st iteration, $k = 1$

$$\mathbf{f}^1 = [-1 \quad 1 \quad -1]$$
$$y_0^1(mT_s) = -y[(m + 1)T_s] + y(mT_s) - y[(m - 1)T_s]$$

Using (2.126), we get:

$$\mathbf{f}^2 = [-0.9641 \quad 1.1099 \quad -0.9720]$$

For 2nd iteration, $k = 2$

$$y_0^2(mT_s) = -0.9641xy[(m + 1)T_s] + 1.1099xy(mT_s) - 0.9720xy[(m - 1)T_s]$$
$$y_0^2(mT_s) = 0.5362$$
$$e_m^2 = 1 - 0.5362 = 0.46$$
$$\mathbf{f}^3 = [-0.9361 \quad 1.1956 \quad -0.9501]$$

For 3rd iteration, $k = 3$

$$y_0^3(mT_s) = -0.9361xy[(m + 1)T_s] + 1.1956xy(mT_s) - 0.9501xy[(m - 1)T_s]$$
$$y_0^3(mT_s) = 0.6345$$
$$e_m^3 = 1 - 0.6345 = 0.37$$
$$\mathbf{f}^4 = [-0.9136 \quad 1.2644 \quad -0.9326]$$

For 4th iteration, k = 4

$$y_0^4(mT_s) = \mathbf{y} \cdot \mathbf{f}^{4^T}$$
$$y_0^4(mT_s) = -0.9136xy[(m+1)T_s] + 1.2644xy(mT_s) - 0.9326xy[(m-1)T_s]$$
$$y_0^4(mT_s) = 0.7136$$
$$e_m^4 = 1 - 0.7136 = 0.29$$
$$\mathbf{f}^5 = [-0.8960 \quad 1.3184 \quad -0.9188]$$

For 5th iteration, k = 5

$$y_0^5(mT_s) = \mathbf{y} \cdot \mathbf{f}^{5^T} = 0.7757$$
$$e_m^5 = 1 - 0.7757 = 0.22$$
$$\mathbf{f}^6 = [-0.8826 \quad 1.3594 \quad -0.9083]$$

For 6th iteration, k = 6

$$y_0^6(mT_s) = \mathbf{y} \cdot \mathbf{f}^{6^T} = 0.8227$$
$$e_m^6 = 1 - 0.7757 = 0.18$$
$$\mathbf{f}^7 = \mathbf{f}^6 + \Delta e_m^6 \mathbf{y}$$
$$\mathbf{f}^7 = [-0.8717 \quad 1.3929 \quad -0.8998]$$

For 7th iteration, k = 7

$$y_0^7(mT_s) = \mathbf{y} \cdot \mathbf{f}^{7^T} = 0.8612$$
$$e_m^7 = 1 - 0.8612 = 0.14$$
$$\mathbf{f}^8 = \mathbf{f}^7 + \Delta e_m^7 \mathbf{y} = [-0.8632 \quad 1.4190 \quad -0.8932]$$

For 8th iteration, k = 8

$$y_0^8(mT_s) = \mathbf{y} \cdot \mathbf{f}^{8^T} = 0.8911$$
$$e_m^8 = 1 - 0.8911 = 0.11$$
$$\mathbf{f}^9 = \mathbf{f}^8 + \Delta e_m^8 \mathbf{y} = [-0.8565 \quad 1.4395 \quad -0.8879]$$

For 9th iteration, k = 9

$$y_0^9(mT_s) = \mathbf{y} \cdot \mathbf{f}^{9^T} = 0.9146$$
$$e_m^9 = 1 - 0.9146 = 0.09$$
$$\mathbf{f}^{10} = \mathbf{f}^9 + \Delta e_m^9 \mathbf{y} = [-0.8510 \quad 1.4562 \quad -0.8836]$$

For 10th iteration, k = 10

$$y_0^{10}(mT_s) = \mathbf{y} \cdot \mathbf{f}^{10^T} = 0.9339$$
$$e_m^{10} = 1 - 0.9339 = 0.07$$
$$\mathbf{f}^{11} = \mathbf{f}^{10} + \Delta e_m^{10} \mathbf{y} = [-0.8468 \quad 1.4693 \quad -0.8803]$$

For 11th iteration, $k = 11$

$$y_0^{11}(mT_s) = \mathbf{y} \cdot \mathbf{f}^{11^T} = 0.9488$$

$$e_m^{11} = 1 - 0.9488 = 0.05$$

$$\mathbf{f}^{12} = \mathbf{f}^{11} + \Delta e_m^{11}\mathbf{y} = [-0.8437 \quad 1.4786 \quad -0.8779]$$

For 12th iteration, $k = 12$

$$y_0^{12}(mT_s) = \mathbf{y} \cdot \mathbf{f}^{12^T} = 0.9595$$

$$e_m^{12} = 1 - 0.9595 = 0.04$$

$$\mathbf{f}^{13} = \mathbf{f}^{12} + \Delta e_m^{12}\mathbf{y} = [-0.8413 \quad 1.4860 \quad -0.8760]$$

For 13th iteration, $k = 13$

$$y_0^{13}(mT_s) = \mathbf{y} \cdot \mathbf{f}^{13^T} = 0.9681$$

$$e_m^{13} = 1 - 0.9681 = 0.03$$

$$\mathbf{f}^{14} = \mathbf{f}^{13} + \Delta e_m^{13}\mathbf{y} = [-0.8395 \quad 1.4916 \quad -0.8746]$$

Summary of the results:

Iteration index k	Error a_m^k
1	0.59
2	0.46
3	0.37
4	0.29
5	0.22
6	0.18
7	0.14
8	0.11
9	0.09
10	0.07
11	0.05
12	0.04
13	0.03

The results above are plotted in Figure 2.24 (see over).

2.3.2   Non-linear equalizers

2.3.2.1   Decision-feedback equalizers

The linear equalizers perform efficiently in channels with low to moderate ISI such as telephone wire channels. However, in channels with severe ISI and null spectrum,

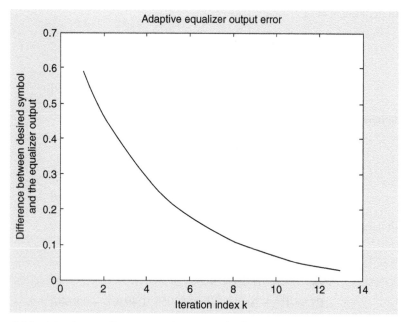

Figure 2.24 *Adaptive equalizer error vs iteration index k.*

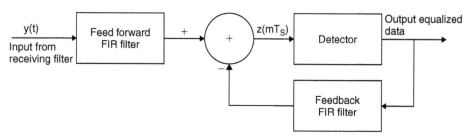

Figure 2.25 *Block diagram of decision feedback equalizer.*

the noise enhancement degrades the probability of error of the channel. Under such operating conditions, the decision-feedback equalizers outperform the linear equalizers' performance.

The decision-feedback equalizer contains two linear FIR filters. The forward filter implemented as *fractionally spaced linear equalizer* with input from the received distorted signal. The feedback filter is implemented as *symbol spaced linear equalizer* where the detected symbol is fed to its input. This feedback loop provides the non-linearity to the equalizer. The block diagram of the decision-feedback equalizer is shown in Figure 2.25.

Let the tap weights of the feed forward filter be denoted by f_n^F with $n = 1, 2, \ldots, N_F$ and f_n^B with $n = 1, 2, \ldots N_B$ for feedback filter. Denote the sampled received signal from the receive filter as y(m T$_S$) then the input to the detector is:

$$z(mT_s) = \text{Feed forward filter output} - \text{feedback filter output}$$

$$= \sum_{n=1}^{N_F} f_n^F y(mT_s - n\tau) - \sum_{n=1}^{N_B} f_n^B a_{m-n} \qquad (2.127)$$

where a$_n$ represents the detected output and $\tau = \frac{T_s}{K}$ is the delay of the fractionally spaced feed forward filter and K is an integer usually K = 2. Work carried out by Proakis (1995) has shown that the probability of error of ML detector using Viterbi algorithm is better by about 4 dB than using decision-feedback equalizers.

2.4 Digital modulation/demodulation schemes used in CDMA systems

In this section, we focus our attention on the modulation system used in CDMA systems currently in service (i.e. the Quadrature Phase Shift Keying (QPSK)). We present the modulation scheme used in IS-95 systems and in the wideband CDMA (3G) systems implemented according to Universal Mobile Telecommunications System (UMTS) standard. The special filtering employed in IS-95 for pulse shaping is presented in Chapter 8. The Root Raised Cosine (RRC) pulse filters used for pulse shaping in UMTS (3G) systems are presented in (2.4) in this chapter.

2.4.1 Quadrature/Offset Phase Shift Keying (QPSK/OQPSK) modulation system

The QPSK modulation is widely employed in many systems such as satellite communications, 3G communications, IS-95 and IS-136 systems using TDMA technology.

The QPSK modulation system converts the digital information to carrier phase changes. The term 'Quadrature' implies that binary information is transported on two orthogonal dimensions, one binary bit on each dimension per each modulation cycle. Since the two dimensions are orthogonal, the information does not interfere with each other even though they occupy the same spectrum. In this sense we can envisage the QPSK modulation system with amplitude A as being two orthogonal Binary Phase Shift Keying systems (BPSK) with identical amplitude $\pm\frac{A}{\sqrt{2}}$, that is BPSK-1 and BPSK-2 and quadrature carriers cos $\omega_c t$ and sin $\omega_c t$, one carrier per dimension as shown in Figure 2.26. In this representation, the QPSK corresponds to an equivalent 4-QAM system.

During each modulation cycle, binary '1' or binary '0' are sent on the individual BPSK carriers as shown in Figure 2.26 where the QPSK symbols are (± 1, y) and (x, ± 1) and E$_b$ is energy per information bit.

Since y is a binary symbol with possible values ± 1, the mapping of the binary bits to the carrier phase may be accomplished in a number of ways. For example, we may map the information [00, 01, 10, 1 1] to carrier phase $\frac{\pi}{4}, \frac{3\pi}{4}, \frac{5\pi}{4}, \frac{7\pi}{4}$. However, the most likely error caused by the noise at the receiver is due to the selection of an adjacent phase to the one transmitted. An error between the first adjacent phases or between the last adjacent phases will only cause a single error in the symbol while an error between the adjacent middle phases will cause two errors in the symbol. To eliminate the possibility of two errors occurring at the same instant, we encode the binary information with Gray code. The Gray code was invented by Frank Gray, a researcher at Bell Labs, in 1953. The Gray code is based on the following algorithm. Let the symbol bits be denoted as b_1, b_2 and Gray encoded bit be denoted by G_k where k = 1, 2 then:

$$G_1 = b_1 \oplus b_2$$
$$G_2 = b_2 \tag{2.128}$$

Using (2.128), the Gray encoded QPSK symbols are given in Table 2.1.

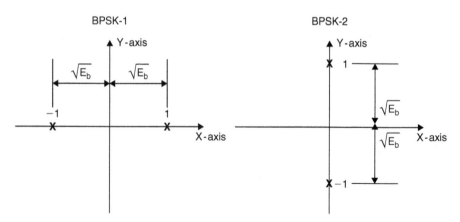

Figure 2.26 *Constellations of two BPSK signals on two orthogonal dimensions.*

Table 2.1 *Gray coded QPSK symbols*

Binary code	Gray code	Carrier phase
0 0	0 0	$\frac{\pi}{4}$
0 1	0 1	$\frac{3\pi}{4}$
1 0	1 1	$\frac{5\pi}{4}$
1 1	1 0	$\frac{7\pi}{4}$

In most QPSK applications, the Gray encoded symbols are mapped to the carrier phase within $\pm\pi$ phase shift anti-clockwise to produce the signal constellation diagram (also called the signal-space diagram) shown in Figure 2.27 which is the combination of the two orthogonal BPSK carriers discussed earlier with E_s energy per QPSK symbol.

The Euclidean distance between the two signal points in the BPSK signal constellation is $2\sqrt{E_b}$ and the Euclidean distance between any two points in the QPSK constellation is $2\sqrt{E_s}\cos\frac{\pi}{4} = \sqrt{2E_s}$. The symbol rate of QPSK system R_s is half the bit rate R_b.

$$R_s = \frac{R_b}{2} \tag{2.129}$$

That is, the symbol duration T_s is given by:

$$T_s = 2T_b \tag{2.130}$$

Since the power in the Inphase channel is equal to the power in the Quadrature channel, each is equal half the total transmitted power P, the energy per symbol E_s on each channel of the QPSK is given as:

$$E_s = \frac{1}{2}PT_s = \frac{1}{2}P2T_b = PT_b = E_b \tag{2.131}$$

As a result, the bit probability of error, which depends on the energy per symbol, is the same for QPSK and BPSK signalling.

Mathematically, we express the QPSK signal waveform $s_{qpsk}(t)$ as:

$$s_{qpsk}(t) = \sqrt{\frac{2E_s}{T_s}}\cos\left(\omega_c t + \frac{\pi(1+2m)}{4}\right) \tag{2.132}$$

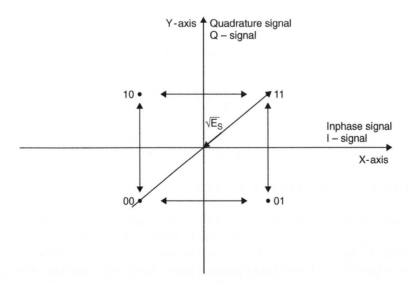

Figure 2.27 *QPSK signal constellation showing allowed phase transitions.*

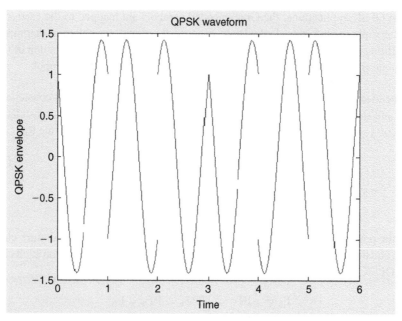

Figure 2.28 *QPSK signal waveform.*

where m = 0, 1, 2, 3. An example of QPSK signal waveform carrying data 110010110110 is shown in Figure 2.28.

Expanding (2.132) we get:

$$s_{qpsk}(t) = \sqrt{\frac{2E_s}{T_s}} \left[\cos\left(\frac{\pi(1+2m)}{4}\right) \cos(\omega_c t) - \sin\left(\frac{\pi(1+2m)}{4}\right) \sin(\omega_c t) \right] \quad (2.133)$$

The Inphase and the Quadrature components of the QPSK signals are:

$$s_{I-qpsk}(t) = \sqrt{\frac{2E_s}{T_s}} \cos\left(\frac{\pi(1+2m)}{4}\right) \quad (2.134)$$

$$s_{Q-qpsk}(t) = \sqrt{\frac{2E_s}{T_s}} \sin\left(\frac{\pi(1+2m)}{4}\right) \quad (2.135)$$

The I and Q components are tabulated in Table 2.2.

It is clear from Table 2.2 that the carrier phase will change by $\frac{\pi}{2}$ at the instant of time the symbol changes from 00 to 01, 01 to 11 or 11 to 10 and this will cause only minor changes in the QPSK envelope since only one bit of the symbol has reversed polarity. On the other hand, transition from 01 to 10 and 00 to 11 causes π phase change and a large change in QPSK envelope. In fact, the envelope has to pass through zero to make the symbol transition.

Table 2.2 *I and Q components*

Carrier degrees	Phase	QPSK symbol	Inphase component	Quadrature component
45	$m=0$	00	$\sqrt{\dfrac{E_s}{T_s}}$	$\sqrt{\dfrac{E_s}{T_s}}$
135	$m=1$	01	$-\sqrt{\dfrac{E_s}{T_s}}$	$\sqrt{\dfrac{E_s}{T_s}}$
225	$m=2$	11	$-\sqrt{\dfrac{E_s}{T_s}}$	$-\sqrt{\dfrac{E_s}{T_s}}$
315	$m=3$	10	$\sqrt{\dfrac{E_s}{T_s}}$	$-\sqrt{\dfrac{E_s}{T_s}}$

A QPSK modulator consists of a serial to parallel converter splitting the input serial data into two parallel channels, the Inphase and the Quadrature channels. The symbols on these two channels multiply the $\cos \omega_c t$ and $\sin \omega_c t$ waveforms using a two balanced multiplier. The two modulated waveforms are combined to form the QPSK signal waveform. Thus, the QPSK modulator is effectively two BPSK modulators arranged in quadrature and each operating at $R_s = \frac{R_b}{2}$.

The other variant of the QPSK used in spread spectrum systems is the Offset Quadrature Phase Shift Keying (OQPSK); also called Staggered Quadrature Phase Shift Keying (SQPSK). The OQPSK is identical to the QPSK system except that the Q channel symbol stream is offset (delayed) by half symbol interval (i.e $\frac{T_s}{2} = T_b$) prior to multiplication by the Quadrature carrier. A schematic for the QPSK modulator is shown in Figure 2.29.

The effect of delaying the symbol stream on the Q channel with respect to the stream on the I channel is to avoid the QPSK envelope passing through the origin, causing a large envelope distortion. When such a distorted signal passes through a non-linear power amplifier, the output contains a large number of frequency components generated by the amplifier non-linearity.

The signal constellation for QPSK and OQPSK are similar except the transitions through zero are removed in the OQPSK constellation as shown in Figure 2.30. Since the I and Q channel symbols are drawn from random input of equally likely binary digits and have rectangular symbol shape of $\pm\sqrt{\frac{E_s}{T_s}}$ given in Table 2.2, the Power Spectral Density (PSD) of QPSK/OQPSK signals G(f) is given by:

$$G(f) = \text{constant} \times \sin c^2(2T_b f)$$

$$= \text{constant} \times \left[\frac{\sin(4\pi f T_b)}{4\pi f T_b}\right]^2 \qquad (2.136)$$

The constant term in (2.136) is positive real number. The PSD given by (2.136) with constant $= 1$ is plotted in Figure 2.31.

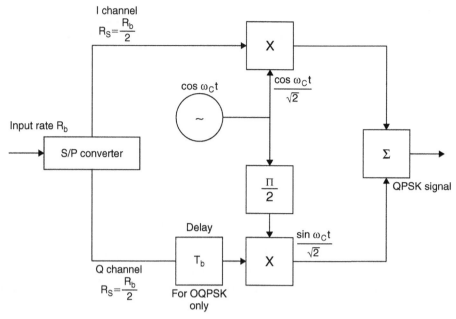

Figure 2.29   *A block diagram for the QPSK/OQPSK modulator.*

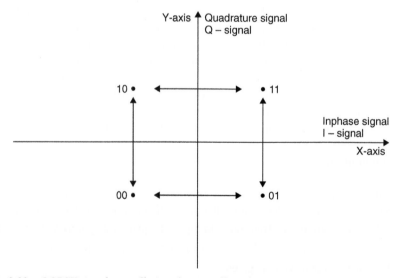

Figure 2.30   *OQPSK signal constellation showing allowed state transitions.*

The spectral efficiency (defined as 'number of bits per Hz') of the QPSK is twice that for BPSK since with the same data rate, the QPSK occupies half the spectral width of a BPSK signal. The QPSK demodulator consists of two main circuits, one for Carrier Recovery (CR) required for carrier coherent demodulation and the other for Symbol Time Recovery (STR) required for symbol time synchronous detection. The received signal plus noise is

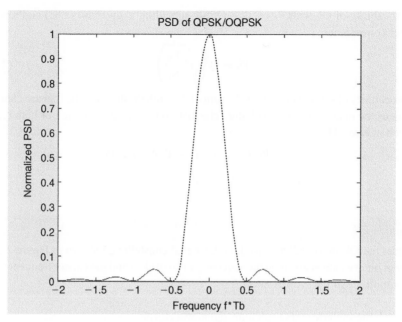

Figure 2.31   *PSD of QPSK/OQPSK signals.*

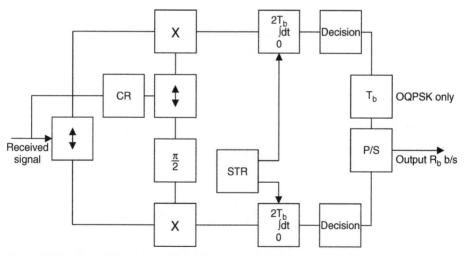

Figure 2.32   *Block diagram for QPSK/OQPSK demodulator.*

applied to the CR circuit generating cos $\omega_c t$ and sin $\omega_c t$ which operates the two balanced multipliers. The output of the balanced multipliers is integrated, sampled and applied to a threshold device for decisions on the output symbols. For the OQPSK demodulator, the Inphase OQPSK channel is delayed by T_b for time synchronization with the Quadrature channel, which is being delayed by the modulator. A block diagram of the QPSK demodulator is shown in Figure 2.32.

The probability of bit error P_b of ideal QPSK signalling is given by (2.83) and repeated here for convenience:

$$P_b = Q\left(\sqrt{2\frac{E_b}{N_0}}\right) \tag{2.137}$$

The probability of symbol error P_s is the sum of the probabilities that the Inphase channel's bit is detected in error, or the Quadrature channel's bit is detected in error, or both bits are detected in error. That is:

$$P_s = P_b(1 - P_b) + (1 - P_b)P_b + P_bP_b$$

$$P_s = 2P_b - P_b^2 \cong 2P_b \tag{2.138}$$

Example 2.8

Consider the QPSK signal described by the signal constellation shown in Figure 2.27. If the input binary data is given by the sequence 1100101100 and the carrier frequency is an integer multiple of the symbol rate:

(a) sketch the Inphase and Quadrature signal waveforms
(b) sketch the QPSK signal waveform

Solution
The two binary streams of the Inphase and Quadrature channels consist of the odd-numbered and even-numbered digits, respectively:

   I-channel stream $d_i = 10110$
   Q-channel stream $d_q = 10010$

Using (2.134) and (2.135), the quadrature waveforms are:

The Inphase waveform is $\sqrt{\dfrac{2E_s}{T_s}}\cos\left(\dfrac{\pi(1 + 2m)}{4}\right)\cos\omega_c t$

The Quadrature waveform is $\sqrt{\dfrac{2E_s}{T_s}}\sin\left(\dfrac{\pi(1 + 2m)}{4}\right)\sin\omega_c t$

where m is the carrier phase shift caused by the symbol amplitude change. We may express the two waveforms as:

   The in phase waveform is $d_i \cos\omega_c t$
   The Quadrature waveform is $d_q \sin\omega_c t$

The QPSK signal waveform is given by the sum of the two waveforms. The Inphase, Quadrature and QPSK signal waveforms are plotted in Figure 2.33. Note the change in QPSK signal amplitude as the symbols change from 11 to 00.

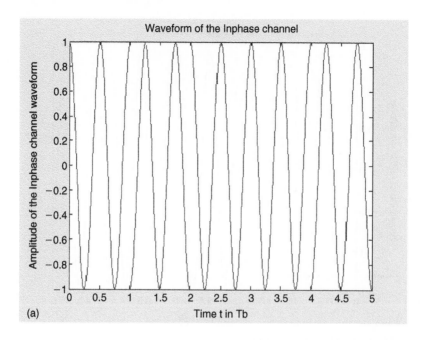

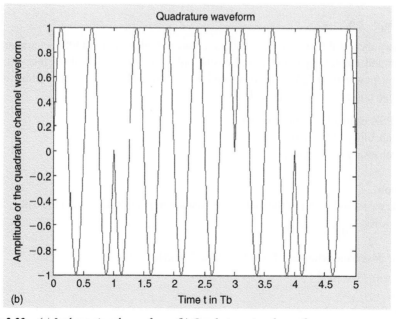

Figure 2.33 *(a) Inphase signal waveform; (b) Quadrature signal waveform.*

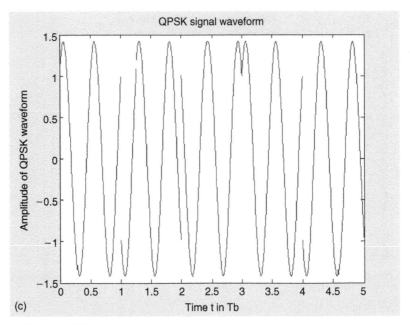

Figure 2.33 *(c) QPSK signal waveform.*

Example 2.9

A binary data at the rate of $R_b = 10^5$ b/s is transmitted using the QPSK communication system and employing a root raised cosine pulse filtering with roll off factor 0.5 with transmit power of 0.75 W. The two-sided power spectral density of the additive white Gaussian noise is 10^{-6} W/Hz. Calculate:

  i. QPSK channel pass band width
 ii. probability of bit error
iii. probability of symbol error

Solution

  i. The pass band width of the root raised cosine filtered QPSK signal $(1 \pm \alpha)R_S$

$$R_s = \frac{R_b}{2} = 0.5 \times 10^5 \text{ s/s}$$

The pass band width $= 1.5 \times 0.5 \times 10^5 = 750 \text{ kHz}$

 ii. $E_b = 0.75 \times 10^{-5}$ J

$$\frac{N_0}{2} = 10^{-6}$$

Thus $\dfrac{E_b}{\dfrac{N_0}{2}} = \dfrac{0.75 \times 10^{-5}}{10^{-6}} = 7.5$

$$P_b = Q(\sqrt{7.5}) = Q(2.74) = 3.0720 \times 10^{-3}$$

iii. $P_s \approx 2P_b = 6.144 \times 10^{-3}$

2.5 RAKE receivers

It is a well-known and theoretically well-studied subject that communication signals through a fading channel are heavily attenuated, and that the information might be lost in deep fade. It would greatly improve the reception if we are able to present the receiver with two or more replicas of the same information signal subject. These replicas would have to be transmitted through independent fading channels so that the probability of all fading at the same time is very small. For example, if the probability that any one signal will fade is 2%, then the probability that three copies, for example, propagated through independent paths fading simultaneously, is reduced to 0.0008%.

We may provide the signal replicas using various techniques. For instance, we can transmit the information signal on L carriers where the separation between adjacent carriers equals or exceeds the coherent bandwidth of the channel. This method is called *frequency diversity*. We can, if we prefer, transmit the information on L time slots where the separation between successive time slots equals or exceeds the coherent time of the channel. This method is called *time diversity*. We may also use one transmit antenna but receive the information signal using multiple receiving antennas. This method is called *space diversity*. Clearly all of these diversity techniques require extensive planning and the skills of specially trained engineers. However, from its very nature, multiple path propagation of wireless signal creates a number of replicas arriving at the receiver at different times. The time it takes a wireless signal to travel from A to B is given by the distance between A and B divided by speed of light c ($= 3 \times 10^8$ m/s). The *delay spread* of the replicas is the time it takes the wireless signal, travelling at speed of light, the longest path minus the shortest path. Thus:

$$\text{the } delay\ spread = \frac{\text{distance of longest path} - \text{distance of shortest path}}{c}$$

The inverse of the *delay spread* is equal the *coherent bandwidth*. We will assume that the symbol duration is much larger than the delay spread so that there is no ISI. However, in a spread spectrum communication operating at high spreading codes, the delay between the multipath components is likely be equal to or greater than one code pulse (chip) duration. For efficient spreading code, the cross-correlation between successive chips has to be low, meaning that the multipath components are generally uncorrelated. Consequently, it is feasible to resolve the strongest uncorrelated components and enable the receiver to achieve L^{th} order diversity. A receiver that collects energy from the received components to provide this diversity is acting somewhat similar to an ordinary *garden rake*. Such a receiver is named by the invertors Price and Green (1958) the *Rake receiver*. The Rake receiver turns a destructive multipath process into a transmission enhancing diversity technique and thus mitigates the fading problem of the wireless channel.

The Rake receiver consists of a number of branches equal to the multipath signal components, each is called *finger*. Each finger attempts to demodulate one path of the composite multipath signal. Modern base station receivers use a maximum of four fingers and mobile stations use a maximum of three fingers (Lee and Miller, 1998). A block diagram of Rake demodulator is shown in Figure 2.34 (Haykin and Moher, 2005).

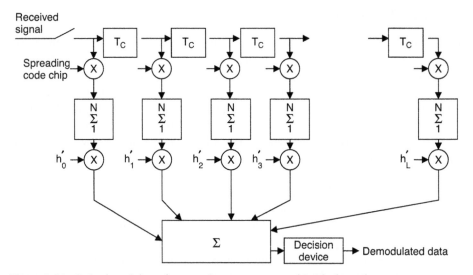

Figure 2.34 *Rake demodulator for spread spectrum over multipath channels.*

In Figure 2.34, h_j' represents the equivalent channel sampled impulse response where $j = 1, 2, \ldots L$, L is the number of resolvable multipath components and N is the number of chips in the spreading code.

The Rake demodulator shown in Figure 2.34 consists of L correlators to detect the L multipath components that have relative delays larger than the chip duration and assumed prior knowledge of amplitudes and relative delays. The input to each correlator is multiplied by the spread spectrum code on a chip by chip basis to despread the received signal; the output of the multiplier is summed up and weighed by the coefficient of the corresponding multipath component channel for maximum ratio combining which achieves optimum combating the fading. The output of the correlators is added up and applied to a decision device for estimating the transmitted bit.

The bit probability of error of Rake receiver over Rayleigh fading channel is given by:

$$P_b = \left(\frac{1-v}{2}\right)^L \sum_{\ell=0}^{L-1} \binom{L-1}{\ell} \left(\frac{1+v}{2}\right)^\ell \tag{2.139}$$

where

$$v = \sqrt{\frac{\overline{\gamma}_c}{1+\overline{\gamma}_c}} \tag{2.140}$$

$\overline{\gamma}_c$ is the average bit energy to noise spectral density per multipath component channel. If $\overline{\gamma}_c$ is the same for all channels, then the average bit energy-to-noise spectral density of transmission channel $\overline{\gamma}_b$ is given by:

$$\overline{\gamma}_b = L\overline{\gamma}_c \tag{2.141}$$

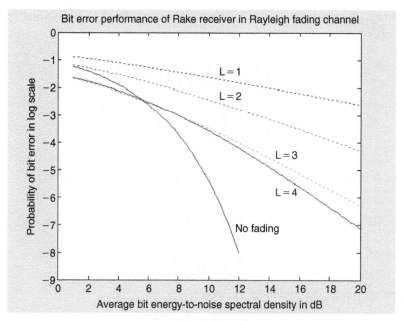

Figure 2.35 *Probability of error of Rake receiver over Rayleigh fading channel.*

Figure 2.35 shows the probability of error for a BPSK modulated spread spectrum system over a Rayleigh fading channel when using a Rake receiver with L = 1, 2, 3 and 4 compared with the probability of error for a BPSK system using a matched filter detector operating over an AWGN channel (no fading). The probability of error versus average bit energy per noise spectral density is plotted.

Example 2.10

Consider the transmission of data over an AWGN channel using a BPSK carrier and a matched filter receiver. The average probability of error measured at the output of the matched filter is 3.3×10^{-3}. If the same average bit energy-to-noise spectral density is kept when the data is transmitted over multipath fading channel and the matched filter is replaced with a 3-finger Rake receiver, what would be the average probability of error of the new system?

Solution

The probability of error at the output of the matched filter receiver is given in terms of the average bit energy-to-noise spectral density as follows:

$$P_b = Q\left(\sqrt{2\frac{E_b}{N_0}}\right)$$

For $\quad P_b = 3.3 \times 10^{-3}, \quad \dfrac{E_b}{N_0} = 3.7$

The average probability of error of Rake receiver is given by (2.139). Substituting $L = 3$ in (2.139), we get:

$$P_b = \left(\frac{1-v}{2}\right)^3 \sum_{\ell=0}^{2} \binom{2}{\ell} \left(\frac{1+v}{2}\right)^\ell$$

$$P_b = \left(\frac{1-v}{2}\right)^3 \left[2 + v + \left(\frac{1+v}{2}\right)^2\right]$$

Now we determine the value v but first we have to calculate γ_c using (2.141).

$$\gamma_c = \frac{\gamma_b}{3} = \frac{3.7}{3} = 1.23$$

$$v = \sqrt{\frac{1.23}{2.23}} = 0.74$$

Substituting v in the expression for P_b, we get:

$$P_b = 7.7 \times 10^{-3}$$

2.6 Channel forward error correction coding

Reliable transmission of information over a digital communication system is required in many practical applications, such as for patient medical recorders, or financial transactions. In fact, ideally, in these applications the transmission must not contain any errors at all. However, in the real-world, the additive Gaussian noise creates errors and the number of errors depends on the ratio of the received power to the additive noise power.

The fundamental concepts of the Shannon (1948) theory are that the transmission of digital information over a noisy channel with small or no errors is possible provided that there is appropriate channel coding, and the data rate is less than or equal to the channel capacity. Channel coding for the purpose of Forward Error Correction (FEC) is accomplished by adding some carefully designed redundant information to the data being transmitted through the channel. In considering the channel coding used in IS-95 CDMA and UMTS wideband systems, the half and one-third rate convolutional codes and turbo codes are of great interest.

In this section, we explore the theory of convolutional coding and explain the optimum algorithm used for decoding, namely the Viterbi algorithm. Convolutional coding and Viterbi decoding are particularly suited to a channel in which the transmitted signal is corrupted, mainly by Additive White Gaussian Noise (AWGN). However, in addition to the additive white noise, wireless communications have other impairments, such as multipath fading, selective frequency fading, multiple access interference. Consequently, although convolutional coding with Viterbi decoding might be useful in dealing with AWGN, it may not be the best technique for wireless communications. Thus, it is appropriate to introduce the more advanced channel coding technology based on turbo coding at this stage.

2.6.1 The convolutional encoder

The basic convolutional encoder consists of a cascade of shift register stages and the associated combinational logic in the form of exclusive OR gates. The computation of the encoded symbols depends not only on the present set of input symbols but on some of the previous input symbols. Therefore, a convolutional encoder generally encodes a given frame of k bits into frames of different sequences of n bits at different times depending on the state of the encoder. The *code rate* of the encoder is defined as $R = \frac{k}{n}$. The shift register consists of K stages where the input binary data is shifted into and along the register k bits at a time. A binary convolutional encoder with k bits input, n bits output and K constraint length is denoted as (n, k, K) encoder. The *constraint length* defines the total span of input binary symbols that influence the encoded output at any time. A conceptual diagram of the convolutional encoder is shown in Figure 2.36.

A convolutional code is defined by kn binary polynomials which constituent the code generator matrix $\mathbf{G}(x)$. For simplicity, consider $k = 1$, the generator matrix is:

$$\mathbf{G}(x) = [\mathbf{g}_1(x), \mathbf{g}_2(x), \dots\dots\dots\dots\dots\dots\dots, \mathbf{g}_n(x)] \tag{2.142}$$

Where
$$\mathbf{g}_i(x) = g_{i0} + g_{i1}x + g_{i2}x^2 + \dots\dots\dots\dots\dots\dots + g_{iK}x^{K-1} \tag{2.143}$$

The code generator polynomials are usually expressed in octal numbers. The octal numbers $(0,1,\dots,7)$ are expressed in three digits, i.e. 57 in octal form is equal to 101 111 in binary form.

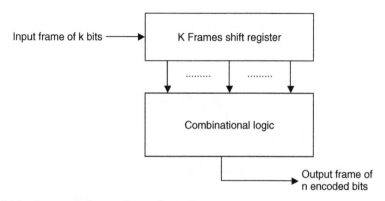

Figure 2.36 *Structure of a convolutional encoder.*

Example 2.11

A convolutional encoder that provides the best error performance in satellite communication systems has the following parameters:

$$\mathbf{G} = [133 \quad 171]$$
$$R = 1/2$$
$$K = 7$$

Determine the structure of the encoder.

Solution

The two octal number are converted to binary forms as:

$$133 = 001\ 011\ 011 = 1\ 011\ 011$$
$$171 = 001\ 111\ 001 = 1\ 111\ 001$$

The generator polynomials are:

$$\mathbf{g}_1(x) = 1 \cdot (x^0) + 0 \cdot (x^1) + 1 \cdot (x^2) + 1 \cdot (x^3) + 0 \cdot (x^4) + 1 \cdot (x^5) + 1 \cdot (x^6)$$

$$\mathbf{g}_2(x) = 1 \cdot (x^0) + 1 \cdot (x^1) + 1 \cdot (x^2) + 1 \cdot (x^3) + 0 \cdot (x^4) + 0 \cdot (x^5) + 1 \cdot (x^6)$$

Denote the input as $i(x)$, the 1st digit is computed from $i(x) \cdot \mathbf{g}_1(x)$. The 2nd digit is computed from $i(x) \cdot \mathbf{g}_2(x)$.

Thus for $i(x) = 101 = 1 + x^2$,

$$1\text{st digit} = (1 + x^2)(1 + x^2 + x^3 + x^5 + x^6) = 10\ 01\ 10\ 00\ 1$$

$$2\text{nd digit} = (1 + x^2)(1 + x + x^2 + x^3 + x^6) = 11\ 00\ 11\ 10\ 1$$

The encoded sequence is 11 01 00 10 11 01 01 00 11

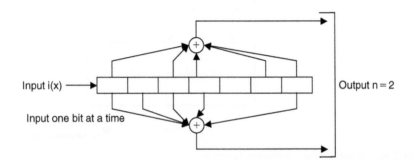

Example 2.12

Draw the convolutional encoder with the following parameters:

$\mathbf{G} = [5\ 7]$
$R = 1/2$
$K = 3$

Compute the encoded bit for $i(x) = 101101$

Solution
The generator polynomials are:

$$g_1(x) = 5 \text{ in octal} = 101 \text{ in binary} = 1 + x^2$$
$$g_2(x) = 7 \text{ in octal} = 111 \text{ in binary} = 1 + x + x^2$$

The structure of the convolutional code is:

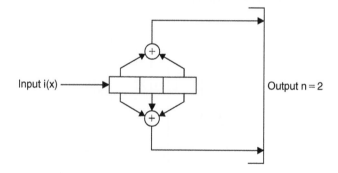

$$i(x) = 101101 = 1 + x^2 + x^3 + x^5$$

1st digit $= i(x) \cdot g_1(x) = (1 + x^2 + x^3 + x^5)(1 + x^2) = 10\ 01\ 10\ 01$

2nd digit $= i(x) \cdot g_2(x) = (1 + x^2 + x^3 + x^5)(1 + x + x^2) = 11\ 00\ 00\ 11$

The output sequence is   11 01 00 10 10 00 01 11

2.6.2 *Convolutional coding representation*

Convolutional codes belong to a group described as tree codes where their coding process can be traced through the coding tree that comprises of nodes and branches. To construct the tree, the encoder is first initialized to start from the all-zero state. Since the input to the encoder is either binary 0 or 1, then at each node two branches emerge, one representing the output for binary 0 input (upper branch) and the other (lower branch) represents the output for binary 1 input. The (2, 1, 3) encoder shown in Figure 2.37 has the *tree diagram* shown in Figure 2.38. For every input bit, the rightmost bit in the register drops out and no longer influences the output. The output bits of the encoder for input 0 or 1 are shown on the tree branches.

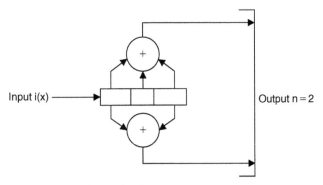

Figure 2.37   *The (2, 1, 3) convolutional code.*

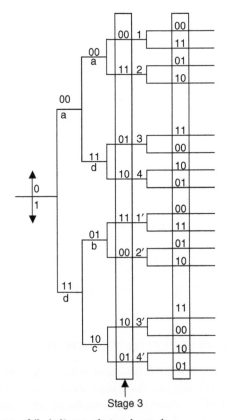

Figure 2.38   *Tree diagram of (2, 1, 3) convolutional encoder.*

A closer inspection of Figure 2.38 reveals that the tree structure repeats itself after the 3rd stage, as the upper and lower halves of the tree structure at the 4th stage are equal to the whole structure at stage 3. These portions of the tree diagram are surrounded by a solid line. Generally, any tree diagram starts repeating itself after K branches. In this sense, node 1 can be connected to node 1′ so as 2 and 2′, 3 and 3′, 4 and 4′. Under ideal circumstances, decoding an encoder output can simply be achieved by tracing back the output sequence on the tree, considering an upper branch represents an input binary 0 and a lower branch corresponds to an input binary 1. The output of the encoder at each stage is determined by the input bit and the bits stored in the two stages on the right. These two stages have four possible states: a = 00, b = 01, c = 10 and d = 11.

An elegant way of representing the information in the tree diagram is the *trellis diagram* shown in Figure 2.39.

One of the important factors in the convolutional coding is the distance properties of the encoded sequence relative to the all-zero sequence. The minimum distance of the code is called the *minimum free distance* and is denoted by d_{free}. The distance distribution of a given convolutional code can be found by using the code *state diagram*, derived from the code trellis diagram, as shown in Figure 2.40. Let us represent every branch in the trellis

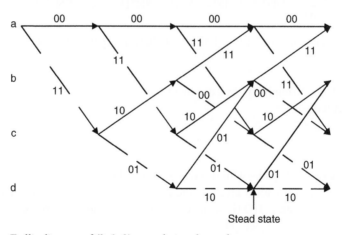

Figure 2.39 *Trellis diagram of (2, 1, 3) convolutional encoder.*

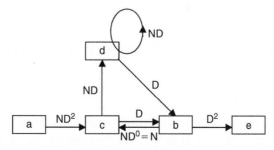

Figure 2.40 *The (2, 1, 3) convolutional code state diagram.*

diagram by $N^k D^j$ where j represents weight of the output for the branch and k the weight of the input to the encoder.

Nodes 'e' and 'a' represent the output and the input nodes, respectively. D represents the Hamming distance of the output sequence from the all-zero branch. The state equations are represented by X_a, X_b, X_c, X_d and X_e such that:

$$\begin{aligned} X_b &= DX_c + DX_b \\ X_c &= ND^2X_a + NDX_b \\ X_d &= NDX_c + NDX_d \\ X_e &= D^2X_b \end{aligned} \qquad (2.144)$$

The transfer function of the convolutional code T(D,N) is given by:

$$T(D, N) = \frac{X_e}{X_a} \qquad (2.145)$$

Substituting (2.144) in (2.145) and simplifying

$$T(D, N)\frac{ND^5}{1 - 2ND} = ND^5 + 2N^2D^6 + 4N^3D^7 + 8N^4D^8 + \cdots + 2^r N^{r+1} D^{r+5} + \cdots \qquad (2.146)$$

The transfer function in (2.146) shows that there are paths with their corresponding distances:

1 path with $d = 5$ (minimum free distance)
2 paths with $d = 6$
4 paths with $d = 7$
etc.

In general there are 2^r paths with distance $(r + 5)$. The number of bits in errors due to choosing the incorrect path is given by the exponent of N which become multiplication factors by differentiating T(D,N) with respect to N. We now differentiate T(D,N) with respect to N and set $N = 1$, we get:

$$\left.\frac{\partial T(D, N)}{\partial N}\right|_{N=1} = D^5 + 2.2D^6 + 3 * 4D^7 + \cdots\cdots + (r + 1) * 2^r D^{r+5} + \cdots\cdots \qquad (2.147)$$

Denote $\beta(d)$ for the number of paths with hamming distance d. In this example, for the path with $d = r + 5$:

$$\beta(d) = (r + 1)2^r \quad \text{for } r = 0, 1, 2, \ldots. \qquad (2.148)$$

The transfer function of the code given in (2.146) shows that the shortest path is of distance 5 which connects state 'a' to 'c' to 'b' to 'e'. Thus, the minimum free distance for this code is $d_f = 5$.

Example 2.13

Consider the convolutional code shown below. Determine the encoded sequence that corresponds to an input sequence [1001].

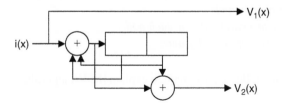

Solution

The forward generator polynomial is $f(x) = 1 + x^2$

The feedback generator polynomial is $g(x) = 1 + x + x^2$

$$V_1(x) = i(x)$$

$$V_2(x) = i(x)\frac{1 + x^2}{1 + x + x^2}$$

Thus:

$$V_1(x) = 1 + x^3 = 1001$$

$$V_2(x) = 1 + x^3\frac{1 + x^2}{1 + x + x^2} = 1111$$

2.6.3 Viterbi decoding algorithm

The optimum convolutional decoding, in the presence of random errors caused by AWGN, is to compare the received sequence with every possible code sequence. This method can achieve a maximum-likelihood performance that minimizes the probability of error for statistically independent and equally likely binary data. However, the computation requirement for this decoding method increases exponentially with the number of paths in the code trellis diagram. Viterbi and Omura (1979) proved that not all of these paths need be considered since paths that are non-optimal (paths with high metrics) at any node can never become optimal in the future and can safely be discarded.

In a binary system, each node in the trellis diagram has 2 paths entering and 2 paths emerging. We compute the 'metric' which is equal to the Hamming distance between the received sequence and each of these paths, i.e. the number of bits that differ between the two sequences. We start by computing the metrics of all possible paths up to the first re-emerging node and keep the path with lowest metric. Since channel errors are random, paths with high metrics at any node can be safely discarded. If v is the number of states in the coder, then there are 2^v paths and v surviving paths.

Considering a single bit-input convolutional code, the Viterbi decoding Algorithm (VA) can be summarized as follows:

- *For each of the 2^ν stored paths, compute the distance between the received frame and the 2 branches at each node.*
- *Select the best (surviving) path for each node.*
- *Updates the metrics after each frame.*

We now consider the following example to explain the VA algorithm.

Example 2.14

Consider the simple (2, 1, 3) covolutional coder shown below:

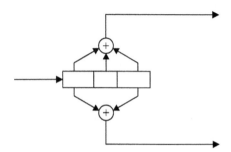

The code has four states: 00, 01, 10, 11 and the trellis diagram is shown in Figure 2.39. Let data **[101100100]** input serially to the encoder and the received data [11 **11** 00 01 **00** 11 11 10 11].

Assume initial path of the start state (state 00) has metric $= 0$. We notice that the given input data produces the following encoded data [11 10 00 01 01 11 11 10 11]. Thus, we can see that the received data has two bits in error. We will show that VA is capable of correcting these two errors.

Define the following variables:

Branch metric: distance between encoder output frame and corresponding received frame.
Initial path metric (i path): Metric at start state.
Final path metric (F path) $=$ Initial path Metric $+$ Branch Metric

Below we identify the number of the start state, the number of the end state, the output digits, the branch metric, the initial path metric, the final path metric and whether the path being considered should be kept or discarded.

1st FRAME – 11 RECEIVED, ENCODED 11

Start state	End state	O/p	Branch	i path	F path	Keep
00	00	00	2	0	2	N
00	**10**	**11**	**0**	**0**	**0**	**Y**

2nd FRAME – 11 RECEIVED, ENCODED 10

Start state	End state	O/p	Branch	i path	F path	Keep
00	00	00	2	2	4	N
00	10	11	0	2	2	Y
10	**01**	**10**	**1**	**0**	**1**	**Y**
10	11	01	1	0	1	N

3rd FRAME – 00 RECEIVED, ENCODED 00

Start state	End state	O/p	Branch	i path	F path	Keep
00	00	00	0	4	4	N
01	00	11	2	1	3	Y
10	01	10	1	2	3	N
11	01	01	1	1	2	Y
00	10	11	2	4	6	N
01	**10**	**00**	**0**	**1**	**1**	**Y**
10	11	01	1	2	3	N
11	11	10	1	1	2	Y

4th FRAME – 01 RECEIVED, ENCODED 01

Start state	End state	O/p	Branch	i path	F path	Keep
00	00	00	1	3	4	N
01	00	11	1	2	3	Y
10	01	10	2	1	3	N
11	01	01	0	2	2	Y
00	10	11	1	3	4	N
01	10	00	1	2	3	Y
10	**11**	**01**	**0**	**1**	**1**	**Y**
11	11	10	2	2	4	N

5th FRAME – 00 RECEIVED, ENCODED 01

Start state	End state	O/p	Branch	i path	F path	Keep
00	00	00	0	3	3	Y
01	00	11	2	2	4	N
10	01	10	1	3	4	N
11	**01**	**01**	**1**	**1**	**2**	**Y**
00	10	11	2	3	5	N
01	10	00	0	2	2	Y
10	11	01	1	3	4	N
11	11	10	1	1	2	Y

6th FRAME – 11 RECEIVED, ENCODED 11

Start state	End state	O/p	Branch	i path	F path	Keep
00	00	00	2	3	5	N
01	**00**	**11**	**0**	**2**	**2**	**Y**
10	01	10	1	2	3	Y
11	01	01	1	2	3	N
00	10	11	0	3	3	Y
10	10	00	2	2	4	N
10	11	01	1	2	3	Y
11	11	10	1	2	3	N

7th FRAME – 11 RECEIVED, ENCODED 11

Start state	End state	O/p	Branch	i path	F path	Keep
00	00	00	2	2	4	N
01	00	11	0	3	3	Y
10	01	10	1	3	4	N
11	01	01	1	3	4	Y
00	**10**	**11**	**0**	**2**	**2**	**Y**
01	10	00	2	3	5	N
10	11	01	1	3	4	Y
11	11	10	1	3	4	N

8th FRAME – 10 RECEIVED, ENCODED 10

Start state	End state	O/p	Branch	i path	F path	Keep
00	00	00	1	3	4	Y
01	00	11	1	4	5	N
10	**01**	**10**	**0**	**2**	**2**	**Y**
11	01	01	2	4	6	N
00	10	11	1	3	4	Y
01	10	00	1	4	5	N
10	11	01	2	2	4	Y
11	11	10	0	4	4	N

9th FRAME – 11 RECEIVED, ENCODED 11

Start state	End state	O/p	Branch	i path	F path	Keep
00	00	00	2	4	6	N
01	**00**	**11**	**0**	**2**	**2**	**Y**
10	01	10	1	4	5	N
11	01	01	1	4	5	Y
00	10	11	0	4	4	N
01	10	00	2	2	4	Y
10	11	01	1	4	5	N
11	11	10	1	4	5	Y

Examining the 9th Frame, it is clear that the branch of the path with the lowest metric starts from state 01 and ends in state 00 shown in the bold row above. Following this branch back to the 1st Frame we have shown the optimum branch in each Frame as the bold row. This optimum path is also shown in the trellis diagram in Figure 2.41.

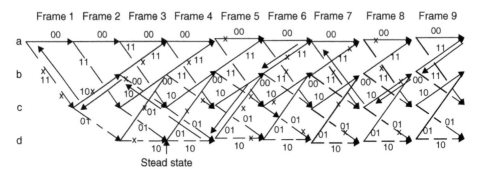

Figure 2.41   *VA decoding Frame by Frame, x denotes selected survivors.*

In summary, the received data has two errors shown in bold digits:

	11 1*1* 00 01 0*0* 11 11 10 11
The encoded data is	11 10 00 01 01 11 11 10 11
The input data to the encoder is	10 11 00 100
The decoded data using VA	10 11 00 100
Thus VA corrected the two errors.	

2.6.4   *Probability of error using VA decoding*

There are two decision methods that can be used to decode the digital information. In the hard decision decoding, the demodulator immediately decides whether the transmitted bit is either binary '0' or binary '1' and feeds the binary digit to the input of the VA decoder. In the soft decision decoding, the received signal is compared to a scale of 8 or 16 levels to decide how close it is to binary '0' or binary '1' and the result of the decision is again fed into the soft input VA decoder. It is widely accepted that soft decision decoding generally outperforms the hard decision decoding considering the upper probability of error bounds.

Let us now consider the probability of error for VA decoding. The probability of error for soft decision VA decoder of antipodal signalling (i.e. binary PSK) in AWGN channel when comparing two paths that differ in d bits is (Viterbi and Omura, 1979):

$$P'_s(d) = Q\left(\sqrt{2\frac{E_s}{N_0}d}\right) \qquad (2.149)$$

where E_S is the symbol energy related to energy per bit E_b as:

$$n\,E_s = k\,E_b$$

Thus
$$E_s = \frac{k}{n}E_b = RE_b \qquad (2.150)$$

Substituting for E_S in (2.149) we get:

$$P_s'(d) = Q\left(\sqrt{2R\frac{E_b}{N_0}d}\right) \qquad (2.151)$$

The average probability of error for soft decision VA decoding is upper-bounded by multiplying the pair-wise probability of error $P_s'(d)$ by β_d (number of paths with Hamming distance d) and sum up the product over all possible values of d. Thus, the probability of bit error, $P_b^s(d)$, for $k = 1$ soft decision VA decoding is given by (Proakis and Salehi, 2002):

$$P_b^s \leq \sum_{d=d_f}^{\infty} \beta_d Q\left(\sqrt{2R\frac{E_b}{N_0}d}\right) \qquad (2.152)$$

where   d   = Hamming distance
d_f = Minimum free distance
β_d = Number of paths with hamming distance d given by (2.148)
R   = Code rate = $\frac{k}{n}$
$\frac{E_b}{N_0}$ = Energy per bit to noise density ratio

Similarly, the probability of bit error, P_b^h, for hard decision VA decoding is given by:

$$P_b^h \leq \sum_{d=d_f}^{\infty} \beta_d P_h'(d) \qquad (2.153)$$

where $P_h'(d)$ is the probability of choosing the incorrect path that has distance d from all-zero path. Now if d is odd, then the probability of selecting the incorrect path occurs when the number of errors in the received sequence is greater than $\frac{d+1}{2}$ and is given by the binomial distribution:

$$P_h'(d) = \sum_{k=\frac{d+1}{2}}^{d} \binom{d}{k} p_1^k(1-p_1)^{d-k} \quad \text{for d odd} \qquad (2.154)$$

Similarly, when d is even, selecting the incorrect path of distance d occurs when the number of errors *equals or exceeds* $\frac{d}{2}+1$. Using the binomial distribution again, the probability of selecting the incorrect path when number of errors exceeds $\frac{d}{2}+1$ is:

$$\sum_{k=\frac{d}{2}+1}^{d} \binom{d}{k} p_1^k(1-p_1)^{d-k} \quad \text{for d even}$$

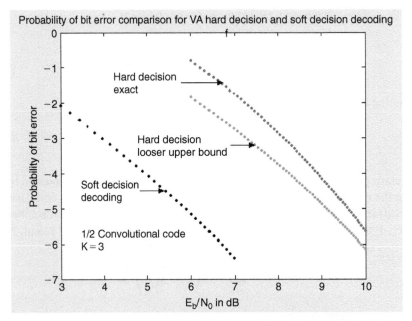

Figure 2.42 *Comparison of Probability for soft decision and hard decision VA decoding.*

The probability of selecting the incorrect path when number of errors equal $\frac{d}{2} + 1$ is:

$$0.5 \binom{d}{d/2} P_h'^{d/2}(1 - P_h')^{d/2}$$

Thus for d even, the pair-wise probability of error is:

$$P_h'(d) = \sum_{k=\frac{d}{2}+1}^{d} \binom{d}{k} p_1^k (1 - p_1)^{d-k} + 0.5 \binom{d}{d/2} P_h'^{d/2}(1 - P_h')^{d/2} \quad \text{for d even} \quad (2.155)$$

Expressions (2.153), (2.155) and (2.154) give the exact probability of bit error for hard decision VA decoded convolutional codes with k = 1. An approximate upper bound on the pair-wise probability of error for any d can be expressed as:

$$P_b^h < \left. \frac{\partial T(D, N)}{\partial N} \right|_{N=1, D=\sqrt{4p(1-p)}} \quad (2.156)$$

The probability of bit error for VA decoding given by (2.152), (2.53) and (2.156) are plotted verses $\frac{E_b}{N_0}$ in dB in Figure 2.42. It can be seen that the soft decision VA decoder outperforms the hard decision decoder by approximately 3 dB for P_b between 10^{-2} and 10^{-6}.

2.6.5 Turbo encoding and decoding

The error correcting capability of convolutional codes is determined in terms of the coding gain for a particular code rate and for certain BER in comparison with the uncoded transmission. The coding gain grows almost linearly with the code memory but the complexity of the decoding grows exponentially with code memory. For these limitations, most practical applications use convolutional codes with a memory of no more than seven. This practical selection of convolutional codes puts an upper limit on the possible achievable coding gain.

In order to obtain high coding gain with moderate decoding complexity, serial concatenation of Reed-Solomon, as the outer code, and convolutional code, as the inner code, is used. This coding structure delivers a large asymptotic gain but with poor convergence towards a theoretical Shannon limit since the inner convolutional decoder makes no use of the outer Reed-Solomon redundant symbols. Furthermore, each constituent code runs with its own clock.

A solution to the problem of getting high asymptotic gain with moderate decoding complexity has inspired many researchers: Gallager (1962) in the 1960s, Tanner (1981) in the early 1980s and Hagenauer and Hoeher (1989) in the late 1980s and many others. Berrou et al. (1993) came up with a proposal for the solution and announced the invention of new codes, called 'Turbo codes'. Then Berrou and Glavieux (1996) published a theory behind these codes in more detail. It took ten years, after the first publication of turbo codes in 1993, for these codes to be implemented in practical systems (Berrou, 2003). Since their invention, turbo codes have been widely studied and the 'turbo' concept is currently being researched for a possible application in other topics such as turbo equalization and turbo multi-user detection.

Turbo codes, which operate very close to the Shannon limit, are recommended by the 3GPP, for use in third generation wireless systems based on UMTS/IMT2000 standards. More precisely, the 3GPP recommend two types of channel coding for use with physical layer of the third generation systems. For voice transmission applications, a 256-state 1/2 rate convolutional encoding with Viterbi algorithm decoding is recommended, while for data transmission, a 2×8-state 1/3 turbo code, decoded by the Max-Log-MAP, is proposed. The number of states processed by the turbo encoder for six iteration (say) is $= 2$ (number of codes) $\times 2$ (double recursions) $\times 8$ (number of states in each code) $\times 6$ (number of iterations) $= 196$ which is less than the number of states processed by the voice decoder.

2.6.6 Turbo code construction

Conventionally, turbo codes are configured using two *Recursive Systematic Convolutional codes* (RSC). We first describe the binary RSC encoder as a convolutional encoder with a feedback loop as shown in Figure 2.43, denoting the feed forward polynomial and the feedback polynomial, h(D) and g(D), respectively. The transfer function of the RSC encoder shown in Fig. 2.43 is given by:

$$G(D) = \frac{h(D)}{g(D)} = \frac{1 + D^2}{1 + D + D^2} \tag{2.157}$$

Turbo codes are flexible block codes in terms of coding rates, which are made up of two RSC constituent encoders C1 and C2, as shown in Figure 2.44. Each of the constituent encoders can be similar to the RSC encoder shown in Figure 2.43. These constituent encoders run with the same clock.

Each block at the output of the turbo encoder contains the original data followed by the parity bits. Considering the turbo encoder shown in Figure 2.44, the binary input data sequence $\mathbf{d}_k = [d_1, d_2, \ldots, d_N]$ goes directly to encoder C1 to generate the first parity stream $\mathbf{y}_k^1$ at times $k = 1, 2, \ldots, N$. The input data is then interleaved before entering the other encoder C2 to generate the second parity stream $\mathbf{y}_k^2$. The encoded block consists of the systematic bits $\mathbf{x}_k = \mathbf{d}_k$ followed by parity bits from encoder C1 and encoder C2, Consequently, the turbo code rate is 1/3. The transmission of the parallel symbols of the encoded sequence is shown in Figure 2.45.

The serial transmission of the encoded sequence is shown in Figure 2.46.

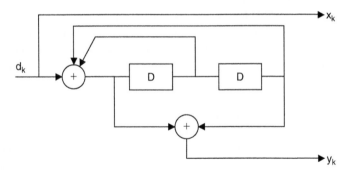

Figure 2.43   *RSC encoder, K = 3, rate 1/2, g1 = 7, g2 = 5.*

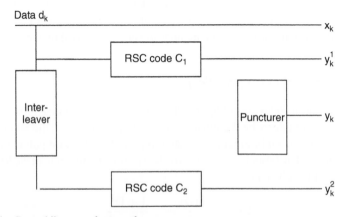

Figure 2.44   *Basic 1/3 rate turbo encoder.*

When a different code rate is required, the turbo encoded blocks are punctured. The *puncturer* periodically deletes selected bits to reduce the coding overhead. Puncturing the x_k data sequence is not allowed since it leads to severe degradation of the turbo code performance. Therefore, puncturing is restricted only to parity bits y_k^1 and y_k^2. For example, if we want to produce 1/2 turbo code from the 1/3 turbo code described above, we may simply delete every other parity bit at C1 and C2 alternately so the unpunctured encoded blocks in Figure 2.45 are now transformed to the punctured blocks shown in Figure 2.47 where the grey cells denote deleted parity bits.

The serial transmission of the punctured 1/2 rate turbo code is given in Figure 2.48.

Time k	0	1	2	3	4	5	6	7
	x_0	x_1	x_2	x_3	x_4	x_5	x_6	x_7
	y_0^1	y_1^1	y_2^1	y_3^1	y_4^1	y_5^1	y_6^1	y_7^1
	y_0^2	y_1^2	y_2^2	y_3^2	y_4^2	y_5^2	y_6^2	y_7^2

Figure 2.45 *Encoded blocks at times k = 0,1, . . . 7 for 1/3 rate turbo encoder.*

k	0			1			2			3			4		
	x_0	y_0^1	y_0^2	x_1	y_1^1	y_1^2	x_2	y_2^1	y_2^2	x_3	y_3^1	y_3^2	x_4	y_4^1	y_4^2

Figure 2.46 *Encoded blocks at times k = 0,1, . . . 7 for 1/3 rate turbo encoder for serial transmission.*

Time k	0	1	2	3	4	5	6	7
	x_0	x_1	x_2	x_3	x_4	x_5	x_6	x_7
	y_0^1		y_2^1		y_4^1		y_6^1	
		y_1^2		y_3^2		y_5^2		y_7^2

Figure 2.47 *Punctured encoded blocks at times k = 0,1, . . . 7 for 1/2 rate turbo encoder.*

k	0		1		2		3		4		5		6	
	x_0	y_0^1	x_1	y_1^2	x_2	y_2^1	x_3	y_3^2	x_4	y_4^1	x_5	y_5^2	x_6	y_6^1

Figure 2.48 *Punctured encoded blocks at times k = 0,1, . . . 7 for 1/2 rate turbo encoder for serial transmission.*

Denote the code rate for C1 and C2 to be R_1^{C1} and R_2^{C2}, respectively, then the global rate R_C of the turbo code produced by these constituent encoders is given by:

$$\frac{1}{R_c} = \frac{1}{R_1^{C1}} + \frac{1}{R_2^{C2}} - 1 \tag{2.158}$$

For example, if encoders C1 and C2 are identical then to construct a 2/3 rate turbo code, the code rate of the constituent RSC encoder is calculated using (2.158):

$$\frac{3}{2} = \frac{2}{R^C} - 1$$

where $\quad R^C = R_1^{C1} = R_2^{C1}$

Thus: $\quad R^C = \dfrac{4}{5}$

2.6.7 Turbo code interleavers

The interleaver is a device used to alter the order of bits or symbols presented at its input to ensure groups of bits or symbols that are close together are spread far apart at its output. Interleavers are widely used after error correct coding and signal mapping in wireless systems to combat fading over a wireless channel. The de-interleaver reverses the interleaver operation at the decoder by breaking up the burst of symbols affected by the fading in order to enable the decoder to effectively estimate the received symbols. In turbo codes, the interleaver basically performs the same functions to increase the Hamming distance of the codeword, thereby reducing the probability of bit error. Generally, turbo codes experience low BER at low $\frac{E_b}{N_0}$ and the BER curve tends to flatten out at moderate to high $\frac{E_b}{N_0}$. The BER flattening out results in the turbo code *error floor*. A high error floor is commonly produced by inadequately designed interleavers that pair low weight code words resulting in a relatively low free distance turbo code.

Interleavers can be of the type *block interleavers* in which data bits are written as one row at a time and read out one column at a time. The de-interleaving is the inverse of this process. However, block interleavers generate large numbers of code words with low weights causing high error floor.

Another type of interleaver is the *random interleavers* which map the input bits to output positions randomly, i.e. according to a pseudo-random number set. These interleavers are simple to implement and have improved BER performance. However, it is almost impossible to guarantee a uniform BER performance for such random processes.

The UMTS interleaver is a complex 3-stage process resulting in 10 or 20 rows. In the first stage, the data bits are written into a rectangular matrix row by row. The dimensions of the matrix are carefully selected as a prime number p. The second stage re-arranges the bit positions within each row by performing an intra-row permutation. The third stage re-arranges the rows themselves by performing an inter-row permutation. Finally the bits are read out column by column. The UMTS interleaving algorithm is capable of generating an excellent bit error pattern for a wide range of interleaver lengths that adequately support voice and data transmission.

2.6.8   Turbo code tail-biting

It is desirable to have the turbo encoders starting and stopping in a known state, preferably in the same state, since this property simplifies the decoding process. A block oriented non-recursive convolutional encoder starting from the zero state returns to the zero state when (K-1) zeros are inserted into the encoder at the end of the data block where K is the constraint length. On the other hand, inserting zeros into the RSC encoders may not return it to the initial zero state because of their recursive nature.

On an optimistic note, any state can be considered as an initial state if a block of k bits returns the RSC encoder to the selected state – as the final state in a process called a tail-biting (Ma and Wolf, 1986) where the trellis is regarded as circular and such encoders are called Circular RSC (CRSC) coders. That is, the encoder starts the encoding at a known state (called the circular state S_C) and ends the encoding in the same state, as we shall see later. Since S_C depends on the contents of the input block, determining S_C requires a pre-encoding operation. We will demonstrate the tail-biting by an example where a look-up table is used for initializing the RSC encoder to the circular state S_C.

Example 2.15

Consider RSC encoder with generator polynomials (5,7) shown in Figure 2 49.

The feed forward polynomial $(5) = 1 + D^2$
The feedback polynomial $(7) = 1 + D + D^2$

Denote the current states, next states and the transition matrix of the encoder as **S**, **S′** and **g**, respectively, where:

$$\mathbf{g}\,\mathbf{S} = \mathbf{S}' \tag{2.159}$$

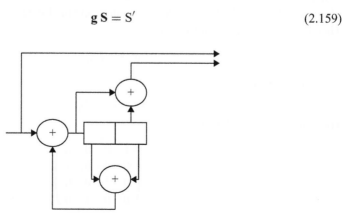

Figure 2.49   *(5,7) RSC encoder.*

Consider initial states $\mathbf{S} = [S_0 \ S_1 \ S_2 \ S_3]^T$, so if the encoder is in state S_0, with an input '0', the next state is S_0'. Similarly, if the encoder is in state S_1 with an input '0', the next state is S_2'. Proceeding with input '0', the initial state and next state are given below:

Initial state	Next state
S_0	S_0'
S_1	S_2'
S_2	S_3'
S_3	S_1'

Substituting in 2.159, we get:

$$\begin{pmatrix} g_{00} & g_{01} & g_{02} & g_{03} \\ g_{10} & g_{11} & g_{12} & g_{13} \\ g_{20} & g_{21} & g_{22} & g_{23} \\ g_{30} & g_{31} & g_{32} & g_{33} \end{pmatrix} \begin{pmatrix} S_0 \\ S_1 \\ S_2 \\ S_3 \end{pmatrix} = \begin{pmatrix} S_0' \\ S_2' \\ S_3' \\ S_1' \end{pmatrix}$$

Therefore matrix $\mathbf{g}$ is

$$\mathbf{g} = \begin{pmatrix} 1 & 0 & 0 & 0 \\ 0 & 0 & 1 & 0 \\ 0 & 0 & 0 & 1 \\ 0 & 1 & 0 & 0 \end{pmatrix} \tag{2.160}$$

Equation 2.159 can be written in another form:

$$\begin{pmatrix} 1 & 0 & 0 & 0 \\ 0 & 0 & 1 & 0 \\ 0 & 0 & 0 & 1 \\ 0 & 1 & 0 & 0 \end{pmatrix} \begin{pmatrix} 0 \\ 1 \\ 2 \\ 3 \end{pmatrix} = \begin{pmatrix} 0 \\ 2 \\ 3 \\ 1 \end{pmatrix} \tag{2.161}$$

Expression (2.161) means that the encoder goes from current state $S = 2$ to state $S' = 3$ with a single binary '0' input. Now consider the encoder with an input which consists of a block on N binary '0' bits and denotes the initial and final states as $\mathbf{S_c}$ and $\mathbf{S_c'}$, respectively. Then (2.159) will be transformed to:

$$\mathbf{g}^N \mathbf{S_c} = \mathbf{S_c'} \tag{2.162}$$

The final states $\mathbf{S_N}$ are:

$$\mathbf{S_N} = \mathbf{S_c} \oplus \mathbf{S_c'} \tag{2.163}$$

For example for $N = 5$, $\mathbf{g}^5$ is

$$\mathbf{g}^5 = \begin{pmatrix} 1 & 0 & 0 & 0 \\ 0 & 0 & 0 & 1 \\ 0 & 1 & 0 & 0 \\ 0 & 0 & 1 & 0 \end{pmatrix}$$

Substituting in (2.162), S'_c becomes

$$\begin{pmatrix} 1 & 0 & 0 & 0 \\ 0 & 0 & 0 & 1 \\ 0 & 1 & 0 & 0 \\ 0 & 0 & 1 & 0 \end{pmatrix} \begin{pmatrix} 0 \\ 1 \\ 2 \\ 3 \end{pmatrix} = \begin{pmatrix} 0 \\ 3 \\ 1 \\ 2 \end{pmatrix}$$

For
$$S_c = \begin{pmatrix} 0 \\ 1 \\ 2 \\ 3 \end{pmatrix}$$

The final states $\mathbf{S_N}$ is calculated using (2.163)

$$\mathbf{S_N} = \begin{pmatrix} 0 \\ 1 \\ 2 \\ 3 \end{pmatrix} \oplus \begin{pmatrix} 0 \\ 3 \\ 1 \\ 2 \end{pmatrix} = \begin{pmatrix} 0 & 0 \\ 0 & 1 \\ 1 & 0 \\ 1 & 1 \end{pmatrix} \oplus \begin{pmatrix} 0 & 0 \\ 1 & 1 \\ 0 & 1 \\ 1 & 0 \end{pmatrix} = \begin{pmatrix} 0 \\ 2 \\ 3 \\ 1 \end{pmatrix}$$

In summary,

For initial states $S_c = \begin{pmatrix} 0 \\ 1 \\ 2 \\ 3 \end{pmatrix}$ the final states $\mathbf{S_N} = \begin{pmatrix} 0 \\ 2 \\ 3 \\ 1 \end{pmatrix}$

Re-arranging the above matrices,

for $\mathbf{S_N} = \begin{pmatrix} 0 \\ 1 \\ 2 \\ 3 \end{pmatrix}$ the initial states $S_c = \begin{pmatrix} 0 \\ 3 \\ 1 \\ 2 \end{pmatrix}$

The look-up table for this example is:

Final states $\mathbf{S_N}$	0	1	2	3
Initial states $\mathbf{S_C}$	0	3	1	2

The above look-up table simply means that if we start the input 5-bit block encoding when the encoder is initialized to state 0 and the final state, say 2, then the circular state is 1, i.e. start the encoding with initial state 1 to finish encoding with final state 1 as well.

We now explore this statement by working through an example. Let the 5-bit block be [1 1 0 1 0]. The trellis diagram for the encoder is shown in Figure 2.50.

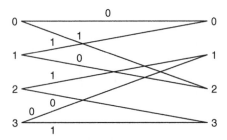

Figure 2.50   *Trellis diagram of (5,7) RSC encoder.*

Using Figure 2.50, the encoder changes state as follows:

Initial state 0, 5-bit block input [1 1 0 1 0],

Input bit	Current state	Next state
1	**0**	2
1	2	1
0	1	2
1	2	1
0	1	**2**

Therefore, final state $S_{NS} = 2$ and from the look-up table the circulation state $S_C = 1$, i.e. if we initialize the encoder with state (0 1), then the final state after completion of block encoding is state 1 as well. Let us investigate this by re-working the above table for initial state 1.

Initial state 1, 5-bit block input [1 1 0 1 0],

Input bit	Current state	Next state
1	**1**	0
1	0	2
0	2	3
1	3	3
0	3	**1**

Example 2.16

Consider the rate 1/3 turbo encoder in Figure 2.44 with RSC encoders C1 and C2 being similar to RSC encoder shown in Figure 2.49. Let the interleaver be described by

$$\Pi = [7, 3, 6, 5, 0, 2, 4, 1]$$

The input to the turbo encoder is

$$\mathbf{d}_k = [1 \quad 1 \quad 0 \quad 1 \quad 0 \quad 0 \quad 1 \quad 0]$$

Determine the encoded bit sequence. Convert the turbo encoder to rate 1/2 and find the encoded bit sequence.

Solution

The interleaver maps the input sequence $\mathbf{d}_k$ to $\mathbf{d}'_k$ as follows:

$$\mathbf{x}_k = \mathbf{d}_k = [1 \quad 1 \quad 0 \quad 1 \quad 0 \quad 0 \quad 1 \quad 0] = [x_0, x_1, x_2, \ldots\ldots\ldots, x_7]$$

The interleaver maps $\mathbf{x}_k$ to $\mathbf{x}'_k$

$$= \mathbf{d}'_k = [0 \quad 1 \quad 1 \quad 0 \quad 1 \quad 0 \quad 0 \quad 1] = [x_7, x_3, x_6, x_5, x_0, x_2, x_4, x_1]$$

Now $\quad \mathbf{x}_k = \mathbf{d}_k = [1 \quad 1 \quad 0 \quad 1 \quad 0 \quad 0 \quad 1 \quad 0]$

Assume both encoders are initialized to state 0, then

$$\mathbf{y}_k^1 = [1 \quad 0 \quad 0 \quad 0 \quad 0 \quad 1 \quad 0 \quad 1] \quad \mathbf{y}_k^2 = [0 \quad 1 \quad 0 \quad 0 \quad 0 \quad 0 \quad 1 \quad 0]$$

Thus the encoded bit sequence is

$$[1 \quad 1 \quad 0 \quad 1 \quad 0 \quad 1 \quad 0 \quad 0 \quad 0 \quad 1 \quad 0 \quad 0 \quad 0 \quad 0 \quad 0 \quad 0 \quad 1 \quad 0 \quad 1 \quad 0 \quad 1 \quad 0 \quad 1 \quad 0]$$

For rate 1/2 turbo encoder with alternate deletion of parity bits, the encoded bit sequence is

$$[1 \quad 0 \quad 1 \quad 1 \quad 0 \quad 0 \quad 1 \quad 0 \quad 0 \quad 0 \quad 1 \quad 1 \quad 1 \quad 0 \quad 1]$$

2.6.9 Turbo decoding

The turbo decoder is made up of two constituent soft-in soft-out decoders. Decoder 1 is associated with encoder C1 and decoder 2 is associated with encoder C2, and each of these decoders is a modified Maximum A Posteriori (MAP) decoder. A block diagram of the turbo decoder is shown in Figure 2.51.

The received signal $\mathbf{r} = \{\mathbf{r}_x, \mathbf{r}_y\}$ is demultiplexed into the received information sequence (systematic sequence) $\mathbf{r}_x$ and the received parity sequence $\mathbf{r}_y$. The latter is de-punctured (if puncturing is used at the encoder) and demultiplexed into $\mathbf{r}_{y1}$ applied to decoder 1 and $\mathbf{r}_{y2}$ which is applied to decoder 2. The received systematic sequence $\mathbf{r}_x$ is applied to decoder 1 and the interleaved version is applied to decoder 2.

The turbo decoding algorithm, based on the MAP decoding algorithm, is commonly known as the BCJR algorithm which was originally proposed in 1974 (Bahl, 1974). We first explore

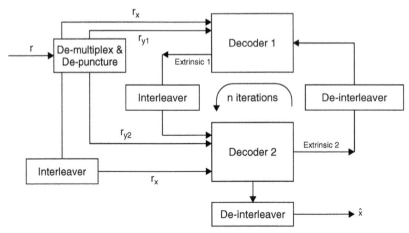

Figure 2.51   *Schematic diagram of turbo decoder.*

the BCJR algorithm in more detail and look at the necessary modification of the BCJR algorithm for turbo decoding.

We now turn our attention to Figure 2.51 to see the output of decoder 1 as a sequence of soft estimates of the transmitted data d_k. These soft estimates are the probabilities of the transmitted d_k conditioned to the received signal r and are called *extrinsic probabilities*. Extrinsic outputs do not have any information that is given to the respective decoders. The extrinsic 1 is interleaved and passed into decoder 2. Decoder 2 takes as its input, beside the interleaved extrinsic 1, the interleaved systematic received data r_x and the parity received sequence, and outputs a set of soft values (extrinsic 2) which is fed back to decoder 1 after de-interleaving.

The interleavers used in the turbo decoder are identical to the interleaver used in the encoders. This process of passing soft information in iterative manner is repeated until the process is converged or until a given number of iteration is completed.

2.6.10   The MAP algorithm

The Viterbi Algorithm (VA), considered in Section 2.6.3, is the conventional algorithm for optimum decoding of the non-systematic convolutional codes. The essence of the VA is rejecting the least likely paths through the trellis at each node (i.e. keeping the most likely paths). The process of path removal at each node usually leaves a single path through the trellis. Consequently, the VA is in fact conducting a hard decision process through the trellis and in this process produces the most likely sequence estimates. Clearly, making hard decisions through the trellis in this way corresponds to loss of valuable information which can be used to improve the decoding process.

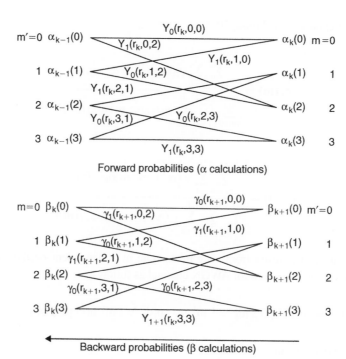

Forward probabilities (α calculations)

Backward probabilities (β calculations)

Figure 2.52 *Forward and backward passes through the turbo encoder trellis.*

Therefore, it is advantageous to compute a set of soft probabilities, about the possible transmitted data, rather than the hard decision values of the data itself given by the VA. It can be concluded then that the best algorithm for the turbo decoding is one that outputs soft decision probabilities on bit by bit of data and should accept soft decision values from the previous iterative decoding process. Such a decoder is generally referred to as a MAP decoder.

The MAP decoder achieves the soft decision decoding by making two passes through the trellis (compared with only one pass in VA). One pass in the forward direction and the other in the backward direction as shown in Figure 2.52, for the (5,7) trellis in Figure 2.51, where $m = 0, 1, \ldots, M - 1$ denotes the state of each decoder.

Let k be the time index and $\alpha_k(m)$ be the probability of the prior observations $\mathbf{r}$ up to time instant $(k - 1)$ with decoder state ending in state m at time k, $\beta_{k+1}(m)$ be the probability of future observations $\mathbf{r}$ at time instant k with decoder starting at state m at time $k + 1$. Let $\gamma_i(r_k, m', m)$, i is the data bit ($i = 0, 1$), be the probability of transition at time k from state m' to state m for each branch, i.e. branch metric.

The values of α and β have to be normalized, since these values are probabilities and the sum of α and β individually has to be one. The normalized values can be calculated using the following expressions (Moon, 2005):

$$\alpha_k(m) = \frac{\sum_{m'=0}^{M-1} \sum_{i=0}^{1} \gamma_i(r_k, m', m)\alpha_{k-1}(m')}{\sum_{m=0}^{M-1} \sum_{m'=0}^{M-1} \sum_{i=0}^{1} \gamma_i(r_k, m', m)\alpha_{k-1}(m')} \qquad (2.164)$$

$$\beta_k(m) = \frac{\sum_{m'=0}^{M-1} \sum_{i=0}^{1} \gamma_i(r_{k+1}, m', m)\beta_{k+1}(m')}{\sum_{m=0}^{M-1} \sum_{m'=0}^{M-1} \sum_{i=0}^{1} \gamma_i(r_{k+1}, m', m)\beta_{k+1}(m')} \qquad (2.165)$$

The probability of transition from state m' to state m is dependent on the transmission channel and can be expressed for AWGN channel as follows:

$$\gamma_i(r_k, m', m) = p(r_{y_k}|d_k = i, m', m) \cdot p(d_k = i|m', m) \cdot p(m|m') \qquad (2.166)$$

where i is the data bit and

$$p(r_{y_k}|d_k = i, m', m) = \frac{1}{\sqrt{2\pi\sigma^2}} \exp\left(-\frac{1}{2\sigma^2}[r_{y_k} - y_k(i, m', m)]^2\right) \qquad (2.167)$$

$$p(r_{x_k}|d_k = i, m', m) = \frac{1}{\sqrt{2\pi\sigma^2}} \exp\left(-\frac{1}{2\sigma^2}[r_{x_k} - x_k(i)]^2\right) \qquad (2.168)$$

Let us consider the MAP algorithm in more depth by working an example.

Example 2.17

Consider the (5,7) RSC encoder in Figure 2.49 with input sequence $d_k = [1011]$. The encoded sequence is formatted into antipodal bits so that binary '1' is mapped to $(+1)$ and binary '0' is mapped to (-1) and transmitted through AWGN channel with channel Signal-to-Noise power (SNR) = 2 dB. The received sequence is:

$$[(0.3605 \quad 1.4200) \quad (-0.8258 \quad 0.2677) \quad (-0.3152 \quad -1.0470) \\ (0.1972 \quad -0.5119)]$$

The decoder contains MAP algorithm. Assuming the encoder starts in state 0, determine the decoded sequence.

Solution

Given that SNR $= 2$ dB $= 1.58$ and assuming the average signal power $= 1$ W.

Therefore, $1.58 = \dfrac{1}{\sigma^2}$ so that $\sigma^2 = 0.63$.

Since the encoder starts in state 0, for time instant $k = 0$, we set:

$$[\alpha_0(0), \alpha_0(1), \alpha_0(2), \alpha_0(3)] = [1, 0, 0, 0]$$

Similarly

$$[\beta_N(0), \beta_N(1), \beta_N(2), \beta_N(3)] = [1, 0, 0, 0]$$

where N is the length of input sequence $\mathbf{d_k}$. The trellis of the (5,7) RSC encoder is:

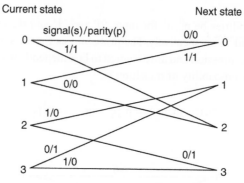

The encoded sequence is:

$$[11 \quad 01 \quad 10 \quad 10]$$

The BPSK mapped encoded sequence is:

$$[1 \quad 1 \quad -1 \quad 1 \quad 1 \quad -1 \quad 1 \quad -1]$$

We add AWGN to get the received sequence:

$$[(0.3605 \ 1.4200) \ (-0.8258 \ 0.2677) \ (-0.3152 \ -1.0470) \ (0.1972 \ -0.5119)]$$

If hard decision is made on the received sequence, the resulting sequence would be:

$$[1 \quad 1 \quad 0 \quad 1 \quad \mathbf{0} \quad 0 \quad 1 \quad 0]$$

Hard decision makes one bit in error shown in bold.

We now proceed with soft decision decoding using MAP algorithm. We start by calculating the α and β probabilities with the encoder in state 0.

Consider an N encoder where $I = 0, 1, \ldots, N$. Each next state combines two paths from two previous states, one for an input '1' and the other for an input '0'. Consequently, calculations of α, β, γ are carried out as shown here.

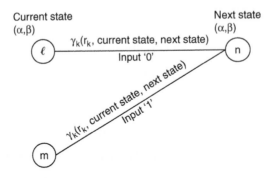

Let the signal probabilities be p_s^0, p_s^1 for input '0' and '1' and the parity probabilities be p_p^0, p_p^1 for input '0' and '1', respectively. The probabilities for the signal and the parity are given by a Gaussian expression and are assumed to be independent as seen by the channel. At the kth instant, the probability of transition is:

$$\gamma_k(r_k, m, m') = p_s^0 p_p^0 \quad \text{or} \quad = p_s^1 p_{p1}$$

Calculations of α:

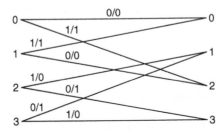

The encoder has four states, namely 0, 1, 2, 3.

At instant k = 0

$\alpha_0(0) = 1$
$\alpha_0(1) = 0$
$\alpha_0(2) = 0$
$\alpha_0(3) = 0$

At instant k = 1, the values of α are:

$\alpha_1(0) = \alpha_0(0) * \gamma_0(r_0, 0, 0) + \alpha_0(1) * \gamma_1(r_0, 1, 0) = \alpha_0(0) * \gamma_0(r_0, 0, 0)$ since $\alpha_0(1) = 0$
$\gamma_0(r_0, 0, 0) = p_s^0 p_p^0$

$$p_s^0 = \frac{1}{\sqrt{2\pi\sigma^2}} \exp\left(-\frac{1}{2\sigma^2}[r_{x,0} - x_0]^2\right)$$

$$r_{x,0} = 0.3605, r_{y,0} = 1.42$$

From the trellis

$x_0 = -1$
$y_0 = -1$

$$p_s^0 = \frac{1}{\sqrt{2\pi * 0.63}} \exp\left(-\frac{1}{2 * 0.63}[0.3605 - (-1)]^2\right) = 0.1157$$

$$p_p^0 = \frac{1}{\sqrt{2\pi * 0.63}} \exp\left(-\frac{1}{2\sigma^2}[r_{y,0} - y_0]^2\right)$$

$$p_p^0 = \frac{1}{\sqrt{2\pi * 0.63}} \exp\left(-\frac{1}{2 * 0.63}[1.42 - (-1)]^2\right) = 0.0048$$

Therefore, $\gamma_0(r_0, 0, 0) = p_s^0 \, p_p^0 = 0.1157 * 0.0048 = 5.5536e - 004 = \alpha_1(0)$

$\alpha_1(1) = \alpha_0(2) * \gamma_1(r_0, 2, 1) + \alpha_0(3) * \gamma_0(r_0, 3, 1) = 0$

$\alpha_1(2) = \alpha_0(0) * \gamma_1(r_0, 0, 2) + \alpha_0(1) * \gamma_0(r_0, 1, 2) = \alpha_0(0) * \gamma_1(r_0, 0, 2)$ since $\alpha_0(1) = 0$

$r_{x,0} = 0.3605, r_{y,0} = 1.42$

From the trellis

$x_0 = 1$
$y_0 = 1$

$$p_s^0 = \frac{1}{\sqrt{2\pi * 0.63}} \exp\left(-\frac{1}{2 * 0.63}[0.3605 - 1]^2\right) = 0.3633$$

$$p_p^0 = \frac{1}{\sqrt{2\pi * 0.63}} \exp\left(-\frac{1}{2\sigma^2}[r_{y,0} - y_0]^2\right)$$

$$p_p^0 = \frac{1}{\sqrt{2\pi * 0.63}} \exp\left(-\frac{1}{2 * 0.63}[1.42 - 1]^2\right) = 0.4370$$

Therefore, $\gamma_1(r_0, 0, 2) = 0.3633 * 0.4370 = 0.1588 = \alpha_1(2)$

$\alpha_3(1) = 0$

Now, since $\alpha_1(0) + \alpha_1(1) + \alpha_1(2) + \alpha_1(3) = 1$, we have to normalize values of α so that the new values are:

$\alpha_1(0) = 0.0035$
$\alpha_1(1) = 0$
$\alpha_1(2) = 0.9965$
$\alpha_1(3) = 0$

Repeating the same procedure for calculating values of α for $k = 2$, 3 and 4, we get the values for received sequence through *AWGN channel* as shown here:

Forward Pass ⟶ Direction of processing

state	α_0	α_1	α_2	α_3	α_4
0	1	0.0035	0.0015	0.0293	0.0297
1	0	0	0.0301	0.0811	0.1433
2	0	**0.9965**	2.4712e-4	0.0665	0.0460
3	0	0	**0.9682**	**0.8231**	**0.7810**

It is worth noting that the forward pass for received sequence *without noise* is:

Forward Pass ⟶ Direction of processing

state	α_0	α_1	α_2	α_3	α_4
0	**1**	0	0	0	0
1	0	0	0	0	0
2	0	**1**	0	0	0
3	0	0	**1**	**1**	**1**

The encoder terminates in state 3, so for received sequence *without noise*, β values are:

Direction of processing ⟵ Backward Pass

state	β_0	β_1	β_2	β_3	β_4
0	**1**	0	0	0	0
1	0	0	0	0	0
2	0	**1**	0	0	0
3	0	0	**1**	**1**	**1**

For received sequence through *AWGN channel*, we calculate β values using the same procedure used for calculating α values previously.

Direction of processing ⟵ Backward Pass

state	β_0	β_1	β_2	β_3	β_4
0	**0.8889**	0.0062	0.0028	0	0
1	0.0106	0.0294	0.1992	0	0
2	0.0862	**0.7408**	0.0684	0.0952	0
3	0.0144	0.2236	**0.7296**	**0.9048**	**1**

The posterior input bit at an instant of time k is decided by computing the sum of all the probabilities for $d_k = 1$ conditioned to $\mathbf{r}$ denoted by $\text{prob}(d_k = 1 | \mathbf{r})$ and the sum of all the probabilities for $d_k = 0$ conditioned to $\mathbf{r}$ denoted by $\text{prob}(d_k = 0|\mathbf{r})$ using the trellis shown below:

For $\text{prob}(d_k = 1|\mathbf{r})$

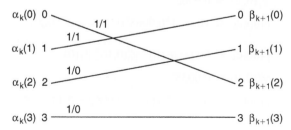

Similarly for $\text{prob}(d_k = 0|\mathbf{r})$

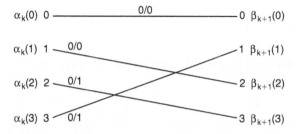

Mathematically we express the probabilities for $d_k = 1$ as:

$$\text{prob}(d_k = 1|\mathbf{r}) = \alpha_k(0) * \gamma_1^k(0,2) * \beta_{k+1}(2) + \alpha_k(1) * \gamma_1^k(1,0) * \beta_{k+1}(0)$$
$$+ \alpha_k(2) * \gamma_1^k(2,1) * \beta_{k+1}(1) + \alpha_k(3) * \gamma_1^k(3,3) * \beta_{k+1}(3)$$

$$(2.169)$$

and the probabilities for $d_k = 0$ as:

$$\text{prob}(d_k = 0|\mathbf{r}) = \alpha_k(0) * \gamma_0^k(0,0) * \beta_{k+1}(0) + \alpha_k(1) * \gamma_0^k(1,2) * \beta_{k+1}(2)$$
$$+ \alpha_k(2) * \gamma_0^k(2,3) * \beta_{k+1}(3) + \alpha_k(3) * \gamma_0^k(3,1) * \beta_{k+1}(1)$$

$$(2.170)$$

where $\gamma_i^k(m', m)$ is the transition probability (branch metric) for encoder binary input $i = 1, 0$, instant k, and encoder state transition from m' to m.

For $k = 0$ we have:

$\alpha_0(0) = 1 \quad \beta_1(0) = 0.0062$
$\alpha_0(1) = 0 \quad \beta_1(1) = 0.0294$
$\alpha_0(2) = 0 \quad \beta_1(2) = 0.7408$
$\alpha_0(3) = 0 \quad \beta_1(3) = 0.2236$

Substituting these values in 2.169 and 2.170, we get

$$\text{prob}(d_k = 1|\mathbf{r}) = \alpha_0(0) * \gamma_1^0(0,2) * \beta_1(2)$$

$$\gamma_1^0(0,2) = 0.9965$$

$$\text{prob}(d_k = 1|\mathbf{r}) = 1*0.9965*0.7804 = 0.7382$$

$$\text{prob}(d_k = 0|\mathbf{r}) = \alpha_0(0) * \gamma_0^0(0,0) * \beta_1(0)$$

$$\gamma_0^0(0,0) = 0.0035$$

$$\text{prob}(d_k = 0|\mathbf{r}) = 1*0.0035*0.0062 = 2.17e - 5$$

Normalized $\text{prob}(d_k = 1|\mathbf{r}) = 0.99997$
Normalized $\text{prob}(d_k = 0|\mathbf{r}) = 0.00003$

Therefore, $\hat{d}_k = 1$

Repeating the calculation for k = 1, 2, 3, we get the following table:

k	0	1	2	3
$\text{prob}(d_k = 1|\mathbf{r})$	**0.99997**	0.0084	**0.9912**	**0.9917**
$\text{prob}(d_k = 0|\mathbf{r})$	0.00003	**0.9916**	0.0088	0.0083
$\hat{d}_k$	1	0	1	1

Thus the MAP decoder output is [1 0 1 1], which is the same as the transmitted sequence.

In the above example, we have used the normalized posterior probabilities, $\text{prob}(d_k = 1|\mathbf{r})$ and $\text{prob}(d_k = 0|\mathbf{r})$, given by (2.169) and (2.170) to provide a maximum likelihood sequence soft estimate of d_k. This has certain computation implications since the dynamic range of α, β and γ is high, signifying the need for high memory and intensive computation. Therefore, it is more efficient to compute the logarithmic values of likelihood probabilities $\Lambda(d_k)$ since logarithmic computation converts multiplication operations to additions.

$$\Lambda(d_k) = \ln \frac{\sum_{m=0}^{M-1} \sum_{m'}^{M-1} \gamma_1^k(r_k, m', m)\alpha_{k-1}(m')\beta_k(m)}{\sum_{m=0}^{M-1} \sum_{m'=0}^{M-1} \gamma_0^k(r_k, m', m)\alpha_{k-1}(m')\beta_k(m)} \tag{2.171}$$

Expression (2.171) represents the probability of d_k being a binary '0' if the sign($\Lambda(d_k)$) is negative otherwise it is binary '1' and can be re-written as the subtraction of two logarithmic terms

$$\Lambda(d_k) = \ln\left[\sum_{m=0}^{M-1}\sum_{m'}^{M-1}\gamma_1^k(r_k, m', m)\alpha_{k-1}(m')\beta_k(m)\right]$$

$$- \ln\left[\sum_{m=0}^{M-1}\sum_{m'=0}^{M-1}\gamma_0^k(r_k, m', m)\alpha_{k-1}(m')\beta_k(m)\right] \quad (2.172)$$

Now if

$$x = \ln y$$

Then

$$y = \exp[\ln(y)] \quad (2.173)$$

Using the variable transformation in (2.173), we can write (2.172) as

$$\Lambda(d_k) = \ln\left[\sum_{m=0}^{M-1}\sum_{m'}^{M-1}\exp[\ln\gamma_1^k(r_k, m', m)]\exp[\ln\alpha_{k-1}(m')]\exp[\ln\beta_k(m)]\right]$$

$$- \ln\left[\sum_{m=0}^{M-1}\sum_{m'=0}^{M-1}\exp[\ln\gamma_0^k(r_k, m', m)]\exp[\ln\alpha_{k-1}(m')]\exp[\ln\beta_k(m)]\right]$$

Thus $\Lambda(d_k) = \ln\left[\sum_{m=0}^{M-1}\sum_{m'}^{M-1}\exp[\ln\gamma_1^k(r_k, m', m) + \ln\alpha_{k-1}(m') + \ln\beta_k(m)]\right]$

$$- \ln\left[\sum_{m=0}^{M-1}\sum_{m'=0}^{M-1}\exp[\ln\gamma_0^k(r_k, m', m) + \ln\alpha_{k-1}(m') + \ln\beta_k(m)]\right]$$

$$(2.174)$$

Expression (2.174) defines the log likelihood probability as natural logarithm of sum of exponential terms as

$$\Lambda(d_k) = \ln\left(\sum_{i=1}^{n}\exp(x_i)\right) - \ln\left(\sum_{j=1}^{n}\exp(y_i)\right)$$

We will simplify the above expression by using the following approximation

$$\ln\left(\sum_{i=1}^{n}\exp(x_i)\right) \approx \max(x_i) \quad i \in 1, 2, \ldots\ldots\ldots, n \quad (2.175)$$

We now simplify (2.174) using (2.175)

$$\Lambda(d_k) \cong \max_{(m,m')}\{\ln\gamma_1^k(r_k, m', m) + \ln\alpha_{k-1}(m') + \ln\beta_k(m)\}$$

$$- \max_{(m,m')}\{\ln\gamma_0^k(r_k, m', m) + \ln\alpha_{k-1}(m') + \ln\beta_k(m)\} \quad (2.176)$$

We know that r_k is made up of information (r_{x_k}) and parity part (r_{y_k}). In order to carry out the iterative decoding we need to separate the log likelihood probability into three components. These are the *extrinsic component* due to each decoder independent of the other decoder, the *priori component* which is derived by the other decoder, and the *systematic component* due to the systematic part of the received signal (r_{x_k}). We now write (2.176) in a form suitable for defining these terms:

$$
\Lambda(d_k) = \max_{(m,m')} \{\ln \gamma_1^k(r_{y_k}, m', m) + \ln \alpha_{k-1}(m') + \ln \beta_k(m)
$$
$$
+ \ln [\text{prob}(r_{x_k}|d_k = 1) \cdot \text{prob}(d_k = 1)]\}
$$
$$
- \max_{(m,m')} \{\ln \gamma_0^k(r_{y_k}, m', m) + \ln \alpha_{k-1}(m') + \ln \beta_k(m)
$$
$$
+ \ln [\text{prob}(r_{x_k}|d_k = 0) \cdot \text{prob}(d_k = 0)]\} \qquad (2.177)
$$

Since r_{x_k} is independent of the encoder states (m, m'), we can take the r_{x_k} terms outside the maximization bracket

$$
\Lambda(d_k) = \max_{(m,m')} \{\ln \gamma_1^k(r_{y_k}, m', m) + \ln \alpha_{k-1}(m') + \ln \beta_k(m)\}
$$
$$
- \max_{(m,m')} \{\ln \gamma_0^k(r_{y_k}, m', m) + \ln \alpha_{k-1}(m') + \ln \beta_k(m)\}
$$
$$
+ \ln[\text{prob}(r_{x_k}|d_k = 1) \cdot \text{prob}(d_k = 1)] - \ln[\text{prob}(r_{x_k}|d_k = 0) \cdot \text{prob}(d_k = 0)]
$$

$$
\Lambda(d_k) = \max_{(m,m')} \{\ln \gamma_1^k(r_{y_k}, m', m) + \ln \alpha_{k-1}(m') + \ln \beta_k(m)\}
$$
$$
- \max_{(m,m')} \{\ln \gamma_0^k(r_{y_k}, m', m) + \ln \alpha_{k-1}(m') + \ln \beta_k(m)\}
$$
$$
+ \ln \text{prob}(r_{x_k}|d_k = 1) + \ln \text{prob}(d_k = 1) - \ln \text{prob}(r_{x_k}|d_k = 0) - \ln \text{prob}(d_k = 0)
$$
$$
\qquad (2.178)
$$

The log of *posterior probability of systematic* bit $\Lambda_{s,k}$ is:

$$
\Lambda_{s,k} = \ln \text{prob}(r_{x_k}|d_k = 1) - \ln \text{prob}(r_{x_k}|d_k = 0) = \frac{2r_{x_k}}{\sigma^2}
$$

The log likelihood ratio of the *priori probability*, $\Lambda_{p,k}$ is:

$$
\Lambda_{p,k} = \ln \text{prob}(d_k = 1) - \ln \text{prob}(d_k = 0) = \ln \frac{\text{prob}(d_k = 1)}{\text{prob}(d_k = 0)}
$$

The *extrinsic probability* $\Lambda_{e,k}$ is:

$$
\Lambda_{e,k} = \max_{(m,m')} \{\ln \gamma_1^k(r_{y_k}, m', m) + \ln \alpha_{k-1}(m') + \ln \beta_k(m)\}
$$
$$
- \max_{(m,m')} \{\ln \gamma_0^k(r_{y_k}, m', m) + \ln \alpha_{k-1}(m') + \ln \beta_k(m)\}
$$

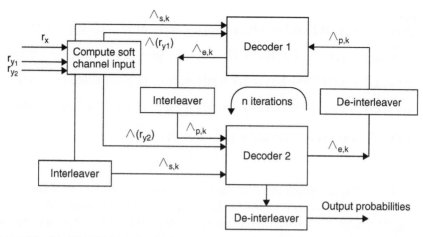

Figure 2.53 *Max-Log-MAP turbo decoder.*

The turbo detection algorithm expressed in (2.178) is known as the *Max-Log-MAP algorithm*. A block diagram of the Max-Log-MAP decoder is shown in Figure 2.53.

The approximation (2.175) used in deriving the Max-Log-MAP algorithm takes into consideration only the maximum exponential term and neglects all other terms. This causes a small $\frac{E_b}{N_0}$ degradation in the turbo decoder processing. A better approximation can be achieved using the '*Jacobian algorithm*':

$$\ln[\exp(x) + \exp(y)] = \max(x, y) + \ln[1 + \exp(-|x - y|)] = g(x, y) \qquad (2.179)$$

where $\ln[1 + \exp(-|x - y|)]$ is the *correction factor* that can be pre-computed for look-up table. The correction factor was ignored in the derivation of the Max-Log-MAP algorithm. We actually want to apply the *Jacobian algorithm* to multiple term summations like the following:

$$\ln[\exp(\delta_1) + \exp(\delta_2) + \cdots \cdots \cdots + \exp(\delta_{n-1}) + \exp(\delta_n)] \qquad (2.180)$$

Using (2.179), the expression in (2.180) is given by:

$$g_n(\delta_n, g_{n-1}(\delta_{n-1}, \cdots \cdots \cdots, g_2(\delta_3, g_1(\delta_1, \delta_2))) \qquad (2.181)$$

We will now show how to calculate (2.181) through the following worked example.

Example 2.18

Calculate the following expression using the *Jacobian algorithm* in (2.181):

$$Y = \ln[\exp(0.23) + \exp(0.72) + \exp(1.32) + \exp(2.3) + \exp(3.7)]$$

Solution

$$g_1(\delta_1, \delta_2) = g_1(0.23, 0.72) = 0.72 + \ln[1 + \exp(-|0.72 - 0.23|)] = 1.233$$

$$g_2(\delta_3, g_1) = g_2(1.32, 1.233) = 1.32 + \ln[1 + \exp(-|1.32 - 1.21)|] = 1.9706$$

$$g_3(\delta_4, g_2) = g_3(1.9706, 2.3) = 2.3 + \ln[1 + \exp(-|2.3 - 1.9706)|] = 2.8419$$

$$g_4(\delta_5, g_3) = g_4(2.8419, 3.7) = 3.7 + \ln[1 + \exp(-|3.7 - 2.8419)|] = 4.0467$$

Calculating Y using a pocket calculator gives $Y = 4.05$.

The turbo decoding algorithm that uses the '*Jacobian algorithm*' for calculating the logarithm of exponential terms is called the *Log-MAP algorithm* to distinguish it from *the Max-Log-MAP algorithm*, which approximates the logarithm of exponential terms using the maximum values. The BER performance verses $\frac{E_b}{N_0}$ in dB for rate 1/2 turbo encoder (37,21) with interleaving 256×256 is shown in Figure 2.54.

2.7 Channel capacity

Shannon's most famous theorem (Shannon, 1948) on channel capacity, emphasizes most brilliantly the interaction between three key parameters in information transmission, namely *channel bandwidth* B, *average received power* S and *noise power spectral density* N_0 at the channel output. The theorem specifies the fundamental limit on the rate of error-free transmission for a power-limited, band-limited channel perturbed by an additive white Gaussian noise. The Shannon limit for channel capacity C can be expressed mathematically as:

$$C = B \log_2\left(1 + \frac{S}{N_0 B}\right) \text{ bits/sec} \tag{2.182}$$

Expressed in terms of $\log_{10}$ (for students who use a pocket calculator), expression (2.182) becomes:

$$C = 3.32B \log_{10}\left(1 + \frac{S}{N_0 B}\right) \text{ bits/sec}$$

The received power (S) can be expressed in terms of the received energy per bit (E_b) and bit duration T_b as:

$$E_b = S T_b$$

Therefore,

$$S = E_b R_b \tag{2.183}$$

Consequently, the channel capacity in (2.182) can be simplified to:

$$\frac{C}{B} = \log_2\left(1 + \frac{E_b}{N_0} \frac{R_b}{B}\right)$$

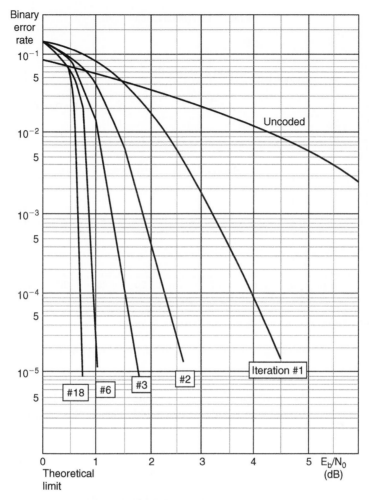

Figure 2.54 *BER given by iterative decoding (p = 1, . . . ,18) of a rate 1/2 encoder, memory = 4, generators $G_1 = 37$, G21 with interleaving 256 × 256 (Figure 9, Reproduced with permission from: Berrou, C. and Glavieux, A. (1996) Near optimum error correcting coding and decoding: Turbo codes, IEEE Transactions on Communications, 44(10), 1261–1271).*

Therefore, we may express $\frac{E_b}{N_0}$ in terms of the bandwidth efficiency $\left(\frac{R_b}{B}\right)$ as:

$$\frac{E_b}{N_0} = \frac{2^{\frac{R_b}{B}} - 1}{\frac{R_b}{B}} \tag{2.184}$$

A plot of *spectral efficiency* or *throughput* $\left(\frac{R_b}{B}\right)$ in b/s/Hz ($\log_{10}$ scale) verses $\frac{E_b}{N_0}$ in dB is shown in Figure 2.55.

The throughput diagram in Figure 2.55 defines the capacity boundary separating two regions. Region-1 supports low BER transmission at a bit rate less or equal to C by

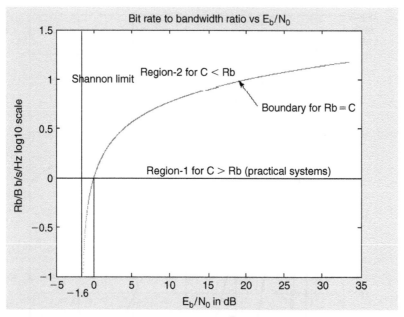

Figure 2.55 *Throughput in bits/Hz (log scale) verses $\frac{E_b}{N_0}$ in dB.*

employing efficient encoding schemes and region-2 for which low BER transmission is unattainable.

Let us now consider the maximum bit rate (i.e. the channel capacity) that can be used for transmission over a channel with an infinite bandwidth. Denote this capacity by C_∞ and define β as:

$$\beta = \frac{S}{N_0 B}$$

Then
$$C = \frac{S}{N_0 \beta} \log_2 (1 + \beta) \qquad (2.185)$$

Now
$$\log_2 (1 + \beta) = \frac{\ln (1 + \beta)}{\ln 2}$$

Therefore
$$C = \frac{S}{N_0 \beta} \frac{\ln (1 + \beta)}{\ln 2}$$

We can expand $\ln (1 + \beta)$ by the following series expansion:

$$\ln (1 + \beta) = \beta - \frac{\beta^2}{2} + \cdots \cdots \text{ and as } B \to \infty, \beta \to 0 \text{ so that}$$

$$C_\infty = \frac{S}{N_0 \beta} \left. \frac{\beta - \frac{\beta^2}{2} + \cdots \cdots}{\ln 2} \right|_{\beta \to 0}$$

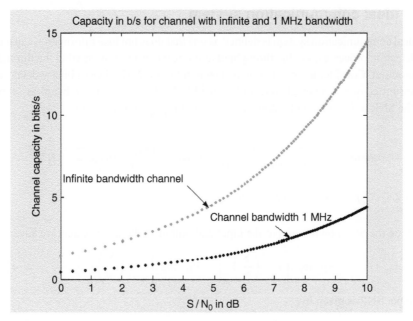

Figure 2.56 *Comparison of channel capacity with infinite and limited bandwidth.*

Thus,
$$C_\infty = \frac{S}{N_0} \frac{1}{\ln 2} = 1.44 \frac{S}{N_0} \qquad (2.186)$$

The channel capacities given by (2.182) and (2.186) are compared in Figure 2.56.

Substituting from (2.183) and choosing $R_b = C_\infty$ then (2.186) becomes:

$$\frac{E_b}{N_0} = 0.69 = -1.6\,\text{dB} \qquad (2.187)$$

The value $\frac{E_b}{N_0} = -1.6\,\text{dB}$ is the theoretical Shannon limit on $\frac{E_b}{N_0}$. The significance of this limit is that for power-bandwidth limited systems when $\frac{E_b}{N_0}$ is less or equal than -1.6 dB, theoretically the BER approaches 50%. This is clearly seen from Figure 2.55 since the channel capacity curve tends toward 0, indicating zero information transfer through the channel. However, above $\frac{E_b}{N_0} = -1.6$ dB the BER is nil or arbitrary low when a suitable coding scheme is used. For throughput of 1 b/s/Hz the Shannon limit on $\frac{E_b}{N_0} = 0$ dB. In practice, achieving the Shannon limit requires a highly complex and long encoding scheme which would be impractical to decode. However, emerging channel code schemes, such as turbo codes, exhibit a BER performance that approaches the theoretical Shannon limits within 1–2 dB.

2.8   Ideal communication system

An ideal communication system is defined as one that does not lose information capacity in the detection processes, i.e. the throughput at the receiver input is equal to the throughput at the output. Consider a typical system shown in Figure 2.57 where a binary data rate R_b is transmitted over AWGN channel of bandwidth B_T. The receiver bandwidth is set to R_b and the SNR at its input and output are $\left(\frac{S}{N}\right)_{in}$ and $\left(\frac{S}{N}\right)_{out}$, respectively.

Figure 2.57   *Typical communication system.*

Applying the Shannon relation to the input and output of the ideal system we have:

$$B_T \cdot \log_2\left[1 + \left(\frac{S}{N}\right)_{in}\right] = R_b \cdot \log_2\left[1 + \left(\frac{S}{N}\right)_{out}\right] \qquad (2.188)$$

The input SNR is given by:

$$\left(\frac{S}{N}\right)_{in} = \frac{S_i}{N_0 \cdot B_T}$$

$$= \left(\frac{S_i}{N_0 \cdot R_b}\right) \cdot \left(\frac{R_b}{B_T}\right) \qquad (2.189)$$

Now denote $\left(\frac{S}{N}\right)_{rec} = \frac{S_i}{N_0 R_b}$ as the receiver SNR. Substituting for $\left(\frac{S}{N}\right)_{in}$ from (2.189) and re-arranging (2.188), we get:

$$\log_2\left[1 + \left(\frac{S}{N}\right)_{rec} \cdot \frac{R_b}{B_T}\right]^{\frac{B_T}{R_b}} = \log_2\left[1 + \left(\frac{S}{N}\right)_{out}\right]$$

Therefore,
$$\left(\frac{S}{N}\right)_{out} = \left[1 + \left(\frac{S}{N}\right)_{rec} \cdot \frac{R_b}{B_T}\right]^{\frac{B_T}{R_b}} - 1 \qquad (2.190)$$

The SNR at the receiver output, $\left(\frac{S}{N}\right)_{out}$, verses the bandwidth expansion factor $\frac{B_T}{R_b}$ for various receiver SNR, $\left(\frac{S}{N}\right)_{rec}$ is plotted in Figure 2.58.

It is clear from Figure 2.58 that the bandwidth expansion factor, $\frac{B_T}{R_b}$, increases the SNR at the receiver output. For example, for bandwidth expansion of 4 the output SNR is 56.6 dB when the receiver SNR at 20 dB corresponds to an increase in output SNR by 36.6 dB. This increment in output SNR is only 23 dB when the receiver SNR is at 15 dB. The conclusion from this discussion is that the receiver performance (BER, resistant to fading and jamming) improves when the transmission bandwidth expands. This is one of the fundamental principles underpinning the spread spectrum techniques introduced in the next chapter. However, no physical system can deliver this theoretical increment in output SNR.

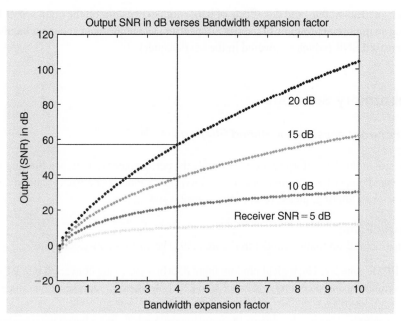

Figure 2.58   *Output SNR verses bandwidth expansion factor for various receiver SNR.*

2.9   Summary

In this chapter we have reviewed the fundamentals of the essential elements in digital communications used for spread spectrum wireless communications. We started out with consideration of the data transmission codes, since these codes have an impact on the bandwidth requirements to indicate the possibility of clock reconstruction from the received codes. The chapter then presented the general theory of N-dimensional signals and their optimum receivers, and used the theory to give detailed analysis of the optimum detection and the probability of error of binary signalling, and gradually introduced the reader to the pulse shaping (raised cosine and root raised cosine) and matched filtering techniques.

The chapter then presented an important element in a system called channel equalization. We explained why we need equalizers, where within the system it is located, and described with thorough analysis three types of equalizers (zero forcing, MMSE and adaptive), and their theoretical performances. This section was followed by consideration of the modulation/demodulation used in the spread spectrum communication systems and their RAKE receivers.

The channel coding was described in detail, including the convolutional codes and its Viterbi decoding algorithm used in IS-95, the MAP decoding algorithm for recursive systematic convolutional codes, the turbo coding and its decoding algorithms the Max-Log MAP and the Log-Map algorithms used in UMTS systems. The chapter then presented the Shannon channel capacity and the Shannon limit. Finally, we looked at the information through an ideal system, and the effects of bandwidth expansions on the output SNR high-lighting an important principle of the spread spectrum technique that exchanges bandwidth for increased SNR (which is covered in the next chapter).

Laboratory Sessions

Laboratory session I: Matched filtering

The aim of this laboratory session is to study the performance of matched filtering for baseband binary transmission operating in AWGN channel. For a background theory of matched filtering, please refer to Sections 2.7 and 2.8.

Instructions:

Construct the simulation model in Figure 2.59. The various subsystems are:

- BPSK encoder: The logical random input data bits $\{0, 1\}$ are converted into BPSK symbols using the simple relation: $b'_i = 2^*b_i - 1$ where b'_i represents bits in (± 1) and b_i in $(0,1)$.
- Upsample: This block converts the input binary symbols into a baseband waveform ready to be transmitted over the channel. The upsampler repeats the symbol 16 times (samples) before filtering.
- Transmit filter: This block is Matlab's subroutine for the Square Root Raised Cosine (SQRC) filter. Choose the roll off factor greater than 0 but less or equal 1.
- AWGN channel: This block adds Additive White Gaussian Noise with a specified signal-to-noise power level. Choose the SNR for your system.
- Receive filter: This block converts the received baseband waveform to a symbol stream. The receive filter is SQRC filter.
- Sampler: This block samples the received waveform and considers the sample point at the 16th sample in the bit duration.

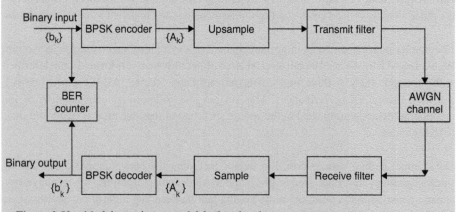

Figure 2.59 *Matlab simulation model for baseband communication system.*

- BPSK decoder: This block decodes the symbol stream using the decision rule for decoding:

$$\hat{b}_i = 1 \quad \text{if } b_i' \geq 0$$

$$= 0 \quad \text{if } b_i' < 0$$

The above decision rule provides Minimum BER over AWGN with matched filtering.
- BER counter: This block counts the number of errors and computes the probability of error.

Note: The system must be tuned to produce zero errors when noise source is switched off. This must be carried out before the start of the simulation.

Monte Carlo simulation

The BER performance simulation method used in this laboratory session is the well-known Monte-Carlo simulation technique and the following steps are used in estimating the BER:

- Choose minimum target BER to be estimated.
- Set the number of symbols in the single simulation run to at least 10/BER.
- Set up the Transmit/Receiver Filter to Square root raised cosine filters, the media to AWGN channel, and calibrate the required noise to a particular receiver operating point.
- Run the simulation and estimate the BER.
- Repeat the simulation run for a specified number of iterations and compute the average BER.
- Repeat the average BER estimation over all the operating points of interest.

Laboratory session II: Signal equalization

In this laboratory, we will design (zero-forcing and MMSE) linear equalizers to remove as much ISI as possible since the equalizer length is finite. For background reading on the subject, the reader is referred to Section 2.3.

Part A: MATLAB simulation of Zero-forcing equalizer with Impulse input:
  (i)   Generate an impulse at the input of the channel.
  (ii) Model the wireless channel by 4th order Butterworth LPF.
  (iii) Plot the channel impulse response.
  (iv) Compute the equalizer tap weights.
  (v)   Use FIR filter of length $(2M+1)$ as an equalizer.
  (vi) Plot the equalized output for $M = 1, 2, 7$.

Part B: MATLAB simulation of MMSE equalizer:
Write and Run the MATLAB code to carry out the following procedure:

   (i) Generate a BPSK (± 1) data of length 1200 bits. Take 4 samples/bit.
  (ii) Model the wireless channel by 4th order Butterworth LPF as in part A.
 (iii) Pass the transmitted signal through the channel and add noise to the signal.
  (iv) Compute equalizer tap weights.
   (v) Use FIR filter of length $(2M + 1) = 7$ as an equalizer.
  (vi) Plot the MSE vs. $\frac{E_b}{N_0}$ in dB.
 (vii) Tabulate the equalizer weights for each $\frac{E_b}{N_0}$ in dB.
(viii) Discuss your results and draw conclusions.

Problems

2.1 A sequence of binary polar data is transmitted at the rate of 9.6 kb/s through AWGN
    channel using rectangular pulse signalling. The SNR at the detector input is 5.6 dB
    and the effective noise bandwidth 12.5 kHz. Assume an ideal centre point detection,
    compute:
     i. $\frac{E_b}{N_0}$ at the detector input
    ii. probability of error of the system
   iii. if the noise bandwidth is set equal to the bit rate, what change in dB in the
        transmitted power is required to keep the same probability of error?
2.2 A sequence of equiprobable unipolar data is transmitted at the rate of 19.2 kb/s through
    AWGN channel. The average transmitted signal power is 0.5 mW and the average
    noise power is 5.5 μW. The resulting errors are 50 bits/s. Calculate the attenuation
    between the transmitter and the receiver.
2.3 A sequence of polar data of amplitude ± 1 is sent through AWGN channel. The
    standard deviation of the noise is 1.6. If the a priori probability of binary bit '1' is
    three times the a priori probability of binary '0', calculate:
     i. optimum detector threshold voltage
    ii. probability of bit error for the system.
2.4 A training sequence **a** is transmitted through AWGN channel and sampled at the input
    of Minimum Mean Square Error (MMSE) equalizer giving the following samples **y**:

$$\mathbf{y} = [0.65 \quad -0.15 \quad 0.4 \quad 0.2 \quad -0.75 \quad -0.25 \quad 0.55 \quad -0.2]$$

    Given the training sequence used is:

$$\mathbf{a} = [1 \quad -1 \quad 1 \quad -1 \quad 1 \quad -1 \quad 1 \quad -1]$$

    Calculate the tap weights of a three-tap MMSE equalizer.
2.5 In a BER experiment, a sequence of binary data, modulated on BPSK carrier, is
    transmitted through two channels: an AWGN channel and the output is fed into a
    matched filter receiver, and through a multipath fading channel where the matched

filter is replaced with 3-figure rake receiver. What would be the required $\frac{E_b}{N_0}$ in dB to keep the BER not more than 2.5×10^{-3} for both channels.

2.6 Consider 1/3 rate convolutional encoder with generator polynomials [5,7,7] and constraint length 3. The encoder is initialized to state 0.

    i. calculate the encoded sequence for an input $i(x) = 1100101$

    ii. construct the encoder trellis

    iii. calculate the encoder distance distribution.

2.7 Consider the 1/2 rate convolutional encoder with generator polynomials [5,7], constraint length 3 and the encoder is initialized to state 0. The data sequence [**101100100**] is input serially to the encoder and the encoded sequence is transmitted through AWGN channel. The received data sequence [11 11 00 01 00 11 11 10 11] is applied to the input of Viterbi algorithm decoder. Assume initial path of the start state (state 00) has metric $= 0$. Compute the estimated data sequence at the output of the Viterbi decoder.

2.8 Consider the rate 1/3 turbo encoder with RSC encoders C1 and C2 being similar to RSC encoder shown in Figure 2.49. Let the interleaver be described by:

$$\Pi = [5, 2, 7, 6, 0, 3, 4, 1]$$

Assume both encoders are initialized to state 0. The input to the turbo encoder is

$$\mathbf{d}_k = [1 \quad 0 \quad 0 \quad 1 \quad 1 \quad 0 \quad 1 \quad 1]$$

Determine the encoded bit sequence. Convert the turbo encoder to rate 1/2 and find the encoded bit sequence.

2.9 Consider the (5,7) RSC encoder in Figure 2.49 with input sequence $\mathbf{d}_k = [1100\ 10]$. The encoded sequence is BPSK modulated with binary bits ± 1 and transmitted through AWGN channel with channel Signal-to-Noise power (SNR) $= 2.2$ dB. The received sequence

$$[1.9231 \quad 0.9708 \quad 1.2541 \quad -0.8644 \quad -1.1449 \quad -0.4366 \quad -1.4567 \quad 2.6947$$
$$0.8941 \quad -0.9116 \quad -0.1719 \quad 1.0460]$$

The received sequence is applied to the input of MAP algorithm decoder.

    i. Calculate the encoded sequence.

    ii. If hard decision is made on the received sequence, what would be the resulting sequence?

    iii. Assuming the encoder starts in state 0, determine the estimated sequence at the output of the MAP decoder.

2.10 Consider an ideal communication system with receiver's SNR $= 10$ dB, and bandwidth expansion factor 5. The bandwidth expansion factor in spread spectrum systems defines the system processing gain. Calculate the output SNR of the ideal system and compare it with (input SNR in dB + expansion factor in dB).

References

Bahl, L., Cocke, J., Jelinek, F. and Raviv, J. (1974) Optimal decoding of linear codes for minimum symbol error rate', *IEEE Transactions on Information Theory*, 20, 284–287.

Berrou, C., Glavieux, A. and Thitimajshima, P. (1993) Near Shannon limit error-correcting coding and decoding: Turbo codes, *Proceedings of ICC '93*, Geneva, Switzerland, pp. 1064–1070, May.

Berrou, C and Glavieux, A. (1996) Near optimum error correcting coding and decoding: Turbo codes', *IEEE Transactions on Communications*, 44(10), 1261–1271.

Berrou, C. (2003) The ten-year-old turbo codes are entering service', *IEEE Communications Magazine*, pp. 110–116, August.

Gallager, R.G. (1962) Low-density parity-check codes, *IRE Transactions on Information Theory*, IT-8, 21–28.

Glover, I.A. and Grant, P.M. (1998) *Digital Communications*, Prentice Hall.

Hagenauer, J. and Hoeher, P. (1989) A veterbi algorithm with soft-decision outputs and its applications, *Proceedings of Globecom '89*, Dallas, TX, pp. 47.11–47.17, Nov.

Haykin, S. (1988) *Digital Communications*, John Wiley & Sons.

Haykin, S. and Moher, M. (2005) *Modern Wireless Communications*, Pearson Prentice Hall, International Edition.

Lee, J.S. and Miller, L.E. (1998) *CDMA Systems Engineering Handbook,* Artech House.

Ma, H.H. and Wolf, J.K. (1986) On tail biting convolutional codes, *IEEE Transactions on Communications*, COM-34(02), 104–111.

Moon, T.K. (2005) *Error Correction Coding*, John Wiley & Sons.

Price, R. and Green, P.E. (1958) A communication technique for multipath channels', *Proc. IRE*, 46, 555–570.

Proakis, J.G. (1995) *Digital Communications*, 3rd edition, McGraw-Hill.

Proakis, J.G. and Salehi, M. (2002) *Communication Systems Engineering*, 2nd edn, Prentice Hall.

Shannon, C.E. (1948) Mathematical theory of communication, *Bell System Technical Journal*, 27(3), 379–423.

Tanner, R.M. (1981) A recursive approach to low complexity codes, *IEEE Transactions on Information Theory*, IT-27, 533–547.

Viterbi, A.J. and Omura, J.K. (1979) *Principles of Digital Communications and Coding*, McGraw-Hill, International student edition.

3

Fundamentals of Spread-Spectrum Techniques

In this chapter we consider the spread-spectrum transmission schemes that demand channel bandwidth much greater than is required by the Nyquist sampling theorem. You will recall from Chapter 2 that the minimum bandpass bandwidth required for data transmission through an ideal channel is equal to the data symbol rate. You will also recall that wideband reception allows a large amount of input noise power to the detector and thus degrades the quality of the detected data. Therefore, the receivers for spread-spectrum schemes have to convert the received wideband signals back to their original narrowband waveforms before detection. This process generates a certain amount of processing gain that can be used to combat radio jamming and interference. We will describe and discuss in detail the properties and methods of generation of the functions used in creating wide spectrum signals. Finally, we consider the multiple access properties of the spread-spectrum systems and outline the analytical model for evaluating the system performance.

3.1 Historical background

There was intensive use of communications warfare during World War II. This technique outlined the ability to intercept and interfere with hostile communications. Consequently, this procedure stimulated a great deal of interest which led to the development of secure communications systems and work in this field was carried out on two fronts. Firstly, development in communication theory initiated encryption schemes (Shannon, 1949) to provide certain information protection. Secondly, work was initiated to harness the development of a new technology. This technology is called the Spread-Spectrum techniques (Scholtz, 1982), which exchanges bandwidth expansion for communications security and targets ranging for military applications.

By the end of the war, the theory of spread-spectrum techniques had developed and its anti-jamming capability had been recognized. Communication systems were developed by the military establishments during the 1960s, using frequency hopping and pseudo-noise spread-spectrum schemes. During this period, a multiple users' pseudo-noise spread-spectrum system was constructed, providing a 16 dB processing gain (Corneretto, 1961). An interesting system was also developed which combines pseudo-noise spread spectrum with Fourier transform (Goldberg, 1981). This is conceptually similar to the contemporary multicarrier spread-spectrum schemes.

Work on spread spectrum during the 1970s prompted commercial use of the spread-spectrum techniques, and theoretical work on spread-spectrum systems revealed the new system's ability to offer multiple access communications at an increased capacity compared to the time division or frequency division schemes of that time (Yue, 1983). The RAKE receiver concept (Price and Green, 1958) was developed to further accelerate the implementation of the systems. By the end of the decade, commercial applications of spread spectrum had become a reality.

The 1980s witnessed the development of the Global System of Mobile Telecommunications (commercially known as GSM) system, and a slow frequency hopping concept from spread-spectrum technique was implemented in the GSM systems to randomize the affects of interference from multiple users accessing the GSM network. The first trial of commercial spread-spectrum system with multiple access capabilities was carried out by Qualcom in the USA in 1993 (Gilhouse et al., 1991). The Qualcom's system was built according to the interim standard IS-95. The first commercial cellular radiophone service based on spread spectrum was inaugurated in Hong Kong in 1995. Korea and the USA soon introduced similar services.

During the 1990s, the spread-spectrum technique was further developed into 'multicarrier techniques' (Fazel, 1993) providing a higher diversity gain against deep fade than a single carrier spread-spectrum system could provide. The spread-spectrum multicarrier technique is based upon low rate data transmission over orthogonal frequency division multiplexing. This scheme generates multiple copies of the conventional spread spectrum; each copy is transmitted on a separate carrier. At the time of writing, many billions of dollars have already been invested in spread-spectrum development for the provision of high data rate for the next generation of communication networks.

3.2   Benefits of spread-spectrum technology

3.2.1   Avoiding interception

In military communications, interception of hostile communications is commonly used for various operations such as identification, jamming, surveillance or reconnaissance. The successful interceptor usually measures the transmitted power in the allocated frequency band. Thus, spreading the transmitted power over a wider band undoubtedly lowers the power spectral density, and thus hides the transmitted information within the background noise. The intended receiver recovers the information with the help of system processing gain generated in the spread process. However, the unintended receiver does not get the advantage of the processing gain and consequently will not be able to recover the information. Because of its low power level, the spread spectrum transmitted signal is said to be a Low Probability of Interception (LPI) signal.

3.2.2 Privacy of transmission

The transmitted information over the spread-spectrum system cannot be recovered without knowledge of the spreading code sequence. Thus, the privacy of individual user communications is protected in the presence of other users. Furthermore, the fact that spreading is independent of the modulation process gives the system some flexibility in choosing from a variety of modulation schemes.

3.2.3 Resistance to fading

In a multipath propagation environment, the receiver acquires frequent copies of the transmitted signal. These signal components often interfere with each other causing what is commonly described as signal fading. The resistance of the spread-spectrum signals to multipath fading is brought about by the fact that multipath components are assumed to be independent. This means that if fading attenuates one component, the other components may not be affected, so that unfaded components can be used to recover the information.

3.2.4 Accurate low power position finding

The distance (range) between two points can be determined by measuring the time in seconds, taken by a signal to move from one point to the other and back. This technique is exploited in the Global Positioning System (GPS). Since the signal travels at the speed of light (3×10^8 metres/sec),

$$\text{Range in metres} = 3 \times 10^8 \, \frac{\text{transit time}}{2}$$

It is clear from the above expression that the accuracy of the transit time measurement determines the ultimate range accuracy.

In practice, the transit time is determined by monitoring the correlation between transmitted and received code sequences. The transit time can be computed by multiplying the code duration by the number of code bits needed to align the two sequences. Clearly, higher resolution requires code symbols to be narrow which means high code rates. Thus, the sequences are selected to provide the required resolution so that if the code sequence has N chips, each with duration T_c seconds, then:

$$\text{Maximum range} = 1.5NT_c \cdot 10^8 \text{ metres}$$

The range resolution requires the chip duration T_c to be small so that sequence chip rate is as high as possible. On the other hand, maximum range requires a long sequence (i.e. N is large) so that many chips are transmitted in a single sequence period.

The GPS system consists of twenty-four satellites orbiting the earth along six orbital planes, spaced 60 degrees apart with nominally four satellites in each orbit. These clusters of satellites provide any user with visibility of five to eight satellites from any point on earth. The position, in 3-D, of a moving receiver and its speed can be measured using signals received from at least four satellites.

GPS provides two services. The precise positioning service uses very long code sequence at a code rate of 10.23 MHz. The standard positioning service, on the other hand, uses a shorter code (1023 bits) at a rate of 1.023 MHz. Each satellite is identified by a different phase of the short code.

3.2.5   Improved multiple access scheme

Multiple access schemes are designed to facilitate the efficient use of a given network resource by a group of users. Conventionally, there are two schemes in use: the Frequency Division Multiple Access (FDMA), and the Time Division Multiple Access (TDMA). In FDMA, the radio spectrum is shared between the users such that a fraction of the channel is allocated to each user at a time. On the other hand, in TDMA, each user is able to access the whole of the spectrum at a unique time slot.

The spread spectrum offers a new network access scheme due to the use of unique code sequences. Users transmit and receive signals with access interference that can be controlled or even minimized. This technique is called Code Division Multiple Access (CDMA) and is considered in more detail in Chapter 6.

3.3   Principles of spread-spectrum communications (Scholtz, 1977)

Digital transmission schemes which provide satisfactory performance and an adequate bit rate can be arranged into two categories.

- In applications like satellite communications, these schemes provide efficient usage of the limited power available.
- In applications such as mobile wireless, where the schemes achieve efficient usage of the limited bandwidth available for the service in demand.

However, both schemes are narrowband and vulnerable to hostile jamming and radio interference. The novelty of the spread-spectrum concept is that it provides protection against such attacks. This concept is based upon exchanging bandwidth expansion for anti-jamming capability.

The bandwidth expansion in spread spectrum is acquired through a coding process that is independent of the message being sent or the modulation being used. The spread spectrum, unlike FM, does not combat interference originated from thermal noise. The trade-off between Signal-to-Noise Ratio (SNR) and data bit rate (or bandwidth) in the spread-spectrum scheme can be demonstrated by the following.

Consider a digital signal transmission over a Gaussian channel occupying a bandwidth B with SNR = 10 dB. A channel coding scheme can be used to receive data with as small an error probability as desired if transmission is carried out at a data bit rate less or equal to the channel capacity (C) defined by the Shannon equation:

$$C = B \cdot \log_2(1 + SNR) \tag{3.1}$$

Substituting for the SNR = 10 dB in equation (3.1) gives the ratio of bit rate to bandwidth:

$$\frac{C}{B} = \log_2(11) = 3.46$$

Now if we reduce the channel SNR to 5 dB (i.e. to 3.16 in ratio), then referring to the bandwidth-efficiency diagram shown in Figure 3.1, the reliable transmission is still possible at the same bit rate but with expanded bandwidth B′ given by:

$$\frac{C}{B'} = \log_2(4.16) = 2.06$$

Now consider $B' = \frac{C}{2.06}$ and $B = \frac{C}{3.46}$ so that the expansion in the bandwidth is given by

$$\frac{B'}{B} = \frac{3.46}{2.06} = 1.73 \quad \text{Thus } B' = 1.73B$$

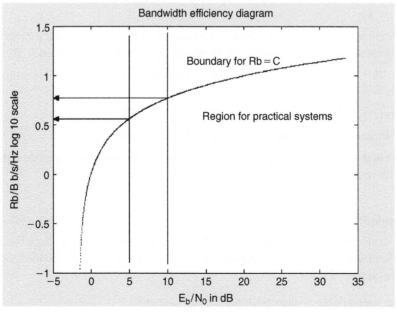

Figure 3.1 *Throughput in bits/Hz (log scale) versus $\frac{E_b}{N_0}$ in dB.*

The original bandwidth has to be expanded by a factor of about 1.73 to compensate for the reduction in the channel SNR.

It is worth noting that increasing the transmission bandwidth will undoubtedly increase the amount of the input noise power in a wideband receiver. But, as we will see, we commonly use a narrowband receiver to limit the amount of the input noise.

Example 3.1

Binary data is transmitted through an Additive White Gaussian Noise (AWGN) channel with $SNR = 3.5\,dB$ and bandwidth B. Channel coding is used to ensure reliable communications. Then:

i. What is the maximum bit rate that can be transmitted?
ii. If the bit rate is increased to 3B, how much must the channel SNR be increased to ensure reliable transmission?

Solution
$SNR = 3.5$ dB (=2.24 in ratio)

i. Channel capacity is given by Shannon equation (3.1):

$$C = B \cdot \log_2(1 + 2.24) = B \cdot \log_2(3.24)$$

$$= B \cdot \frac{\log_{10}(3.24)}{\log_2(2)} = 1.7B$$

Note the maximum bit rate for binary transmission that can be achieved with no errors in an ideal channel (no noise) is 2B. In this example the bit rate is about 1.7B.

ii. $C = 3B = B\log_2(1 + SNR)$ where SNR represents the channel's new signal-to-noise ratio.

Thus $(1 + SNR) = 2^3 = 8$, therefore, $SNR = 7 = 8.45$ dB

The increase in the channel $SNR = 8.45 - 3.5 = 4.95$ dB.

Note in this case, the bit rate is greater than 2B and the transmission of the data over the channel is multi-level but the symbol rate is still 2B.

Example 3.2

Binary data is transmitted at the rate of R_b bits/sec over a channel occupying a bandwidth B and the channel $SNR = 3$ dB. If the data bit rate is increased to $2.65R_b$ and the bandwidth is increased to 1.75B:

i. What would be the channel SNR for the new system?
ii. What channel bandwidth is required to keep the same channel signal-to-noise ratio?

Solution

  i. Substitute the SNR in equation (3.1):

$$SNR = 3\,dB = 2 \text{ in ratio}$$

So that for the first case: $R_b = B\log_2(1 + 2)$ and for the second case:

$2.65 \cdot R_b = 1.75B\log_2(1 + SNR)$ where SNR is the channel SNR for the new system.

$$\frac{1}{2.65} = \frac{B \cdot \log_2(3)}{1.75B \cdot \log_2(1 + SNR)}$$

$$= \frac{\log_{10}(3)}{1.75 \log_{10}(1 + SNR)}$$

Therefore, $1.75 \log_{10}(1 + SNR) = 2.65 \log_{10}(3)$. This gives

$$SNR = 4.28(=6.3\,dB).$$

  ii. If the channel signal-to-noise ratio is kept at 3 dB, the expanded bandwidth (B′) is computed from $\frac{1}{2.65} = \frac{B}{B'}$.

Thus $B' = 2.65B$ compared with 1.75B in the first case.

The spread-spectrum concept has developed from the principle of Shannon theorem. If data is transmitted at a rate of R_b over a channel occupying a bandwidth much greater than R_b, Shannon theorem indicates that reliable communications can be achieved at a reduced SNR. However, if the transmitted power is kept fixed, even though the power density is substantially reduced, a surplus in the SNR is generated and can be used to combat interference and jamming. This surplus in SNR is called processing gain.

The spreading of the energy is achieved by phase modulating the input data with the user code sequence. The modulation reduces the high power density of the original data to a low level shown in Figure 3.2(a).

A simple Matlab code is written to compare the power spectral density of 6 data symbols with power spectral density of the same data symbols spread using Gold sequence number 7 of length 31 and is shown in Figure 3.2(b). The spreading process generates enough processing gain to protect the transmission from hypothetical jammer employing a narrow band tone as shown in Figure 3.2(c).

The received signal has to be converted into the original narrowband to limit the amount of input noise accompanying the wideband reception. The conversion is performed at the receiver with the aid of a locally generated code sequence causing the spread spectrum to collapse. Moreover, the de-spreading process is accompanied with spreading of the jamming power into background noise as shown in Figure 3.2(d). Thus, de-spreading the wanted signal is accompanied by reduction of the impact of jamming attack on the data transmission.

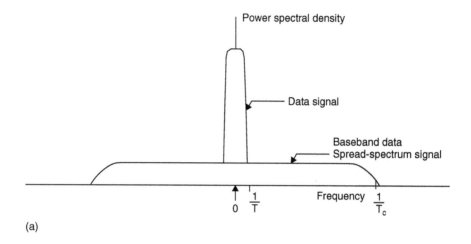

(a)

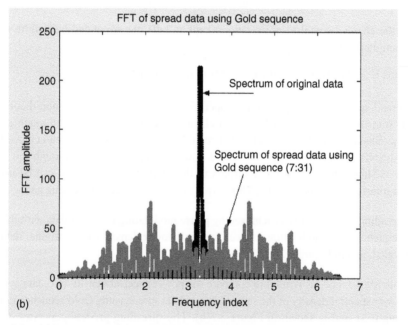

(b)

Figure 3.2 *(a) Power spectral density of data signal before and after spreading; (b) Power spectral density of spread-spectrum signal using Gold code sequence (7:31) generated by Matlab.*

3.4 Most common types of spread-spectrum systems

Two spread-spectrum systems are widely employed in the provision of reliable communications: the Direct Sequence Spread Spectrum (DS-SS), and the Frequency Hopped Spread Spectrum (FH-SS) systems. The DS-SS system executes the spreading of the data energy in real time by phase modulating the data with a high rate code sequence. On the other hand, the FH-SS scheme performs the energy spreading in the frequency domain.

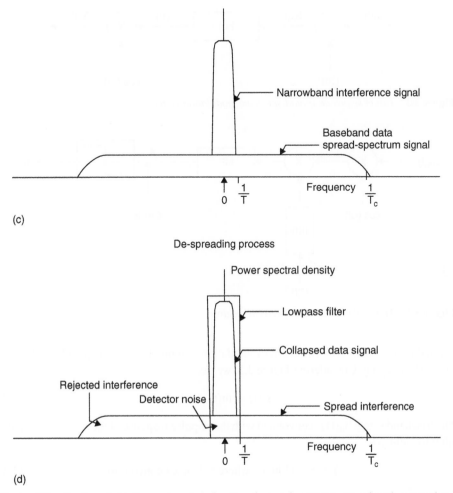

Figure 3.2 *Continued (c) Power spectral density of spread-spectrum signal with narrowband jamming signal; (d) Power spectral density of signal and interference after the despreading process.*

The latter is accomplished by forcing the narrowband carrier to jump pseudo-randomly from one frequency slot to the next according to the state of the code sequence in use.

Furthermore, a hybrid of both schemes can be developed to improve the processing gain compared to what is obtainable from a single scheme. The emphasis in this textbook is on the DS-SS systems and their applications in wireless communications.

3.4.1 DS-SS systems

A block diagram of the modulator that generates DS-SS signals is shown in Figure 3.3.

The binary data m(t) is first multiplied by the high rate code sequence to acquire the energy spreading. The baseband signal $S_n(t)$ is filtered to confine energy within the bandwidth

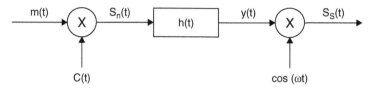

Figure 3.3   *Direct sequence spread-spectrum modulation system.*

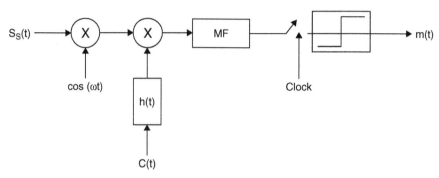

Figure 3.4   *Matched filter spread-spectrum receiver.*

defined by the code rate. The carrier modulation commonly used in spread spectrum is phase shift keying. Considering Figure 3.3, we get:

$$S_n(t) = m(t) \cdot C(t) \tag{3.2}$$

The baseband signal $S_n(t)$ is convoluted with the impulse response of the spectrum-shaping filter to yield $y(t)$:

$$y(t) = S_n(t) {}^* h(t) \quad \text{where * denotes convolution} \tag{3.3}$$

The bandpass signal $\qquad\qquad S_S(t) = [S_n(t) {}^* h(t)] \cdot \cos \omega_C t \tag{3.4}$

A basic block diagram of the matched filter receiver is shown in Figure 3.4. The received bandpass signal $S_S(t)$ is converted to an equivalent complex lowpass signal $A(t)$ by mixing with a locally generated coherent carrier. The lowpass spread spectrum is caused to collapse by multiplying by a locally generated in-phase copy of the transmitted code sequence. The de-spread signal B is matched filtered and sampled.

The complex lowpass signal $\qquad\qquad A(t) = S_S(t) \cdot \cos \omega_C t \tag{3.5}$

The de-spread signal $\qquad\qquad B(t) = A(t) \cdot [C(t) {}^* h(t)] \tag{3.6}$

The output of the matched filter $\qquad\qquad D(T) = \displaystyle\int\limits_{(K-1)T}^{KT} B(t) \cdot dt \tag{3.7}$

The receiver decodes the data according to the following rule:

$$D(T) > 0 \text{ decode binary '1' otherwise decode binary '0'}.$$

Example 3.3

A binary data stream of 4 digits [1011] is spread using an 8-chip code sequence $C(t) =$ [01 10 10 01]. The spread data phase modulates a carrier using binary phase shift keying. The transmitted spread-spectrum signal is exposed to interference from a tone at the carrier frequency but with 30 degrees phase shift. The receiver generates an in-phase copy of the code sequence and a coherent carrier from a local oscillator.

i. Determine the baseband transmitted signal.
ii. Express the signal received. Ignore the background noise.
iii. Assuming negligible noise, determine the detected signal at the output of the receiver.

Solution

i. Let the data stream be denoted as $m(t)$. The baseband spread-spectrum data $m_S(t)$ can be represented as:

$$m_S(t) = m(t) \cdot C(t) = [01\ 10\ 10\ 01, 10\ 01\ 01\ 10, 01\ 10\ 10\ 01, 01\ 10\ 10\ 01]$$

Since the data is transmitted as binary PSK, we map $0 \to 1$ and $1 \to -1$.

ii. The baseband spread-spectrum signal, $m_S(t)$, now modulates a carrier at frequency ω_C and the transmitted signal, $m_t(t)$, is given by:

$$m_t(t) = m_S(t) \cdot \cos \omega_C t.$$

The received signal $m_r(t)$ comprised the baseband signal $m_t(t)$, the interfering tone $I(t)$, and additive white noise $n(t)$. However in this example we ignore the noise so that signal plus interference is:

$$m_r(t) = m_t(t) + I(t)$$

The interfering signal is a sinusoidal waveform at frequency ω_c with 30 degrees phase shift:

$$I(t) = \cos(\omega_C t + 30)$$

Thus, the received signal $m_r(t) = m_t(t) + \cos(\omega_C t + 30)$

iii. The front end stage of the receiver mixes the received signal $m_r(t)$ with the local oscillator by multiplying $m_r(t)$ by the reference carrier, $(\cos \omega_C t)$ to compose the baseband signal, $m_b(t)$. Therefore:

$$\begin{aligned} m_b(t) &= m_t(t) \cdot \cos \omega_C t + \cos(\omega_C t + 30) \cdot \cos \omega_C t \\ &= 0.5 m_S(t)[1 + \cos 2\omega_C t] + 0.5[\cos 30 + \cos(2\omega_C t + 30)] \end{aligned}$$

Assume that $2\omega_C$ is removed by filtering and the signal level adjusted to unit by amplification then:

$$m_b(t) = m_S(t) + \cos 30$$

The next stage in the detection provokes the collapse of the spread spectrum into its original narrowband data. The de-spread signal $m_d(t)$ is given by multiplying $m_b(t)$ by the locally generated code sequence, that is:

$$m_d(t) = m_b(t) \cdot C(t) = [m_S(t) + \cos 30] \cdot C(t)$$
$$= m(t) \cdot C(t) \cdot C(t) + 0.866C(t)$$

Now $C(t) \cdot C(t)$ is a constant which can be normalized to one. The detector samples the de-spread signal at the code sequence rate and adds the samples to be compared with a threshold level. The summation of the sample of $C(t)$ when sampled at the code rate is

$$\sum_{k=0}^{7} C(kT_c) = -1 + 1 + 1 - 1 + 1 - 1 - 1 + 1 = 0$$

Therefore $m_d(t) = m(t)$
The output of the receiver is [10 11].

The quadrature spread-spectrum modulator, shown in Figure 3.5(a), comprises two orthogonal binary modulators similar to the one just described. The input data is demultiplexed into two parallel streams. Data transported on the in-phase channel is spread by the code sequence $C_i(t)$ and data on the quadrature channel is spread by the code sequence $C_q(t)$. The two parallel channels are combined to modulate a main RF carrier.

The quadrature spread-spectrum receiver consists of two binary matched filter receivers as shown in Figure 3.5(b). The detection of the data is carried out by each channel separately in a method identical to the one described for the binary channel.

3.4.2 Frequency hopping spread-spectrum system

Frequency hopping entails the transmission carrier frequency hopping between available channels within the spread-spectrum band. A narrow spectral band and an individual carrier frequency at the centre of the band define each transmitted channel. Successive carrier frequencies are chosen in accordance with the pseudo-random phases of the spreading code sequence. There are two widely used FH schemes: (1) Fast frequency hopping where one complete, or a fraction of the data symbol, is transmitted within the duration between carrier hops. Consequently, for a binary system, the frequency hopping rate may exceed the data bit rate. (2) On the other hand, in a slow frequency hopping system, more than one symbol is transmitted in the interim time between frequency hops.

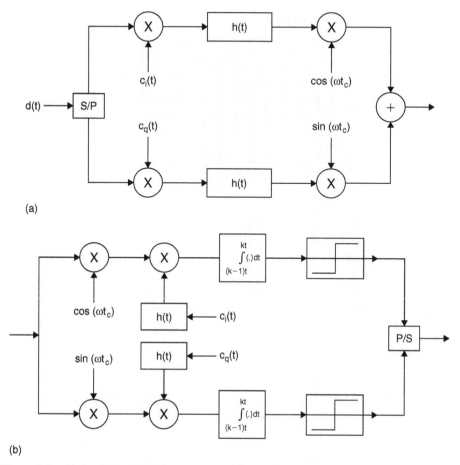

(a)

(b)

Figure 3.5 *(a) Quadrature spread-spectrum modulator; (b) Quadrature spread-spectrum receiver.*

Figure 3.6 illustrates how the carrier frequency hops with time. Let time duration between hops be T_h and data bit duration be denoted by T_b, then:

$$T_h \leq T_b \quad \text{for fast hopping} \tag{3.8}$$

$$T_h > T_b \quad \text{for slow hopping} \tag{3.9}$$

The basic FH modulation system, depicted in Figure 3.7(a), comprises a digital phase or frequency shift keying modulator and a frequency synthesizer. The latter generates carrier frequencies according to the pseudo-random phases of the spreading code sequence that is then mixed with the data carrier to originate the FH signal.

In the basic FH receiver, shown in Figure 3.7(b), the received FH signal is first filtered using a wideband bandpass filter and then mixed with a replica of the FH carrier. The mixer output is applied to the appropriate demodulator. A coherent demodulator may be used when a PSK carrier is received.

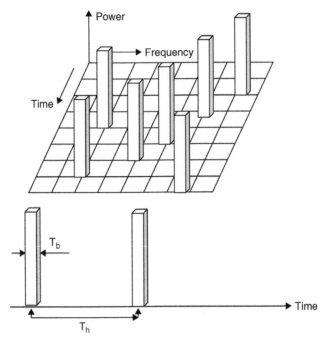

Figure 3.6 *Carrier frequency hopping from one frequency to another.*

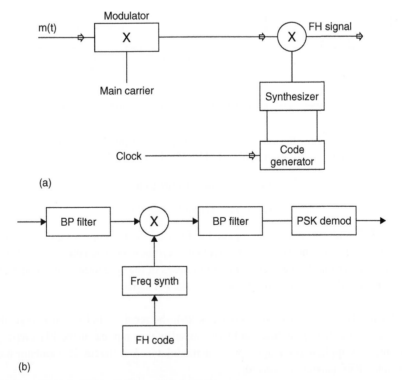

Figure 3.7 *(a) Basic FH modulator; (b) Basic FH receiver.*

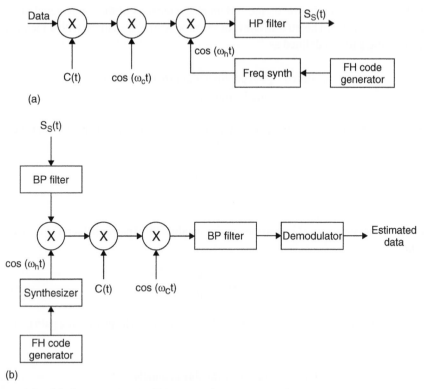

Figure 3.8 *(a) Direct sequence/Frequency hopping spread-spectrum transmitting system; (b) DS/FH spread-spectrum hybrid receiving system.*

3.4.3 Hybrid DS/FH systems

In special applications such as anti-jamming work, there may be a need for a hybrid system using both the DS and FH spread-spectrum schemes. A hybrid system is shown conceptually in Figure 3.8. Two code sequences are employed in this system. The DS/FH hybrid modulation system is shown in Figure 3.8(a); first code sequence is used to generate the DS-SS signal as described previously. The resulting signal is linearly modulated on a hopping carrier frequency generated by a frequency synthesizer according to the second code sequence. A replica of the hopping carrier is generated locally at the receiver using a coherent hopping code sequence.

The DS/FH hybrid receiver is shown in Figure 3.8(b) where the received signal is filtered and mixed with the hopping frequency and the output of the mixer is de-spread using the DS code.

3.5 Processing gain

Digital signal transmission is normally preceded by signal processing such as filtering, modulation and coding. At the receiver, processing like matched filtering and detection is used to recover the data.

In each of these processing methods, certain characteristics of the input signal are being modified or amplified. The effectiveness of the processor is measured with a factor called the processing gain G_p defined as:

$$\frac{\text{Modified signal parameter at processor output}}{\text{Signal parameter at input}}$$

In spread-spectrum systems, the parameter of interest is the signal spectrum at the input (B_b) and the spectrum of the output (B_s). Thus:

$$G_p = \frac{B_s}{B_b} \tag{3.10}$$

Thus the processing gain (G_p) expresses the bandwidth expansion factor. For a DS-SS system:

$$G_p = \frac{R_c}{R_b} \tag{3.11}$$

where R_c is the code sequence rate and R_b is the data bit rate. The processing gain generated by FH-SS system is:

$$G_p = \text{number of available channels} = N \tag{3.12}$$

Example 3.4

A speech conversation is transmitted by a DS-SS system. The speech is converted to PCM using an anti-aliasing filter with a cut-off frequency of 3.4 kHz and using 256 quantization levels. It is anticipated that the processing gain should not be less that 23 dB.

i. Find the required chip rate.
ii. If the speech was transmitted by an FH-SS system, what would be the number of hopping channels?

Solution

i. Sampling the speech at the Nyquist frequency generates $2 \times 3.4 = 6.8$ k samples/sec. We encode these samples using 256 quantization levels. Thus each sample is represented by n bits where

$$256 = 2^n$$

Thus $n = 8$

The PCM bit rate $= R_b = n \times 6.8 = 54.4$ k bits/sec

$$\text{Processing gain} = 23\text{dB} = 199.53 = G_p = \frac{R_c}{R_b}$$

Substituting for R_b gives $R_c = 10854.2\,\text{k chip/sec}$.

ii. Applying the definition of processing gain to transmission over an FH-SS system, we get:

$$G_p = \frac{B_s}{B_b} = N$$

Therefore the number of FH channels $= N \approx 200$.

3.6 Correlation functions (Sarwate and Pursley, 1980)

The interaction and the interdependence between two time (or frequency) varying signals are defined by the correlation function derived from the comparison of the two signals. The comparison of a signal with itself is described as the autocorrelation function. On the other hand, the cross-correlation is a measure of similarity between two autonomous signals. The correlation processing forms the basis upon which optimum detection algorithms in digital communication systems are derived.

Consider two binary sequences {a} and {b} with elements $\widehat{a}_n$ and $\widehat{b}_n$ that can be real or complex such that:

$$\{a\} = \{\widehat{a}_0, \widehat{a}_1, \widehat{a}_2, \ldots\ldots, \widehat{a}_{N-1}\} \tag{3.13}$$

$$\{b\} = \{\widehat{b}_0, \widehat{b}_1, \widehat{b}_2, \ldots\ldots, \widehat{b}_{N-1}\} \tag{3.14}$$

In the analysis, we assume the two sequences to be periodic with long period N and $0 \le n \le N - 1$. The reason behind this assumption is that, while code sequences in practical CDMA systems have long period N, they are in essence considered pseudo-random.

Two correlation functions of interest when considering spread-spectrum communications: periodic correlation function and aperiodic correlation function. Each is designed for a specific application.

3.6.1 Periodic correlation function

The periodic correlation function $R_{a,b}(\tau)$ of N-element sequences {a} and {b} is defined by:

$$R_{a,b}(\tau) = \sum_{n=0}^{N-1} \widehat{a}_n \cdot \widehat{b}_{n+\tau}^* \tag{3.15}$$

where $\widehat{b}_i^*(\tau)$ denotes the complex conjugate of $\widehat{b}_i(\tau)$. When $\widehat{a}_n = \widehat{b}_n$, $R_a(\tau)$ represents the Periodic Auto-Correlation Function [PACF] and with $\widehat{a}_n \neq \widehat{b}_n$, $R_{a,b}(\tau)$ describes the Periodic Cross-Correlation Function [PCCF]. The normalized correlation function is given by:

$$R_{a,b}(\tau)\big|_{norm} = \frac{R_{a,b}(\tau)}{N} \tag{3.16}$$

Now we consider the equation (3.15) for the periodic correlation $R_{a,b}(\tau)$ in more detail by expanding the summation:

$$R_{a,b}(\tau) = \widehat{a}_0 \cdot \widehat{b}_\tau^* + \widehat{a}_1 \cdot \widehat{b}_{\tau+1}^* + \widehat{a}_2 \cdot \widehat{b}_{\tau+2}^* + \cdots\cdots\cdots + \widehat{a}_{N-1-\tau} \cdot \widehat{b}_{N-1}^* + \widehat{a}_{N-\tau} \cdot \widehat{b}_N^*$$
$$+ \widehat{a}_{N-\tau+1} \cdot \widehat{b}_{N+1}^* + \widehat{a}_{N-\tau+2} \cdot \widehat{b}_{N+2}^* + \cdots\cdots\cdots + \widehat{a}_{N-1} \cdot \widehat{b}_{N-1+\tau}^* \tag{3.17}$$

$$= \widehat{a}_0 \cdot \widehat{b}_\tau^* + \widehat{a}_1 \cdot \widehat{b}_{\tau+1}^* + \widehat{a}_2 \cdot \widehat{b}_{\tau+2}^* + \cdots\cdots\cdots + \widehat{a}_{N-1-\tau} \cdot \widehat{b}_{N-1}^*$$
$$+ \widehat{a}_{N-\tau} \cdot \widehat{b}_{N\cdot\mathrm{mod}\,N}^* + \widehat{a}_{N-\tau+1} \cdot \widehat{b}_{(N+1)\cdot\mathrm{mod}\,N}^* + \widehat{a}_{N-\tau+2} \cdot \widehat{b}_{(N+2)\cdot\mathrm{mod}\,N}^*$$
$$+ \cdots\cdots\cdots + \widehat{a}_{N-1} \cdot \widehat{b}_{(N-1+\tau)\cdot\mathrm{mod}\,N}^* \tag{3.18}$$

Since the code sequences are assumed periodic with period $= N$ then:

$$\widehat{b}_{(N+i)\cdot\mathrm{mod}\,N}^* = \widehat{b}_i^*$$

Thus $R_{a,b}(\tau)$ in equation (3.18) is simplified to:

$$R_{a,b}(\tau) = \widehat{a}_0 \cdot \widehat{b}_\tau^* + \widehat{a}_1 \cdot \widehat{b}_{\tau+1}^* + \widehat{a}_2 \cdot \widehat{b}_{\tau+2}^* + \cdots\cdots\cdots + \widehat{a}_{N-1-\tau} \cdot \widehat{b}_{N-1}^*$$
$$+ \widehat{a}_{N-\tau} \cdot \widehat{b}_0^* + \widehat{a}_{N-\tau+1} \cdot \widehat{b}_1^* + \widehat{a}_{N-\tau+2} \cdot \widehat{b}_2^* + \cdots\cdots\cdots + \widehat{a}_{N-1} \cdot \widehat{b}_{\tau-1}^* \tag{3.19}$$

Now let us define the functions $R'_{a,b}(\tau)$ and $R''_{a,b}(\tau)$ as:

$$R'_{a,b}(\tau) = \widehat{a}_0 \cdot \widehat{b}_\tau^* + \widehat{a}_1 \cdot \widehat{b}_{\tau+1}^* + \widehat{a}_2 \cdot \widehat{b}_{\tau+2}^* + \cdots\cdots\cdots + \widehat{a}_{N-1-\tau} \cdot \widehat{b}_{N-1}^* \tag{3.20}$$

$$R''_{a,b}(\tau) = \widehat{a}_{N-\tau} \cdot \widehat{b}_0^* + \widehat{a}_{N-\tau+1} \cdot \widehat{b}_1^* + \widehat{a}_{N-\tau+2} \cdot \widehat{b}_2^* + \cdots\cdots\cdots + \widehat{a}_{N-1} \cdot \widehat{b}_{\tau-1}^* \tag{3.21}$$

Comparing equations (3.20) and (3.21) with (3.19), we have:

$$R_{a,b}(\tau) = R'_{a,b}(\tau) + R''_{a,b}(\tau) \tag{3.22}$$

Computation of $R'_{a,b}(\tau)$ and $R''_{a,b}(\tau)$ is carried out in the method explained in the following example. We start by sketching sequence {a} and {b} when aligned with zero time shifts ($\tau = 0$).

$\widehat{b}_0^*$	$\widehat{b}_1^*$	$\widehat{b}_2^*$	...	...	...	...	...	...	...	...	...	...	$\widehat{b}_{N-2}^*$	$\widehat{b}_{N-1}^*$
$\widehat{a}_0$	$\widehat{a}_1$	$\widehat{a}_2$	...	...	...	...	...	...	...	...	...	...	$\widehat{a}_{N-2}$	$\widehat{a}_{N-1}$

Now we delay sequence {a} by $\tau \neq 0$ relative to sequence {b} and only integer values of τ are considered. In practice, the shift τ can take on any real value. The sequences would look as in the following sketch.

$\widehat{b}_0^*$	$\widehat{b}_1^*$	$\widehat{b}_2^*$	. . . . .	$\widehat{b}_\tau^*$	$\widehat{b}_{\tau+1}^*$				$\widehat{b}_{N-2}^*$	$\widehat{b}_{N-1}^*$
$\widehat{a}_{N-\tau}$			$\widehat{a}_{N-1}$	$\widehat{a}_0$	$\widehat{a}_1$	$\widehat{a}_2$	. . . . . .	. . . . . .		$\widehat{a}_{N-\tau-1}$

The term $R'_{a,b}(\tau)$ is given by equation (3.20) as the summation of the product of the corresponding elements of the two sequences. Elements of sequence {b} have indices limited to $\tau \leq n \leq N-1$ and sequence {a} elements have indices limited to $0 \leq n \leq N-\tau-1$ as shown in the following sketch:

$\widehat{b}_\tau^*$	$\widehat{b}_{\tau+1}^*$		. . . . . .	$\widehat{b}_{N-2}^*$	$\widehat{b}_{N-1}^*$
$\widehat{a}_0$	$\widehat{a}_1$	$\widehat{a}_2$	. . . . . .	$\widehat{a}_{N-\tau-2}$	$\widehat{a}_{N-\tau-2}$

The term $R''_{a,b}(\tau)$ is given by equation (3.21) as the summation of the product of the sequence elements shown in following sketch:

$\widehat{b}_0^*$	$\widehat{b}_1^*$			. . . . . .	$\widehat{b}_{\tau-2}^*$	$\widehat{b}_{\tau-1}^*$
$\widehat{a}_{N-\tau}$	$\widehat{a}_{N-\tau+1}$	...	. . . . . .	. . . . . .	$\widehat{a}_{N-2}$	$\widehat{a}_{N-1}$

The periodic autocorrelation of the spreading code sequences plays an important role in the time tracking of the spread-spectrum system code sequence, as we will see in Chapter 5.

Example 3.5

Sequences {a} and {b}, each with period $N = 15$, are given by:

$$\{a\} = \{1, 1, 1, -1, 1, 1, -1, -1, 1, -1, 1, -1, -1, -1, -1\}$$
$$\{b\} = \{1, -1, -1, -1, -1, 1, -1, -1, -1, -1, 1, -1, -1, -1, -1\}$$

Find the periodic autocorrelation and cross-correlation functions of the sequences.

Solution

Periodic autocorrelation functions with shift right is shown in the following sketches:

Sequence {a}

1	2	3	4	5	6	7	8	9	10	11	12	13	14	15	$R_a(\tau)$
1	1	1	−1	1	1	−1	−1	1	−1	1	−1	−1	−1	−1	15
−1	1	1	1	−1	1	1	−1	−1	1	−1	1	−1	−1	−1	−1
−1	−1	1	1	1	−1	1	1	−1	−1	1	−1	1	−1	−1	−1
−1	−1	−1	1	1	1	−1	1	1	−1	−1	1	−1	1	−1	−1
−1	−1	−1	−1	1	1	1	−1	1	1	−1	−1	1	−1	1	−1
1	−1	−1	−1	−1	1	1	1	−1	1	1	−1	−1	1	−1	−1
−1	1	−1	−1	−1	−1	1	1	1	−1	1	1	−1	−1	1	−1
1	−1	1	−1	−1	−1	−1	1	1	1	−1	1	1	−1	−1	−1
−1	1	−1	1	−1	−1	−1	−1	1	1	1	−1	1	1	−1	−1
−1	−1	1	−1	1	−1	−1	−1	−1	1	1	1	−1	1	1	−1
1	−1	−1	1	−1	1	−1	−1	−1	−1	1	1	1	−1	1	−1
1	1	−1	−1	1	−1	1	−1	−1	−1	−1	1	1	1	−1	−1
−1	1	1	−1	−1	1	−1	1	−1	−1	−1	−1	1	1	1	−1
1	−1	1	1	−1	−1	1	−1	1	−1	−1	−1	−1	1	1	−1
1	1	−1	1	1	−1	−1	1	−1	1	−1	−1	−1	−1	1	−1
1	1	1	−1	1	1	−1	−1	1	−1	1	−1	−1	−1	−1	15
−1	1	1	1	−1	1	1	−1	−1	1	−1	1	−1	−1	−1	−1

Sequence {b}

1	2	3	4	5	6	7	8	9	10	11	12	13	14	15	$R_b(\tau)$
1	−1	−1	−1	−1	1	−1	−1	−1	−1	1	−1	−1	−1	−1	15
−1	1	−1	−1	−1	−1	1	−1	−1	−1	−1	1	−1	−1	−1	3
−1	−1	1	−1	−1	−1	−1	1	−1	−1	−1	−1	1	−1	−1	3
−1	−1	−1	1	−1	−1	−1	−1	1	−1	−1	−1	−1	1	−1	3
−1	−1	−1	−1	1	−1	−1	−1	−1	1	−1	−1	−1	−1	1	3
1	−1	−1	−1	−1	1	−1	−1	−1	−1	1	−1	−1	−1	−1	3
−1	1	−1	−1	−1	−1	1	−1	−1	−1	−1	1	−1	−1	−1	3
−1	−1	1	−1	−1	−1	−1	1	−1	−1	−1	−1	1	−1	−1	3
−1	−1	−1	1	−1	−1	−1	−1	1	−1	−1	−1	−1	1	−1	3
−1	−1	−1	−1	1	−1	−1	−1	−1	1	−1	−1	−1	−1	1	3
1	−1	−1	−1	−1	1	−1	−1	−1	−1	1	−1	−1	−1	−1	15
−1	1	−1	−1	−1	−1	1	−1	−1	−1	−1	1	−1	−1	−1	3
−1	−1	1	−1	−1	−1	−1	1	−1	−1	−1	−1	1	−1	−1	3
−1	−1	−1	1	−1	−1	−1	−1	1	−1	−1	−1	−1	1	−1	3
−1	−1	−1	−1	1	−1	−1	−1	−1	1	−1	−1	−1	−1	1	3
1	−1	−1	−1	−1	1	−1	−1	−1	−1	1	−1	−1	−1	−1	3
−1	1	−1	−1	−1	−1	1	−1	−1	−1	−1	1	−1	−1	−1	3

Periodic cross-correlation function

1	2	3	4	5	6	7	8	9	10	11	12	13	14	15	$R_{a,b}(\tau)$
1	1	1	-1	1	1	-1	-1	1	-1	1	-1	-1	-1	-1	
1	-1	-1	-1	-1	1	-1	-1	-1	-1	1	-1	-1	-1	-1	7
-1	1	-1	-1	-1	-1	1	-1	-1	-1	-1	1	-1	-1	-1	-1
-1	-1	1	-1	-1	-1	-1	1	-1	-1	-1	-1	1	-1	-1	-1
-1	-1	1	-1	-1	-1	-1	1	-1	-1	-1	-1	1	-1	-1	-1
-1	-1	-1	-1	1	-1	-1	-1	-1	1	-1	-1	-1	-1	1	-1
1	-1	-1	-1	-1	1	-1	-1	-1	-1	1	-1	-1	-1	-1	7
-1	1	-1	-1	-1	-1	1	-1	-1	-1	-1	1	-1	-1	-1	-1
-1	-1	1	-1	-1	-1	-1	1	-1	-1	-1	-1	1	-1	-1	-1
-1	-1	-1	1	-1	-1	-1	-1	1	-1	-1	-1	-1	1	-1	-1
-1	-1	-1	-1	1	-1	-1	-1	-1	1	-1	-1	-1	-1	1	-1
1	-1	-1	-1	-1	1	-1	-1	-1	-1	1	-1	-1	-1	-1	7
-1	1	-1	-1	-1	-1	1	-1	-1	-1	-1	1	-1	-1	-1	-1
-1	-1	1	-1	-1	-1	-1	1	-1	-1	-1	-1	1	-1	-1	-1
-1	-1	-1	1	-1	-1	-1	-1	1	-1	-1	-1	-1	1	-1	-1
-1	-1	-1	-1	1	-1	-1	-1	-1	1	-1	-1	-1	-1	1	-1
1	-1	-1	-1	-1	1	-1	-1	-1	-1	1	-1	-1	-1	-1	7

3.6.2 Aperiodic correlation function

The aperiodic correlation function between sequence $\{a\}$ and $\{b\}$ is defined by $C_{a,b}(\tau)$ where:

$$C_{a,b}(\tau) = \sum_{n=0}^{N-1-\tau} \widehat{a}_n \cdot \widehat{b}^{*}_{n+\tau} \quad 0 \leq \tau \leq N-1 \tag{3.23}$$

$$= \sum_{n=0}^{N-1+\tau} \widehat{a}_{n-\tau} \cdot \widehat{b}^{*}_{n} \quad 1-N \leq \tau \leq 0 \tag{3.24}$$

$$= 0 \quad |\tau| \geq N$$

Again if $\widehat{a}_n = \widehat{b}_n$, the expression $C_{a,b}(\tau)$ represents the Aperiodic Auto-Correlation Function [AACF]. When $\widehat{a}_n \neq \widehat{b}_n$, the expression defines the Aperiodic Cross-Correlation Function [ACCF]. The significance of [ACCF] becomes evident when we consider the access interference in multi-user spread-spectrum system in Chapter 6.

Let us focus our attention for now on the aperiodic cross-correlation, $C_{a,b}(\tau - N)$, between $\{a\}$ and $\{b\}$ when $\{a\}$ is time shifted to the left by $(\tau - N)$ with respect to $\{b\}$ such that:

$$C_{a,b}(\tau - N) = \sum_{n=0}^{N-1+(\tau-N)} \widehat{a}_{n-(\tau-N)} \cdot \widehat{b}^{*}_{n} \quad 1-N \leq \tau \leq 0 \tag{3.25}$$

$$= \sum_{n=0}^{\tau-1} \widehat{a}_{n-(\tau-N)} \cdot \widehat{b}_n^*$$

$$= \widehat{a}_{N-\tau} \cdot \widehat{b}_0^* + \widehat{a}_{N-\tau+1} \cdot \widehat{b}_1^* + \widehat{a}_{N-\tau+2} \cdot \widehat{b}_2^* + \cdots\cdots + \widehat{a}_{N-1} \cdot \widehat{b}_{\tau-1}^* \qquad (3.26)$$

Now compare equation (3.26) with equation (3.21), where we have just proved that:

$$C_{a,b}(\tau - N) = R''_{a,b}(\tau) \qquad (3.27)$$

Similarly, we can show that the aperiodic cross-correlation $C_{a,b}(\tau)$ is given as:

$$C_{a,b}(\tau) = R'_{a,b}(\tau) \qquad (3.28)$$

Thus, the periodic cross-correlation that has been defined in the previous section can be expressed in terms of the aperiodic cross-correlation as:

$$R_{a,b}(\tau) = C_{a,b}(\tau - N) + C_{a,b}(\tau) \qquad (3.29)$$

Example 3.6

Consider the sequences given in Example 3.5. Calculate the aperiodic correlation functions.

Solution

The AACF for sequence $\{a\}$

1	2	3	4	5	6	7	8	9	10	11	12	13	14	15	$C_a(\tau)$
1	1	1	-1	1	1	-1	-1	1	-1	1	-1	-1	-1	-1	15
	1	1	1	-1	1	1	-1	-1	1	-1	1	-1	-1	-1	0
		1	1	1	-1	1	1	-1	-1	1	-1	1	-1	-1	1
			1	1	1	-1	1	1	-1	-1	1	-1	1	-1	2
				1	1	1	-1	1	1	-1	-1	1	-1	1	1
					1	1	1	-1	1	1	-1	-1	1	-1	0
						1	1	1	-1	1	1	-1	-1	1	1
							1	1	1	-1	1	1	-1	-1	-2
								1	1	1	-1	1	1	-1	1
									1	1	1	-1	1	1	-2
										1	1	1	-1	1	-1
											1	1	1	-1	-2
												1	1	1	-3
													1	1	-2
														1	-1
											1	1	1	-1	-2
										1	1	1	-1	1	-3
									1	1	1	-1	1	1	-2

The AACF for sequence {b}

1	2	3	4	5	6	7	8	9	10	11	12	13	14	15	$C_b(\tau)$
1	−1	−1	−1	−1	1	−1	−1	−1	−1	1	−1	−1	−1	−1	15
	1	−1	−1	−1	−1	1	−1	−1	−1	−1	1	−1	−1	−1	4
		1	−1	−1	−1	−1	1	−1	−1	−1	−1	1	−1	−1	3
			1	−1	−1	−1	−1	1	−1	−1	−1	−1	1	−1	2
				1	−1	−1	−1	−1	1	−1	−1	−1	−1	1	1
					1	−1	−1	−1	−1	1	−1	−1	−1	−1	10
						1	−1	−1	−1	−1	1	−1	−1	−1	3
							1	−1	−1	−1	−1	1	−1	−1	2
								1	−1	−1	−1	−1	1	−1	1
									1	−1	−1	−1	−1	1	0
										1	−1	−1	−1	−1	5
											1	−1	−1	−1	2
												1	−1	−1	1
													1	−1	0
														1	−1
													1	−1	0
												1	−1	−1	1
											1	−1	−1	−1	2
										1	−1	−1	−1	−1	5
									1	−1	−1	−1	−1	1	0
								1	−1	−1	−1	−1	1	−1	1
							1	−1	−1	−1	−1	1	−1	−1	2
						1	−1	−1	−1	−1	1	−1	−1	−1	3
					1	−1	−1	−1	−1	1	−1	−1	−1	−1	10
				1	−1	−1	−1	−1	1	−1	−1	−1	−1	1	1
			1	−1	−1	−1	−1	1	−1	−1	−1	−1	1	−1	2
		1	−1	−1	−1	−1	1	−1	−1	−1	−1	1	−1	−1	3
	1	−1	−1	−1	−1	1	−1	−1	−1	−1	1	−1	−1	−1	4

The ACCF between sequences {a} and {b}

1	2	3	4	5	6	7	8	9	10	11	12	13	14	15	$C_{a,b}(\tau)$
1	1	1	−1	1	1	−1	−1	1	−1	1	−1	−1	−1	−1	
1	−1	−1	−1	−1	1	−1	−1	−1	−1	1	−1	−1	−1	−1	7
	1	−1	−1	−1	−1	1	−1	−1	−1	−1	1	−1	−1	−1	0
		1	−1	−1	−1	−1	1	−1	−1	−1	−1	1	−1	−1	1
			1	−1	−1	−1	−1	1	−1	−1	−1	−1	1	−1	2

(Continued)

1	2	3	4	5	6	7	8	9	10	11	12	13	14	15	$C_{a,b}(\tau)$
				1	−1	−1	−1	−1	1	−1	−1	−1	−1	1	1
					1	−1	−1	−1	−1	1	−1	−1	−1	−1	8
						1	−1	−1	−1	−1	1	−1	−1	−1	1
							1	−1	−1	−1	−1	1	−1	−1	0
								1	−1	−1	−1	−1	1	−1	3
									1	−1	−1	−1	−1	1	0
										1	−1	−1	−1	−1	5
											1	−1	−1	−1	2
												1	−1	−1	1
													1	−1	0
														1	−1

3.6.3 Even and odd cross-correlation functions

Another classification of the correlation functions frequently used is the even and odd correlation functions. These functions can be defined in terms of the periodic and aperiodic functions as we shall show in this section. Consider data $m_a(t)$ that is spread using code sequence $a(t)$ and data $m_b(t)$ that is spread using code sequence $b(t)$. Transmission of $m_b(t)$ is delayed by (τ) relative to transmission of data $m_a(t)$. The receiver is synchronized to code sequence $a(t)$ so that the received signal $r(t)$ can be expressed as:

$$r(t) = m_a(t) \cdot a(t) + m_b(t - \tau) \cdot b(t - \tau). \tag{3.30}$$

The correlation of $r(t)$ with code sequence $a(t)$ during the kth symbol of data $m_a(t)$ is given by:

$$y(kT) = \frac{1}{T} \int_{kT}^{(k+1)T} m_a(t) \cdot a(t) \cdot a(t) \cdot dt + \frac{1}{T} \int_{kT}^{(k+1)T} m_b(t - \tau) \cdot b(t - \tau) \cdot a(t) \cdot dt$$

$$= \frac{1}{T} m_a(kT) \int_{kT}^{(k+1)T} a(t) \cdot a(t) \cdot dt + \frac{1}{T} \int_{kT}^{kT+\tau} m_b(t - \tau) \cdot b(t - \tau) \cdot a(t) \cdot dt$$

$$+ \frac{1}{T} \int_{kT+\tau}^{(k+1)T} m_b(t - \tau) \cdot b(t - \tau) \cdot a(t) \cdot dt \tag{3.31}$$

Now $\frac{1}{T} \int_{kT}^{(k+1)T} a(t) \cdot a(t) \cdot dt =$ autocorrelation function of code sequence $a(t)$ at zero timeshift $= R_a(0)$. The 2nd and 3rd terms in the expression (3.31) are illustrated in Figure 3.9.

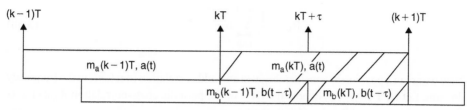

Figure 3.9 *Second and third terms for equation (3.31).*

Now considering the sketch in Figure 3.9, we have:

$$\frac{1}{T} \cdot \int_{kT}^{kT+\tau} m_b((k-1)T) \cdot b(t-\tau) \cdot a(t) \cdot dt + \frac{1}{T} \cdot \int_{kT+\tau}^{(k+1)T} m_b(kT) \cdot b(t-\tau) \cdot a(t) \cdot dt$$

$$= m_b(k-1)T \cdot C_{a,b}(\tau-N) + m_b(kT) \cdot C_{a,b}(\tau)$$

Thus

$$y(kT) = m_a(kT) \cdot R_a(0) + m_b(k-1)T \cdot C_{a,b}(\tau-N) + m_b(kT) \cdot C_{a,b}(\tau) \qquad (3.32)$$

If $m_b(kT) = m_b(k-1)T$, then

$$y(kT) = m_a(kT) \cdot R_a(0) + m_b(kT)[C_{a,b}(\tau-N) + C_{a,b}(\tau)] \qquad (3.33)$$

But when $m_b(kT) = -m_b(k-1)T$,

$$y(kT) = m_a(kT) \cdot R_a(0) + m_b(kT)\left[C_{a,b}(\tau-N) - C_{a,b}(\tau)\right] \qquad (3.34)$$

Thus, the even cross-correlation $R_{a,b}(\tau)$ and the periodic cross-correlation are the same, that is:

$$R_{a,b}(\tau) = C_{a,b}(\tau-N) + C_{a,b}(\tau) \qquad (3.35a)$$

The odd cross-correlation $\widehat{R}_{a,b}(\tau)$ is defined as:

$$\widehat{R}_{a,b}(\tau) = C_{a,b}(\tau-N) - C_{a,b}(\tau) \qquad (3.35b)$$

Similarly, the even and odd autocorrelation functions can be expressed in terms of the aperiodic autocorrelation function as follows:

$$R_a(\tau) = C_a(\tau-N) + C_a(\tau) \qquad (3.36a)$$

$$\widehat{R}_a(\tau) = C_a(\tau-N) - C_a(\tau) \qquad (3.36b)$$

Let the discrete Fourier transform (DFT) of the periodic code sequence {a} and {b} be sequence {A} and {B}, respectively, such that:

$$A_k = \frac{1}{N} \sum_{i=0}^{N-1} a_i \, e^{-j \cdot 2\pi \cdot \frac{ki}{N}} \qquad (3.37)$$

$$B_k = \frac{1}{N} \sum_{i=0}^{N-1} b_i\, e^{-j \cdot 2\pi \cdot \frac{ki}{N}} \tag{3.38}$$

Now we are in a position to consider the even and odd correlation functions in the frequency domain. Let DFT $(R_{a,b}(\cdot))$ denote the DFT of the periodic cross-correlation $R_{a,b}(\tau)$. It is shown in Sarwate and Pursley (1980) that:

$$\text{DFT}\,(R_{a,b}(k)) = N \cdot A_{-k} \cdot (B_{-k})^* \tag{3.39}$$

$$\text{DFT}\,(R_a(k)) = N \cdot |A_{-k}|^2 \tag{3.40}$$

Example 3.7

Consider the following two Walsh-Hadamard sequences, each with period $N = 8$, where sequences $\{a\}$ and $\{b\}$ are given by:

$\{a\} = \{010\ 1001\}$
$\{b\} = \{0000\ 1111\}$

i. Find the even and odd cross-correlation.
ii. Find the DFT of the periodic cross-correlation and express it in terms of the DFT of both sequences.

Solution
i. The even cross-correlation $R_{a,b}(\tau)$ is given by equation (3.35a):

$$R_{a,b}(\tau) = C_{a,b}(\tau - N) + C_{a,b}(\tau)$$

The odd cross-correlation $\widehat{R}_{a,b}(\tau)$ is given by equation (3.35b):

$$\widehat{R}_{a,b}(\tau) = C_{a,b}(\tau - N) - C_{a,b}(\tau)$$

The periodic cross-correlation is given by:

$$R_{a,b}(\tau) = \sum_{n=0}^{N-1} \hat{a}_n \cdot \hat{b}_{n+\tau}^*$$

We use the convention: binary '1' $= +1$ and '0' $= -1$.

For $\tau = 0$

-1	1	1	-1	1	-1	-1	1
-1	-1	-1	-1	1	1	1	1
1 $+$	(-1) $+$	(-1) $+$	1 $+$	1 $+$	(-1) $+$	(-1) $+$	1

$R_{a,b}(0) = 0$

For $\tau = 1$

-1	1	1	-1	1	-1	-1	1
1	-1	-1	-1	-1	1	1	1

$(-1)\ +\ (-1)\ +\ (-1)\ +\ 1\ +\ (-1)\ +\ (-1)\ +\ (-1)\ +\ 1$

$R_{a,b}(1) = -4$

For $\tau = 2$

-1	1	1	-1	1	-1	-1	1
1	*1*	-1	-1	-1	-1	1	1

$(-1)\ +\ 1\ +\ (-1)\ +\ 1\ +\ (-1)\ +\ 1\ +\ (-1)\ +\ 1$

$R_{a,b}(2) = 0$

It can be shown that:

τ	0	1	2	3	4	5	6	7
$R_{a,b}(\tau)$	0	-4	0	4	0	4	0	-4

The aperiodic cross-correlation, $C_{a,b}(\tau)$, is computed as follows:

For $\tau = 0$

-1	1	1	-1	1	-1	-1	1
-1	-1	-1	-1	1	1	1	1

$1\ +\ (-1)\ +\ (-1)\ +\ 1\ +\ 1\ +\ (-1)\ +\ (-1)\ +\ 1$

$C_{a,b}(0) = 0$

For $\tau = 1$

-1	1	1	-1	1	-1	-1	1
	-1	-1	-1	-1	1	1	1

$(-1)\ +\ (-1)\ +\ 1\ +\ (-1)\ +\ (-1)\ +\ (-1)\ +\ 1$

$C_{a,b}(1) = -3$

For $\tau = 2$

-1	1	1	-1	1	-1	-1	1
		-1	-1	-1	-1	1	1

$(-1)\ +\ 1\ +\ (-1)\ +\ 1\ +\ (-1)\ +\ 1$

$C_{a,b}(2) = 0$

It can be shown that:

τ	0	1	2	3	4	5	6	7
$C_{a,b}(\tau)$	0	-3	0	3	0	1	0	-1

The odd cross-correlation is calculated as follows:

τ	0	1	2	3	4	5	6	7
$R_{a,b}(\tau)$	0	-4	0	4	0	4	0	-4
$C_{a,b}(\tau)$	0	-3	0	3	0	1	0	-1
$C_{a,b}(\tau - N)$	0	-1	0	1	0	3	0	-3
$\hat{R}_{a,b}(\tau)$	0	2	0	-2	0	2	0	-2

ii. The Fast Fourier Transform (FFT) of $R_{a,b}(\tau)$ is:

τ	0	1	2	3	4
$FFT(R_{a,b})$	0	-1.414	0	1.414	0

The FFT of sequence {a} is:

τ	0	1	2	3	4
$FFT(a)$	0	$0.104 - j0.25$	0	$-0.604 + j0.25$	0

and the FFT of sequence {b} is:

τ	0	1	2	3	4
$FFT(b)$	0	$-0.25 + j0.604$	0	$-0.25 + j0.104$	0

Therefore, $N.A_{-k} \cdot (B_{-k})^* = 8$.

0    $(0.104 + j0.25)(-0.25 + j0.604)$ 0 $(-0.604 - j0.25)(-0.25 + j0.104)$ 0

$=$

0    $-1.416 + j0.002528$ 0 $1.416 - j0.002528$ 0

Therefore, (FFT) of $R_{a,b}(\tau) \approx N \cdot A_{-k} \cdot (B_{-k})^*$

3.6.4   The Merit Factor (Golay, 1982)

The Merit Factor (M_F) is defined by the ratio of the energy of the in-phase autocorrelation ($C_a(0)$) to the total energy of the out-of-phase autoscorrelation ($C_a(\tau)$); that is:

$$M_F = \frac{|C_a(0)|^2}{2 \sum_{\tau=1}^{N-1} |C_a(\tau)|^2} \tag{3.41}$$

The Merit Factor provides an insight into the behaviour of the sequence autocorrelation function, such that we can use M_F as an indicator to improve the design of code sequences. Ideally, sequences used in spread spectrum should exhibit large in-phase autocorrelation and zero (or very small) out-of-phase autocorrelation components. Consequently, such sequences enjoy very large Merit Factor. However, in practice such sequences do not necessarily have acceptable cross-correlation properties. Thus, the design of code sequences is based upon a compromise between providing low cross-correlation and a large Merit Factor.

Example 3.8

Compute the Merit Factor of sequence {a} where:

$$\{a\} = \{001100000101011\}$$

Solution

The Merit Factor (M_F) can be computed using equation (3.41) as follows:

For $\tau = 0$

−1	−1	1	1	−1	−1	−1	−1	−1	1	−1	1	−1	1	1
−1	−1	1	1	−1	−1	−1	−1	−1	1	−1	1	−1	1	1

$1 + 1 + 1 + 1 + 1 + 1 + 1 + 1 + 1 + 1 + 1 + 1 + 1 + 1 + 1$

$C_a(0) = 15$

For $\tau = 1$

−1	−1	1	1	−1	−1	1	−1	−1	1	−1	1	−1	1	1
	−1	−1	1	1	−1	−1	−1	−1	−1	1	−1	1	−1	1

$1 + (-1) + 1 + (-1) + 1 + 1 + 1 + 1 + (-1) + (-1) + (-1) + (-1) + (-1) + 1$

$C_a(1) = 0$

For $\tau = 2$

−1	−1	1	1	−1	−1	−1	−1	−1	1	−1	1	−1	1	1
		−1	−1	1	1	−1	−1	−1	−1	−1	1	−1	1	−1

$(-1) + (-1) + (-1) + (-1) + 1 + 1 + 1 + (-1) + 1 + 1 + 1 + 1 + (-1)$

$C_a(2) = 1$

Therefore, the aperiodic autocorrelation is:

τ	0	1	2	3	4	5	6	7	8	9	10	11	12	13
$C_a(\tau)$	15	0	1	−2	1	0	1	0	−1	−2	−1	2	1	−2

Thus, Merit Factor is given by:

$$M_F = \frac{|15|^2}{2 \cdot [0 + 1 + 4 + 1 + 0 + 1 + 0 + 1 + 4 + 1 + 4 + 1 + 4 + 1]} = 4.89$$

3.6.5 Interference rejection capability

Interference can be caused by an external transmitter tuned to a frequency within the passband of the intended receiving equipment, possibly with the same modulation and with enough power to override any signal at the intended receiver. There are many other types of signal interferences such as interference from random noise, random radio pulse, sweep-through, and stepped tones. Radio interference limits the effectiveness of the communication equipments.

Consider a spread-spectrum system transmitting information signal m(t) between two fixed points. Further, assume that the transmission is being exposed to a jamming signal, j(t). The channel noise and the interfering signal are assumed to be uncorrelated. The received signal r(t) can be expressed as:

$$r(t) = s_s(t) + j(t) + n(t) \tag{3.42}$$

where signal $s_s(t)$ is given by:

$$s_s(t) = m(t) \cdot C(t) \cdot \cos \omega_c t \tag{3.43}$$

where C(t) is the spreading code sequence. The reference signal used by the matching filter receiver is given by:

$$r_{ref} = C(t - \tau) \cdot \cos(\omega_c t + \theta) \tag{3.44}$$

where τ represents the phase delay between transmitted and locally generated sequences and θ is the carrier phase shift. The *signal component* at the matched filter output is:

$$s_0(t) = \int_0^{T_b} s_s(t) \cdot r_{ref}(t) \cdot dt \tag{3.45}$$

$$= \int_0^{T_b} m(t) \cdot C(t) \cdot \cos(\omega_c t) \cdot C(t - \tau) \cdot \cos(\omega_c t + \theta) \cdot dt$$

$$= \int_0^{T_b} m(t) \cdot C(t) \cdot C(t - \tau) \cdot \left[\frac{\cos \theta + \cos(2\omega_c t + \theta)}{2} \right] \cdot dt$$

$$= \int_0^{T_b} m(t) \cdot C(t) \cdot C(t - \tau) \cdot \left[\frac{\cos \theta}{2} \right] \cdot dt + \int_0^{T_b} m(t) \cdot C(t) \cdot C(t - \tau) \cdot \frac{\cos(2\omega_c t + \theta)}{2} \cdot dt$$

$$\tag{3.46}$$

To simplify the analysis, choose ω_c to be integer multiple of the data rate $\left(\frac{1}{T_b} \right)$ and τ is an integer number of chips so that:

$$\int_0^{T_b} m(t) \cdot C(t) \cdot C(t - \tau) \cdot \frac{\cos(2\omega_c t + \theta)}{2} \cdot dt = 0 \tag{3.47}$$

Substituting equation (3.47) in equation (3.46), we get:

$$s_0(t) = \int_0^{T_b} m(t) \cdot C(t) \cdot C(t - \tau) \cdot \left[\frac{\cos \theta}{2}\right] \cdot dt \qquad (3.48)$$

Now during each symbol interval, the input data m(t) is either $+1$ or -1 so that equation (3.48) simplifies to:

$$s_0(t) = \pm \cos \theta \cdot R_C(\tau) \qquad (3.49)$$

where $R_C(\tau)$ is the autocorrelation function of the code sequence C(t) at time shift τ and defined by:

$$R_C(\tau) = \int_0^{T_b} C(t) \cdot C(t - \tau) \cdot dt \qquad (3.50)$$

The output *noise component* is given by:

$$n_0(t) = \int_0^{T_b} n(t) \cdot r_{ref}(t) \cdot dt \qquad (3.51)$$

The *interference component* at matched filter receiver output is:

$$j_0(t) = \int_0^{T_b} j(t) \cdot r_{ref}(t) \cdot dt \qquad (3.52)$$

It is always useful to make the analysis more generic and so we will now consider power analysis of the receiver output rather than proceed with time domain analysis of the matched filter outputs as given by equations (3.49), (3.51) and (3.52) since further time domain analysis of these expressions requires specifications of the signal, interference and reference used.

The noise considered has white spectral density and zero mean value. Let the one-sided noise power density at the input of the receiver be N_0 in W/Hz. Clearly, the noise power at the output of the matched filter depends only on the noise spectral density at its input and the receiver bandwidth. The noise power is normally independent of the code chip rate. Let the bandwidth of the narrowband receiver be B_b, so the noise power output is:

$$\frac{N_0}{2} B_b \qquad (3.53)$$

Let the interference power at the input of the matched filter be J, and assume it is uniformly distributed across the spread-spectrum bandwidth B_S. Consequently, we can assume the average interference power spectral density to be $\frac{J}{B_s}$. It follows that the interference power at the receiver output is $\frac{J}{B_s}B_b$. Since we assume the noise to be independent of the interference, the total noise output power is the addition of output noise power and output interference power.

Let the received signal power be P_r with the receiver providing unit power gain. The ratio of output signal power to noise power, $(SNR)_0$ is expressed as:

$$(SNR)_0 = \frac{P_r}{\dfrac{N_0}{2}B_b + \dfrac{J}{B_s}B_b} \tag{3.54}$$

Substituting for the processing gain $G_p = \frac{B_s}{B_b}$ gives the output signal-to-noise ratio as:

$$(SNR)_0 = \frac{G_p \cdot P_r}{\dfrac{N_0}{2}B_s + J} \tag{3.55}$$

The signal power to noise power ratio at the input of the receiver is:

$$(SNR)_i = \frac{P_r}{\dfrac{N_0}{2}B_s + J} \tag{3.56}$$

Therefore, combining equation (3.55) with (3.56) we get:

$$(SNR)_0 = G_p \cdot (SNR)_i \tag{3.57}$$

The interference rejection capability of the spread-spectrum system can be evaluated in terms of the jamming margin, M_j, which is defined as the level of interference (jamming) that the system is able to tolerate and still maintain a specified level of performance such as specified bit error rate even though the signal-to-noise ratio is <1.

Let Loss be system losses between transmitter and receiver in dB, then jamming margin, M_j, is defined by:

$$M_j(dB) = -(SNR)_i(dB) - L(dB) \tag{3.58}$$

But $\qquad\qquad (SNR)_0(dB) = G_p(dB) + (SNR)_i(dB) \tag{3.59}$

Combining equation (3.58) with (3.59) we get:

$$M_j(dB) = G_p(dB) - (SNR)_0(dB) - L(dB) \tag{3.60}$$

The above equation (3.60) indicates that, under ideal conditions, the desired signal can be recovered with minimum distortion provided that there is enough processing gain to eradicate the effects of the jamming signal.

The most effective form of jamming against a DS spread-spectrum system is tone jamming at the centre of the spread-spectrum band (Dixon, 1994). We will not delve into details of various schemes available in the literature to combat jamming, but it is important to inspire the reader with two efficient schemes that can be employed when the processing gain is not sufficient to remove the effects of jamming: a closed loop scheme to cancel the effects of interference described in Mowbray et al. (1992) and a phase lock loop proposed to acquire and subtract the jamming tone is described in Abu-Rgheff et al. (1989).

Example 3.9

A message is transmitted at bit rate of 9.6 kb/s using a direct sequence spread-spectrum system. The clock rate of the code sequence is 512 Mb/s. The receiver reference carrier is off by 12 degrees and the code synchronization error is 30%. The one-sided noise density (N_0) at the receiver input is 10^{-18} W/Hz. The received signal power is 2.3×10^{-14} W.

i. Find the signal-to-noise ratio at receiver output $(SNR)_0$ in dB
ii. What is the maximum obtainable $(SNR)_0$ in dB?
iii. Find the synchronization error that would reduce $(SNR)_0$ by 1.5 dB when carrier phase coherence has been achieved.

Solution

i. Recall from previous analysis, the output signal $s_0(t)$ is given by equation (3.49) as:

$$s_0(t) = m(t) \cdot \cos \theta \cdot R_c(\tau)$$

Thus, received power $P'_r = P_r \cdot \cos^2 \theta \cdot R_c^2(\tau)$. In the ideal case, the autocorrelation, $R_c(\tau)$, is shown in Figure 3.10.

In the region between 0 and $\pm T_c$, the autocorrelation function $R_c(\tau)$ can be expressed in terms of τ as:

$$R_c(\tau) = \left[1 - \frac{|\tau|}{T_c} \right]$$

Given $\dfrac{|\tau|}{T_c} = 30\%$, then $R_c(\tau) = 0.70$

Therefore, $P'_r = 2.3 \times 10^{-14} \times \cos^2 12 \times (0.7)^2 = 1.08 \times 10^{-14}$

We calculate the output SNR from equation (3.56) but with $J = 0$ since no interference is applied:

$$(SNR)_0 = \frac{P'_r}{\dfrac{N_0}{2} \times B_b} = \frac{1.08 \times 10^{-14}}{\dfrac{10^{-18}}{2} \times 9600} = 2.25 = 3.5 \, dB$$

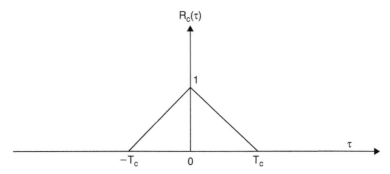

Figure 3.10 *Signal-to-noise ratio at receiver output (SNR)$_0$ in dB.*

ii. The maximum possible (SNR)$_0$ is evaluated when system employed coherent reference carrier and the code synchronization error is zero.

$$(SNR)_0|_{max} = \frac{P_r}{\frac{N_0}{2} \times B_b} = \frac{2.3 \times 10^{-14}}{\frac{10^{-18}}{2} \times 9600} = 4.79 = 6.8 \, dB$$

iii. The new value for (SNR)$_0$ is $6.8 - 1.5 = 5.3 \, dB = 3.39$

$$(SNR)_0 = 3.39 = \frac{P'_r}{\frac{N_0}{2} \times B_b} = \frac{P'_r}{\frac{10^{-18}}{2} \times 9600}$$

Thus $P'_r = 16.27 \times 10^{-15} \, W = 2.3 \times 10^{-14} \times R_c^2(\tau)$

Thus $R_c^2(\tau) = 0.707$ giving $R_c(\tau) = 0.84 = 1 -$ synchronization error percent.

Therefore synchronization error $= 16\%$.

3.7 Performance of spread-spectrum systems (Pursley, 1977)

The performance of a spread-spectrum system is measured in terms of the Bit Error Rate (BER), a quantity related to the theoretically computed probability of detection error.

The main parameter that influences the performance of the system when the link shared by a group of users is the level of interference generated by the multiple user-access, due to the fact that signals from individual users in spread-spectrum transmission co-exist within the same frequency band. The transmission from the users can take one of two possible modes: the first mode is when transmissions from individual users are coordinated in time and frequency and this mode results in a synchronous spread-spectrum system. Furthermore, if the synchronous signals are generated using orthogonal sequences (i.e. the cross-correlation between any pair is zero), then such signals do not interfere with

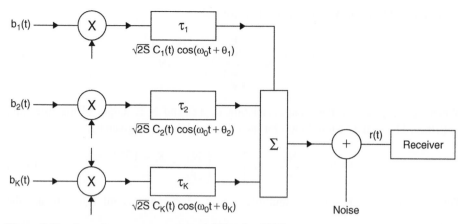

Figure 3.11 *Spread-spectrum system model (Pursley 1977).*

each other. Consequently, the channel thermal noise exclusively determines the performance of the system. The second mode is when the transmissions are uncoordinated and results in an asynchronous system and the multiple access interference has to be considered when evaluating system performance.

The performance of an asynchronous spread-spectrum system is analysed by Pursley (1977) using the basic model shown in Figure 3.11. It has been shown (Mowbray and Grant, 1992) that the distribution of the multiple access interference generated by a large group of users simultaneously accessing the network, tends to be Gaussian.

Let us consider the Pursley model with K users such that data $d_k(t)$ generated by k^{th} user is transmitted using code sequence $C_K(t)$, at a time delay τ_K and carrier phase offset θ_K relative to the intended user. Pursley has shown that for a large community of users ($N \gg 1$), the worst-case probability of error P_{max} is given by:

$$P_{max} \leq 1 - \Phi\left(\left[1 - (K-1)\left(\frac{2C_c}{N}\right)\right] \cdot \sqrt{\frac{2E_b}{N_0}}\right) \qquad (3.61)$$

Where $\frac{N_0}{2}$ is channel two-sided thermal noise spectral density, E_b is the energy per data bit, and $\Phi(\cdot)$ is the zero mean, unit variance Gaussian distribution, C_c is the maximum magnitude of the aperiodic cross-correlation given by:

$$C_c = \max\{|C_{k,i}(\tau)|\} \quad 1 - N \leq \tau \leq N - 1 \qquad (3.62)$$

$|C_{k,i}(\tau)|$ is the magnitude of the cross-correlation between code sequences that belong to users k and i. Furthermore, Pursley has shown that the amount of multiple access interference, Q_a, from $(K - 1)$ other active users is:

$$Q_a = \frac{1}{6 \cdot N^3} \cdot \sum_{\substack{k = 1 \\ k \neq i}}^{K} r_{k,i} \approx \frac{(K - 1)}{3N} \tag{3.63}$$

Where $r_{k,i}$ is given by the cross-correlation $C_{k,i}(\tau)$ and N is length of the spreading sequence. The signal-to-noise ratio for the ith channel, SNR_i, is given by:

$$SNR_i = \left[\frac{N_0}{2E_b} + Q_a\right]^{-1} \approx \left[\frac{N_0}{2E_b} + \frac{K - 1}{3N}\right]^{-1} \tag{3.64}$$

For a single user, $K = 1$, the multiple access interference is nil $(Q_a = 0)$ and the

$$SNR = \frac{E_b}{\dfrac{N_0}{2}}.$$

The probability of error P_e is given by:

$$P_e = Q(\sqrt{SNR}) \quad \text{single user} \tag{3.65}$$

$$= Q(\sqrt{SNR_i}) \quad \text{K users} \tag{3.66}$$

Example 3.10

Three users share a spread-spectrum transmission link. The system performance is to be evaluated while the users are sharing the link. The code sequences for the users are chosen from the family of Walsh-Hadamard sequences of period $N = 64$. The code sequences for the active users are:

{x} is sequence number 7
{y} is sequence number 11
{g} is sequence number 32

{x} = {0110 1001 0110 1001 0110 1001 0110 1001 0110 1001 0110 1001 0110 1001 0110 1001}

{y} = {0110 0110 1001 1001 0110 0110 1001 1001 0110 0110 1001 1001 0110 0110 1001 1001}

{g} = {0000 0000 0000 0000 0000 0000 0000 0000 1111 1111 1111 1111 1111 1111 1111 1111}

The signal power to channel thermal noise is 3 dB.

  i. Find the probability of bit error assuming transmission is conducted on an asynchronous link.
 ii. What would be the worst-case error probability of the asynchronous transmission?
iii. Find the probability of error when users are transmitting on a synchronous link.

Solution

i. Signal power to channel thermal noise ratio $= 3\,dB = 2$ in ratio

Thus $\dfrac{2E_b}{N_0} = 2$

$N = 64$

$K = 3$

$$SNR_i = \left(\frac{1}{2} + \frac{2}{3 \times 64}\right)^{-1} = 1.959$$

$$P_e = Q(\sqrt{1.959}) = Q(1.4) = 0.081$$

ii. The highest magnitude of the cross-correlation between any pair of the active users can be shown to be 4.

$$C_c = 4$$
$$N = 64$$
$$K = 3$$

Thus

$$P_{max} = 1 - \Phi\left(\left[1 - \frac{2 \times 2 \times 4}{64}\right] \times \sqrt{2}\right)$$
$$= 1 - \Phi(0.75 \times \sqrt{2}) = 1 - \Phi(1.0607)$$
$$= Q(1.0607)$$

Thus

$$P_{max} = Q(1.0607) = 0.145$$

iii. When transmission is synchronous, the cross-correlation is zero since code sequences are orthogonal at zero time shift. Therefore:

$$P_e = Q(\sqrt{2}) = Q(1.414) = 0.074$$

3.8 Summary

A brief historical background to the development of the spread-spectrum technique and the deployment of a commercial system that implements the spread-spectrum technique was given in Section 3.1. The benefits acquired in the use of such a system were discussed in Section 3.2. The principles of the spread-spectrum techniques were explained in terms of Shannon bandwidth efficiency diagram in Section 3.3. We have also shown the trade-off between the system bandwidth and the channel signal-to-noise power ratio so that a reduction in SNR can be compensated by an increase in the system bandwidth.

After a brief description of the most common types of spread-spectrum systems in Section 3.4, our attention was focused on the direct sequence spread-spectrum system model.

One of the most exciting phenomenon's at the heart of this technique, namely, the processing gain that is introduced through the process of spread–despread scheme was treated in depth in Section 3.5.

Since a spread-spectrum system employs bandwidth exorbitantly, the frequency spectrum is generally shared by a number of users leading to a spread-spectrum system based on CDMA operation.

The other important phenomenon, namely the correlations between different code sequences used by users accessing the system, was treated in more detail in Section 3.6 and information from this section will be cited in forthcoming sections in Chapters 5 and 6.

One of the spread-spectrum factors that are exposed to ongoing research is the choice of family of the spreading code sequences for multiple users. Conventionally this research is carried out by exhaustive computer searches. One of the parameters used for benchmark comparisons of the outcome of the search is the Merit factor which was discussed in Section 3.6.

An important facet of the spread-spectrum technique is its ability to resist interference/jamming so that the desired signal can be recovered with minimum distortion provided that there is enough processing gain to eradicate the effects of the interfering signal. This anti-jamming scheme was defined in terms of the jamming margin in Section 3.6. Finally, a scheme for evaluating the performance of the spread-spectrum system in terms of bit error rate based on Purley's model was introduced in Section 3.7.

Laboratory session III: Introduction to spread-spectrum techniques

The aim of this laboratory is to present a laboratory-intensive understanding of the spread-spectrum techniques and of the main concepts involved in such systems. These concepts are fundamental to direct-sequence multiple access systems such as IS-95 and the wideband code division multiple access used in 3G cellular networks. For background information, please refer to the material in this chapter.

Part A: Simulation of system parameters

Write and RUN MATLAB code to model the following system:
  i. Generate random data of (± 1).
  ii. Generate user code sequence from 64×64 Walsh matrix.
  iii. Perform the spreading process on the data using one out of the 64 Walsh code sequences.
  iv. Use FFT, overlay the spectrum of the baseband data and the spread-spectrum signal in the frequency domain.
  v. Measure the processing gain from FFT plot and compare this value with computed value of $G_p = N = 64$.

vi. Add Gaussian noise to the system at specified $\frac{E_b}{N_0}$ setting ($=3$ dB).

vii. Measure the BER for $\frac{E_b}{N_0}$ between 1 to 8 dB in steps of 0.5 dB.

viii. Plot BER vs. $\frac{E_b}{N_0}$ in dB.

Part B: Evaluation of system BER performance

In this part of the laboratory, the signature waveforms are chosen from 31×31 Gold code matrix where each user uses 31-chip Gold sequence. The focus is on the distribution of the multiple access interference and its impact on the system BER performance.

1. Gold Sequence matrix

i. Generate 31×31 Gold code matrix using MATLAB code.

ii. Extend the M-file to compute the Autocorrelation function for shift delay (τ) between 0 to 31 chip in a step of one chip duration.

iii. Plot the magnitude of the Autocorrelation function versus τ.

iv. Modify the M-file in (ii) to compute the cross-correlation function between code sequence 1 and code sequence 4 for shift delay (τ) between 0 to 31 chip in a step of one chip duration.

v. Plot the magnitude of the cross-correlation function versus τ.

2. Distribution of multiple access interference

i. Write a MATLAB code to generate transmitter signal for k users sharing the spread-spectrum channel.

ii. Set $k = 31$ so you have one desired user (say user 1) and interference from 30 users.

iii. Compute the interfering signal for each transmission cycle defined by: Interference = estimated data − transmitted data.

iv. Run (iii) for 4000 data symbols.

v. Plot the distribution (Histogram) of the multiple access inference.

3. Simulation of BER of multiple access spread-spectrum system

i. Modify MATLAB in 2.i to vary k from 1 up to 31 in a step on 1.

ii. Measure BER of the system for each value of k. Assume input noise power set to zero.

iii. Plot BER versus number of users.

iv. Plot theoretical BER, calculated from equations in Section 3.7, on the same graph of (iii).

v. Comment of your results.

Problems

3.1 Binary data is transmitted at the rate of 16.6 kb/s over a Gaussian channel that has a signal-to-noise power ratio (SNR) equal to 2.7 dB. Channel coding is employed to ensure error-free data detection with a matched filter receiver. If the channel SNR is reduced by 1.5 dB, find the new bandwidth required to keep the transmission rate constant.

3.2 A direct sequence spread-spectrum system is transmitting binary data at a bit rate of 9.6 kb/s and code sequence chip rate of 1.22 Mc/s. At the input of the matched filter receiver, the received power is 3.7 μW and the noise spectral density N_0, is 0.65 μW. Find the signal-to-noise ratio at the input of the receiver.

3.3 A frequency hopping spread-spectrum system is transmitting binary data at the rate of 4.8 kb/s. The system operates with 4 hops per data bit and final frequency is multiplied by 8. The processing gain is 36 dB. Find the:
  (a) number of hopping channel frequencies used. Assume this number to be a power of 2.
  (b) bandwidth of the frequency-hopped signal.

3.4 A hybrid system, consisting of a direct sequence and a frequency hopping spread-spectrum system, is transmitting binary data at the rate of 9.6 kb/s and chip rate 1.228 Mc/s. The number of hopping frequencies used is 1024 and the system hops once every time a single symbol is being transmitted. Find the:
  (a) system processing gain.
  (b) bandwidth of the system.

3.5 A group of 23 users are sharing a direct sequence spread-spectrum link. Each user transmits at a binary bit rate of 9.6 kb/s using a binary PSK carrier. If the users transmit at equal power and the additive Gaussian channel noise is to be ignored, what chip rate of the code sequence is required to ensure a bit error probability not greater than 10^{-3}.

3.6 A group of 13 equal-power users are sharing a direct sequence spread-spectrum channel and transmitting binary data at the rate of 12.6 kb/s on a binary PSK carrier. The system is operating at a chip rate of 1.2 Mc/s.
  (a) Compute the processing gain.
  (b) Find the ratio of energy per bit to power spectral density of the access interference $\left(\frac{E_b}{J_0}\right)$.
  (c) If the number of users is to double, keeping the output signal-to-noise ratio (SNR) fixed, how much should the processing gain be increased?

3.7 A group of 27 users is sharing a direct sequence spread-spectrum channel. The signal to Gaussian noise power ratio is 2.5 dB. Each user is using a code sequence of period 128 and chip rate of 1.22 Mc/s. Find the:
  (a) maximum magnitude of cross-correlation allowed if the worst-case probability of error is not to exceed 10^{-3}.
  (b) average probability of error.
  (c) increase in processing gain required to keep the probability of error equal to the value determined by the channel noise only.

3.8 Binary data is transmitted at the bit rate of 9.6 kb/s over a spread-spectrum channel that employs a code sequence of chip rate 1.228 Mc/s. The received signal at the input of a matched filter receiver is 3.1 μW. The noise power spectral density, N_0, is 2.9 µW/Hz, and the signal-to-noise ratio at the receiver output, $(SNR)_0$, is 27 dB. Find the:

(a) ratio of energy per bit to jamming power spectral density, $\frac{E_b}{J_0}$, assuming that the jamming power spectral density is uniform over the system bandwidth.

(b) jamming margin of the system.

3.9 Consider the following two Walsh-Hadamard sequences {A} and {B}, each of period N = 8 where:

$$\{A\} = \{0101\ 1010\}$$

$$\{B\} = \{0011\ 1100\}$$

Find the:

(a) periodic autocorrelation and periodic cross-correlation of both sequences.

(b) aperiodic autocorrelation and the aperiodic cross-correlation of both sequences.

(c) even and odd cross-correlation.

3.10 Consider the following two Walsh-Hadamard sequences, each with period N = 8 where sequences {a} and {b} are given by:

$$\{a\} = \{0011\ 0011\}$$

$$\{b\} = \{0110\ 0110\}$$

Find the:

(a) merit factor for each sequence.

(b) DFT of the cross-correlation and thus show that it can be expressed in terms of the DFT of both sequences.

References

Abu-Rgheff, M.N.A., Al-Shamma, N.K., Suliman, A.N. and Gailani, W.A. (1989) Cancellation of a sweep CW interfere in direct sequence spread-spectrum signals, *Proceedings of MELECON '89*, pp. 399–401, Lisbon, Portugal, April.

Corneretto, A. (1961) Spread-spectrum com system uses modified PPM, *Electronic Design,* June.

Dixon, R.C. (1994) Spread-spectrum systems with commercial applications', 3rd edition, pp. 177–189, John Wiley & Sons.

Fazel, K. (1993) Performance of CDMA/OFDM for mobile communication system, *Proceedings of the IEEE International Conference on Universal Personal communications*, Oct, 975–979.

Gilhouse, K.S., Jacobs, I.M., Padovani, R., Viterbi, A.J., Weaver, L.A. and Wheatley, C.E. (1991) On the capacity of a cellular CDMA system, *IEEE Transactions on Vehicle Technology*, VT-40, 303–312.

Golay, M.J.E. (1982) The Merit factor of long low autocorrelation binary sequences, *IEEE Transactions on Information Theory*, IT-28(3), 543–549.

Goldberg, B. (1981) Applications of statistical communication theory, *IEEE Communications Magazine*, Vol. 19, July, pp. 26–33.

Mowbray, R.S. and Grant, P.M. (1992) 'Wide band coding for uncoordinated multiple access communication', *IEE Electronic & Communication Engineering Journal*, 351–361.

Mowbray, R.S., Pringle, R.D. and Grant, P.M. (1992) 'Increased CDMA system capacity through adaptive co-channel interference regeneration and cancellation', *IEE Proceedings – I*, 139(5), 515–524.

Price, R. and Green, P.E. (1958) A communication technique for multipath channels, *Proceedings IRE*, 46, 555–570.

Pursley, M.B. (1977) 'Performance evaluation for phase-coded spread-spectrum multiple access communication', *IEEE Transactions on Communications*, COM-25(8), 795–802.

Sarwate, D.V. and Pursley, M.B. (1980) Cross-correlation properties of pseudo random and related sequences, *Proceedings IEEE*, 68(5), 593–619.

Scholtz, R.A. (1982) The origin of spread-spectrum communications, *IEEE Transactions on Communications*, COM-30(5), 822–854.

Scholtz, R.A. (1977) The spread-spectrum concept, *IEEE Transactions on Communications*, 25(8), 748–755.

Shannon, C.E. (1949) Communication theory of secrecy systems, *Bell Systems Technical Journal*, 28, 656–715.

Yue, O.C. (1983) Spread-spectrum mobile radio 1977–1982, *IEEE Transactions on Vehicle Technology*, VT-32, 98–105.

4

Pseudo-Random Code Sequences for Spread-Spectrum Systems

4.1 Introduction

In the previous chapter we learnt that essential to the spread-spectrum system is the spreading of the data energy at the transmitter, and the collapsing of the spreading (in a process known as de-spreading) at the receiver. The operations at the transmitter and the receiver generate the required processing gain without which the system could not combat jamming and interference. The spreading code sequences, used in the system, have special properties which will be introduced in this chapter. These code sequences have an important role to play in the spread-spectrum technique and have to be chosen carefully for efficient communication systems.

Each code sequence used in the spread-spectrum communications must easily be distinguishable from a time shifted version of itself, in order to enable the code acquisition and tracking and, therefore, the synchronization of the system. Furthermore, when multiple users are accessing the system for services, each code sequence assigned to a user must be distinguishable from every other user code sequence in the set and ideally should generate little or no interference to other users sharing the channel.

The code sequences are pseudo-random sequences but in practice they are generally periodic. Two types of these sequences can be used in spread-spectrum applications: the binary code sequence and the non-binary (also known as complex) code sequence. The elements of the binary sequence are made up of real numbers ± 1. These are most commonly applied in spread-spectrum communications, for ease of generation using widely available binary logic circuits. However, there has also been some interest in generating complex sequences through exhaustive computer searching, such as quadric phase and poly phase sequences which have low correlation properties.

195

The bit error probability performance of spread-spectrum communications depends on the correlation properties of the code sequence used since the latter define the amount of interference generated from multiple access. The technique for generating code sequences should be aimed at a large family of sequences in order to accommodate a number of users and, with an impulse-type autocorrelation which enhances the system synchronization and possibly with low cross-correlation functions, to reduce multiple access interference.

We start this chapter with basic binary algebra in Sections 4.2, 4.3 and 4.4. This is followed by block level structure of the commonly used binary pseudo-random sequence generators. The binary sequences are called *maximal-length sequences* or simply *m-sequences*. Their decimation and the preferred pairs of the m-sequence are discussed, together with other sequences widely used in spread spectrum such as the Gold, Kassami and Walsh sequences, in detail in Section 4.5. Finally, Section 4.6 introduces the complex sequences such as quadric phase, poly phase sequences, and the whole chapter is summarized in Section 4.7.

4.2 Basic Algebra concepts

We start this section with a basic definition of terms used in the numbers theory such as number set, group and field. The reader who may be unfamiliar with these terms may then feel at ease reading the section.

An algebraic *set* of M elements is defined by an array of M real or complex numbers acted upon by an operator for addition (and its inverse, subtraction) and division or a multiplication (and its inverse, division). The set is said to be a *closed set* if the algebraic operations (addition and division or multiplication) on the original set, yield a new element already existing in the same set. A *group* is a set of elements acted upon by an operator for addition (additive group) or multiplication (multiplicative group). A *ring* is a set of elements operated upon by addition and multiplication. A *field* is defined as a ring with every element in the ring (except zero) having an inverse.

A field is therefore a system that has two operations both with inverses and each element in the field (except zero) having an inverse. A field can be real or complex, depending on the type of its constituting elements being real or complex numbers. An infinite field has a theoretically infinite number of elements.

A field with a finite number of elements, M, is called a Galois (pronounced as 'Gal-Wah') field and is denoted GF(M). Generally, finite fields only exist when M is prime or M is the power of a prime, i.e. $M = P^m$ when m is integer. Galois field GF(M) has M elements with index $0, 1, 2, \ldots, M - 1$. The simplest Galois field uses modulo 2 arithmetic and is denoted GF(2) with elements drawn from $\{0, 1\}$ which is also called a binary field.

The elements of a given set must satisfy precise rules to become a constituent member of the field. Let e_i, e_j, e_k, e_n be any four elements of a set. The field requires the set to possess the following properties:

1 Commutative property

$$e_i + e_j = e_j + e_i$$
$$e_i \cdot e_j = e_j \cdot e_i \tag{4.1}$$

2 Associative property

$$e_i + (e_j + e_k) = (e_i + e_j) + e_k$$
$$e_i \cdot (e_j \cdot e_k) = (e_i \cdot e_j) \cdot e_k \tag{4.2}$$

3 Distributed property

$$e_i \cdot (e_j + e_k) = e_i \cdot e_j + e_i \cdot e_k$$
$$(e_i + e_j) \cdot (e_k + e_n) = e_i \cdot e_k + e_i \cdot e_n + e_j \cdot e_k + e_j \cdot e_n \tag{4.3}$$

4 Inverse property

$$e_i + e_i^* = 0 \quad \text{additive inverse} \tag{4.4}$$
$$e_i \cdot e_i^{-1} = 1 \quad e_i \neq 0 \quad \text{multiplication inverse} \tag{4.5}$$

where e_i^* and e_i^{-1} denote inverse elements.

5 Closure property

$$e_i + e_j \in \text{GF}(\cdot)$$
$$e_i \cdot e_j \in \text{GF}(\cdot) \tag{4.6}$$

Subtraction of one element from any other element in the given field is performed by adding the additive inverse of the element while division by an element (other than zero) is accomplished by multiplying by the multiplicative inverse.

Example 4.1

Consider a set of binary elements drawn from $\{0, 1\}$. If the basic algebraic operations (addition/subtraction and multiplication/division) are acted upon each pair of the set, find the resulting elements.

Solution

1 Mod-2 addition

+	0	1
0	A1	A2
1	B1	B2

$A1 = 0; \quad A2 = 1; \quad B1 = 1; \quad B2 = 0 \ (\text{mod } 2)$

2 Mod-2 subtraction

−	0	1
0	A1	A2
1	B1	B2

Here we apply the additive inverse such that:
$$0 = -0$$
$$1 = -1$$
Therefore, $A1 = 0$; $A2 = 1$; $B1 = 1$; $B2 = 0$ (mod 2)

3 Mod-2 multiplication

×	0	1
0	A1	A2
1	B1	B2

$A1 = 0$; $A2 = 0$; $B1 = 0$; $B2 = 1$

4 Mod-2 division

÷	0	1
0	A1	A2
1	B1	B2

Here we use the multiplicative inverse:
$A1 = ?$; $A2 = ?$; $B1 = 0$; $B2 = 1$

4.3 Arithmetic of binary polynomial

Galois's theory is based on algebra articulated by a young French mathematician (Evariste Galois [1811–1832]). Galois fields are extensively used in applications such as cryptography, error correcting codes and random number generators.

Galois field binary polynomials are seen as mathematical equivalents *of Linear Feed-Back Shift Registers* (LFSRs) in sequence generator design, using widely available logic gates such as X-OR, AND, and OR gates. This subject will be dealt with in more detail in Section 4.5.

In most applications of Galois Field GF(M), M is chosen as a prime number, i.e. *M is a positive integer not divisible, without a remainder, by any positive integer other than itself and one*. The elements in the field are integer numbers. Furthermore, M corresponds to an integral power of a prime number. A field $GF(2^m)$ is considered to be an extension of the binary field GF(2), so to generate a field $GF(2^m)$, we extend the constituent elements $\{0, 1\}$ of GF(2) using a primitive element 3. If $3 \in GF(2^m)$, then under multiplication $3.3 = 3^2$, $3.3^2 = 3^3, \ldots, 3^m - 1$, are also elements of $GF(2^m)$. Therefore, the elements of $GF(2^m)$ are:

$$GF(2^m) = \{0, 1, \beta, \beta^2, \beta^3, \ldots \ldots, \beta^m - 1. \qquad (4.7)$$

Consider a polynomial P(x) of degree m and coefficients that are chosen from elements of the field GF(p) such that:

$$p(x) = \sum_{i=0}^{m} h_i \cdot x^i = h_0 + h_1 x + h_2 x^2 + \cdots \cdots \qquad (4.8)$$

For example, the polynomial $[1 + x^2]$ is a second order with coefficients drawn from the binary field (i.e. $h_0 = 1$, $h_1 = 0$, $h_2 = 1$). On the other hand, the polynomial $[1 + x + 3x^3]$ is a third order polynomial with coefficients drawn from the field $GF(2^2)$ (i.e. $h_0 = 1$, $h_1 = 1$, $h_2 = 0$, $h_3 = 3$).

Example 4.2

Consider the polynomials $P_1(x)$ and $P_2(x)$ such that:

$$P_1(x) = 1 + x + x^3$$
$$P_2(x) = x + x^2 + x^3$$

Evaluate the following mathematical expressions:

  i. $P_1(x) + P_2(x)$
 ii. $P_1(x) - P_2(x)$
iii. $P_1(x) \times P_2(x)$
 iv. $\dfrac{P_1(x)}{P_2(x)}$

Solution
Since both polynomials have binary coefficients, we can use binary arithmetic in the analysis.

  i. $P_1(x) + P_2(x) = 1 + x + x^3 + x + x^2 + x^3 = 1 + x^2$ since $x + x = 0$ and $x^3 + x^3 = 0$
 ii. $P_1(x) - P_2(x) = 1 + x^2$ Binary addition and subtraction are the same.
iii. $P_1(x) P_2(x) = (1 + x + x^3)(x + x^2 + x^3)$
$$= x + x^2 + x^3 + x^2 + x^3 + x^4 + x^4 + x^5 + x^6$$
$$= x + x^5 + x^6$$

iv. $\dfrac{P_1(x)}{P_2(x)}$ Using long division, we get:

$$
\begin{array}{r}
1 \\ \hline
x^3+x^2+x \,\big)\, \overline{\,x^3+x+1\,} \\
x^3+x^2+x \\ \hline
1+x+x+x^2+x^3+x^3 = x^2+1
\end{array}
$$

Thus $\dfrac{P_1(x)}{P_2(x)} = 1 + \dfrac{x^2+1}{x^3+x^2+x}$

Example 4.3

Consider polynomials $P_1(x)$ and $P_2(x)$ with coefficients drawn from Galois field GF(3) such that:

$$P_1(x) = x + 2x^2 + x^3$$
$$P_2(x) = 1 + 2x + x^2$$

Evaluate the following mathematical expressions:

i. $P_1(x) + P_2(x)$
ii. $P_1(x) - P_2(x)$
iii. $P_1(x) \times P_2(x)$
iv. $\dfrac{P_1(x)}{P_2(x)}$

Solution

The arithmetic rules for Galois field GF(3) are:

Addition:

+	0	1	2
0	0	1	2
1	1	2	0
2	2	0	1

Subtraction:
Since $2 + 1 = 0$, then $-0 = -0$, $-1 = 2$ and $-2 = 1$

−	0	1	2
0	0	1	2
1	2	0	1
2	1	2	0

Multiplication:

This is a straight forward arithmetic operation.

×	0	1	2
0	0	0	0
1	0	1	2
2	0	2	1

Division:

This is accomplished by multiplying by the multiplicative inverse.

$1^{-1} = 1$ and $2^{-1} = 2$

÷	0	1	2
0	?	?	?
1	0	1	2
2	0	2	1

i. $P_1(x) + P_2(x) = x + 2x^2 + x^3 + 1 + 2x + x^2 = 1 + x^3$

ii. $P_1(x) - P_2(x) = x + 2x^2 + x^3 - 1 - 2x - x^2 = 2 + 2x + x^2 + x^3$

iii. $P_1(x) \times P_2(x) = (x + 2x^2 + x^3) \times (1 + 2x + x^2)$

$= x + 2x^2 + x^3 + 2x^2 + 4x^3 + 2x^4 + x^3 + 2x^4 + x^5$

$= x + 4x^2 + 6x^3 + 4x^4 + x^5$

$= x + x^2 + x^4 + x^5$

iv. $\dfrac{P_1(x)}{P_2(x)} = x$

An irreducible polynomial is a polynomial that cannot be factored into non-trivial polynomials over the same field. For example, polynomial f(x) is said to be irreducible if it cannot be expressed as a product of at least two non-trivial polynomials in the f(x) field. An irreducible polynomial of degree m is primitive if it divides $[1 + x^n]$ for which the smallest positive integer $n = 2^m - 1$.

In general, a primitive polynomial exists for every irreducible polynomial with positive integer m. The addition of any two elements of $GF(2^m)$ is defined as mod-2 addition of two binary polynomials. A multiplication of two elements of $GF(2^m)$ is referred to as modulo-h(x) multiplication where h(x) is a primitive polynomial of order m.

Example 4.4

Find the elements that arise from the addition and multiplication of each pair of elements of the polynomials in $GF(2^2)$.

Solution

The polynomials in GF(2^2) have 4 elements and each may be represented by a binary polynomial of order m $-$ 1 where m $=$ 2. These elements can be expressed as binary digits: 00, 01, 10, 11 and in binary polynomial as: 0, 1, x, 1 $+$ x.

Addition:

$+$	0	1	x	1 + x
0	**0**	**1**	**x**	**1 + x**
1	**1**	**0**	**1 + x**	**x**
x	**x**	**1 + x**	**0**	**1**
1 + x	**1 + x**	**x**	**1**	**0**

Multiplication:

To carry out the multiplication, we choose the following primitive polynomial, h(x) of degree 2:

$$h(x) = 1 + x + x^2$$

Now the mathematical operations that we need to evaluate are:

$$x \times x \bmod[h(x)] = (1 + x)$$

$$(1 + x) \times x \bmod[h(x)] = (x + x^2)\bmod[h(x)] = 1$$

$$(1 + x) \times (1 + x)\bmod[h(x)] = (1 + x + x + x^2)\bmod[h(x)] = x$$

$\times$	0	1	x	1 + x
0	**0**	**0**	**0**	**0**
1	**0**	**1**	**x**	**1 + x**
x	**0**	**x**	**1 + x**	**1**
1 + x	**0**	**1 + x**	**1**	**x**

4.4   Computing elements of GF(2^m)

We now explain how to find the elements in GF(2^m) by a worked example. Consider a primitive polynomial p(x_i) of degree m which is primitive over GF(2^m). Since p(x_i) has a finite number of roots, if we associate these roots as elements of GF(2^m), this will ensure that GF(2^m) is a finite field. A polynomial of degree m has m roots, x_i, that satisfies the equation:

$$p(x_i) \approx 0 \tag{4.9}$$

Let us consider the elements in GF(2^3) using the primitive polynomial P(x) $= 1 + x + x^3$.

The polynomial in $GF(2^3)$ has 8 elements. Since β is the root of $P(x)$ then:

$$P(\beta) = 1 + \beta + \beta^3 = 0$$

Therefore, $1 + \beta = \beta^3$

The elements due to power of β are:

$$\beta^0, \beta, \beta^2, \beta^3, \beta^4, \beta^5, \beta^6, \beta^7$$

Therefore, the elements are:

$$0, 1, \beta, \beta^2, \ldots, \beta^7, \quad \text{where :}$$

$$\beta^3 = 1 + \beta,$$

$$\beta^4 = \beta\beta^3 = \beta(1 + \beta) = \beta + \beta^2,$$

$$\beta^5 = \beta\beta^4 = \beta^2(1 + \beta) = \beta^2 + \beta^3 = 1 + \beta + \beta^2$$

$$\beta^6 = \beta^3 \cdot (1 + \beta) = (1 + \beta) \cdot (1 + \beta) = \beta^3 + \beta^4 = 1 + \beta + \beta + \beta^2 = 1 + \beta^2$$

$$\beta^7 = \beta(1 + \beta^2) = \beta + \beta^3 = 1 + \beta + \beta = 1$$

4.5 Binary pseudo-random sequences

4.5.1 *Generation of binary pseudo-random sequences*

We have hinted in Section 4.3 that shift registers, with linear feedback, can be used in the implementation of binary code sequence generators. We now examine this proposal in more detail. Our knowledge of the Galois field algebra will help us to understand the involved algebraic analysis. To explain the basic concept of the sequence generators, consider the simple feedback shift registers shown in Figure 4.1 where the initial states of the r-stage shift registers are $(a_{r-1}, a_{r-2}, \ldots, a_1, a_0)$ and the feedback function $f(x_0, x_1, \ldots, x_{r-1})$ is a binary function.

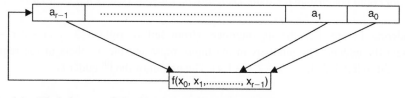

Figure 4.1 *Block diagram shift registers with linear feedback.*

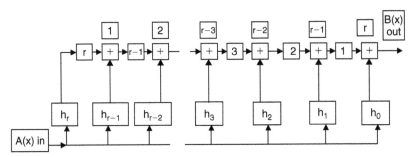

Figure 4.2 *Block diagram of a general sequence generator (Peterson et al., 1995).*

At each clock pulse, the content of each register is shifted to the next register on the left or right. For example, for shift right, the next state shown in Figure 4.1 is shown as:

a_r	....	a_2	a_1

Where $a_r = f(a_0, a_1, \ldots \ldots \ldots, a_{r-1})$.

We now show that the feedback shift register circuit performs the multiplication of Galois polynomials. Consider the block diagram of a general sequence generator depicted in Figure 4.2 where we use the following symbols:

j	j^{th} stage shift register # j
$+$	Modulo-2 adder
h_i	Modular-2 multiplier

Let x denote the time delay of a unit clock duration and x^j denote the time delay of j such units. The input A(x) specifies the initial states of the registers and are denoted by the sequence $(a_{r-1}, a_{r-2}, \ldots, a_1, a_0)$.

The generator connections can be expressed by polynomial h(x) where:

$$h(x) = h_0 + h_1 \cdot x + h_2 \cdot x^2 \ldots \ldots \ldots \ldots \ldots h_r \cdot x^r \qquad (4.10)$$

The coefficients h_i are such that a connection is present if $h_i = 1$ and no connection is present if $h_i = 0$. It is worth noting that there is no output from this generator for the first r clock time shifts and that once all registers are loaded with zeros (i.e. A(x) = 0), the generator could not change it's state. *Therefore an all-zero state is not allowed.*

Considering Figure 4.2, the top numbers (from left to right $1, 2, \ldots, r-2, r-1, r$) represent the adders. The numbers in the figure represent the numbers of the registers (from right to left $1, 2, 3, \ldots, r-2, r-1, r$). The output of the jth adder is:

$$x: B_j(x) = B_{j-1}(x)x + A(x)h_{r-j}$$

Let us start the iteration from right to the left:

$$B(x) = B_r(x)\cdot$$
$$= B_{r-1}(x) \cdot x + A(x) \cdot h_0$$
$$= B_{r-2}(x) \cdot x^2 + A(x) \cdot xh_1 + A(x) \cdot h_0$$
$$= B_{r-3}(x) \cdot x^3 + A(x)x^2h_2 + A(x) \cdot xh_1 + A(x) \cdot h_0$$

$$\bullet$$
$$\bullet$$
$$\bullet$$
$$\bullet$$
$$\bullet$$

$$= B_1(x) \cdot x^{r-1} + A(x)x^{r-2}h_{r-2} + A(x) \cdot x^{r-1}h_{r-1} + \cdots\cdots\cdots + A(x) \cdot h_0$$
$$= A(x)x^r h_r + A(x)x^{r-1}h_{r-1} + A(x)x^{r-2}h_{r-2} + A(x) \cdot x^{r-1}h_{r-1} + \cdots + A(x) \cdot h_0$$
$$= \sum_{j=0}^{r} [A(x) \cdot x^j] \cdot h_j$$

Thus
$$B(x) = A(x) \cdot h(x) \tag{4.11}$$

Equation (4.11) expresses the generator output $B(x)$ as a product of two polynomials, i.e. the input polynomial $A(x)$ and the generator connections polynomial $h(x)$.

The maximum period of the binary sequence generated by the r-stage shift register is limited to $2^r - 1$. A binary sequence which achieves this maximum period is called *maximal-length sequence* or simply *m-sequence*. Primitive polynomials which can be used to connect the feedback are given in Table 4.1. It must be emphasized that the period of the generated sequence depends on the choice of $h(x)$ and only connections based on these primitive polynomials are capable of generating sequences of length $2^r - 1$.

Having demonstrated that feedback shift register circuits are capable of performing multiplication of Galois polynomials, we now consider the block diagram depicted in Figure 4.3 to show that these circuits can also perform division of Galois polynomials. We use the same symbols in Figure 4.2.

Applying (4.11) to Figure 4.3a, we get:

$$B(x) = A_1(x) \cdot h(x) + A_2(x) \cdot k(x) \tag{4.12}$$

Suppose that $k_0 = 0$ so that the connection between multiplier k_0 and the corresponding adder is disconnected and that $A_2(x)$ is taken from the output, i.e. $A_2(x) = B(x)$. Let us define the polynomial $g(x)$ such that $k(x) = g(x) + 1$.

Table 4.1 *Primitive polynomials for m-sequence generator connections*

Number of SR stages r	Sequence length $N = 2^r - 1$	Number of m-sequences	Polynomial generator h(x)
2	3	1	$1 + x + x^2$
3	7	2	$1 + x + x^3$ $1 + x^2 + x^3$
4	15	2	$1 + x + x^4$ $1 + x^3 + x^4$
5	31	6	$1 + x^2 + x^5$ $1 + x^3 + x^5$ $1 + x + x^2 + x^3 + x^5$ $1 + x + x^2 + x^4 + x^5$ $1 + x + x^3 + x^4 + x^5$ $1 + x^2 + x^3 + x^4 + x^5$
6	63	6	$1 + x + x^6$ $1 + x + x^3 + x^4 + x^6$ $1 + x^5 + x^6$ $1 + x + x^2 + x^5 + x^6$ $1 + x^2 + x^3 + x^5 + x^6$ $1 + x + x^4 + x^5 + x^6$
7	127	18	$1 + x + x^7$ $1 + x^3 + x^7$ $1 + x + x^2 + x^3 + x^7$ $1 + x^4 + x^7$ $1 + x^2 + x^3 + x^4 + x^7$ $1 + x + x^2 + x^5 + x^7$ $1 + x + x^3 + x^5 + x^7$ $1 + x^3 + x^4 + x^5 + x^7$ $1 + x + x^2 + x^3 + x^4 + x^5 + x^7$ $1 + x^6 + x^7$ $1 + x + x^3 + x^6 + x^7$ $1 + x + x^4 + x^6 + x^7$ $1 + x^2 + x^4 + x^6 + x^7$ $1 + x^2 + x^5 + x^6 + x^7$ $1 + x + x^2 + x^3 + x^5 + x^6 + x^7$ $1 + x^4 + x^5 + x^6 + x^7$ $1 + x + x^2 + x^4 + x^5 + x^6 + x^7$ $1 + x^2 + x^3 + x^4 + x^5 + x^6 + x^7$
8	255	16	$1 + x^2 + x^3 + x^4 + x^8$ $1 + x + x^3 + x^5 + x^8$ $1 + x^2 + x^3 + x^5 + x^8$ $1 + x^2 + x^3 + x^6 + x^8$ $1 + x + x^2 + x^3 + x^4 + x^6 + x^8$ $1 + x + x^5 + x^6 + x^8$

(Continued)

Table 4.1 *Continued*

Number of SR stages r	Sequence length $N = 2^r - 1$	Number of m-sequences	Polynomial generator h(x)
			$1 + x^2 + x^5 + x^6 + x^8$
			$1 + x^3 + x^5 + x^6 + x^8$
			$1 + x^4 + x^5 + x^6 + x^8$
			$1 + x + x^2 + x^7 + x^8$
			$1 + x^2 + x^3 + x^7 + x^8$
			$1 + x^3 + x^5 + x^7 + x^8$
			$1 + x + x^6 + x^7 + x^8$
			$1 + x + x^2 + x^3 + x^6 + x^7 + x^8$
			$1 + x + x^2 + x^5 + x^6 + x^7 + x^8$
			$1 + x^2 + x^4 + x^5 + x^6 + x^7 + x^8$
9	511	48	$1 + x^4 + x^9$ (example)
10	1023	60	$1 + x^3 + x^{10}$ (example)

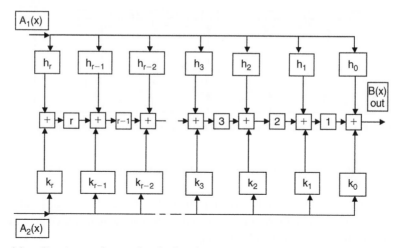

Figure 4.3a *Two-input polynomial multiplier (Peterson et al., 1995).*

Substituting in (4.12):

$$B(x) = A_1(x) \cdot h(x) + B(x) \cdot [g(x) + 1]$$

Add $B(x)[1 + g(x)]$ to both sides of the above expression:

$$B(x) + B(x) + B(x) \cdot g(x) = A_1(x) \cdot h(x) + B(x) \cdot g(x) + B(x) + B(x) + B(x)\, g(x)$$

Therefore, $B(x) \cdot g(x) = A_1(x) \cdot h(x)$

Thus

$$B(x) = A_1(x)\frac{h(x)}{g(x)} \tag{4.13}$$

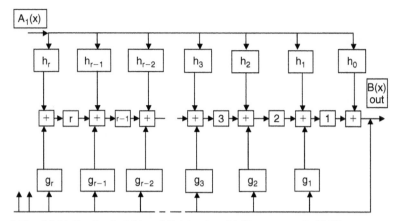

Figure 4.3b   *Multiplication by h(x) and division by g(x) (Peterson et al., 1995).*

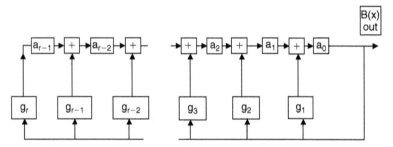

Figure 4.4   *Galois linear feedback shift register sequence generator (Peterson et al., 1995).*

Implementing these changes, Figure 4.3a will be transformed to Figure 4.3b.

Examining (4.13), it is clear that the circuit in Figure 4.3b divides h(x) by g(x) and multiply the result by $A_1(x)$. Generally $A_1(x)$ sets the initial state of the registers contents and can be represented by a finite polynomial given by:

$$A_1(x) = a_0 + a_1 \cdot x + a_2 \cdot x^2 + \cdots\cdots\cdots\cdots\cdots a_{r-1} \cdot x^{r-1} \qquad (4.14)$$

When the registers are loaded with sequence $A_1(x)$, the circuit in Figure 4.3b can be simplified to that shown in Figure 4.4. The loading process takes r time units and while the registers are loading, the output B(x) is zero for these r time units as has been mentioned previously. Therefore, B(x) starting at time r is:

$$B(x) = \frac{A_1(x)}{g(x)} \qquad (4.15)$$

There are two methods of constructing LFSR sequence generators for a given generator polynomial (Hanzo et al., 2003). The generator shown in Figure 4.4 follows *Galois feedback implementation* where the output bits are feedback into the shift registers according to the connection polynomial. The Galois implementation is commonly used when high speed sequences are required since feedback is not delayed. The other implementation is known as *Fibonacci feedback generator* and is shown in Figure 4.5 (see later). The Fibonacci generator can output several delayed versions of the same sequence at the output of each shift register with no additional hardware.

The maximum possible period for an LFSR with the generator connection defined by the polynomial g(x) is computed as follows:

Given the generator polynomial g(x), formulate $x^r \cdot g(x^{-1})$ then find the smallest integer N such that $x^N + 1$ is divisible by $x^r \cdot g(x^{-1})$ where N is the maximum period of the sequence. We now consider an example to explain the calculation.

Example 4.5

Consider the sequence generator shown in Figure 4.4 with the generator polynomial g(x) is given by:

$$g(x) = 1 + x^2 + x^3 + x^4$$

Assume the initial load of the register be 0001.

i. Find the output periodic sequence.
ii. What would the maximum possible period be?

Solution
Using the procedure described in this section, we have:

$$A_1(x) = 1$$

$$B(x) = \frac{A_1(x)}{g(x)} = \frac{1}{1 + x^2 + x^3 + x^4}$$

Using long division, we get:

$$B(x) = 1 + x^2 + x^3 + x^7 + x^9 + x^{10} + x^{14} + \cdots\cdots\cdots\cdots\cdots\cdots$$

The binary sequence at the output of generator shown in Figure 4.4 is:

$$B(x) = 1011000\ \mathbf{1011000}\ 101\ldots\ldots$$

The output B(x) is a periodic of 1011000 with a period N = 7.

We may check the length of the sequence using the minimum integer method described above for which $x^N + 1$ is divisible by $x^r \cdot k(x^{-1})$ as described above.

Now $x^r \cdot g(x^{-1}) = x^4(1 + x^{-2} + x^{-3} + x^{-4}) = x^4 + x^2 + x + 1$

$$\frac{x^N + 1}{x^r \cdot g(x^{-1})} = \frac{x^N + 1}{x^4 + x^2 + x + 1}$$

The minimum integer value for N is 7, that is:

$$\frac{x^N + 1}{x^r \cdot g(x^{-1})} = \frac{x^7 + 1}{x^4 + x^2 + x + 1} = x^3 + x + 1$$

Note that the maximum possible period for the output sequence given by this sequence generator is $2^4 - 1 = 15$.

Example 4.6

Find the sequence at the output of the generator shown in Figure 4.5 with polynomial given by:

$$g(x) = 1 + x^2 + x^3 + x^4$$

Assume the initial load is 0001.

Solution

Consider the initial load 0001 then:

$A_1(x) = 1$ which is the first out of the generator.

Let the equivalent initial load for generator in Figure 4.5 be expressed as

$$A_1'(x) = a_0' + a_1' \cdot x + \cdots \cdots \cdots + a_{r-1}' \cdot x^{r-1}$$

now $A_1(x) + b_r \cdot x^r + b_{r+1} \cdot x^{r+1} + \cdots \cdots \cdots = \frac{A_1'(x)}{k(x)}$ Equate the first r-coefficients, we get:

$$A_1(x) \cdot g(x) = a_0' + a_1' \cdots \cdots \cdots \cdots \cdots + a_{r-1}' \cdot x^{r-1}$$

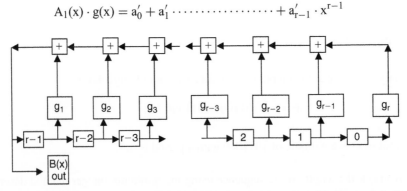

Figure 4.5   *Fibonacci linear feedback shift register sequence generator (Peterson et al., 1995).*

Therefore, $1 + x^2 + x^3 + x^4 = a_0' + a_1' \cdot x + a_2' \cdot x^2 + a_3' \cdot x^3$

So that: $a_0' = 1 \; a_1' = 0 \; a_2' = 1 \; a_3' = 1$

Therefore, $A_1'(x) = 1 + x^2 + x^3$

$$B(x) = \frac{x^{-r} \cdot A_1'(x)}{g(x)} = x^{-4} \cdot \frac{1 + x^2 + x^3}{1 + x^2 + x^3 + x^4}$$

$$= x^{-4} \cdot [1 + x^4 + x^6 + x^7 + x^{11} + x^{13} + x^{14} + x^{18} + \cdots \cdots]$$

$$= x^{-4} + 1 + x^2 + x^3 + x^7 + x^9 + x^{10} + x^{14} + \cdots \cdots$$

The output polynomial $B(x)$ is the same as that generated from generator shown in Figure 4.4 in the previous example.

4.5.2 Maximal-length sequences (m-sequences)

The m-sequences have found numerous applications in digital communication systems, including spread-spectrum systems. These sequences have a maximum period $N = 2^r - 1$ for an r-stage LFSR generator connected according to a primitive binary polynomial of degree r selected from Table 4.1.

The salient features of the m-sequences are their two-valued autocorrelation functions which are optimal, with the absence of any side-lobe peaks. This is the key parameter which determines the probability of detection and false alarm, during code acquisition and tracking, as we will discover in the material presented in Chapter 5.

The periodic cross-correlation function between any pair of m-sequences of the same period can be relatively large. However, the peak values depend on the sequences chosen and their respective phases. To reduce interference, it is desirable to constrain these peak values to a minimum. All m-sequences of the same length can be derived from each other by a process of proper decimation discussed in Section 4.5.3.

A list of the peak magnitude for the periodic cross-correlation between pairs of m-sequences for $3 \leq r \leq 12$ is shown in Table 4.2.

The m-sequences have the following well-known properties discussed in Sarwate and Pursley's (1980) paper:

i. There are exactly N non-zero sequences representing the N different phases of the m-sequence. If the m-sequence is $\mathbf{x} = (x_0, x_1, x_2, \ldots, x_{N-1})$, then the non-zero sequences are $(x_1, x_2, x_3, \ldots, x_{N-1}, x_0)$, $(x_2, x_3, x_4, \ldots, x_{N-1}, x_0, x_1)$, $(x_3, x_4, x_5, \ldots, x_{N-1}, x_0, x_1, x_2)$, etc.

Table 4.2 *Peak periodic cross-correlation between a pair of m-sequences*

r	$N = 2^r - 1$	Number of m-sequences	Peak cross-correlation
3	7	2	5
4	15	2	9
5	31	6	11
6	63	6	23
7	127	18	41
8	255	16	95
9	511	48	113
10	1023	60	383
11	2047	176	287
12	4095	144	1407

Source: Proakis, 1995

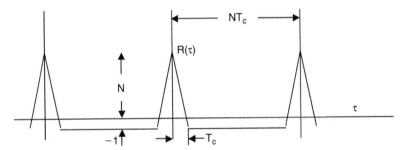

Figure 4.6 *Auto-correlation function for an m-sequence.*

ii. Shift-and-add property of the m-sequences suggests that the modulo-2 sum of an m-sequence and any phase shifted version of itself is another phase of the same m-sequence.

iii. The Hamming weight of an m-sequence is $\left(\frac{N+1}{2}\right)$. This is because the number of ones in an m-sequence is $\left(\frac{N+1}{2}\right)$. The number of zeros is of course $\left(\frac{N-1}{2}\right)$.

iv. The periodic autocorrelation function of an m-sequence is a two-valued function given by

$$R(\tau) = N \quad \text{for } \tau = jN \qquad (4.16)$$
$$= -1 \quad \text{for } \tau \neq jN$$

where j is any integer. A plot of the autocorrelation for an m-sequence with chip duration T_c and time period NT_c is shown in Figure 4.6.

v. A run is defined as a set of identical symbols within the m-sequence. The length of the run is equal to the number of these symbols in the run. For any m-sequence generated by r-stage shift registers, it has the following statistics:

　1 run of ones of length r
　1 run of zeros of length r − 1

1 run of ones and one run of zeros of length r − 2

2 runs of ones and 2 runs of zeros of length r − 3

4 runs of ones and 4 runs of zeros of length r − 4

8 runs of ones and 8 runs of zeros of length r − 5

●

●

●

2^{r-3} runs of ones and 2^{r-3} of zeros of length 1.

For example the m-sequence 000100110101111 contains a total of eight runs as follows: one run of four 1s, one run of three 0s, one run of two 0s, two runs of s single 1, two runs of a single 0.

4.5.3   Decimation of m-sequences

The application of code sequences in spread-spectrum communications necessitates the generation of large sets of codes with highly peaked autocorrelation and minimum cross-correlation. In Section 4.5.1 we considered electronic circuits to generate these sequences. We now look at the generation of number sequences by decimating a single sequence. As usual, we start by defining the basic element involved in the decimation process.

Consider sequence $u = u_0, u_1, u_2, u_3, \ldots$, then sequence v, constructed by taking every q^{th} bit of the nonzero elements of sequence u denoted by u(q) and said to be the *decimation by q of u* where q is a positive integer, that is:

$$v = u(q) \tag{4.17}$$

where $v = u_0, u_q, u_{2q}, u_{3q}, \ldots\ldots\ldots\ldots\ldots$.

If u has a period N and is generated by LFSR with generator connection polynomial h(x) then u(q) has period N_v

$$N_v = \frac{N}{\gcd(N, q)} \tag{4.18}$$

The sequence u(q) can be generated using LFSR with generator polynomial $\hat{h}(x)$ whose roots are the qth powers of the roots of h(x).

The term gcd(N, q) denotes the *greatest common divisor* of N and q. For example, the gcd(23, 7) = 1 because the two numbers can only be divisible by 1, the gcd(81, 45) = 9 since 45 is divisible by 3, 5 and 9 and 81 is divisible by 3 and 9.

The decimation of an m-sequence *may or may not yield another m-sequence*. When the decimation yields an m-sequence, it is called *proper decimation* and if gcd(N, q) = 1, sequence v = u(q) has a period N. Proper decimation guarantees that sequence v = u(q) is an m-sequence and the polynomial $\hat{h}(x)$ is primitive. Clearly, there are N possible sequences that correspond to the N phases of m-sequence u.

The decimation of any phase of sequence u will give a certain phase of v. In general, regardless of which of the m-sequences generated by h(x) we choose to decimate, the result will be an m-sequence generated by $\hat{h}(x)$.

The characteristic sequence $\tilde{u}$ of m-sequence u is such that $\tilde{u} = \tilde{u}(2)$. Since the m-sequence u is periodic, we only need to consider values of less than or equal to $[N-1]$, that is $u[q] = u\,[q \cdot \mathrm{mod}\,N]$. When proper decimation is achieved by odd integer q, then $u[2^j q]$ represents different phases of the same m-sequence $u(q)$. Let the m-sequence u be generated by polynomial h(x) such that:

$$h(x) = h_0 x^r + h_1 x^{r-1} + \cdots\cdots\cdots\cdots\cdots\cdots\cdots\cdots + h_{r-1} x + h_r$$

$$(4.19)$$

Decimating u by $q = \frac{1}{2}[N-1]$ will generate the reciprocal polynomial of h(x) that is $\hat{h}(x)$ where:

$$\hat{h}(x) = h_r x^r + h_{r-1} x^{r-1} + \cdots\cdots\cdots\cdots\cdots\cdots\cdots + h_1 x + h_0$$

$$(4.20)$$

Example 4.7

A primitive polynomial h(x) of degree 5, given by the octal number 45, is used to generate an m-sequence u. Decimation of u by 3 generates the m-sequence 75 and decimation by 5 produces the m-sequence 67. Consider every possible decimation in the range $1 \leq q \leq N-1$, find the m-sequences that can be formulated by each decimation.

Solution
Sequence u, given by polynomial 45, has period $N = 2^5 - 1 = 31$. The decimation of u will then take place for values of q in the range:

$$1 \leq q \leq 30$$

Sequence u is generated by primitive polynomial $h_0(x)$ given by the octal number 45 and is represented by [100101] in binary so that $h_0(x)$ is given by:

$$h_0(x) = 1 \cdot x^0 + 0 \cdot x^1 + 1 \cdot x^2 + 0 \cdot x^3 + 0 \cdot x^4 + 1 \cdot x^5$$

Therefore

$$h_0(x) = 1 + x^2 + x^5$$

Decimation of u by $2^j q$ where $j \geq 0$ with $q = 1$, that is 1, 2, 4, 8, 16 produces different phases of u.

Decimating the sequence u by 3 ($q = 3$) generates an m-sequence with primitive polynomial $h_3(x)$ given by the octal number 75 which is equivalent to [111101] in binary. Using the same procedure used for representing $h_0(x)$, we can express $h_3(x)$ as:

$$h_3(x) = 1 + x^2 + x^3 + x^4 + x^5$$

Decimating the sequence u by $2^j q$ where $j \geq 0$ and $q = 3$ results in phases of m-sequence given by $h_3(x)$, that is 3, 6, 12, 24, 17. Now decimating the sequence u by 17 is the same as decimating the same sequence by 48 since the sequence is periodic with period 31. Therefore, $48 - N = 17$ where $N = 31$.

Decimating the sequence u by 5 ($q = 5$) generates an m-sequence with primitive polynomial $h_5(x)$ given by the octal number 67 is equivalent to [110111] in binary where:

$$h_5(x) = 1 + x + x^2 + x^4 + x^5$$

Similarly, decimating 5, 10, 20, 9, 18 produces the m-sequence given by polynomial 67. The last two decimations are given by:

$$40 - N = 9 \quad \& \quad 80 - 2N = 18.$$

Consider decimating u by 7. This decimation will generate the same primitive polynomial as decimating by 14, 28, 25, 19. Note that decimation by $25 = 56 - N$ and decimation by $19 = 112 - 3N$ generates the same m-sequences.

Now decimation by 14 is equivalent to decimation by $14 + N = 45$ which is the same as decimating u(3) by 15. Decimation by $15 = \frac{N-1}{2}$ results in an m-sequence generated by the reciprocal of polynomial 75. The octal number 75 is [111101] in binary and the reciprocal polynomial $h_7(x)$ is given by [101111] that is the octal number 57.

$$h_7(x) = 1 + x + x^2 + x^3 + x^5$$

Consider decimating u by $q = 11$. The same primitive polynomial is used when decimation by 22, 13, 26, 21. Now $13 = 44 - N$, $26 = 88 - 2N$, $21 = 176 - 5N$.

Now the decimation by 13 is equivalent to decimating by $13 + 2N = 75$ which is the same as decimating u(5) by 15. Thus the m-sequence is produced by the reciprocal polynomial 67. The octal number $67 = [110111]$ in binary and the reciprocal polynomial is given by [111011] which is 73 in octal number format. Thus, the primitive polynomial $h_{11}(x)$ corresponding to decimation by 11 is:

$$h_{11}(x) = 1 + x + x^3 + x^4 + x^5$$

Lastly, consider decimating u by 15. The same polynomial corresponds to decimation by 30, 29, 27, 23. Note that 29 is equivalent to $60 - N$, 27 is $120 - 3N$ and 23 is $240 - 7N$. The primitive polynomial is the reciprocal polynomial 45. That is, octal number 45 is [100101] in binary and the reciprocal polynomial $h_{15}(x)$ is [101001], which is 51 in octal format.

$$h_{15}(x) = 1 + x^3 + x^5$$

4.5.3.1 Summary of the sequences

The decimation of u generates a total of six m-sequences for primitive polynomials of degree 5. These m-sequences have the following primitive polynomials:

$$h_0(x) = 1 + x^2 + x^5$$ generates m-sequence u.
$$h_3(x) = 1 + x^2 + x^3 + x^4 + x^5$$ generates u(3)
$$h_5(x) = 1 + x + x^2 + x^4 + x^5$$ generates u(5)
$$h_7(x) = 1 + x + x^2 + x^3 + x^5$$ generates u(7)
$$h_{11}(x) = 1 + x + x^3 + x^4 + x^5$$ generates u(11)
$$h_{15}(x) = 1 + x^3 + x^5$$ generates u(15)

4.5.4 Preferred pairs of m-sequences

According to property (iv) in Section 4.5.2, the periodic autocorrelation of m-sequence is a two-valued function. However, the cross-correlation between two m-sequences generated by two different primitive polynomials can be three-valued, four-valued, or possibly many-valued. It is possible to choose a pair of m-sequences which has a three-valued cross-correlation function. These two chosen m-sequences are called the *preferred pair*. The designated pair could be selected as the m-sequence u and its decimated version v = u(q) using the decimation process discussed in Example 4.7. The preferred pairs that have period $N (= 2^r - 1)$ must satisfy the following conditions:

i. $r \neq 0 \bmod 4$, that is n is odd or r = 2 mod 4   (4.21)
    Where n is the degree of the primitive polynomial and r could not take on such values as: 4, 8, 12, 16, 20, …. That is, r = 2, 6, 10, 14, 18,…, etc. These values of r give odd values for $N (= 3, 63, 1023, 16383, 262143, \ldots$, etc.).
ii. v = u (q) q is odd given by:

$$q = \begin{matrix} 2^k + 1 \\ \text{or} \\ 2^{2k} - 2^k + 1 \end{matrix}$$ (4.22)

where k is given by property (iii).

iii. $\gcd(r, k) = \begin{matrix} 1 & \text{for n odd} \\ 2 & \text{for } r = 2 \bmod 4 \end{matrix}$ (4.23)

We have shown in Section 4.5.3 how to find the gcd of two numbers. It is clear that because $r \neq 0 \bmod 4$, N is not a power of 2. Typical values for k are (1, 2). These values of k make q = 3, 5, 13. The preferred pairs of m-sequences have three-valued cross-correlation function defined as $[-1, -t(r), t(r) - 2]$ where

$$t(r) = 1 + 2^{\frac{r+1}{2}} \text{ for r odd}$$ (4.24)

$$t(r) = 1 + 2^{\frac{r+2}{2}} \text{ for } r = 2 \bmod 4$$ (4.25)

Table 4.3 *Maximum cross-correlation associated with preferred pair of m-sequences*

r	1	2	3	5	6	7	10
t(r)	3(4.24)	5 (4.25)	5 (4.24)	9 (4.24)	15(4.25)	15(4.24)	64(4.25)
Cross-correlation	$-1, -3, 1$	$-1, -5, 3$	$-1, -5, 3$	$-1, -9, 7$	$-1, -15, 13$	$-1, -15, 13$	$-1, -64, 63$

Let us compute typical values of the cross-correlation for an assumed m-sequences with $r = 1, 2, 3, 5, 6, 7$, and 10. Using (4.24) and (4.25), t(r) and maximum cross-correlations are given in Table 4.3.

A collection of m-sequences where the property of each pair in the set is a preferred pair is called a *connected set*. The largest possible connected set is called *Maximal connected set*. The size of this set, M_n, is important in applications such as multiple users' spread-spectrum systems.

Example 4.8

Consider the m-sequence u generated by a primitive polynomial of degree $n = 5$ as given in Example 4.7. Construct the maximal connected set of preferred pairs of m-sequences produced by the decimation of u. What is the size of this set?

Solution
A preferred pair of m-sequences must satisfy conditions i, ii, iii as stated in Section 4.5.4. Condition (i) is being satisfied since n in this example is odd.

Consider the pair [u, u(3)] , where $q = 3$ (odd) and $k = 1$ so that $\gcd(r,k) = 1$.

Therefore [u, u(3)] are a preferred pair.

The pair [u, u(5)] gives $q = 5$ (odd) and $k = 2$ so that $\gcd(r,k) = 1$, therefore, [u, u(5)] is another preferred pair.

The m-sequence u(5) is another phase for m-sequence u(9) as we proved in Example 4.7. The pair [u(3), u(5)] produces the same cross-correlation as the pair [u(3), u(9)] so that [u(3), u(5)] is a preferred pair of m-sequences.

Following a similar procedure, we can show that the following pairs of m-sequences are preferred pairs [u(3), u(5)], [u(5), u(15)] and [u(15), u(3)].

Considering all of the possible pairs, we can present a graphical plot with each preferred pair connected by a line as shown in Figure 4.7.

It can be seen from the plot that $M_5 = 3$. Considering all of the m-sequences generated by primitive polynomials of degree n, where $3 \leq n \leq 16$, the size of the maximal connected

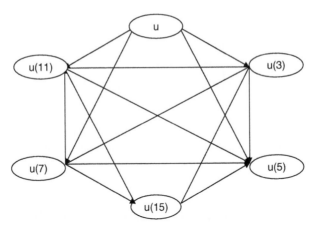

Figure 4.7   *Set of preferred pairs of m-sequence.*

Table 4.4   *Maximum connected size of m-sequences*

Number of SR stages r	Number of m-sequences	Maximal connected set size M_n
3	2	2
4	2	0
5	**6**	**3**
6	6	2
7	18	6
8	16	0
9	48	2
10	60	3
11	176	4
12	144	0
13	630	4
14	756	3
15	1800	2
16	2048	0

set is a small fraction of the number of m-sequences as given by Fan and Darnell (1995) (see Table 4.4).

4.5.5   Gold sequences

If [u, v] is any preferred pair of m-sequences generated by primitive polynomials h(x) and $\hat{h}(x)$ and each of degree n and period $N = 2^n - 1$, then a set of Gold sequences G[u, v] can be generated by $u \oplus v$ where $\oplus$ represents module-2 addition. Taking into consideration

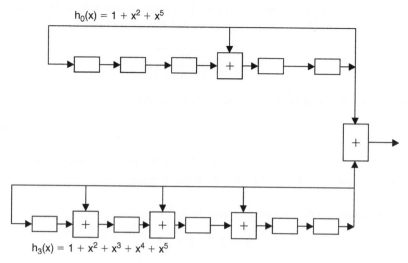

Figure 4.8 *Block diagram of Gold generator.*

the N possible phases of the sequences, we can define the set G[u, v] as:

$$u, v, u \oplus v, u \oplus Tv, u \oplus T^2 v, u \oplus T^3 v, \ldots\ldots\ldots\ldots\ldots\ldots\ldots\ldots, u \oplus T^{N-1}v$$

(4.26)

Where $T^i v$ represents m-sequence v phase shifted by i symbols with $i = 0, 1, 2, \ldots, N - 1$.

The Gold set of sequences contains $N + 2$ sequences and is generated by polynomial given by $h(x) \hat{h}(x)$. A typical Gold generator, shown in Figure 4.8, can be constructed using the preferred pair of m-sequences u[u, u(3)] from Example 4.7 where:

$$h_0(x) = 1 + x^2 + x^5 \qquad \text{gives m-sequence u.}$$
$$h_3(x) = 1 + x^2 + x^3 + x^4 + x^5 \quad \text{gives u(3)}$$

Considering the set of Gold sequences, the out-of-phase autocorrelation of any sequence in the set and the cross-correlation between any pair in the set have three-valued correlation functions given by $[-1, -t(r), t(r) - 2]$ as mentioned previously. However, Gold sequences have cross-correlation (-1) for many offsets of the preferred pair of m-sequence. It turns out that attaching '0' to the original Gold sequences will eliminate the cross-correlation. In fact, simple zero-padding to the Gold sequences can originate 2^r code sequences which have *zero cross-correlation* between them. These code sequences are called '*orthogonal Gold codes*'.

It should be noted that the literature presents an earlier definition for the set of Gold sequences as G[u, v] where $v = u[t(r)]$. At present, it has been accepted that u and v should be any preferred pair of m-sequences.

The lower bound on the peak cross-correlation (Φ_{max}) between any pair of binary sequences of period N in a set of M sequences is given by Welch bounds (Welsh, 1974) as:

$$\Phi_{max} \geq N\sqrt{\frac{M-1}{NM-1}} \tag{4.27}$$

For large values of N and M, Φ_{max} can be approximated as:

$$\Phi_{max} \approx \sqrt{N} \tag{4.28}$$

This lower bound is commonly taken as a bench mark for the cross-correlation between a set of binary sequences when computing the multiple access interference.

For a Gold sequence with a reasonably large value for m,

$$N = 2^m - 1 \approx 2^m \tag{4.29}$$

Thus the lower bound on Gold sequences Φ_{max} is:

$$\Phi_{max} \approx 2^{\frac{m}{2}} \tag{4.30}$$

The maximum cross-correlation between the preferred sequences of a Gold sequence is:

$$2^{\frac{m+1}{2}} + 1 \text{ m is odd, i.e. } \sqrt{2} \text{ times lower bound} \tag{4.31}$$

$$2^{\frac{m+2}{2}} + 1 \text{ m is even, i.e. } 2 \text{ times lower bound} \tag{4.32}$$

Example 4.9

Four sequences, 7-bits each, are all chosen from a set of Gold sequences. Compute the cross-correlation between each pair and suggest a format that eliminates this cross-correlation. Assume the zero time shift between the sequences.

$G_1 = [01\ 01\ 00\ 0]$
$G_2 = [00\ 10\ 01\ 0]$
$G_3 = [10\ 00\ 00\ 1]$
$G_4 = [11\ 11\ 01\ 1]$

Solution

Using the expressions for cross-correlation function derived in Section 4.5.5, we now replace logic '0' with '-1' and the cross-correlations are given below:

$$R_{1,2}(0) = -1, \quad R_{1,3}(0) = -1, \quad R_{1,4}(0) = -1$$
$$R_{2,3}(0) = -1, \quad R_{2,4}(0) = -1$$
$$R_{3,4}(0) = -1$$

Now let us zero-pad the sequences, we get the following sequences:

$G_1 = [01\ 01\ 00\ 00]$
$G_2 = [00\ 10\ 01\ 00]$
$G_3 = [10\ 00\ 00\ 10]$
$G_4 = [11\ 11\ 01\ 10]$

It can be shown that the cross-correlation between any pair of these sequences is zero.

4.5.6 Kasami sequences

A set of Kasami sequences can be generated using two different procedures described below:

i. Generating a small set of Kasami sequences.
   Starting with the m-sequence u generated by a primitive polynomial $h_u(x)$ with period $N = 2^n - 1$ where n is an even number, we can generate a sequence v using primitive polynomial $h_v(x)$ by decimating u by $2^{\frac{n}{2}} + 1$; that is:

$$v = u(2^{\frac{n}{2}} + 1) \tag{4.33}$$

It has been proven (Fan and Darnell, 1995) that v is an m-sequence with period derived as follows:

$$\text{period} = \frac{N}{\gcd[N, (2^{\frac{n}{2}} + 1)]} \tag{4.34}$$

$$= \frac{2^n - 1}{\gcd[(2^n - 1), (2^{\frac{n}{2}} + 1)]} = \frac{(2^{\frac{n}{2}} - 1)(2^{\frac{n}{2}} + 1)}{\gcd[(2^{\frac{n}{2}} - 1)(2^{\frac{n}{2}} + 1), (2^{\frac{n}{2}} + 1)]}$$

$$\text{Period} = N_v = \frac{(2^{\frac{n}{2}} - 1)(2^{\frac{n}{2}} + 1)}{(2^{\frac{n}{2}} + 1)} = 2^{\frac{n}{2}} - 1 \tag{4.35}$$

The small set of Kasami sequences is generated by the primitive polynomial $h(x) = h_u(x)$ $h_v(x)$ using a module addition of u with all possible phases of v; that is:

$$\{u, u \oplus v, u \oplus Tv, \ldots\ldots\ldots\ldots\ldots\ldots\ldots\ldots\ldots, u \oplus T^{N_v}v\} \tag{4.36}$$

The set contains $2^{\frac{n}{2}}$ sequences, each of period N and with three-valued correlation function $[-1, -s(n), s(n)-2]$ where

$$s(n) = 2^{\frac{n}{2}} + 1 \tag{4.37}$$

The maximum magnitude of correlation acquired is s(n) and it is approximately one half of the maximum magnitude value achieved by Gold set.

Table 4.5 *Comparison between Kasami and Gold sequences*

	Small set of Kasami	Large set of Kasami	Gold
Period of individual sequence	$2^n - 1$	$2^n - 1$	$2^n - 1$
Size of set	$2^{\frac{n}{2}}$	$2^{\frac{n}{2}}(2^n + 1)$	$2^n + 1$
Values of n	even	2 mod 4 or 0 mod 4	odd or 2 mod 4
Max correlation between any pair	$2^{\frac{n}{2}} + 1$	$2^{\frac{n+2}{2}} + 1$	$2^{\frac{n+2}{2}} + 1$

ii. Generating a large set of Kasami sequences.

Consider the following m-sequences: sequence u is generated by primitive polynomial $h_u(x)$ of degree n and has a period N; sequence v is the decimation of u by s(n), i.e. $v = u[s(n)]$ generated by the primitive polynomial $h_v(x)$ of degree $\frac{n}{2}$ and has period $2^{\frac{n}{2}} - 1$ and $w = u[t(r)]$ generated by a polynomial $h_w(x)$ of degree n with period N where t(r) is given by (4.24) and (4.25) where n is even. Then the large set of Kasami sequences $K_L(u)$ is generated by primitive polynomial $h(x) = h_u(x) \, h_v(x) \, h_w(x)$ and is given by:

$$K_L(u) = u \oplus v \oplus w \qquad (4.38)$$

and has a period $N = 2^n - 1$. The size of $K_L(u)$ is $2^{\frac{n}{2}}(2^n + 1)$ for $n \equiv 2 \bmod 4$, and $2^{\frac{n}{2}}(2^n + 1) - 1$ for $n \equiv 0 \bmod 4$.

The correlation function for $K_L(u)$ is many-valued with values chosen from the set $\{-1, -t(r), t(r) - 2, -s(n), s(n) - 2\}$. The maximum magnitude of correlation is t(r).

It is interesting to compare the Kasami sequences with the Gold sequences and such a comparison is given in Table 4.5.

For example, for n = 6 (i.e. 6-stage LFSR generator), the length of Kasami sequences is 63 bits, the size of the small set is 8 sequences; the size of the large set is 520 sequences and the size of Gold set for the 6-stage LFSR generator is 65. For the same 6-stage LFSR generator, the maximum magnitudes of the cross-correlation between these sequences are as follows: 9 for the Kasami small set, 17 for the Kasami large set, and 17 for the Gold set.

4.5.7 Walsh sequences

Walsh code sequences are obtained from the Hadamard matrix which is a square matrix where each row in the matrix is orthogonal to all other rows, and each column in the matrix is orthogonal to all other columns. The Hadamard matrix H_n is generated by starting with zero matrix and applying the Hadamard transform successively. Each column or row in the Hadamard matrix corresponds to a Walsh code sequence of length n. Orthogonality between codes in the Hadamard matrix is defined such that the cross-correlation values, associated with zero offset between the pair of sequences, is zero.

The Hadamard transform is defined as

$$\mathbf{H_n} = [0] \tag{4.39}$$

$$\mathbf{H_{2n}} = \begin{array}{|c|c|} \hline \mathbf{H_n} & \mathbf{H_n} \\ \hline \mathbf{H_n} & \overline{\mathbf{H}}_n \\ \hline \end{array} \tag{4.40}$$

Thus, $n = 1$ and we get:

$$\mathbf{H_2} = \begin{array}{|c|c|} \hline 0 & 0 \\ \hline 0 & 1 \\ \hline \end{array} \tag{4.41}$$

Repeating the Hadamard transform again for $n = 2$, we get $\mathbf{H_4}$ as:

$$\mathbf{H_4} = \begin{array}{|c|c|c|c|} \hline 0 & 0 & 0 & 0 \\ \hline 0 & 1 & 0 & 1 \\ \hline 0 & 0 & 1 & 1 \\ \hline 0 & 1 & 1 & 0 \\ \hline \end{array} \tag{4.42}$$

Repeating the Hadamard transform again for $n = 4$, we get:

$$\mathbf{H_8} = \begin{array}{|c|c|c|c|c|c|c|c|} \hline 0 & 0 & 0 & 0 & 0 & 0 & 0 & 0 \\ \hline 0 & 1 & 0 & 1 & 0 & 1 & 0 & 1 \\ \hline 0 & 0 & 1 & 1 & 0 & 0 & 1 & 1 \\ \hline 0 & 1 & 1 & 0 & 0 & 1 & 1 & 0 \\ \hline 0 & 0 & 0 & 0 & 1 & 1 & 1 & 1 \\ \hline 0 & 1 & 0 & 1 & 1 & 0 & 1 & 0 \\ \hline 0 & 0 & 1 & 1 & 1 & 1 & 0 & 0 \\ \hline 0 & 1 & 1 & 0 & 1 & 0 & 0 & 1 \\ \hline \end{array} \tag{4.43}$$

Matrix (4.43) consists of 8 Walsh codes, each of length 8 bits. It is worth noting that in matrices $(n = 2)$, $(n = 4)$, $(n = 8)$, *and it is true for all Hadamard matrices*, the first column (or row) has all zero sequences and the second column (or row) has alternating sequences of '0' and '1'. The interesting property of the matrix is that any column (or row) differs from any other column (or row) in exactly $\frac{N}{2}$ positions. Hardware implementation is shown in Figure 4.9 comprising of a clock, three AND gates, and two toggle flip flops and OR gate.

The edge-triggered Toggle Flip Flop (T-FF) is a standard JK flip flip, with both J and K connected to high, and each stage divides the input clock by 2. The input to the generator is eight bits from the clock 01010101, so the output from the first T-FF is 00110011 and from the second T-FF is 00001111. The binary variables u_2 u_1 u_0 represent a Walsh code

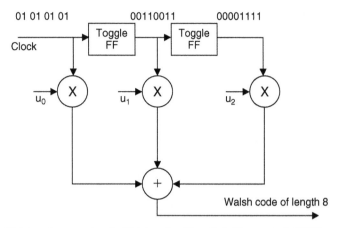

Figure 4.9   *Walsh code generator for Walsh code of length 8 (Hanzo et al., 2003).*

index as shown in this table below:

$u_2\,u_1\,u_0$	Walsh code
000	00000000
001	01010101
010	00110011
011	01100110
100	00001111
101	01011010
110	00111100
111	01101001

(4.44)

It can be seen that to generate a Walsh code of length 16 bits, we need to use three T-FFs, four ANDs and one OR gate. In general, for a Walsh code of length n, we use n–1 toggled flip flips, n AND gates and one multi-input OR gate.

4.5.8   Multi-rate orthogonal codes

In applications such as data transmission using CDMA spread-spectrum systems, there is a need to use orthogonal codes that enable variable data rate transmission. Consequently, for constant chip rate, the spreading factor (i.e. spreading gain) has to be varied. These codes are described in the literature as multi-rate orthogonal codes. An orthogonal code generator is shown in Figure 4.10 and comprises of a Walsh code generator, similar to the one shown in Figure 4.9, and the orthogonal Gold code generator.

In Figure 4.10, R_c is the chip rate for the orthogonal Gold code generator and L_g is the length of the orthogonal Gold code, $\frac{R_c}{L_g}$ is a chip rate for the Walsh code generator.

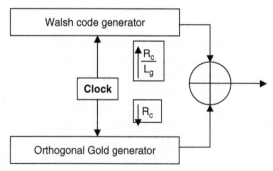

Figure 4.10 *Multi-rate orthogonal code generator (Hanzo et al., 2003).*

The maximum length of the Walsh code L_w is given by:

$$L_w = \frac{G_{p\,max}}{L_g}$$

where $G_{p\,max} = \frac{R_c}{R'_b}$ and R'_b is the lowest bit rate used in the transmission. The Walsh generator is clocked once every L_g Gold chips. Walsh codes with indices selected between 0 to $L_w - 1$ can be used for data transmission at the rate R'_b. However, for data rate $2^k R'_b$, Walsh codes have to be selected carefully to guarantee the orthogonality between the multi-rate codes, and this is better explained by considering the following example.

Example 4.10

Consider the multi-rate orthogonal Gold code generator shown in Figure 4.10 and let $R_c = 128\,kc/s$ and $R'_b = 8\,kb/s$ and $R_b = 16\,kb/s$. Let the length of the orthogonal Gold code $L_g = 8$.

1. Find the code matrix at the generator output.
2. Find the codes that are not orthogonal.

Solution

The maximum spreading factor (spreading gain) $= \dfrac{R_c}{R'_b} = \dfrac{128}{8} = 16$

Walsh code length $L_w = \dfrac{\text{Max. Spreading factor}}{L_g} = \dfrac{16}{8} = 2$

Thus

$$\mathbf{H}_2 = \begin{array}{|c|c|} \hline 0 & 0 \\ \hline 0 & 1 \\ \hline \end{array}$$

Therefore, the length of each of the codes generated from the multi-rate code is $L_w L_g = 2 \times 8 = 16$ and the multi-rate code matrix M_{16} is a square-shaped (16×16) matrix given by:

$$\mathbf{M}_{16} = \begin{array}{|c|c|} \hline G_8 & G_8 \\ \hline G_8 & \bar{G}_8 \\ \hline \end{array}$$

Substituting for $\mathbf{G}_8$, we get:

$$\mathbf{M}_{16} =$$

1	0	1	1	1	0	0	0	1	0	1	1	1	0	0	0
0	1	0	1	0	0	0	0	0	1	0	1	0	0	0	0
1	1	0	0	1	1	0	0	1	1	0	0	1	1	0	0
0	0	1	0	0	1	0	0	0	0	1	0	0	1	0	0
1	0	0	0	0	0	1	0	1	0	0	0	0	0	1	0
0	1	1	0	1	0	1	0	0	1	1	0	1	0	1	0
1	1	1	1	0	1	1	0	1	1	1	1	0	1	1	0
0	0	0	1	1	1	1	0	0	0	0	1	1	1	1	0
1	0	1	1	1	0	0	0	0	1	0	0	0	1	1	1
0	1	0	1	0	0	0	0	1	0	1	0	1	1	1	1
1	1	0	0	1	1	0	0	0	0	1	1	0	0	1	1
0	0	1	0	0	1	0	0	1	1	0	1	1	0	1	1
1	0	0	0	0	0	1	0	0	1	1	1	1	1	0	1
0	1	1	0	1	0	1	0	1	0	0	1	0	1	0	1
1	1	1	1	0	1	1	0	0	0	0	0	1	0	0	1
0	0	0	1	1	1	1	0	1	1	1	0	0	0	0	1

Consider sequences $\mathbf{M}_{16}(1)$ and $\mathbf{M}_{16}(9)$ to work out the cross-correlation associated with them given as:

0	1	0	1	0	0	0	0	0	1	0	1	0	0	0	0
0	1	0	1	0	0	0	0	1	0	1	0	1	1	1	1
1	+1	+1	+1	+1	+1	+1	+1	−1	−1	−1	−1	−1	−1	−1	−1

Clearly, this cross-correlation function has an amplitude zero for zero offset between the sequences indicating these two codes are orthogonal. Similarly, we can show each row in $\mathbf{M}_{16}$ is orthogonal to other rows.

When data rate is doubled (i.e. $R_b = 16$ kb/s), the number of chips per data bit is halved which makes codes $\mathbf{M}_{16}(i)$ and $\mathbf{M}_{16}(i+8)$ not orthogonal where $i = 0, 1, \ldots, 7$ since both are expanded from $\mathbf{G}_8(i)$. Thus for doubling the rate, the number of orthogonal sequences available is halved. Consequently, for transmission at double bit rate, two independent orthogonal codes have to be used.

4.6 Complex sequences

Although only binary code sequences are used in CDMA spread-spectrum systems (IS-95) standards, research carried out on systems using complex codes shows improved bit error rates compared with those using binary Gold codes. Consequently, future wireless systems may benefit from using complex code sequences, although there are challenges in terms of finding large size code sequences with correlation properties suitable for multi-user applications.

4.6.1 Quadriphase sequences

The elements of the quadriphase sequences correspond to the complex qth root of unity and a large family of these sequences can be generated as we will show in this section. Each quadriphase sequence has q complex elements. Such sequences match the M-level phase modulation schemes used in spread-spectrum systems so that the carrier magnitude and phase are modulated by complex valued code sequence.

In its basic form, the quadriphase sequence can be acquired using any two binary sequences $\{a_n\}$ and $\{b_n\}$, each of period N and which are combined into a quadriphase sequence $\{c_n\}$ of period N given by:

$$c_n = \frac{1}{2}(1 + j) \cdot a_n + \frac{1}{2}(1 - j) \cdot b_n \qquad (4.45)$$

Conversely, any quadriphase sequence of length N can be decomposed into two binary sequences, each of length N. The maximum cross-correlation of the quadriphase set is usually lower than the maximum cross-correlation between constituent binary sequences by $\sqrt{2}$ which corresponds to a 3 dB improvement in signal to interference ratio (Fan and Darnell, 1995).

Quadriphase sequences, which can be used as signature sequences in multi-user spread-spectrum systems, have been studied extensively in the literature. Given the size of the literature on this topic, it is not necessary to cover all of the work on quadriphase sequences in this section. Therefore, with apology to researchers whose work is not discussed here, we proceed with selected samples of papers published on this topic and that cover properties such as sequence length, family size and correlation functions suitable for application in spread-spectrum systems. We start with Krone and Sarwate (1984) who have presented methods of construction that obtain sets of quadriphase sequences from sets of binary sequences. A set of $2(q + 1)$ sequences can be constructed, each with period $N = (q - 1)$ and maximum periodic cross-correlation and periodic out-of-phase autocorrelation bounded by $3\sqrt{q} + 5$. Implementation of these methods is not discussed in Krone and Sarwate (1984) and the performance of spread-spectrum systems using the quadriphase code sequences is an open problem.

Kumar et al. (1996) presented design methods for large families of quadriphase sequences with low correlation, and three sets of families of quadriphase sequences are proposed. Each sequence in each family has length N given by:

$$N = 2^r - 1 \qquad (4.46)$$

The sequences can be generated using two multistage LFSR generators – one is binary (mod 2 arithmetic) and the other is quaternary (mod 4 arithmetic). The family size M, the maximum periodic correlation C_{max}, for family sets S(0), S(1), and S(2) are given in Table 4.6.

Consider family set S(1) with r = 5 (number of shift registers), the length of each sequence N = 31 and the family size M = 31^2 = 3 × 31 + 2 = 1056. The maximum magnitude of the periodic correlation C_{max} = 12.31.

The following primitive polynomial of degree 5 is used for the connection of the binary LFSR

$$h(x) = 1 + x^2 + x^3 + x^4 + x^5 \tag{4.47}$$

Sequences in the family set S(1) can be generated using the LFSR shown in Figure 4.11.

The desired quadriphase sequence $\{â_n\}$ is obtained by the mapping relation:

$$â_n = j^{a_n} \quad j = \sqrt{-1} \tag{4.48}$$

Table 4.6   *Family size of polyphase sequences (Kumar et al., 1996)*

Family set	Family size M	C_{max}
S(0)	N + 2	$\sqrt{N+1} + 1$
S(1)	$\geq N^2 + 3N + 2$	$2\sqrt{N+1} + 1$
S(2)	$\geq N^3 + 4N^2 + 5N + 2$	$4\sqrt{N+1} + 1$

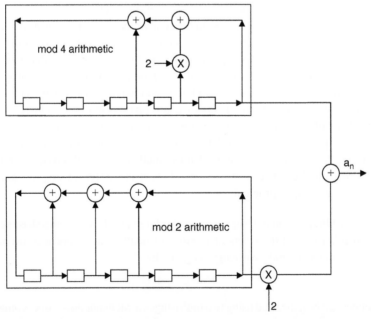

Figure 4.11   *LFSR quadriphase generator using mod 4 arithmetic except for the shift register at the bottom which is implemented using mod 2 arithmetic (Kumar et al., 1996).*

Example 4.11

Consider the quadriphase generator in Figure 4.11 with $M = 5$. The shift registers are loaded with the following initial states: $(a_{n-1}, a_{n-2}, a_{n-3}, a_{n-4}, a_{n-5}) = (1,0,0,0,0)$. Generate the quadriphase sequences.

Solution

Consider the mod 4 arithmetic generator with initial states 10000 shown below:

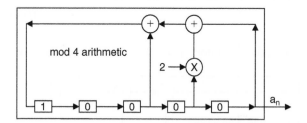

Clearly each register is a 2-bit register with a clock twice as fast as the binary register and all computations are mod 4 arithmetic. We will consider the binary LFSR generator clock as a reference clock. Let us compute the mod 4 output sequence.

At the first clock pulsed, the generator output '0' and the generator states change to:

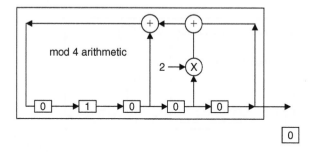

At the 2nd clock pulse, the generator output '0' and its states change to:

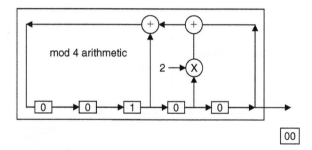

At the 3rd clock pulse, the generator output '0' and its states change to:

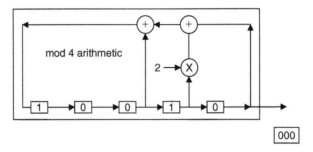

000

At the 4th clock pulse, the generator output '0' and its states change to:

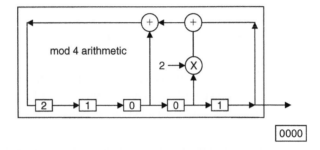

0000

At the 5th clock pulse, the generator output '1' and its states change to:

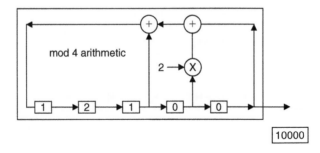

10000

At the 6th clock pulse, the generator output '0' and its states change to:

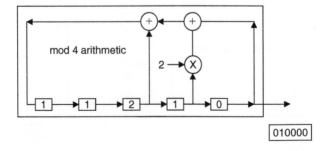

010000

At the 7th clock pulse, the generator output '0' and its states change to:

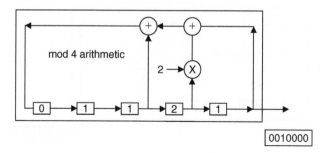

0010000

At the 8th clock pulse, the generator output '1' and its states change to:

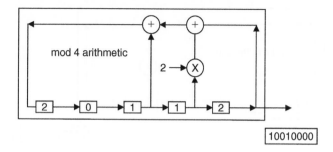

10010000

At the 9th clock pulse, the generator output '2' and its states change to:

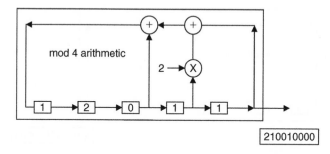

210010000

At the 10th clock pulse, the generator output '1' and its states change to:

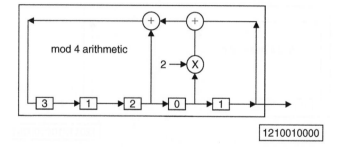

1210010000

At the 11th clock pulse, the generator output '1' and its states change to:

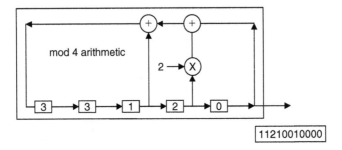

11210010000

At the 12th clock pulse, the generator output '0' and its states change to:

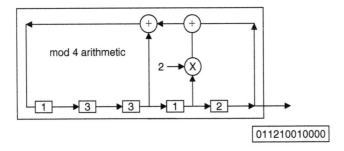

011210010000

At the 13th clock pulse, the generator output '2' and its states change to:

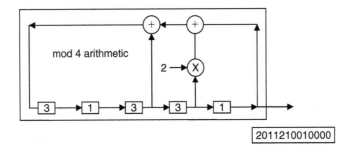

2011210010000

At the 14th clock pulse, the generator output '1' and its states change to:

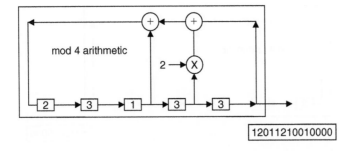

12011210010000

At the 15th clock pulse, the generator output '3' and its states change to:

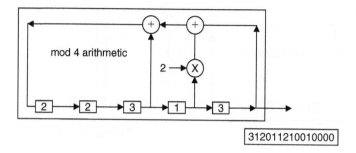

312011210010000

At the 16th clock pulse, the generator output '3' and its states change to:

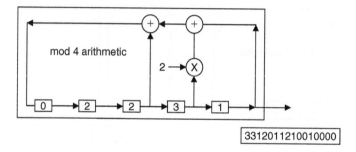

3312011210010000

At the 17th clock pulse, the generator output '1' and its states change to:

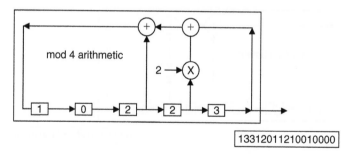

13312011210010000

At the 18th clock pulse, the generator output '3' and its states change to:

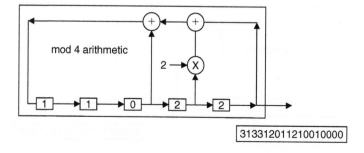

313312011210010000

At the 19th clock pulse, the generator output '2' and its states change to:

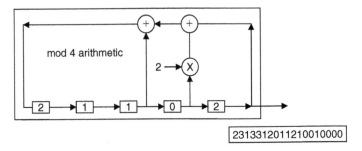

2313312011210010000

At the 20th clock pulse, the generator output '2' and its states change to:

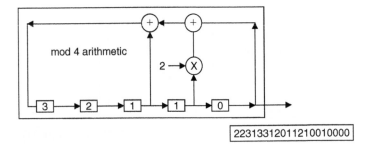

22313312011210010000

At the 21st clock pulse, the generator output '0' and its states change to:

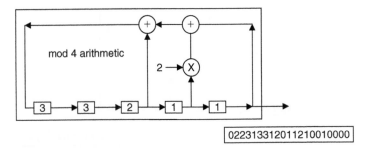

022313312011210010000

At the 22nd clock pulse, the generator output '1' and its states change to:

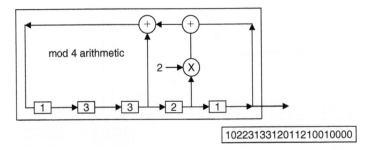

1022313312011210010000

At the 23rd clock pulse, the generator output '1' and its states change to:

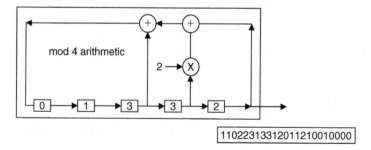

110223133120112100100000

At the 24th clock pulse, the generator output '2' and its states change to:

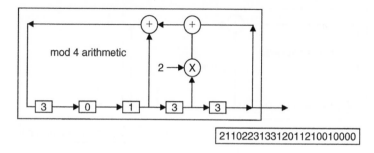

2110223133120112100100000

At the 25th clock pulse, the generator output '3' and its states change to:

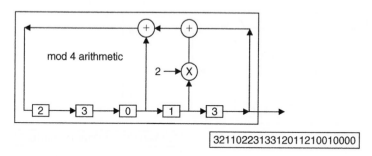

32110223133120112100100000

At the 26th clock pulse, the generator output '3' and its states change to:

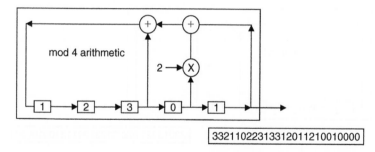

332110223133120112100100000

At the 27th clock pulse, the generator output '1' and its states change to:

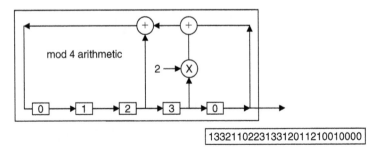

1332110223133120112100 10000

At the 28th clock pulse, the generator output '0' and its states change to:

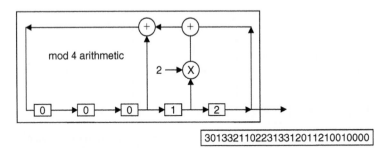

01332110223133120112100 10000

At the 29th clock pulse, the generator output '3' and its states change to:

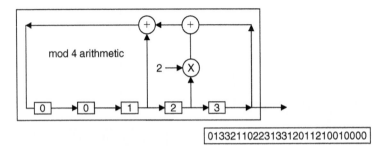

301332110223133120112100 10000

At the 30th clock pulse, the generator output '2' and its states change to:

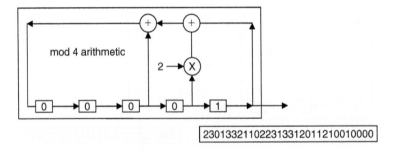

2301332110223133120112100 10000

At the 31st clock pulse, the generator output '1' and its states change to:

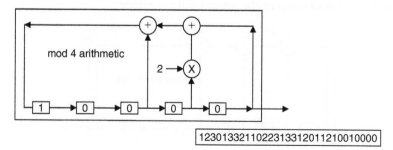

1230133211022313312011210010000

At the 32nd clock pulse, the generator output '0' and its states change to:

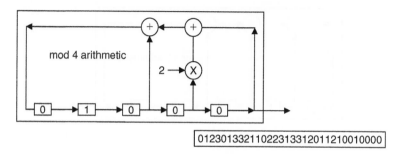

01230133211022313312011210010000

At the 33rd clock pulse, the generator output '0' and its states change to:

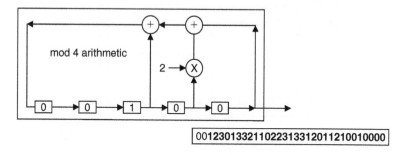

001230133211022313312011210010000

Similarly, it can be shown that the output is:

$- - - - - - - - - - - - - - - -$0010000**1230133211022313312011210010000**

Consequently, the output periodic mod 4 sequence is of length (period) 31:

1230133211022313312011210010000

Now consider the binary LFSR generator in Figure 4.11. The shift registers are initialized with 10000 as shown below. The multiplication of the output binary sequence by 2 corresponds to a shift of the output sequence one place to the left that increases its value by a power of 2. Let us compute the output binary sequence.

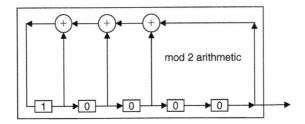

At the first clock pulsed, the generator output '0' and the generator states change to:

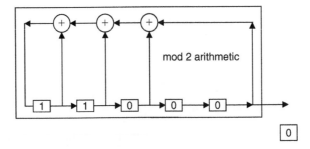

At the 2nd clock pulse, the generator output '0' and its states change to:

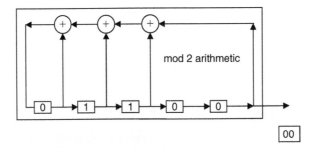

At the 3rd clock pulse, the generator output '0' and its states change to:

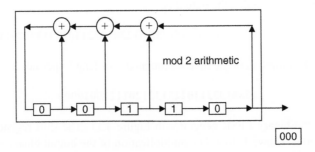

At the 4th clock pulse, the generator output '0' and its states change to:

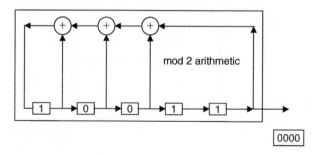

0000

At the 5th clock pulse, the generator output '1' and its states change to:

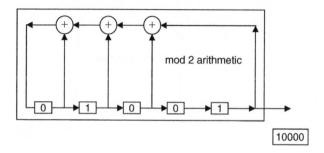

10000

At the 6th clock pulse, the generator output '1' and its states change to:

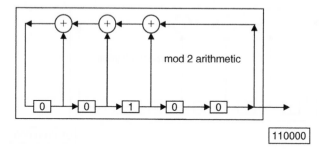

110000

At the 7th clock pulse, the generator output '0' and its states change to:

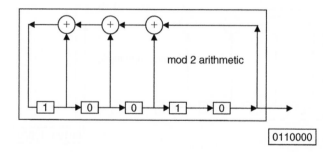

0110000

At the 8th clock pulse, the generator output '0' and its states change to:

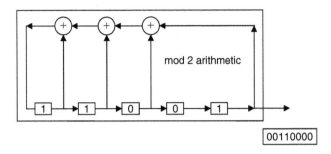

00110000

At the 9th clock pulse, the generator output '1' and its states change to:

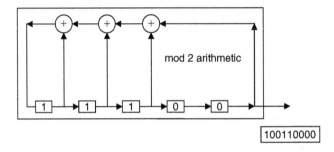

100110000

At the 10th clock pulse, the generator output '0' and its states change to:

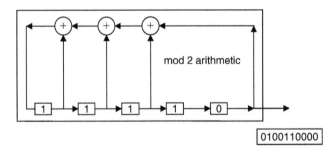

0100110000

At the 11th clock pulse, the generator output '0' and its states change to:

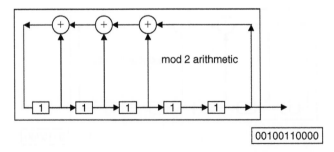

00100110000

At the 12th clock pulse, the generator output '1' and its states change to:

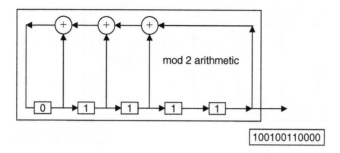

100100110000

At the 13th clock pulse, the generator output '1' and its states change to:

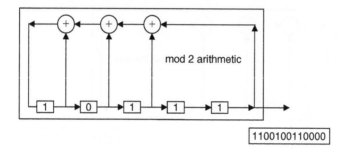

1100100110000

At the 14th clock pulse, the generator output '1' and its states change to:

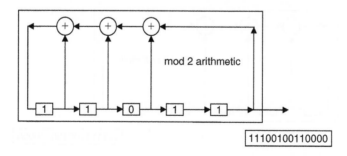

11100100110000

At the 15th clock pulse, the generator output '1' and its states change to:

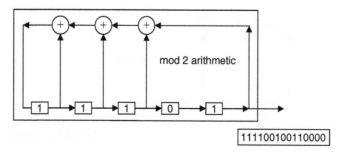

111100100110000

At the 16th clock pulse, the generator output '1' and its states change to:

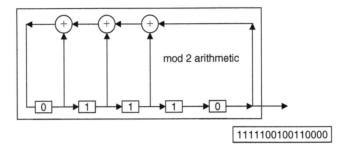

1111100100110000

At the 17th clock pulse, the generator output '0' and its states change to:

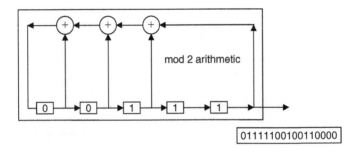

01111100100110000

At the 18th clock pulse, the generator output '1' and its states change to:

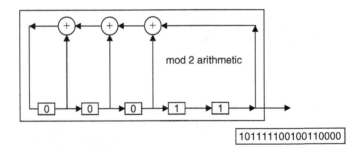

101111100100110000

At the 19th clock pulse, the generator output '1' and its states change to:

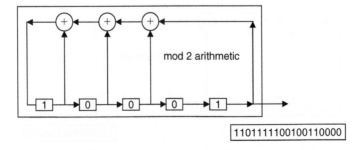

1101111100100110000

At the 20th clock pulse, the generator output '1' and its states change to:

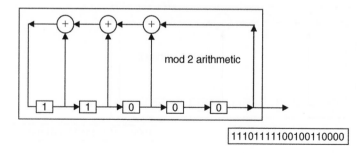

111011111100100110000

At the 21st clock pulse, the generator output '0' and its states change to:

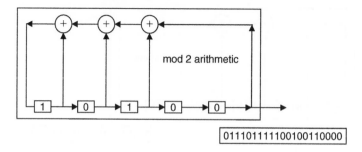

011101111100100110000

At the 22nd clock pulse, the generator output '0' and its states change to:

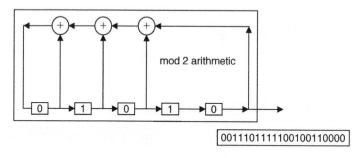

001110111110100100110000

At the 23rd clock pulse, the generator output '0' and its states change to:

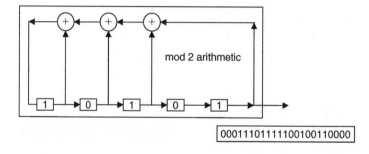

000111011111100100110000

At the 24th clock pulse, the generator output '1' and its states change to:

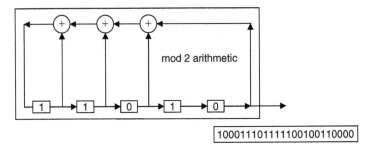

1000111011111100100110000

At the 25th clock pulse, the generator output '0' and its states change to:

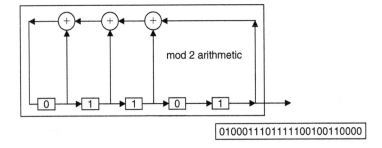

0100011101111100100110000

At the 26th clock pulse, the generator output '1' and its states change to:

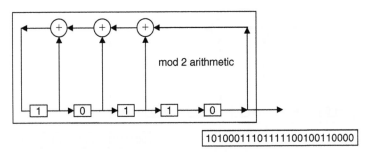

1010001110111110010011000

At the 27th clock pulse, the generator output '0' and its states change to:

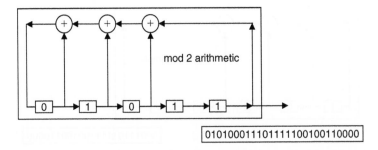

0101000111011111100100110000

At the 28th clock pulse, the generator output '1' and its states change to:

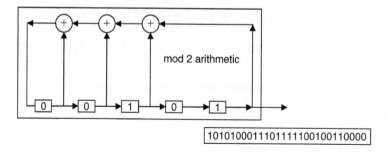

1010100011101111100100110000

At the 29th clock pulse, the generator output '1' and its states change to:

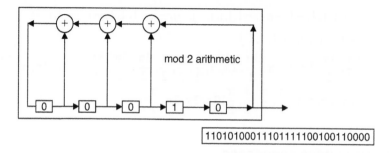

1101010001110111100100110000

At the 30th clock pulse, the generator output '0' and its states change to:

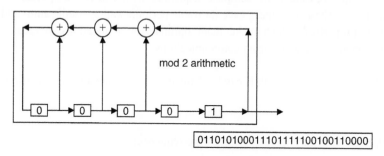

0110101000111011111100100110000

At the 31st clock pulse, the generator output '1' and its states change to:

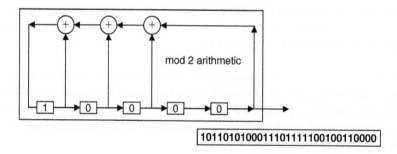

1011010100011101111100100110000

The binary LFSR revert to its initial conditions and the binary sequence is:

$$10110101000111011111100100110000$$

Now shift left one place to account for multiplication by 2 and we get the sequence after the multiplier:

$$01101010001110111111001001100001$$

Now add the two sequences to get:

$$1300230211133320023012211110001$$

The desired sequence is obtained using (4.48):

$$
\begin{aligned}
\{\hat{a}_n\} = {} & j, -j, 1, 1, -1, -j, 1, -1, j, j, j, -j, -j, -j, -1, 1, 1, -1, -j, 1, j, \\
& -1, -1, j, j, j, j, 1, 1, 1, j \\
& \frac{\pi}{2}, \frac{3\pi}{2}, 2\pi, 2\pi, \pi, \frac{3\pi}{2}, 2\pi, \pi, \frac{\pi}{2}, \frac{\pi}{2}, \frac{\pi}{2}, \frac{3\pi}{2}, \frac{3\pi}{2}, \frac{3\pi}{2}, \pi, 2\pi, 2\pi, \pi, \frac{3\pi}{2}, \\
& 2\pi, 2\pi, \pi, \pi, \frac{\pi}{2}, \frac{\pi}{2}, \frac{\pi}{2}, \frac{\pi}{2}, 2\pi, 2\pi, 2\pi, \frac{\pi}{2}
\end{aligned}
$$

4.6.2 Polyphase sequences

Polyphase sequences are finite complex sequences with constant amplitude and time discrete phases. These sequences have desirable periodic and aperiodic auto- and cross-correlation properties, which make them optimal for applications in radar, system identification and spread-spectrum communications.

A polyphase sequence of length N can be defined by N complex elements $\hat{a}_i$ given by:

$$\hat{a}_i = a_i\, e^{j\theta_i} \tag{4.49}$$

where θ_i represents the N phases given by the sequence:

$$\{\theta_0, \theta_1, \theta_2, \dots\dots\dots\dots\dots \theta_{N-1}\} \tag{4.50}$$

$j = \sqrt{-1}$ and a_i is the magnitude of ith element in the complex polyphase sequence. The sequence symbols are the complex N^{th} root of unity where N is a prime number for finite Galois field as demonstrated in Section 4.6.1.

The literature is rich with works on polyphase sequences proposing their use as signature sequences in multi-user spread-spectrum systems but most of the proposed techniques are still open for research; especially the practical issues concerning the implementation of these sequences and the performance of the multi-user spread-spectrum systems which use the polyphase code sequences.

A polyphase sequence with elements of magnitude 1 is called a *Baker sequence* if the maximum magnitude of all side lobes of their aperiodic autocorrelation function is less than or equal to 1. Generally, optimization algorithms are used to search for these sequences (Krone and Sarwate, 1984). However, the number of polyphase Baker sequences discovered is 31, each of length 31 (Fan and Darnell, 1995) and because of their small family size, they have minimal use in multiple access communication system applications.

Frank polyphase sequences with a relatively large set size ($= \sqrt{N} - 1$) are denoted as $F = \{\hat{a}^{(1)}, \hat{a}^{(2)}, \ldots\ldots\ldots, \hat{a}^{(r)}, \ldots\ldots\ldots, \hat{a}^{(q-1)}\}$. Each sequence is of length $N = q^2$. The elements of the sequence are the qth root of unity and are given by Frank (1963):

$$\hat{a}^{(r)} = \{a_0^{(r)}, a_1^{(r)}, \ldots\ldots\ldots\ldots, a_{N-1}^{(r)}\} \tag{4.51}$$

$$a_n^{(r)} = \hat{a}_{iq+k}^{(r)} = e^{j\frac{2\pi}{q}rik} = \alpha^{rik} \tag{4.52}$$

where $\alpha = e^{j\frac{2\pi}{q}}$ and $0 \leq n \leq N-1, i < q$, q is any integer, and $\gcd(r, q) = 1$. It has been shown in Frank (1963) that the periodic ACF and CCF of Frank sequences are as follows:

$$\begin{aligned} ACF(\tau) &= N \quad \text{for } \tau = 0 \\ &= 0 \quad \text{for } \tau \neq 0 \end{aligned} \tag{4.53}$$

$$\text{and } CCF(\tau) = \sqrt{N} \quad \text{for q is odd.} \tag{4.54}$$

Chu (1972) polyphase sequences exist for every integer N. The sequences $\{a_n^{(r)}\}$ of length N and for ($0 \leq n \leq N$) are defined by:

$$a_n^{(r)} = e^{j\frac{\pi}{N}r(1+n)n+mn} \quad \text{for N odd} \tag{4.55}$$

$$= e^{j\frac{\pi}{N}rn^2+mn} \quad \text{for N even} \tag{4.56}$$

where m is any integer and $\gcd(r,N) = 1$.

It has been shown (Chu, 1972) that the periodic ACF and CCF of the Zadoff-Chu sequences are as follows:

$$\begin{aligned} ACF(\tau) &= N \quad \text{for } \tau = 0 \\ &= 0 \quad \text{for } \tau \neq 0 \end{aligned} \tag{4.57}$$

$$\text{and } CCF(\tau) = \sqrt{N} \quad \text{for } \gcd(r-s, N) = 1 \text{ and N is odd.} \tag{4.58}$$

A comprehensive treatment of the complex sequences and the behaviour of the correlation functions is given in Fan and Darnell (1995), Chapters 7 to 11.

The Frank-Zadoff-Chu (FZC) polyphase sequences have excellent correlation properties. The maximum out-of-phase periodic autocorrelation and the maximum periodic cross-correlation (C_{max}) of the Gold, quadriphase, and ZFC sequences are given in Lam and Ozluturk (1992) as:

$$C_{max} = \begin{cases} \sqrt{2(N+1)} + 1 \text{ Gold sequences} \\ \sqrt{N + \sqrt{2(N+1)} + 2} \text{ Quadriphase sequences} \\ \sqrt{N} \text{ FZC sequences} \end{cases} \tag{4.59}$$

where N is the length of the sequence. For example, where $N = 128$, the maximum correlations are 17.06, 14.00, and 11.31 for the three classes of sequences. However, the degradation in system performance does not depend on periodic correlation but on the odd correlation as well. Chu (1972) and Lam and Ozluturk (1992) discuss how multi-user spread-spectrum systems with FZC code sequences exhibit a slightly better BER performance under large SNR values, i.e. show lower multiple access interference.

Oppermann and Vucetic (1997) presented a new family of complex valued sequences for use in multi-user CDMA spread-spectrum systems which offer a choice of a wide range of correlation properties. The ith element of the Oppermann-Vucetic sequence μ_i is defined by:

$$\mu_i = (-1)^{Mi} \, e^{\frac{j\pi(M^m i^p + i^n)}{N}} \quad 1 \leq i \leq N, 1 \leq M \leq N \tag{4.60}$$

where $j = \sqrt{-1}$, and M is an integer that is relatively prime to N. The variables m, n, and p are any real numbers that specify the sequence set. However, the sequences in the family are constant amplitude complex numbers, each sequence in the set differs only in phase terms. When $p = 1$, each sequence in the set will have the same autocorrelation function magnitude. When N is prime, the maximum set size for any combination of parameters of $N - 1$ will be achieved.

The performance of CDMA systems using sequence (4.57) is superior to any other known large set of sequences (Oppermann and Vucetic, 1997). However, methods of constructing these sequences are open problems for research. Oppermann (1997) has identified a sub-family of spreading sequences within the sequences presented in Oppermann and Vucetic (1997), which offer larger sets of orthogonal sequences.

An interesting class of polyphase sequences for prime length (N) greater than 3 is suggested in Fan et al. (1994). The size of the set is equal N and r^{th} polyphase sequence $a^{(r)}$ is defined as:

$$a^{(r)} = \{a_0^{(r)}, a_1^{(r)}, \ldots \ldots, a_{N-1}^{(r)}\} \tag{4.61}$$

$$a_n^{(r)} = \alpha^{\frac{n(n+1)(n+2)}{6} + rn}$$

$$\alpha = e^{\frac{j2\pi v}{N}} \quad 0 \leq n \quad r < N \tag{4.62}$$

where α is the Nth root of unity and r, n and v are integers relatively prime to N. It was shown in Fan and Darnell (1995) that the periodic autocorrelation is a two-valued function given by:

$$|R_{r,r}(\tau)| = N \quad \tau = 0$$
$$= \sqrt{N} \quad \tau \neq 0 \qquad (4.63)$$

and the periodic cross-correlation is a two-valued function given by:

$$|R_{r,s}(\tau)| = 0 \quad \tau = 0$$
$$= \sqrt{N} \quad \tau \neq 0 \qquad (4.64)$$

However, neither the implementation of these sequences nor their performance in a multi-user spread-spectrum system were discussed in Fan et al. (1994).

4.7 Summary

In this chapter we have presented the theory and implementation of the pseudo-random code sequences commonly used in spread-spectrum systems. We started by introducing basic algebra concepts that are used in the sequence analysis in Section 4.2, followed by the arithmetic of the binary polynomials in Section 4.3. Computation of the elements of Galois's field, important in the generation of the sequences, was presented in Section 4.4, while Section 4.5 presented the generation and decimations of the most common PN-sequences, called the m-sequences. These sequences have many applications in coding theory, spread-spectrum techniques and cryptography. Methods using two preferred m-sequences for deriving the Gold and Kasami sequences that have reduced cross-correlation between any pair in the set were also presented in Sections 4.5.5 and 4.5.6. Orthogonal sequences such as the Walsh-Hadamard and orthogonal Gold sequences are used for logical chanallization and were described in Sections 4.5.7 and 4.5.8.

Recent interest in complex code sequences such as quadriphase and polyphase sequences were considered in Section 4.6 where we examined in detail the use of quadriphase and polyphase sequences as users' signatures.

Problems

4.1 Consider polynomials $P_1(x)$ and $P_2(x)$ with coefficients drawn from Galois field GF(4) such that:

$$P_1(x) = 2x + 3x^2 + x^4$$
$$P_2(x) = 1 + 3x^3 + x^4$$

Evaluate the following mathematical expressions:
 i. $P_1(x) + P_2(x)$
 ii. $P_1(x) - P_2(x)$

   iii. $P_1(x) \cdot P_2(x)$

   iv. $\dfrac{P_1(x)}{P_2(x)}$

4.2 Find the elements that arise from the addition of each pair of elements of the polynomials in $GF(2^3)$.

4.3 Find the elements in $GF(2^3)$ using the primitive polynomial $P(x) = 1 + x^2 + x^3$.

4.4 Consider the *Galois feedback implementation* of the sequence generator shown in Figure 4.4 with the generator polynomial $g(x)$ given by:

$$g(x) = 1 + x^3 + x^4$$

Assume the initial load of the register be 0001.

   i. Find the output periodic sequence.

   ii. What would the maximum possible period be?

4.5 Find the sequence at the output of the Fibonacci generator shown in Figure 4.5 with polynomial given by:

$$g(x) = 1 + x + x^4$$

Assume the initial load is 0001.

4.6 Show that the m-sequence obtained from the primitive polynomial $(1 + x + x^4)$ is [00 01 00 11 01 01 11 1].

4.7 A sequence u generated by a primitive polynomial of degree 6 is decimated by $q = 27$ to generate sequence $\upsilon = u(q)$. What would be the period of sequence υ?

4.8 An m-sequence u of period $N = 15$ is decimated by $q = 7$ to generate sequence $\upsilon = u(7)$. Given $u = [00\quad 10\quad 01\quad 10\quad 10\quad 11\quad 11\quad 0]$. What would be sequence υ?

4.9 Four sequences, 7-bits each, are chosen from a set of Gold sequences. Compute the cross-correlation between each pair for time shift (τ) between the sequences to be 0, 1, 2, 3, 4, 5 and suggest a format that eliminates the cross-correlation for zero time shift.

$$G_1 = [01\ 10\ 10\ 1]$$

$$G_2 = [00\ 10\ 01\ 0]$$

$$G_3 = [11\ 00\ 11\ 0]$$

$$G_4 = [10\ 11\ 10\ 0]$$

4.10 Consider the primitive polynomial $h(x)$ of degree $m = 4$ which is dividing $(x^{2^m - 1} - 1)$. The shift registers are loaded with the following initial states: $(a_{n-1}, a_{n-2}, a_{n-3}, a_{n-4}) = (1, 0, 0, 0)$. Generate the quadriphase sequences using mod 4 arithmetic LFSR generator.

References

Chu, D.C. (1972) Polyphase codes with good periodic correlation properties, *IEEE Transactions on Information Theory*, 18, 531–532.

Fan, P. and Darnell, M. (1995) *Sequence Design for Communications Applications*, Publisher: Taunton: Research Studies Press Ltd, page 118.

Fan, P.Z., Darnell, M. and Honary, B. (1994) New class of polyphase sequences with two-valued auto- and cross-correlation functions, *Electronics Letters*, 30(13), 1031–1032.

Frank, R.L. (1963) Polyphase codes with good non-periodic correlation properties, *IEEE Transactions on Information Theory*, IT-9, 43–45.

Hanzo, L., Munster, M., Choi, B. J. and Keller, T. (2003) *OFDM and MC-CDMA for Broadband Multi-user Communications, WLANs and Broadcasting*, John Wiley & Sons.

Krone, S. M. and Sarwate, D.V. (1984) Quadriphase sequences for spread-spectrum multiple access communication, *IEEE Transactions on Information Theory*, IT-30(3), 520–529.

Kumar, P.V., Helleseth, T., Calderbank, A.R. and Hammons, A.R. (1996) Large families of quaternary sequences with low correlation, *IEEE Transactions on Information Theory*, 42(2), 579–592.

Lam, A.W. and Ozluturk, F.M. (1992) Performance bound for DS/SSMA communications with complex signature sequences, *IEEE Transactions on Communications*, 40(10), 1607–1614.

Oppermann, I. and Vucetic, B.S. (1997) Complex spreading sequences with a wide range of correlation properties, *IEEE Transactions on Communications*, 45(3), 365–375.

Oppermann, I. (1997) Orthogonal complex-valued spreading sequences with a wide range of correlation properties, *IEEE Transactions on Communications*, 45(11), 1379–1380.

Ozluturk, F.M., Tantaratana, S. and Lam, A.W. (1995) Performance of DS/SSMA Communications with MPSK signalling and complex signature sequences, *IEEE Transactions on Communications*, 43(2/3/4), Feb/Mar/April.

Peterson, R.L., Ziemer, R.E. and Borth, D.E. (1995) *Introduction to Spread-Spectrum Communications*, Prentice Hall Inc.

Proakis, J.G. (1995) *Digital Communications*, McGraw-Hill International Editions.

Sarwate, D.V. and Pursley, M.B. (1980) Cross-correlation properties of pseudorandom and related sequences, *Proceedings of the IEEE*, 68(5), 593–619.

Welch, L.R. (1974) Lower bounds on the maximum cross-correlation of signals, *IEEE Transactions on Information Theory*, IT-20, 397–399.

5

Time Synchronization of Spread-Spectrum Systems

5.1 Introduction

The reader may recall from the previous chapter that the data in the received spread-spectrum signal cannot be efficiently recovered by a conventional wideband receiver. This was because the amount of noise power input to the detector overwhelms the useful information causing total erroneous output. Consequently, reliable data detection can only be carried out after the received spread-spectrum is converted to its original narrowband equivalent.

The prime task of the receiver is therefore to generate a local replica of the received spreading code to re-modulate or de-spread the incoming signal. When the local code phase is time synchronized to the received code to within a fraction of chip code phase offset, the received spread-spectrum signal collapses in bandwidth and reverts to its original narrowband form and the conventional detection techniques can be applied to recover the data.

The process of synchronizing the local code and the received code is commonly achieved in two stages: initially, the two code signals are aligned in phase to uncertainty less than one chip duration through a process called *code acquisition* or coarse synchronization. In other words, the acquisition is aligning the unknown phase of the received code with the known phase of the local code generated at the receiver. The acquisition process is random and the randomness is due to many sources such as initial uncertainty about the code phase offset; channel fading and interference; unknown carrier phase and frequency offset such as Doppler; and the presence of additive white Gaussian noise.

Once the incoming code is acquired, a verification process attests the correct code phase which is continuously maintained by a closed loop tracking system. However, if for some reason the tracking system has gone out of lock, the acquisition system will be re-activated

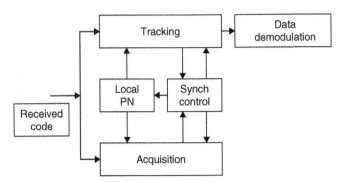

Figure 5.1   *Direct sequence spread-spectrum code time synchronization.*

in order to acquire the incoming code (Pickholtz et al., 1982) and the tracking system takes over again to maintain code synchronization as depicted in Figure 5.1.

In addition, the receiver local carrier must be synchronized to the received carrier for coherent demodulation. However, the frequency-time synchronization necessitates the search over a two dimension (2D) uncertainty region containing the frequency and code phase. Considering a typical code length and the spectrum width of the received signal, the 2D search process is likely to be very complex. The effects of frequency error on the acquisition process are discussed in Section 5.6. In general, as long as the initial carrier frequency error and phase offset are relatively small, the degradation to code synchronization is small and non-coherent code synchronization can be employed.

This chapter starts by considering the time synchronization of a single user spread-spectrum system operating in AWGN environment. The code phase acquisition of such a system is considered in more depth in Sections 5.2–5.6. A brief consideration of the optimum synchronizer is given in Section 5.2.1, followed by sub-optimum acquisition systems using both parallel and serial search in Sections 5.2.2 and 5.2.3. The performance of these systems is analysed in terms of the average and variance acquisition times for given probabilities of false alarm and detection and the analysis is presented in Section 5.3. Other means of acquisition, such as sequential (Section 5.4), and matched filtering (Section 5.5) are also included in this chapter. The effects of frequency errors on the code acquisition are presented in Section 5.6.

The tracking loops operating in the AWGN environment are considered in Sections 5.7 and 5.8 where we start with consideration of the optimum tracking loop in Section 5.7.1. After we have outlined the complexity of the optimum tracker, we move on to look at alternative sub-optimum tracking loops such as delay locked loops (Sections 5.7.2–5.7.7) and the τ-dither loops (Section 5.8). We have then looked at the problem of time synchronization (both acquisition and tracking) in mobile fading channels where timing errors are mainly due to multipath and multi-user interference. This subject is a hot research topic in both academia and industry.

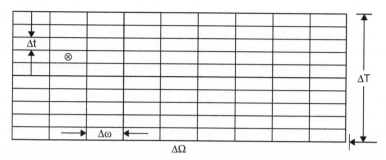

Figure 5.2 *Received signal frequency–code phase uncertainty region. ⊗ Denotes correct received frequency-phase.*

5.2 Code acquisition

5.2.1 Optimum acquisition

Consider the received signal which has carrier frequency with uncertainty $\Delta\Omega$ radians/second and has a received code delay (phase) with uncertainty of ΔT seconds. Let us subdivide the $\Delta\Omega$–ΔT region into a large number of uncertainty or cells of smaller regions ($\Delta\omega$–Δt) (Selin and Tuteur, 1963) as shown in Figure 5.2. For each uncertainty cell, the received signal is de-spread using the receiver local code with hypothesized code phase and carrier frequency. If the hypotheses are correct, the received signal collapses in bandwidth and the signal presence is detected using an energy detector at the de-spreader output. Under such circumstances, the synchronizer is said to acquire the received code phase.

However if the hypotheses are incorrect, a new phase-frequency cell in the $\Delta\Omega$–ΔT uncertainty region is tested. This method will be repeated until the system acquires the received code phase. Clearly, cells in the uncertainty region can be tested serially or in parallel. The synchronizer achieves *minimum acquisition time* when cells are tested in parallel but such a system is never practically implemented due to its excessive hardware requirement.

5.2.2 Sub-optimum acquisition system

Since the 2D search system is too complex to implement in spread-spectrum systems, we now look for a practical acquisition system which achieves sub-optimal performance through a compromise between acquisition time and implementation complexity. One likely possibility for reducing the hardware complexity is to assume a negligible frequency uncertainty. This assumption will develop the 2D search in Figure 5.2 into a simple 1D search system along the code phase as shown in Figure 5.3.

The spread-spectrum synchronization is extensively discussed in the available literature, with various systems of much simpler hardware complexity than the optimum synchronizer. The main differences between these systems are in the type of detectors used and in the search algorithms which act on the detectors' output to verify the code phase acquisition. The detectors can be grouped into either coherent or non-coherent according to the information available concerning the carrier phase offset. A block diagram for the synchronization of a coherent detector used for QPSK system is shown in Figure 5.4.

Non-coherent detectors are most widely used in a mobile environment characterized by Doppler effects and multipath propagation since accurately estimating the carrier frequency error and the carrier phase offset are quite difficult in such environment. The de-spreading of the received signal usually takes place prior to the carrier synchronization. Basic detectors typically comprise bandpass filters centred at the nominal carrier frequency, followed by a square law envelop detector, an integrate-dump circuit, and a decision device.

When carrier frequency error and the carrier phase offset are both accurately known, then a coherent detector can be used in the search system. This typically consists of a lowpass filter implemented as integrate and dump circuit followed by a simple decision device or Bayes detector.

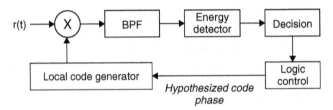

Figure 5.3   *Sub-optimum acquisition system.*

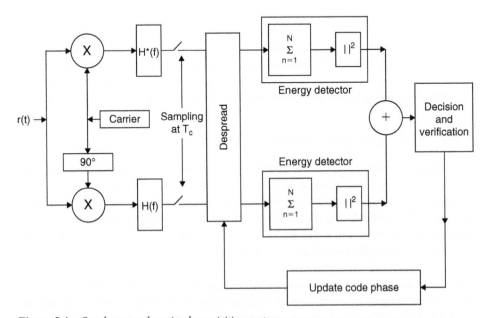

Figure 5.4   *Quadrature sub-optimal acquisition system.*

The integration time used during the detection can be a single fixed integration (dwell) time, multiple dwell time or variable dwell time. Furthermore, acquisition decisions in the fixed single dwell detectors are made on the basis of single fixed time observation of the received signal utilizing partial or full code period duration. The multiple dwell detectors make their decisions based on many observations with various verification techniques that can immediately reject code phase positions as soon as any of the dwell outputs fail below the threshold. Otherwise, decisions are based on majority logic on all the threshold tests. The variable dwell time detectors are similar to the single dwell detector, although the integration time is not fixed but is continuously variable – as in the sequential detection systems which will be discussed in more detail in Section 5.4.

The correlation detectors used in the acquisition systems can relatively reduce system hardware complexity and improve the decision rate. In the serial search, these detectors carry out *active* correlation where the continuously running local code waveform is correlated with the received code waveform and integrated for an integer number of chip duration. Consequently, the active correlation detectors suffer from a basic limitation on the search rate since the local code phase is updated only at an integer number of chip intervals. On the other hand, the search rate can be significantly increased (up to chip rate) when *passive* correlation, such as a matched filter detector, is used. The matched filter acquisition system will be discussed in more detail in Section 5.5.

5.2.3 Search strategies

Conventionally, either serial or parallel search algorithms are employed to search the uncertainty region and to acquire the code phase. In the serial search algorithm, all possible code phases are tested one by one (i.e. serially) as shown in Figure 5.5 to verify the acquisition of the correct code phase where T_i is the integration time. The average time for the serial search to acquire the correct code phase is relatively large and a common scheme used to shorten this acquisition time is to use a two-stage detection system. The first stage is set to a low threshold and short integration time so that a relatively large incorrect code phase can be rejected abruptly. The second stage is designed with an appropriate dwell time and proper decision threshold to reduce the overall acquisition time.

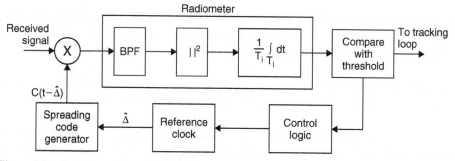

Figure 5.5 *Serial search loop block diagram.*

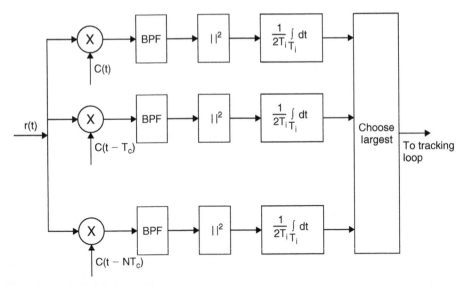

Figure 5.6   *Parallel search circuit.*

The parallel search algorithm tests all possible code phases simultaneously (as shown in Figure 5.6) resulting in a much smaller average acquisition time compared to serial search algorithm. However, the complexity of the parallel search is higher than the serial search system.

All acquisition systems, irrespective of the type of detectors and search algorithms used, are susceptible to various types of errors – the most dominant are the *false alarm errors* and the *miss errors*. The false alarm occurs when the detector output exceeds the threshold for an incorrect code phase, while the miss error occurs when the detector output falls below the threshold for a correct code phase. As a practical rule, the design of an efficient acquisition system is based on a compromise between *small average acquisition time* and *small false alarm probability* together with appropriate detection probability.

5.3   Analysis of serial acquisition system in AWGN channel

Analysis of a serial acquisition system can be carried out using two methods: the direct method, which employs *classical statistical approach*, and the circular state diagram, which is based on the *signal flow graph approach*.

We will focus on the statistical method since it gives a detailed insight of the performance analysis. However, we will first start with a brief description of the signal flow approach.

The application of a signal flow graph to the acquisition problem of the spread-spectrum systems dates back to the late 1970s (Holmes and Chen, 1977) but only developed in the early 1980s (Polydoros and Weber, 1984; Polydoros and Simon, 1984). Consider two

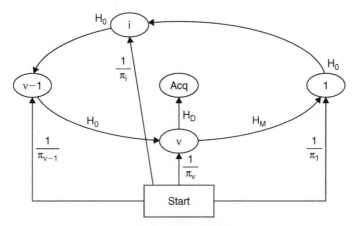

Figure 5.7a *The Circular state diagram of the flow graph for Serial search acquisition system.*

hypotheses: H_D denotes the phases of the received code sequence and local code sequences are misaligned by less than a code chip; H_0 denotes the alternative hypothesis, that the relative alignment is greater than a chip.

The acquisition system can be represented by a $v + 2$ state flow graph, where $v - 1$ of these states corresponds to the cells belonging to hypothesis H_0 while one state (the vth) corresponds to H_D. The remaining two states are the False Alarm (FA) state which can be reached from any of the $v - 1$ states corresponding to offset cells (H_0) and the correct Acquisition (ACQ) state which can be directly reached from the vth state (H_D). The v states are indexed in a circular arrangement as shown in Figure 5.7a where the ith state $(i = 1, 2, \ldots, v - 1)$ corresponding to the ith offset code position to the right of the correct synchronization position H_D.

The search process occurs at any one of the v states in the code phase uncertainty region. The probability of selecting state j is π_j for $j = , 1, 2, \ldots, v$. For total uncertainty, a uniform distribution is commonly assumed, i.e. $\pi_j = \frac{1}{v}$ for $j = 1, 2, \ldots, v$. The worst case occurs when the search starts with $\pi_1 = 1$ and $\pi_j = 0$ for $j \neq 1$ corresponding to an initial relative code phase location furthest from the correct synchronization position. The search starts at the 'start' state.

Let us assume the receiver selects code phase i^{th} state and uses the search loop in Figure 5.5 to evaluate that code phase. The evaluation process dismisses the code phase as incorrect or a false alarm occurs. In the case of dismissal, the search moves to evaluate the code phase in the $(i + 1)^{th}$ state. In the false alarm case, the search also moves to search the $(i + 1)^{th}$ state with a delay incurred by the false alarm detection. The two possible events between the ith state and the $(i + 1)^{th}$ state are represented by function $H_0(z)$ in Figure 5.7b. The integration time is T_1 and probability of dismissal of the incorrect code phase $(1 - P_{fa})$. Assume the penalty for false alarm be T_{fa} so that the time for a false alarm is $T_1 + T_{fa}$.

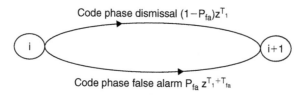

Figure 5.7b   *Single dwell state transition.*

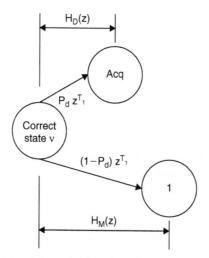

Figure 5.7c   *State transition at correct state.*

The function $H_0(z)$ for single dwell detector

$$H_0(z) = (1 - P_{fa})z^{T_1} + P_{fa}z^{T_1+T_{fa}}$$

This search process is repeated in each code phase within the uncertainty region until it reaches the correct phase. The branch between the correct state and 'Acq' state is represented by function $H_D(z)$ while $H_M(z)$ represents the path from the correct state to state 1. If the detector fails to detect the correct code phase at state v the search advances to state 1 as shown in Figure 5.7c, the search process is repeated and the search may continue indefinitely. Clearly the state diagram is circular.

We now determine the mean acquisition time by defining the function $H(z)$, which represents all paths from the start state to the 'Acq' state (Peterson et al., 1995).

$$H(z) = \sum_{i=1}^{v} H_i(z) \tag{5.1}$$

Now all paths that search for the correct code phase start with branch labelled π_i includes $v - i$ branches from the ith state to the vth state are:

$$h_{i0}(z) = \pi_i[H_0(z)]^{v-i}H_D(z)$$

The path that corresponds to a single miss detection is:

$$h_{i1}(z) = \pi_i[H_0(z)]^{\nu-i}H_D(z)\{H_M(z)[H_0(z)]^{\nu-1}\}$$

For k missed detections

$$h_{ik}(z) = \pi_i[H_0(z)]^{\nu-i}H_D(z)\{H_M(z)[H_0(z)]^{\nu-1}\}^k$$

Now
$$H_i(z) = \sum_{k}^{\infty} h_{ik}(z) \tag{5.2}$$

$$H_i(z) = \pi_i H_D(z)[H_0(z)]^{\nu-i} \sum_{k=0}^{\infty} [H_0(z)]^{(\nu-i)k}[H_M(z)]^k$$

But
$$\sum_{k=0}^{\infty} [H_0(z)]^{(\nu-i)k}[H_M(z)]^k = \frac{1}{1 - H_M(z)[H_0(z)]^{\nu-i}}$$

Therefore,
$$H_i(z) = \pi_i H_D(z)[H_0(z)]^{\nu-i}\frac{1}{1 - H_M(z)[H_0(z)]^{\nu-i}} \tag{5.3}$$

Consequently the desired function H(z) is:

$$H(z) = \frac{H_D(z)}{1 - H_M(z)[H_0(z)]^{\nu-1}} \sum_{i=1}^{\nu} \pi_i[H_0(z)]^{\nu-i} \tag{5.4}$$

The mean acquisition time can be expressed as:

$$E\{T_{acq}\} = \overline{T}_{acq} = \frac{dH(z)}{dz}\bigg|_{z=1} \tag{5.5}$$

5.3.1 Statistical analysis of the mean and variance acquisition time for serial acquisition system

The purpose of this analysis is to derive closed form mathematical representations for the false alarm and the detection probabilities to compute the average and variance of the acquisition time for the specific synchronization system.

We now consider the generic serial acquisition system shown in Figure 5.3 in the analysis with a 1D uncertainty search region for the correct code phase. In this search scheme, the search system evaluates the phase cells in the uncertainty region one after the other until the cell which achieves the correct code acquisition is found. The time taken by the system to find the correct cell, the acquisition time T_{acq}, is a random parameter with a mean $\overline{T}_{acq}$ and variance σ_{acq}^2. Obviously, we are interested in minimizing the overall mean acquisition time $\overline{T}_{acq}$ which depends on the probability of detection (P_d) and the probability of false alarm (P_{fa}) in the test; dwell (integration) time in cell T_i, and the SNR of the received spread-spectrum signal. In the sequel, we assume that the frequency error is insignificant; so that no frequency uncertainty exists. At a later stage, we investigate the degradation in $\overline{T}_{acq}$ when frequency error is taken into account.

Let us proceed with the serial search of the phase uncertainty region that contains N_u chips represented by N cells with the correct cell having an index n. Furthermore, let us assume that the acquisition is completed after j missed detections of the correct cell, k false alarms in all incorrect cells. We will associate a penalty time of T_{fa} seconds ($T_{fa} \gg T_i$) with a false alarm. The penalty time associated with a miss is NT_i since, if we encounter a miss, the correct phase cannot be detected until the next cycle. The total number of cells tested for this event is $(n + j N)$ and the total acquisition time for this event defined by $T_{acq}(n, j, k)$ is given by:

$$T_{acq}(n, j, k) = (n + jN)T_i + k\,T_{fa} \tag{5.6}$$

The number of correct cells is $(j + 1)$ and the number of incorrect cells is K which is the penalty caused by a false alarm in the verification, as multiple of chip intervals i.e. $T_{fa} = K\,T_c$ and K is:

$$K = (n - 1) + j(N - 1) \tag{5.7}$$

Let the correct cells be distributed uniformly within the uncertainty so that the probability of the correct phase in the nth cell is $\left(\frac{1}{N}\right)$. The probability of j missed detections followed by a correct detection is:

$$P_d(1 - P_d)^j \tag{5.8}$$

The probability of k false alarms with the K incorrect cells is:

$$P_{fa}^k(1 - P_{fa})^{K-k} \tag{5.9}$$

Now there are $\binom{K}{k}$ arrangements of k false alarms in K cells. Thus the probability of the correct phase position at the nth cell with j misses and k false alarms $P(n, j, k)$ is:

$$P(n, j, k) = \frac{1}{N}\ \ P_d(1 - P_d)^j\ \ \binom{K}{k}\ \ P_{fa}^k(1 - P_{fa})^{K-k} \tag{5.10}$$

The mean acquisition time $\overline{T}_{acq}$ is:

$$\overline{T}_{acq} = \sum_{n=0}^{N-1}\sum_{j=0}^{\infty}\sum_{k=0}^{K-1} T_{acq}(n, j, k)\,P(n, j, k) \tag{5.11}$$

Substituting for $T_{acq}(n, j, k)$ given by (5.6) and $P(n, j, k)$ given in (5.10), it is shown in Peterson et al. (1995) that (5.11) can be simplified to:

$$\overline{T}_{acq} = (N - 1)(T_i + T_{fa}P_{fa})\left(\frac{2 - P_d}{2\,P_d}\right) + \frac{T_i}{P_d} \tag{5.12}$$

The variance of $T_{acq}(n, j, k)$ can be expressed as:

$$\sigma^2_{acq} = E(T^2_{acq}) - (\overline{T}_{acq})^2 \tag{5.13}$$

where

$$E(T^2_{acq}) = \sum_{n=0}^{N-1} \sum_{j=0}^{\infty} \sum_{k=0}^{K-1} T^2_{acq}(n, j, k) \, P(n, j, k) \tag{5.14}$$

It is shown in Peterson et al. (1995) that (5.13) can be simplified to:

$$\sigma^2_{acq} = \left[\frac{N^2 - 1}{12} - \frac{(N-1)^2}{P_d} + \frac{(N-1)^2}{P_d^2} \right] (T_i + T_{fa} P_{fa})^2$$

$$+ (2N - 1)\frac{1 - P_d}{P_d^2} T_i^2 + 2(N - 1)\frac{1 - P_d}{P_d^2} T_i T_{fa} P_{fa}$$

$$- (N - 1)\frac{2 - P_d}{2 P_d} T_{fa}^2 P_{fa}^2 + (N - 1)\frac{2 - P_d}{2 P_d} T_{fa}^2 P_{fa} \tag{5.15}$$

For a practical case, $N \gg 1$, and for a well-designed acquisition system, we can assume $(1 - P_d) \ll 1$, and $P_{fa} \ll 1$. With these assumptions, the variance can be expressed as:

$$\sigma^2_{acq} \approx (T_i + T_{fa} P_{fa})^2 N^2 \left(\frac{1}{12} - \frac{1}{P_d} + \frac{1}{P_d^2} \right) \tag{5.16}$$

For the ideal case when $P_d = 1$, $P_{fa} = 0$, and $N \gg 1$, the variance becomes:

$$\sigma^2_{acq} = \frac{(T_i N)^2}{12} \tag{5.17}$$

Equation (5.15) expresses the variance of a random variable uniformly distributed over the range $[0, (T_i + T_{fa} \, P_{fa})N]$.

Example 5.1

A BPSK direct sequence spread-spectrum system transmits data using a spreading code clock of 1.8822 MHz and the spreading code is a maximum length sequence generated using a shift register of length 6. Assume the carrier frequency is a priori known and the acquisition system uses a single dwell serial search in step of $0.5 T_c$. Given the probability of detection of $P_d = 0.75$, probability of false alarm $P_{fa} = 10^{-3}$ and the false alarm penalty $72 T_i$ where T_i is the fixed integration time. Calculate the:

 i. average acquisition time
 ii. variance of the acquisition time
iii. variance of the acquisition time given by (5.16)
iv. variance of the acquisition time for the ideal case.

Solution

i. The chip rate $= 1.8822$ c/s

The chip duration $$T_c = \frac{1}{1.8822 \times 10^6} = 0.53\,\mu s$$

Therefore $$T_i = \frac{T_c}{2} = 0.27\,\mu s$$

Assume the uncertainty region restricted to within the code length (N_u)

$$N_u = 2^6 - 1 = 63 \text{ chip}$$

Number of cells $= N = 2 \times 63 = 126$.

The false alarms penalty time $T_{fa} = 72 \times 0.27\,\mu s = 19.44\,\mu s$

Substituting in (5.12), we get

$$\bar{T}_{acq} = \frac{(N-1)\left(1 + \frac{T_{fa}}{T_i}P_{fa}\right)(2 - P_d) + 2}{2P_d}T_i$$

$$\bar{T}_{acq} = \frac{125\left(1 + \frac{19.44 \times 10^{-6}}{0.27 \times 10^{-6}}10^{-3}\right)(2 - 0.75) + 2}{2 \times 0.75} \times 0.27 \times 10^{-6}$$

$$\bar{T}_{acq} = \frac{125\left(1 + \frac{19.44}{0.27}10^{-3}\right) \times 1.25 + 2}{1.5} \times 0.27 \times 10^{-6}$$

$$\bar{T}_{acq} = 30.51\mu s$$

ii. The variance of the acquisition time is calculated as follows:

$$\sigma^2_{acq} = \left[\frac{N^2 - 1}{12} - \frac{(N-1)^2}{P_d} + \frac{(N-1)^2}{P_d^2}\right](T_i + T_{fa}P_{fa})^2$$

$$+ (2N - 1)\frac{1 - P_d}{P_d^2}T_i^2 + 2(N - 1)\frac{1 - P_d}{P_d^2}T_iT_{fa}P_{fa}$$

$$- (N - 1)\frac{2 - P_d}{2P_d}T_{fa}^2P_{fa}^2 + (N - 1)\frac{2 - P_d}{2P_d}T_{fa}^2P_{fa}$$

$$\sigma_{acq}^2 = \left[\frac{126^2 - 1}{12} - \frac{(126-1)^2}{0.75} + \frac{(126-1)^2}{0.75^2} \right] (0.27 * 10^{-6}$$

$$+ \, 19.44 * 10^{-6} \times 10^{-3})^2 + (2*126 - 1)\frac{1 - 0.75}{0.75^2}(0.27*10^{-6})^2$$

$$+ \, (2*126 - 1)\frac{1 - 0.75}{0.75^2}(0.27*10^{-6*}19.44*10^{-6} \times 10^{-3})$$

$$- \, (126 - 1)\frac{2 - 0.75}{2*0.75}(19.44*10^{-6})^{2*}(10^{-3})^2$$

$$+ \, (126 - 1)\frac{2 - 0.75}{2 \times 0.75}(19.44*10^{-6})^{2*}(10^{-3})$$

$$\sigma_{acq}^2 = 7.4064 \times 10^{-10}\text{s}^2$$

iii. We use the following expression

$$\sigma_{acq}^2 \approx (T_i + T_{fa}\,P_{fa})^2 N^2 \left(\frac{1}{12} - \frac{1}{P_d} + \frac{1}{P_d^2} \right)$$

Substituting in the above expression we get:

$$\sigma_{acq}^2 \approx (0.27 * 10^{-6} + 19.44 * 10^{-6} * 10^{-3})^2 (126)^2 \left(\frac{1}{12} - \frac{1}{0.75} + \frac{1}{(0.75)^2} \right)$$

$$\sigma_{acq}^2 \approx 7.0196 \times 10^{-10}\;\text{s}^2$$

iv. For the ideal case we use the following expression:

$$\sigma_{acq}^2 = \frac{(T_i N)^2}{12}$$

Substitute for the integration time and number of cells

$$\sigma_{acq}^2 = \frac{(0.27 \times 10^{-6} \times 126)^2}{12}$$

$$\sigma_{acq}^2 = 9.6447 \times 10^{-11}\;\text{s}^2$$

5.3.2 The Doppler effect on code acquisition

The presence of code Doppler affects the code rate and smears the relative code phase difference between received and locally generated codes during the acquisition dwell time of the integrator, which has the effect of increasing or reducing the probability of detection, depending on the code phase and the algebraic sign of the Doppler rate. The code Doppler also affects the effective code phase update. Clearly, when the code phase shift caused by the Doppler over a single dwell time is equal to the phase update, the average search rate is reduced to zero causing the search time to increase greatly.

In the following analysis we follow the treatment in Holmes and Chen (1977). Let Δf denote the code Doppler in chips/sec. The mean phase update in chips μ' is the summation of: the search step size in fractions of a chip $\left(\frac{\Delta T_c}{T_c}\right)$ in the absence of Doppler, the code phase shift during the dwelling time $(\Delta f\, T_i)$, and the code phase shift during the hit verification caused by false alarm $(\Delta f\, T_{fa}\, P_{fa})$ so that:

$$\mu' = \frac{\Delta T_c}{T_c} + \Delta f\, T_i + \Delta f\, T_{fa}P_{fa}$$
$$= \mu + \Delta f[T_i + T_{fa}P_{fa}] \tag{5.18}$$

where $\mu = \frac{\Delta T_c}{T_c}$, typical value for $\mu \approx 0.5$ chip. For $N \gg 1$, we can be approximate (5.12) to:

$$\overline{T}_{acq}\,|\text{no code Doppler} \approx \frac{\left(1 + \dfrac{T_{fa}}{T_i}P_{fa}\right)(2 - P_d)}{2P_d}NT_i \tag{5.19}$$

Now substituting $N = \frac{N_\mu}{\mu}$ in equation (5.19), we get:

$$\overline{T}_{acq}\,|\text{no code Doppler} = \frac{\left(1 + \dfrac{T_{fa}}{T_i}P_{fa}\right)(2 - P_d)}{2P_d\mu}N.uT_i \tag{5.20}$$

Consider the acquisition process when code Doppler is present and substitute μ' for μ in (5.20):

$$\overline{T}_{acq}\,|\text{code Doppler present} = \frac{\left(1 + \dfrac{T_{fa}}{T_i}P_{fa}\right)(2 - P_d)}{2P_d(\mu + \Delta f[T_i + T_{fa}P_{fa}])}N_\mu T_i$$

Substitute for N_μ

$$\overline{T}_{acq}\,|\text{code Doppler present} = \frac{\left(1 + \dfrac{T_{fa}}{T_i}P_{fa}\right)(2 - P_d)}{2P_d(\mu + \Delta f[T_i + T_{fa}P_{fa}])}N_\mu T_i$$

Simplify the above expression

$$\overline{T}_{acq}\,|\text{code Doppler present} = \frac{\left(1 + \dfrac{T_{fa}}{T_i}P_{fa}\right)(2 - P_d)}{2P_d\left(1 + \dfrac{\Delta f[T_i + T_{fa}P_{fa}]}{\mu}\right)}NT_i$$

Thus the mean acquisition time in the presence of code Doppler is:

$$\overline{T}_{acq}\,|\text{code Doppler present} = \frac{\overline{T}_{acq}\,|\text{no code Doppler}}{\left(1 + \dfrac{\Delta f[T_i + T_{fa}P_{fa}]}{\mu}\right)} \tag{5.21}$$

When code Doppler is present, the variance of the acquisition time can be obtained from (5.16) by substituting $N = \frac{N_u}{\mu'}$:

$$\sigma_{acq}^2 \,|\text{code Doppler present} \approx (T_i + T_{fa}P_{fa})^2 \frac{N_u^2}{\mu'^2}\left(\frac{1}{12} - \frac{1}{P_d} + \frac{1}{P_d^2}\right) \qquad (5.22)$$

Substituting for μ' in (5.22):

$$\sigma_{acq}^2 \,|\text{code Doppler present} \approx (T_i + T_{fa}P_{fa})^2 \frac{N_u^2}{\left[\mu + \Delta f T_i\left(1 + \frac{T_{fa}}{T_i}P_{fa}\right)\right]^2}$$

$$\times \left(\frac{1}{12} - \frac{1}{P_d} + \frac{1}{P_d^2}\right) \qquad (5.23)$$

Simplifying (5.23):

$$\sigma_{acq}^2 \,|\text{code Dopler present} \approx (T_i + T_{fa}P_{fa})^2 \frac{N_u^2}{\mu^2\left[1 + \frac{\Delta f T_i}{\mu}\left(1 + \frac{T_{fa}}{T_i}P_{fa}\right)\right]^2}$$

$$\times \left(\frac{1}{12} - \frac{1}{P_d} + \frac{1}{P_d^2}\right)$$

Thus the variance of the acquisition time in the presence of code Doppler is (Simon et al., 1985):

$$\sigma_{acq}^2 \,|\text{code Doppler present} \approx \frac{\sigma_{acq}^2 \,|\text{no code Doppler}}{\left[1 + \frac{\Delta f T_i}{\mu}\left(1 + \frac{T_{fa}}{T_i}P_{fa}\right)\right]^2} \qquad (5.24)$$

Clearly, since Δf can be either positive or negative, the code Doppler can either speed up or slow down the acquisition search.

Example 5.2

Reconsider the spread-spectrum in Example 5.1 where the received code delay uncertainty is $\pm 11.157\,\mu s$. The system is subjected to code Doppler of 4 kc/s. Calculate each of the following:

i. mean acquisition time without code Doppler
ii. mean acquisition time with code Doppler
iii. variance of the acquisition time without code Doppler
iv. variance of the acquisition time with code Doppler.

Solution

i. Let the uncertainty time be Δt and step size be ΔT

$$\Delta t = \pm 11.157\,\mu s$$

$$\Delta T = 0.5 T_c = 0.27\,\mu s$$

$$N_u = \frac{\Delta t}{T_c} = \Delta t R_c = 22.314\,\mu s \times 1.8822\,\text{Mc/s} = 42\ \text{chips}$$

Number of cells to be tested $= N = 42 \times 2 = 84$

$$\overline{T}_{acq}\,|\text{no code Doppler} = (N-1)(T_i + T_{fa}P_{fa})\left(\frac{2-P_d}{2\,P_d}\right) + \frac{T_i}{P_d}$$

$$= (84-1)(0.27 \times 10^{-6} + 19.44 \times 10^{-6} \times 10^{-3})$$

$$\times \left(\frac{2-0.75}{1.5}\right) + \frac{0.27 \times 10^{-6}}{0.75}$$

$$= 83 \times (0.29 \times 10^{-6})(0.83) + 0.36 \times 10^{-6}$$

$$= 20.3381\,\mu s.$$

ii. The mean acquisition time with Doppler is:

$$\frac{\overline{T}_{acq}\,|\text{no code Doppler}}{\overline{\overline{T}}_{acq}\,|\text{code Doppler present}} = \left(1 + \frac{\Delta f[T_i + T_{fa}\,P_{fa}]}{\mu}\right)$$

$$\mu = \frac{0.5 T_c}{T_c}$$

Substitute values for parameters in the above expression:

$$\left(1 + \frac{\Delta f[T_i + T_{fa}P_{fa}]}{\mu}\right) = \left(1 + \frac{4 \times 10^3 \times [0.27 \times 10^{-6} + 19.44 \times 10^{-6} \times 10^{-3}]}{0.5}\right)$$

$$\frac{\overline{T}_{acq}\,|\text{no code Doppler}}{\overline{\overline{T}}_{acq}\,|\text{code Doppler present}} = 1.0023$$

$$\overline{T}_{acq}|\text{code Doppler present} = \frac{\overline{T}_{acq}\,|\text{no code Doppler}}{1.0023} = \frac{20.338\,\mu s}{1.0023} = 20.29\,\mu s$$

iii. Variance of the acquisition time without code Doppler is:

$$\sigma_{acq}^2 \approx (T_i + T_{fa}P_{fa})^2 N^2 \left(\frac{1}{12} - \frac{1}{P_d} + \frac{1}{P_d^2}\right)$$

$$= (0.27 \times 10^{-6} + 19.44 * 10^{-6} * 10^{-3})^2 * 84^2 * \left(\frac{1}{12} - \frac{1}{0.75} + \frac{1}{0.75^2}\right)$$

$$= 3.1198 * 10^{-10}\,s^2$$

iv. Variance of the acquisition time with code Doppler is:

$$\frac{\sigma_{acq}^2 | \text{no code Doppler}}{\sigma_{acq}^2 | \text{code Doppler present}} = \left[1 + \frac{\Delta f\, T_i}{\mu}\left(1 + \frac{T_{fa}}{T_i}p_{fa}\right)\right]^2$$

$$\frac{\sigma_{acq}^2 | \text{no code Doppler}}{\sigma_{acq}^2 | \text{code Doppler present}} = \left[1 + \frac{4 \times 10^3 * 0.27 \times 10^{-6}}{0.5}\left(1 + \frac{19.44*10^{-6}}{0.27*10^{-6}} \times 10^{-3}\right)\right]^2$$

$$= 1.0046$$

Thus $\sigma_{acq|\text{code Doppler present}}^2 = 3.1198 * 10^{-10}\text{s}^2 / 1.0046 = 3.1055 * 10^{-10}\ \text{s}^2.$

5.3.3 Probabilities of detection and false alarm

Consider the generic serial acquisition system shown in Figure 5.5. The radiometer detects the received energy in a bandwidth $B \approx \frac{1}{T}$ where T is data symbol interval by squaring the amplitude of output of the BPF and integrating the output of the squarer over time duration T_i. The decision statistic, z, is given by:

$$z = \frac{1}{T_i}\int_0^{T_i} \left|[r(t)C^*(t - \hat{\Delta})]\right|^2 dt \qquad (5.25)$$

Where $\hat{\Delta}$ is an integer multiple of the chip duration T_c. Assuming no data modulation is being received by the acquisition system and S is signal power, the received signal is:

$$r(t) = \sqrt{2S}\, C(t - \Delta) + n(t) \qquad (5.26)$$

Substituting (5.26) in (5.25), we get:

$$z = \frac{1}{T_i}\left|\int_0^{T_i}\left[\sqrt{2S}\, C(t - \Delta) + n(t)C^*(t - \hat{\Delta})\right]\right|^2 dt$$

$$z = \frac{1}{T_i}\left|\int_0^{T_i}\left[\sqrt{2S}\, C(t - \Delta)C^*(t - \hat{\Delta})\right] + [n(t)C^*(t - \hat{\Delta})]\right|^2 dt \qquad (5.27)$$

If $\hat{\Delta} \neq \Delta$, a very small power passes through the bandpass filter since the bandwidth of the filter is $B << \frac{1}{T_c}$. The detector indicates the absence of the signal in this test. However, when $\hat{\Delta} \approx \Delta$, i.e. the hypothesized code phase matches the actual phase, the spectrum bandwidth collapses and most of the signal power passes through the filter and hence the energy detector indicates the presence of the signal. The power spectrum of $C(t - \Delta)C^*(t - \Delta)$ contains an impulse with power $R_{auto}^2(\tau)$ where $R_{auto}^2(\tau)$ is the autocorrelation of the spreading sequence.

We denote $[n(t)C * (t - \hat{\Delta})] = \tilde{n}(t)$ which is zero mean Gaussian process with power spectral density $(PSD) = N_0$ for $|f| < \frac{B}{2}$ and zero otherwise. Equation (5.27) now simplifies as:

$$z = \frac{1}{T_i} \left| \int_0^{T_i} \sqrt{2S} + \tilde{n}(t) \right|^2 dt \tag{5.28}$$

Considering the hypotheses H_0 and H_1, the levels of the decision statistic z are:

$$H_0 : z = \frac{1}{T_i} \int_0^{T_i} |\tilde{n}(t)|^2 dt$$

$$H_D : z = \frac{1}{T_i} \int_0^{T_i} \left| \sqrt{2S} + \tilde{n}(t) \right|^2 dt \tag{5.29}$$

Let us now set the decision threshold to γ so that:

if $z > \gamma$, decide H_1, otherwise if $z \leq \gamma$, decide H_0.

Let the output of the squarer in the radiometer be y(t). When signal is present at the squarer output, y(t) has a non-central chi-squared pdf $(P(y)_{nc})$ with two degrees of freedom (Lee and Miller, 1998; Ward, 1965)

$$P(y)_{nc} = \frac{1}{2\sigma^2} \exp\left[- \left(SNR + \frac{y}{2\sigma^2} \right) \right] I_0\left(2\sqrt{\frac{ySNR}{2\sigma^2}} \right) \tag{5.30}$$

Where $I_0(x)$ is the zeroth order modified Bessel function of the first kind. In the absence of signal, the output y(t) distribution reduces to the central chi-squared pdf $(P(y)_c)$ with two degrees of freedom where $I_0(0) = 1$

$$P(y)_c = \frac{1}{2\sigma^2} \exp\left[-\frac{y}{2\sigma^2} \right] \tag{5.31}$$

The integrator output is sampled at T intervals so that $t = k\,T$. The sampled output can be approximated by a summation over these sampled values:

$$z = \frac{1}{T_i} \int_0^{T_i} y(t)dt \approx \frac{1}{T_i'} \sum_{k=0}^{T_i'-1} y(kT) \tag{5.32}$$

where $\quad T_i' = \frac{T_i}{T}$.

We now normalize the pdf with respect to $(2\sigma^2)$ so that normalized y(kT) is $y_k' = \frac{y(kT)}{2\sigma^2}$ and for large T_i', we can express (5.30) as:

$$P(y_k')_{nc} = \exp\left[- (SNR + y_k') \right] I_0(2\sqrt{SNRy_k'}) \tag{5.33}$$

Similarly (5.31) becomes:

$$P(y_k')_c = \exp[-y_k'] \tag{5.34}$$

Now let us make:

$$z' = \sum_{k=0}^{T_i'-1} y_k' \quad \text{so that} \quad z' = z\frac{T_i'}{2\sigma^2}.$$

For large T_i', we can approximate z' by Gaussian distributed with:

$$E(z') = T_i'E(y')$$
$$\text{Var}(z') = \sigma_z^2 = T_i'\sigma_{y'}^2$$

If the signal is present (Simon et al., 1995),

$$E(y') = 1 + \text{SNR}$$
$$\text{Var}(y') = \sigma_{y'}^2 = 1 + 2\text{SNR}$$

If the signal is absent,

$$E(y') = 1$$
$$\text{Var}(y') = \sigma_{y'}^2 = 1$$

Thus if the signal is present,

$$E(z') = T_i'[1 + \text{SNR}] \tag{5.35}$$

$$\text{Var}(z') = \sigma_{z'}^2 = T_i'[1 + 2\text{SNR}] \tag{5.36}$$

Similarly if signal is absent,

$$E(z') = T_i' \tag{5.37}$$

$$\text{Var}(z') = \sigma_{z'}^2 = T_i' \tag{5.38}$$

Using Gaussian assumption for z' when signal is absent, the probability of false alarm is:

$$P_{fa} = \int_{\gamma'}^{\infty} \frac{1}{\sqrt{2\pi T_i'}} \exp\left[-\frac{(z'-T_i')^2}{2T_i'}\right] dz' = Q\left(\frac{\gamma'-T_i'}{\sqrt{T_i'}}\right) = Q(\beta) \tag{5.39}$$

where $Q(\beta)$ is the complementary cumulative probability distribution function of a zero mean, unit variance Gaussian random variable. The normalized threshold γ' is:

$$\gamma' = \gamma\frac{T_i'}{2\sigma^2}$$

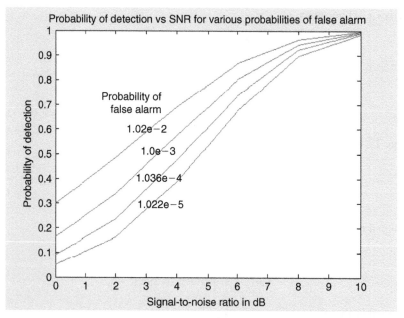

Figure 5.8 *Probability of detection vs SNR in dB for various probabilities of false alarm.*

The variable β is:

$$\beta = \frac{\gamma' - T_i'}{\sqrt{T_i'}}$$

Similarly, the detection probability is:

$$P_d = \int_{\gamma'}^{\infty} \frac{1}{\sqrt{2\pi T_i'(1 + 2SNR)}} \exp\left[-\frac{(z' - T_i'(1 + 2SNR))^2}{2T_i'(1 + 2SNR)}\right] dz'$$

$$P_d = Q\left(\frac{\beta - SNR\sqrt{T_i'}}{\sqrt{1 + 2\,SNR}}\right)$$

$$(5.40)$$

The miss probability P_m is:

$$P_m = 1 - P_d$$

The Receiver Operating Characteristic (ROC) of the radiometer is a plot showing the relationship between the detection probability P_d and the false alarm probability P_{fa} as shown in Figure 5.8. Clearly, as the SNR increases the probability of detection improves rapidly tending to unity for high SNR.

Example 5.3

A spread-spectrum acquisition system uses a serial search scheme. The synchronizer integrates the squarer output for interval of 0.1 s and the synchronizer output is sampled at intervals of 0.05 s. The probability of false alarm is 0.064255. Given that the received SNR is 3.5 dB, calculate the probability of detection.

Solution
From (5.39) the probability of false alarm is:

$$P_{fa} = Q(\beta) = 0.064255$$

Thus $\beta = 1.52$:

$$T_i' = \frac{T_i}{T} = \frac{0.1}{0.05} = 20$$

SNR $= 3.5$ db $= 2.24$

The probability of detect ion is given by (5.40):

$$P_d = Q\left(\frac{\beta - SNR\sqrt{T_i'}}{\sqrt{1 + 2SNR}}\right)$$

Substitute in the above expression we get

$$P_d = Q\left(\frac{1.52 - 2.24 \times \sqrt{20}}{\sqrt{1 + 2 \times 2.24}}\right)$$

$P_d = 0.7591$

5.4 Sequential detection acquisition system

The search algorithm discussed in Section 5.3 acquires the correct code phase using single fixed integration for a given threshold level. Such an algorithm is incapable of quickly dismissing a false phase cell or extending the integration time during phase search in a given cell. Indeed, the algorithm does not make use of the additional information that could be available, such as: whether the threshold statistic is close to, greater or smaller than the threshold level.

Sequential detection suggests the decision in a given code phase cell by using two or more sequences in a successive order. Consider a detection process with two thresholds A and B such that A > B. If the decision statistic >A, the signal is declared present and the test ends; if the decision statistic <B, the signal is declared absent and the test also ends. However, if B< decision statistic <A, no decision is made about the presence or absence of the signal and the test continues by extending the integration time.

The development of single dwell detection into sequential algorithm is similar in concept to the evolution of a hard decision into a soft decision decoding in digital signalling. The similarity is clear when one considers the fixed integration time with a single threshold level used in the search algorithm in Section 5.3, with the single threshold level in the hard decision decoding on one hand and the extended integration time with the multiple threshold levels in the sequential detection with multiple quantization levels in the soft decision decoding on the other. The comparison of data decoding and search algorithms can be extended to system performance. While soft decision decoding improves bit error rate in data transmission, the sequential detection improvement is evident in shorter mean acquisition times.

In sequential detection, the integration time is increased in discrete steps until the test fails and the false phase position is dismissed in a short time. However, the latter process is usually accompanied by a longer verification process. To explain how the sequential detection reduces the acquisition time, let us consider the sequential detection shown in Figure 5.9 (Simon et al., 1985) which operates like a single dwell detector except for a negative bias (b) and a negative threshold Γ.

The detector dismisses the incorrect code phase position whenever the average integrator output $\overline{Z}(t)$ falls below Γ. Consider signal plus noise input to the detector, the integrator output:

$$Z(t) = \int_0^t [(s(t) + n(t))^2 - b]dt \tag{5.41}$$

Expand the integrand in (5.41):

$$Z(t) = \int_0^t [s^2(t) + 2s(t)n(t) + n^2(t) - b]dt$$

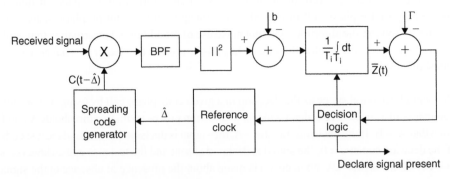

Figure 5.9 *Sequential detector acquisition system.*

Tasking the time average of $Z(t)$:

$$\overline{Z}(t) = \int_0^t [E(s^2(t)) + E(n^2(t)) - b]dt \qquad (5.42)$$

The average of $[s(t) \times n(t)]$ is zero since signal is uncorrelated with noise. The average signal and noise powers are:

$$E(n^2(t)) = N = N_0B \qquad (5.43)$$

$$E(s^2(t)) = S \qquad (5.44)$$

S and N are signal and noise power, respectively. Let the average signal-to-noise ratio be γ:

$$\gamma = \frac{S}{N} \qquad (5.45)$$

Therefore, for signal plus noise input, the average integrator output:

$$\overline{Z}(t) = N_0B(1 + \gamma)t - bt \qquad (5.46)$$

For noise only input, the integrator output:

$$\overline{Z}_n(t) = (N_0B - b)t \qquad (5.47)$$

We choose the bias b to be between the integrator signal plus noise output and noise only output such that:

$$N_0B < b < N_0B(1 + \gamma) \qquad (5.48)$$

The average integrator output $\overline{Z}(t)$ is plotted in Figure 5.10 where you can easily see that $\overline{Z}_n(t)$ will always be negative for noise only input and the detector can quickly reject the noise only input and speed up the acquisition process assuming high signal-to-noise ratio.

Clearly, choosing bias (b) according to (5.48) causes the integrator output to increase linearly when a signal is present and deceases linearly when a signal is absent. If we choose threshold Γ around zero volt, the code phase cell is quickly dismissed when integrator output is below Γ. The negative side of this algorithm, however, is that the detector dismisses the true code phase when the signal is corrupted by *severe fading distortion* such that the signal-to-noise ratio is extremely low.

Consider a sequential detection acquisition system with double threshold comparison (Γ_1, Γ_2) such that $\Gamma_1 > \Gamma_2$. The probabilities of detection and false alarm are given in Simon et al. (1985).

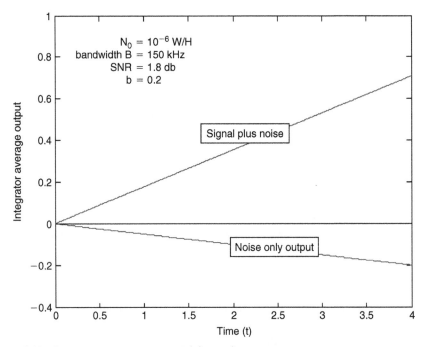

Figure 5.10   *Integrator average output with bias voltage present.*

$$P_d \approx \frac{e^{\Gamma_1} - e^{\Gamma_1 + \Gamma_2}}{e^{\Gamma_1} - e^{\Gamma_2}} \tag{5.49}$$

$$P_{fa} \approx \frac{1 - e^{\Gamma_2}}{e^{\Gamma_1} - e^{\Gamma_2}} \tag{5.50}$$

If we are given P_d and P_{fa}, we can determine the two threshold levels:

$$\Gamma_1 = \ln \frac{P_d}{P_{fa}} \tag{5.51}$$

$$\Gamma_2 = \ln \frac{1 - P_d}{1 - P_{fa}} \tag{5.52}$$

Example 5.4

A sequential detection acquisition system double threshold is designed so that the probability of false alarm $P_{fa} = 10^{-2}$ and the probability of detection is 0.85.

i. calculate the threshold levels Γ_1 and Γ_2.
ii. keeping threshold Γ_2 fixed but increasing Γ_1 to 6.5, what would be the probabilities of detection and false?

Solution

i. Using (5.51) and (5.52)

$$\Gamma_1 = \ln \frac{P_d}{P_{fa}} = \ln \frac{0.85}{10^{-2}} = \ln 85 = 4.4427$$

$$\Gamma_2 = \ln \frac{1 - P_d}{1 - P_{fa}} = \ln \frac{1 - 0.85}{1 - 0.01} = \ln 0.1515 = -1.8872$$

ii. $\Gamma_1 = \ln \dfrac{P_d}{P_{fa}} = 6.5$

$$\Gamma_2 = \ln \frac{1 - P_d}{1 - P_{fa}} = -1.8972$$

Thus
$$e^{6.5} = \frac{P_d}{P_{fa}} = 665.1416$$
$$e^{-1.8972} = \frac{1 - P_d}{1 - P_{fa}} = 0.1500$$

$$P_d = 665.1416 P_{fa}$$

$$1 - 665.1416 P_{fa} = 0.15(1 - P_{fa})$$

Therefore $\quad P_{fa} = 1.3 \times 10^{-3}$

$$P_d = 0.8647$$

5.5 Matched filter acquisition system

In serial search, the received spreading code sequence plus noise is multiplied by continuously running local reference spreading code sequence and, after the removal of the possible modulation using square envelope detection, the output is integrated to make an acquisition decision. The process leading to the acquisition test is known as *active correlation*. Consequently, a *new* set of $\frac{T_i}{T_c}$ chips from the reference code is used in each acquisition test. This means that, if the test fails, the code phase is updated only every T_i-second intervals. The search rate can be significantly increased by using a matched filter.

The reader can recall from Chapter 2 (Section 2.2.6) that matched filtering is basically passive correlation which maximizes the signal-to-noise ratio at its output when the input signal is embedded in additive white Gaussian noise. The received signal continuously slides the *stationary* (stored) spreading code $C(t)$ until the two code sequences are in synchronism. The output of the matched filter is applied to the input of the square law envelope detector and tested against a threshold. The maximum output occurs when the system acquires the correct code phase. The matched filter acquisition system is shown in Figure 5.11 when a perfect coherent system is used.

The received spreading code plus noise is sampled and each sample is digitized to one bit. The digitized samples are compared with the reference code sequence stage by stage

Figure 5.11 *Baseband matched filter acquisition system.*

generating a '+1' if the two stages match and a '−1' if they don't. Summing the resulting set of '+1' and '−1' generates the code phase decision.

We proceed with the analysis by setting the threshold at λ. As before, we denote the output of the square law detection by Z so that if $Z > \lambda$, the hypothesized code phase $\hat{\delta}$ matches the received code phase δ to within $\pm\frac{T_c}{2}$ (hypothesis H_D). On the other hand, if $Z \leq \lambda$, then hypothesized phase is incorrect (hypothesis H_0). Therefore, we are essentially dealing with a binary hypothesis problem.

$$\text{Hypothesis } H_0: \hat{\delta} \neq \delta \tag{5.53}$$

$$\text{Hypothesis } H_D: \hat{\delta} \approx \delta \tag{5.54}$$

Furthermore, we assume that under hypothesis H_0 the signal contribution to Z is zero, so that:

$$H_0: Z = |\tilde{n}(t)|^2 \tag{5.55}$$

$$H_D: Z = |\sqrt{2S} + \tilde{n}(t)|^2 \tag{5.56}$$

The false alarm occurs with probability P_{fa} given by:

$$P_{fa} = \Pr(Z > \lambda \,|\text{no signal present}) \tag{5.57}$$

Similarly, the miss probability is given by:

$$P_m = \Pr(Z \leq \lambda \,|\text{signal present}) \tag{5.58}$$

Now when there is no signal present (hypothesis H_0), Z is a random variable with central chi-square distribution such that:

$$P_{fa} = \exp\left(-\frac{\lambda}{2\sigma^2}\right) \tag{5.59}$$

where σ^2 is the noise variance at the output of the matched filter. When signal is received (hypothesis H_D), Z is non-central chi-square distributed (Proakis, 1995; Lee and Miller, 1998) so that probability of detection, and for a single test P_d, is given by:

$$P_d \leq Q\left(2\sqrt{N_i\frac{E_c}{N_0}}, \sqrt{-2\ln P_{fa}}\right) \tag{5.60}$$

where E_c is the signal energy per chip, N_0 is the noise spectral density, and $N_i = \frac{T_i}{T_c} =$ the number of code chips integrated. The double argument function in the right-hand side

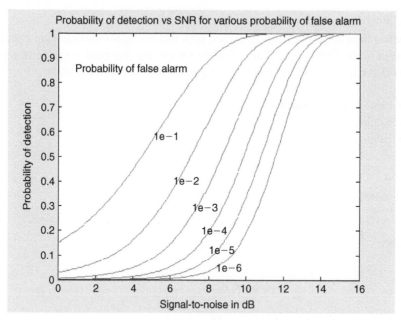

Figure 5.12 *Probability of detection vs SNR for various probability of false alarm.*

of (5.60) is called the *Marcum's Q-function*. The Marcum's Q-function should not be confused with the Gaussian Q-function shown in (5.39). The Marcum's Q-function has been tabulated, but the table is not widely available so it is useful to use the approximation in terms of the Gaussian Q-function.

$$
\begin{aligned}
Q(\alpha, \beta) &\approx Q\left(\sqrt{\beta^2 - \frac{1}{2}} - \sqrt{\alpha^2 + \frac{1}{2}}\right) && \text{for } \beta^2 > \frac{1}{2} \\
&\approx Q(\beta - \alpha) && \text{for } \alpha \gg 1, \beta \gg 1
\end{aligned}
\tag{5.61}
$$

The Marcum's Q-function is given by the following integral:

$$
Q(\alpha, \beta) = \int_{\beta}^{\infty} x e^{-\frac{x^2 + \alpha^2}{2}} I_0(\alpha x) dx
\tag{5.62}
$$

From (5.62), the miss probability is given by:

$$
P_m = 1 - P_d
\tag{5.63}
$$

The Receiver Operating Characteristic (ROC) curves of P_d verses P_{fa} for various values of signal-to-noise ratio are shown in Figure 5.12.

Example 5.5

A matched filter acquisition system is used such that the signal-to-noise ratio per chip at the output of the filter is 3 dB. Given that the ratio of the threshold to the noise variance 13, the chip rate is 1.2288 Mc/s and the integration time is 10 μs, calculate:

  i. probability of false alarm
 ii. probability of detection
iii. probability of miss

Solution

Signal-to-noise ratio per chip $= 3\,\text{dB} = 2 = 2\dfrac{E_c}{N_0}$

Therefore $\dfrac{E_c}{N_0} = 1$

$$T_c = \frac{1}{1.2288 \times 10^6} = 8.1380 \times 10^{-7}\,\text{s}$$

$$N_i = \frac{T_i}{T_c} = \frac{10^{-5}}{8.1380 \times 10^{-7}} = 12.2880\,\text{chip}$$

  i. $P_{fa} = \exp\left(-\dfrac{\lambda}{2\sigma^2}\right) = \exp\left(-\dfrac{13}{2}\right) = 1.5 \times 10^{-3}$

 ii. $P_d \leq Q\left(2\sqrt{N_i\dfrac{E_c}{N_0}},\ \sqrt{-2\ln P_{fa}}\right)$

$\alpha = 2\sqrt{N_i\dfrac{E_c}{N_0}} = 2\sqrt{12.288} = 7.0108$

$\beta = \sqrt{-2\ln P_{fa}} = \sqrt{-2 \times \ln(1.5 \times 10^{-3})} = 3.6062$

$$Q(\alpha, \beta) \approx Q\left(\sqrt{\beta^2 - \frac{1}{2}} - \sqrt{\alpha^2 + \frac{1}{2}}\right) \quad \text{for } \beta^2 > \frac{1}{2}$$

Therefore $P_d \leq Q\left(2\sqrt{N_i\dfrac{E_c}{N_0}},\ \sqrt{-2\ln P_{fa}}\right) \approx Q\left(\sqrt{3.6062^2 - \dfrac{1}{2}} - \sqrt{7.0108^2 + 0.5}\right)$

$$= Q(-3.5102) = 1 - Q(3.5102)$$

$$= 0.9998$$

iii. probability of miss $P_m = 1 - P_d = 2 \times 10^{-4}$

5.6 Effects of frequency errors on the acquisition detector performance

So far we have considered the effects of code Doppler on the acquisition. However, a frequency offset (Δf_c) may exist between receiver carrier and the locally generated carrier which affects the acquisition. This frequency offset originates by the instabilities of the oscillators in the transmitter/receiver circuits.

Consider a matched filter acquisition system. The degradation (loss ratio) in the signal-to-noise ratio at the output of the matched filter due to this frequency error can be expressed as L_D:

$$L_D = \frac{1}{\sin c^2(\Delta f_c T_i)} \tag{5.64}$$

The effect of frequency offset can be taken into account by modifying the noise variance σ^2.

$$\sigma_D^2 = L_D \sigma^2 \tag{5.65}$$

Hence

$$\sigma_D^2 = \frac{\sigma^2}{\sin c^2(\Delta f_c T_i)} \tag{5.66}$$

Taking the frequency error into account, the new probability of a false alarm is given by (5.39) and (5.59) after substituting σ_D^2 for σ^2, the new probability of detection is given (5.40) and (5.60) after substituting σ_D^2 for σ^2, as well as a new value of SNR which can be approximated as follows.

The received signal-to-noise ratio at the MF output when there is a frequency offset $=$ $SNR|_{new}$ where:

$$SNR|_{new} = \frac{E_b}{\frac{N_0}{2}} = \frac{E_b}{\sigma_D^2} \tag{5.67}$$

Substituting for σ_D^2 in (5.67)

$$SNR|_{new} = \frac{E_b}{\sigma^2}(\sin c(\Delta f_c T_i))^2 \tag{5.68}$$

$$SNR|_{new} = SNR(\sin c(\Delta f_c T_i))^2 \tag{5.69}$$

Example 5.6

Consider the matched filter acquisition system in Example 5.5. If the instability of the frequency of the oscillators in the transmitter and the receiver causes frequency error of 12750 Hz calculate the new:

  i. probability of false alarm
 ii. probability of detection
iii. probability of miss

Solution

The loss ratio $L_D = \dfrac{1}{\sin c^2(\Delta f_c T_i)} = \dfrac{1}{\sin c^2(12750 * 10^{-5})} = 1.0552$

Therefore $\sigma_D^2 = L_D\sigma^2 = 1.0552\,\sigma^2$

$$\text{SNR}|_{\text{new}} = \text{SNR}(\sin c(\Delta f_c T_i))^2 = \text{SNR} \times \frac{1}{L_D} = 0.9477\,\text{SNR}$$

i. The new probability of false alarm is given by

$$(P_{\text{fa}})_{\text{new}} = \exp\left(-\frac{\lambda}{2\sigma_D^2}\right) = \exp\left(-\frac{\lambda}{2L_D\sigma^2}\right) = \exp\left(-\frac{\lambda}{2\sigma^2}\right)^{\frac{1}{L_D}}$$

Therefore $(P_{\text{fa}})_{\text{new}} = (P_{\text{fa}})^{\frac{1}{L_D}} = (1.5 \times 10^{-3})^{0.9477} = 2.1 \times 10^{-3}$

ii. The new probability of detection is calculated as follows

$\dfrac{E_c}{N_0} = \dfrac{E_b}{NN_0} = \dfrac{1}{N}\dfrac{\text{SNR}}{2}$, where N is the number of chips per spreading code sequence and spreading is achieved by whole sequence per data symbol.

$$\alpha_D^2 = 4N_i\left(\frac{E_c}{N_0}\right)_D = 4N_i\frac{1}{N}\frac{(\text{SNR})_D}{2}$$

$$\alpha^2 = 4N_i\left(\frac{E_c}{N_0}\right) = 4N_i\frac{1}{N}\frac{(\text{SNR})}{2}$$

Therefore
$$\frac{\alpha_D^2}{\alpha^2} = \frac{(\text{SNR})_D}{\text{SNR}} = 0.9477$$

$$\alpha_D^2 = 0.9477\alpha^2 = 46.5807$$

$$\beta_D^2 = -2\ln(P_{\text{fa}})_D = -2\ln(2.1 \times 10^{-3}) = 10.8478$$

Thus $\quad Q(\alpha, \beta) \approx Q\left(\sqrt{\beta^2 - \frac{1}{2}} - \sqrt{\alpha^2 + \frac{1}{2}}\right) = Q(-3.6447) = 1 - Q(3.6447) = 0.9999$

iii. Probability of miss $P_m = 1 - 0.9999 = 10^{-4}$

5.7 Code tracking in AWGN channels

Having acquired the received code phase within less than one chip, the receiver has to track any changes in the code phase using code phase tracking loops. Code phase tracking loops are identical in operation to the conventional phase locked loops used for carrier phase tracking. The only difference in operation is that code tracking loops track the *timing delay error* between the received code and the locally generated code while the conventional tracking loops track the *phase error* between the received carrier and the reference carrier generated locally. Tracking the delay errors is based on the correlation between the received code and two different replicas of the received code: one is an *early version* and the other is *late version* of the locally generated code.

Code tracking loops can be grouped in several ways: coherent loops that make use of an available carrier phase and non-coherent loops that do not. Loops that use two independent correlators are known as full-time early–late loops and loops that share a single correlator known as Tau-dither or τ-dither early–late loops.

The aim of the tracking loops is to achieve reasonably low tracking delay jitter. We start by considering various spread-spectrum tracking loops operating in the presence of AWGN in Sections 5.7 and 5.8 and develop our analysis to take care of multipath fading and multi-user interference in CDMA systems in Section 5.9. First, however, we start with the optimum code tracker in the next section.

5.7.1 Optimum code tracking

It is well-known by radar specialists that an accurate estimate of delay, between transmitted signal and the return signal reflected from a target, is obtained by matched filtering the received signal and a locally generated reference signal. The matched filter is described in Chapter 2, Section 2.2.6. The technique is *optimum* in AWGN channel that the matched filter maximizes the signal-to-noise ratio at its output and, consequently, the measurement is the maximum likelihood estimate of the delay.

This technique is extended by Spilker and Magill (1961) in the early 1960s for arbitrary wideband signal, such as direct sequence spread-spectrum corrupted by additive Gaussian noise. The proposed optimum code tracking loop is shown in Figure 5.13. It consists of the multiplier, to form the product of the received code, and the derivative, with respect to time, of the locally generated code. The output of the multiplier is averaged by a lowpass filter to extract the dc component related to the delay error. The filter output is used to control the delay of the differentiated locally generated code waveform to maximize the cross-correlation between received and the locally generated code.

The optimum code tracking loop was originally proposed for radar applications. However, as pointed out in Spilker and Magill (1961), the maximum likelihood estimate requires a

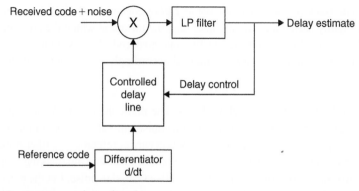

Figure 5.13 *Optimum code tracking loop.*

non-realizable loop filter. Furthermore, generating the impulse function for the derivative of the locally generated code reference is not an easy task and, therefore, the optimum tracker is not used in modern systems. Our interest in this type of code tracker is due to the fact that modern (sub-optimal) tracking loop arrangements are usually approximation of this loop.

Example 5.7

Consider an m-sequence of $N = 7$ representing the reference locally generated code

$$001 \; 0111$$

Denote the received code signal plus noise by $s(t - \delta_r)$ and the reference code by $s(t - \delta_\ell)$ where δ_r and δ_ℓ are the delays in the received and reference codes, respectively. Plot the:

  i. signal waveform $s(t - \delta_r)$
  ii. signal waveform $s(t - \delta_\ell)$
  iii. derivative of ii
  iv. multiplier output

Solution
We will assume that $\delta_\ell > \delta_r$

  i. The received signal waveform is shown in Figure 5.14a.
  ii. The reference code waveform is shown in Figure 5.14b.
  iii. The derivative of the reference code is ± 2 at the waveform transition points.
  iv. Since the normalized delay difference $\left(\frac{\delta_r - \delta_\ell}{T_c} \right)$ is < 0, the output of the multiplier
     $s(t - \delta_r) \frac{d}{dt}[s(t - \delta_\ell)]$ is positive and so as the input to the delay line and thus reducing
     the delay difference to zero.

5.7.2 Baseband early–late tracking loop

The early–late tracking loop, also known as *Delay-Locked Loop* (DLL), examines samples taken slightly earlier and slightly later than the instant at which the cross-correlation between the received code and the locally generated code is maximum. After comparing these instants, the code phase is adjusted accordingly (Davenport and Root, 1958; Meyr, 1976).

We first consider the coherent tracking loops. The non-coherent tracking loops are considered in Sections 5.7.5 to 5.7.7. The signal-to-noise ratio at the receiver input has to be high enough to permit the generation of a coherent carrier reference. Furthermore, we will assume the received signal contains *no data modulation* so that the tracking loop input is the code spreading waveform corrupted by additive Gaussian noise. Let the received code signal power and phase be P and t_d, respectively; $\hat{t}_d$ be the estimate of t_d and that both t_d

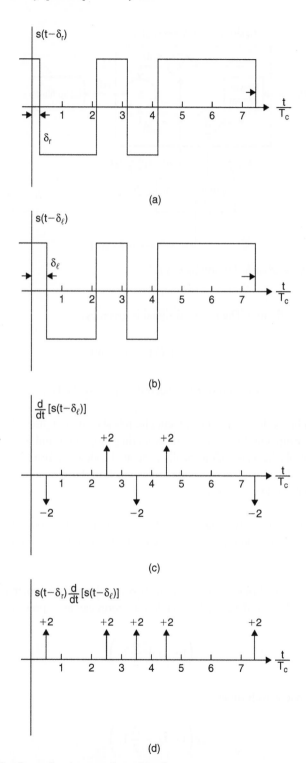

Figure 5.14 *Waveform of optimum code tracking loop.*

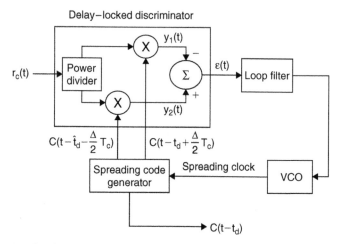

Figure 5.15   *Baseband early–late tracking loop.*

and $\hat{t}_d$ may vary with time. The received signal is given by:

$$r_c(t) = \sqrt{P}\, C(t - t_d) + n(t) \tag{5.70}$$

where n(t) is Gaussian white noise with spectral density $\frac{N_0}{2}$ W/Hz.

The conceptual block diagram of the tracking loop is shown in Figure 5.15 which consists of a phase discriminator followed by an averaging loop filter and a Voltage Oscillator (VCO) that controls the phase of the locally generated code waveform. The received signal $r_c(t)$ is power divided equally between the two channels of the discriminator and correlated with the early/late code waveforms.

Our analysis proceeds in two stages. First, we determine the *loop performance in a noise-free environment* and then we evaluate the *effects of the noise on the tracking loop performance*.

Assume the total time difference between the two correlators is Δ which is restricted to a range of $0 \le \Delta \le T_c$ then the early local code waveform can be expressed as:

$$c\left(t - \hat{t}_d + \frac{\Delta}{2}T_c\right) \tag{5.71}$$

The late local code waveform as:

$$c\left(t - \hat{t}_d - \frac{\Delta}{2}T_c\right) \tag{5.72}$$

5.7.3 Baseband early–late tracking loop in noiseless channels

Denote the delay error $(t_d - \hat{t}_d)$ normalized with respect to chip duration T_c as δ:

$$\delta = \frac{t_d - \hat{t}_d}{T_c} \tag{5.73}$$

Since the received power is divided equally between the two arms of the tracking loop, we divided the received signal by $\sqrt{2}$. The early correlator output, $y_1(t)$ is:

$$y_1(t) = \sqrt{\frac{P}{2}}C(t - t_d)\,C\left(t - \hat{t}_d + \frac{\Delta}{2}T_c\right) \tag{5.74}$$

and the late correlator output, $y_2(t)$ is:

$$y_2(t) = \sqrt{\frac{P}{2}}C(t - t_d)\,C\left(t - \hat{t}_d - \frac{\Delta}{2}T_c\right) \tag{5.75}$$

The discriminator output $\varepsilon\,(t)$ is:

$$\varepsilon(t) = y_2(t) - y_1(t)$$

Thus:

$$\varepsilon(t) = \sqrt{\frac{P}{2}}C(t - t_d)\,C\left(t - \hat{t}_d - \frac{\Delta}{2}T_c\right) - \sqrt{\frac{P}{2}}C(t - t_d)\,C\left(t - \hat{t}_d + \frac{\Delta}{2}T_c\right) \tag{5.76}$$

The error signal, $\varepsilon\,(t)$, is composed of a dc component which is used in the tracking loop, and a time carrying component called a *code self-noise*. The self-noise signal occurs at frequencies much higher than the relatively narrowband tracking loop and we can ignore it without jeopardizing the correctness of the analysis.

The dc component of the error signal $\varepsilon_{dc}\,(t)$ is given by the time average of $\varepsilon\,(t)$.

$$\varepsilon_{dc}(t) = \sqrt{\frac{P}{2}}D_\Delta(t_d, \hat{t}_d) \tag{5.77}$$

Therefore,

$$\sqrt{\frac{P}{2}}D_\Delta(t_d, \hat{t}_d) = \frac{1}{NT_c}\sqrt{\frac{P}{2}}\int\limits_{-\frac{NT_c}{2}}^{\frac{NT_c}{2}} C(t - t_d)\left[C\left(t - \hat{t}_d - \frac{\Delta}{2}T_c\right) - C\left(t - \hat{t}_d + \frac{\Delta}{2}T_c\right)\right]dt \tag{5.78}$$

where NT_c is period of the code waveform. Recall the definition of the autocorrelation of the code waveform in Section 3.6:

$$R_c(t_d - \hat{t}_d - \frac{\Delta}{2}T_c) = \frac{1}{NT_c}\sqrt{\frac{P}{2}} \int_{-\frac{NT_c}{2}}^{\frac{NT_c}{2}} C(t - t_d)\,C\left(t - \hat{t}_d - \frac{\Delta}{2}T_c\right)dt \qquad (5.79)$$

$$R_c(t_d - \hat{t}_d + \frac{\Delta}{2}T_c) = \frac{1}{NT_c}\sqrt{\frac{P}{2}} \int_{-\frac{NT_c}{2}}^{\frac{NT_c}{2}} C(t - t_d)\,C\left(t - \hat{t}_d + \frac{\Delta}{2}T_c\right)dt \qquad (5.80)$$

Thus, substituting (5.79) and (5.80) in (5.78) we get:

$$D_\Delta(t_d, \hat{t}_d) = R_c\left(t_d - \hat{t}_d - \frac{\Delta}{2}T_c\right) - R_c\left(t_d - \hat{t}_d + \frac{\Delta}{2}T_c\right)$$

Substituting (5.84) for δ in the above expression we get:

$$D_\Delta(\delta) = R_c\left(\delta - \frac{\Delta}{2}\right)T_c - R_c\left(\delta + \frac{\Delta}{2}\right)T_c \qquad (5.81)$$

The function $D_\Delta(\delta)$ is known as the *S-curve characteristics of the tracking loop*. It can be seen from the S-Curve that $D_\Delta(\delta)$ is linearly related to δ around $\delta = 0$. This region is commonly selected as the normal operating region for the tracking loop. The S-curve ($D_\Delta(\delta)$ vs. δ) is plotted in Figure 5.16, $\Delta = 0.5$, 1.5 for m-sequence of period $N = 31$.

The autocorrelation functions, $R_c(\tau)$, used in plotting these curves are:

$$R_c\left[\left(\delta - \frac{\Delta}{2}\right)T_c\right] = 1 - \left(\delta - \frac{\Delta}{2}\right)\left(1 + \frac{1}{N}\right) \quad \text{for } \frac{\Delta}{2} \le \delta \le 1 + \frac{\Delta}{2}$$

$$R_c\left[\left(\delta + \frac{\Delta}{2}\right)T_c\right] = 1 + \left(\delta - \frac{\Delta}{2}\right)\left(1 + \frac{1}{N}\right) \quad \text{for } -1 + \frac{\Delta}{2} \le \delta \le \frac{\Delta}{2}$$

$$R_c\left[\left(\delta - \frac{\Delta}{2}\right)T_c\right] = -\frac{1}{N} \quad \text{for } 1 + \frac{\Delta}{2} \le \delta \le (N - 1)$$

$$R_c\left[\left(\delta + \frac{\Delta}{2}\right)T_c\right] = -\frac{1}{N} \quad \text{for } -1 + \frac{\Delta}{2} \le \delta \le -(N - 1)$$

5.7.4   Baseband early–late tracking loop in AWGN channel

Following a similar analysis that leads to the error signal at the discriminator output $\varepsilon(t)$ in the previous section, we now add AWG noise $n'(t)$ to the loop analysis. The error signal at the discriminator output becomes:

$$\varepsilon(t) = \sqrt{\frac{P}{2}}\left[C(t - t_d)\,C\left(t - \hat{t}_d - \frac{\Delta}{2}T_c\right) - C(t - t_d)\,C\left(t - \hat{t}_d + \frac{\Delta}{2}T_c\right)\right] + \left[\frac{1}{\sqrt{P}}n'(t)\right]$$

$$(5.82)$$

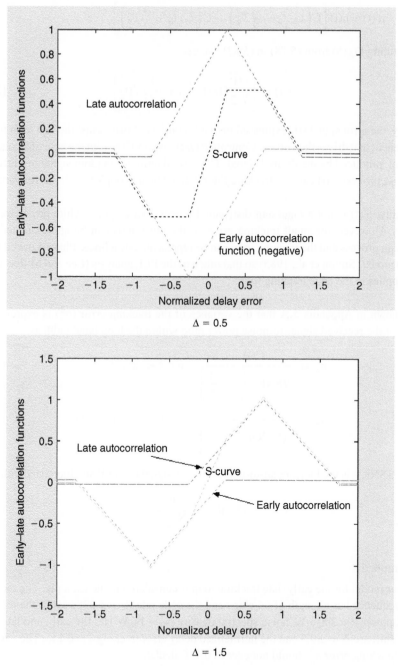

Figure 5.16 *S-curve characteristics for m-sequence with period N = 31.*

where $\quad n'(t) = n(t)\left[C\left(t - \hat{t}_d - \dfrac{\Delta}{2}T_c\right) - C\left(t - \hat{t}_d + \dfrac{\Delta}{2}T_c\right)\right]$

Substituting $D_\Delta(\delta)$ from (5.78) in (5.82), we get:

$$\epsilon(t) = \left[\sqrt{\frac{P}{2}}D_\Delta(t_d, \hat{t}_d) + \frac{1}{\sqrt{P}}n'(t)\right] \tag{5.83}$$

Clearly, the error signal $\epsilon(t)$ expressed in (5.83) consists of two terms: the first term is a dc component that drives the Voltage Controlled Oscillator (VCO) as in the case of a noiseless channel, produced by the desired signal. The second term is a random signal produced by the noise process $n'(t)$ causing tracking jitter defined by the loop's tracking error variance.

Normally, $n'(t)$ is not a Gaussian distributed process but acquires white power spectral density. However, for small tracking error, δ, the noise $n'(t)$ can be approximated as a Gaussian process and the tracking loop can be represented by a linear Phase-Locked Loop (PLL) model (Simon et al., 1985) so that much of the PLL analysis (Lee and Miller, 1998) also applies to the code tracking loop.

It is shown in Appendix 5.A that the variance of the tracking error (σ_δ^2) is expressed in terms of the received signal-to-noise power ratio within the loop bandwidth as:

$$\sigma_\delta^2 = \frac{\Delta}{2(\text{SNR})_\ell\left[1 + \dfrac{1}{N}\right]} \quad \text{for } 0 < \Delta < 1 \tag{5.84}$$

$$\sigma_\delta^2 = \frac{1}{2(\text{SNR})_\ell\left[1 + \dfrac{1}{N}\right]} \quad \text{for } 1 \leq \Delta \leq (N-1) \tag{5.85}$$

Where $(\text{SNR})_\ell$ is the loop signal-to-noise ratio for signal power P and loop bandwidth B_ℓ:

$$(\text{SNR})_\ell = \frac{P}{\dfrac{N_0}{2}B_\ell} \tag{5.86}$$

Example 5.8

A linear model for the early–late tracking loop is considered in the example. The received code sequence has a length of 64 chips and the code rate $R_c = 1.2288\,\text{Mc/s}$, the received signal power $P = 200\,\text{mW}$, noise spectral density $N_0 = 1\,\mu\text{W/Hz}$, the early and late code phase shift are $\Delta = \pm 0.5$ and the normalized delay error tracking $\delta = 0.1T_c$. The variance of the tracking error σ_δ^2 should not exceed 0.1. Calculate:

   i. noise spectral density $S_n'(f)$
  ii. two-sided power spectral density $S_n''(f)$
 iii. slop of the S-curve at $\delta = 0\ D_\Delta(\delta)$
 iv. tracking loop signal-to-noise ratio $(\text{SNR})_\ell$

Solution

$$R_c = 1.2288 \, \text{Mc/s}$$

$$T_c = \frac{10^{-6}}{1.2288} = 0.8138 \, \mu s$$

$$N = 64$$

i. From (5.A20).

$$S_{n'}(f) = \Delta N_0 \left[1 + \frac{1}{N} \right] \quad \text{for } 0 < \Delta < 1$$

Substitute in the corresponding values:

$$S_{n'}(f) = 0.5 \times 10^{-6} \left(1 + \frac{1}{64} \right) = 5.0781 \times 10^{-7} \, \text{W/Hz}$$

ii. $S_{n''}(f) = \dfrac{\Delta N_0}{4P \left[1 + \dfrac{1}{N} \right]} \quad \text{for } 0 < \Delta < 1$

Substitute values in the above equation:

$$S_{n''}(f) = \frac{0.5 \times 10^{-6}}{4 \times 0.2 \times \left[1 + \dfrac{1}{64} \right]} = 6.1538 \times 10^{-7}$$

iii. The slop of the S-curve at $\delta = 0$ $D_\Delta(\delta)$ is given by (5.A5)

$$D_\Delta(\delta) = 2 \left(1 + \frac{1}{N} \right) \delta \quad 0 < \Delta < 1$$

$$D_\Delta(\delta) = 2 \left(1 + \frac{1}{64} \right) \times 0.1 \times 0.8138 \times 10^{-6} = 1.6530 \times 10^{-7}$$

iv. The tracking loop signal-to-noise ratio $(\text{SNR})_\ell$ expressed in terms of the tracking error σ_δ^2 in (5.A26)

$$\sigma_\delta^2 = \frac{\Delta}{2(\text{SNR})_\ell \left[1 + \dfrac{1}{N} \right]} \quad \text{for } 0 < \Delta < 1$$

$$0.1 = \frac{0.5}{2(\text{SNR})_\ell \left[1 + \dfrac{1}{64} \right]}$$

$$(\text{SNR})_\ell = 2.4615 = 3.9120 \, \text{dB}$$

5.7.5 Noncoherent early–late tracking loop

The previous section considers the analysis of a coherent tracking loop when the loop is tracking a received signal with no data modulation (code sequence only). Furthermore, we assumed that a coherent carrier reference is available at the receiver to extract and correlate the received code with the locally generated code. Conventionally, the carrier reference is recovered from the received signal. However, considering the extremely low signal-to-noise ratio at which practical spread-spectrum system is operating, the recovery of this coherent reference is extremely difficult. Consequently, the reference carrier is locally generated. Furthermore, since there will most likely be a phase difference between the received and the local reference carrier, we genuinely expect the received signal input to the tracking loop to contain some data modulation. A simple phase discriminator that avoids these difficulties applies energy detection which is insensitive to carrier phase or data modulation. The non-coherent delay-track loop is illustrated in Figure 5.17.

The received signal from AWGN channel is data and spreading code modulated carrier:

$$r(t) = \sqrt{P}\, C(t - t_d) \cos\left[\omega_c t + \theta_d(t - t_d) + \phi\right] + n(t) \qquad (5.87)$$

where P, t_d, ω_c, $\theta_d(t)$, ϕ, and $n(t)$ are the received power, transmission delay, carrier radian frequency, carrier phase representing transmitted data, carrier random phase ϕ uniformly distributed over $(0, 2\pi)$, and zero mean white Gaussian noise process respectively. The band-limited received noise $n(t)$ has the two-sided power spectral density of $\frac{N_0}{2}$ W/Hz.

$$n(t) = \sqrt{2}[n_i(t) \cos \omega_c t - n_q(t) \sin \omega_c t]$$

where $n_i(t)$, $n_q(t)$ are the inphase and quadrature zero mean white Gaussian noise process. The two code waveforms are given by (5.88) and (5.89):

$$C_{early} = C\left(t - \hat{t}_d + \frac{\Delta}{2} T_c\right) \qquad (5.88)$$

$$C_{late} = C\left(t - \hat{t}_d - \frac{\Delta}{2} T_c\right) \qquad (5.89)$$

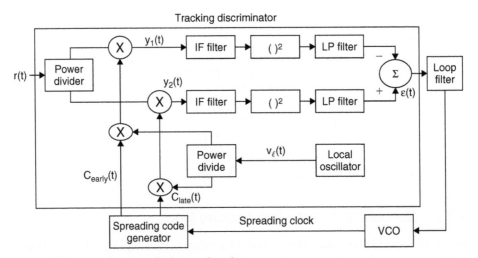

Figure 5.17   *Noncoherent early–late tracking loop.*

The reference local oscillator output is:

$$v_\ell(t) = 2\sqrt{2} \cos\left[(\omega_c - \omega_{IF})t + \phi'\right]$$

Therefore, assuming unity gain phase detectors, their outputs are:

$$y_1(t) = C\left(t - \hat{t}_d + \frac{\Delta}{2}T_c\right)\sqrt{2}\cos\left[(\omega_c - \omega_{IF})t + \phi'\right]$$

$$\times \left[\sqrt{\frac{P}{2}}C(t - t_d)\cos\left(\omega_c t + \theta(t - t_d) + \phi\right) + n(t)\right] \quad (5.90)$$

Similarly,

$$y_2(t) = C\left(t - \hat{t}_d - \frac{\Delta}{2}T_c\right)\sqrt{2}\cos\left[(\omega_c - \omega_{IF})t + \phi'\right]$$

$$\times \left[\sqrt{\frac{P}{2}}C(t - t_d)\cos\left(\omega_c t + \theta(t - t_d) + \phi\right) + n(t)\right] \quad (5.91)$$

5.7.6 Noncoherent early–late noiseless tracking loop

In this section, we analyse the noiseless tracking loop and assume that the bandwidth of the IF bandpass filters is such that the data passes undistorted. The product of the received code and the early–late modulated code waveforms are $\bar{y}_1(t)$ and $\bar{y}_2(t)$.

$$\bar{y}_1(t) = \sqrt{PE}\left\{C\left(t - \hat{t}_d + \frac{\Delta}{2}T_c\right)C(t - t_d)\right\}\cos\left[\omega_{IF}t + \theta_d(t - t_d) + (\phi - \phi')\right] \quad (5.92)$$

$$\bar{y}_2(t) = \sqrt{PE}\left\{C\left(t - \hat{t}_d - \frac{\Delta}{2}T_c\right)C(t - t_d)\right\}\cos\left[\omega_{IF}t + \theta_d(t - t_d) + (\phi - \phi')\right] \quad (5.93)$$

Using the definition of autocorrelation function of the code sequence, we can re-write (5.92) and (5.93) as:

$$\bar{y}_1(t) = \sqrt{P}\,R_c\left(\delta + \frac{\Delta}{2}\right)T_c\cos\left[\omega_{IF}t + \theta_d(t - t_d) + (\phi - \phi')\right] \quad (5.94)$$

$$\bar{y}_2(t) = \sqrt{P}\,R_c\left(\delta - \frac{\Delta}{2}\right)T_c\cos\left[\omega_{IF}t + \theta_d(t - t_d) + (\phi - \phi')\right] \quad (5.95)$$

The output of the non-coherent phase discriminator, $\varepsilon(\delta)$ is given by:

$$\varepsilon(t) = \left([\bar{y}_2(t)]^2 - [\bar{y}_1(t)]^2\right)_{\text{LPfiltered}} \quad (5.96)$$

Now $\quad [\bar{y}_2(t)]^2 = PR_c^2\left(\delta - \frac{\Delta}{2}\right)T_c\left(\cos\left[\omega_{IF}t + \theta_d(t - t_d) + (\phi - \phi')\right]\right)^2$

But $\quad (\cos\left[\omega_{IF}t + \theta_d(t - t_d) + (\phi - \phi')\right])^2 = \frac{1}{2}(\cos\left[2\omega_{IF}t + 2\theta_d(t - t_d) + 2(\phi - \phi')\right]) + \frac{1}{2}$

After LP filtering, it becomes:

$$(\cos [\omega_{IF}t + \theta_d(t - t_d) + (\phi - \phi')])^2 = \frac{1}{2}$$

Thus at LP filters:

$$[\overline{y}_2(t)]^2 = \frac{1}{2}PR_c^2\left(\delta - \frac{\Delta}{2}\right)T_c$$

$$[\overline{y}_1(t)]^2 = \frac{1}{2}PR_c^2\left(\delta + \frac{\Delta}{2}\right)T_c$$

Substituting for $\overline{y}_1(t)$ and $\overline{y}_2(t)$ in (5.96), we get:

$$\varepsilon(\delta) = \frac{P}{2}\left[R_c^2\left(\delta - \frac{\Delta}{2}\right)T_c - R_c^2\left(\delta + \frac{\Delta}{2}\right)T_c\right]$$

Thus,

$$\varepsilon(\delta) = \frac{P}{2}D_\Delta(\delta) \qquad (5.97)$$

Where

$$D_\Delta(\delta) = \left[R_c^2\left(\delta - \frac{\Delta}{2}\right)T_c - R_c^2\left(\delta + \frac{\Delta}{2}\right)T_c\right] \qquad (5.98)$$

Example 5.9

A plot of the S-curve for the non-coherent noiseless early–late tracking loop when $N = 31$ and $\Delta = 1.5$ is shown in Figure 5.18.

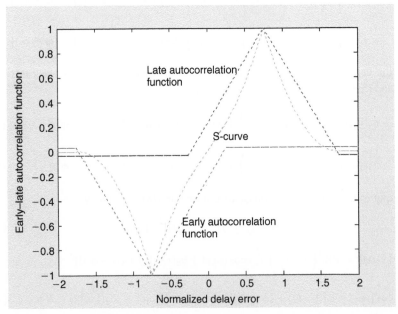

Figure 5.18   *S-curve for the noncoherent noiseless early–late tracking loop. $N = 31$ and $\Delta = 1.5$.*

5.7.7 Noncoherent early–late tracking loop in AWGN channel

Considering Figure 5.17, the noise input to the IF filter in the early arm of the discriminator is:

$$n_{early}(t) = c\left(t - \hat{T}_d + \frac{\Delta}{2}T_c\right)n(t) \tag{5.99}$$

Similarly, the noise input to the IF filter in the late arm of the discriminator is:

$$n_{late}(t) = c\left(t - \hat{T}_d - \frac{\Delta}{2}T_c\right)n(t) \tag{5.100}$$

where $n(t)$ is given by

$$n(t) = \sqrt{2}[n_i(t)\cos \omega_c t - n_q(t)\sin \omega_c t] \tag{5.101}$$

where $n_i(t)$, $n_q(t)$ are the inphase and quadrature zero mean white Gaussian noise process.

Define the noise output of the bandpass filters by $n'_1(t)$ and $n'_2(t)$ and the impulse response of the IF filters as $h_{bp}(t)$.

$$n'_1(t) = \int_{-\infty}^{\infty} n_{early}(\lambda)h_{bp}(t - \lambda)d\lambda$$

$$n'_1(t) = \int_{-\infty}^{\infty} c\left(\lambda - \hat{t}_d + \frac{\Delta}{2}T_c\right)n(\lambda)h_{bP}(t - \lambda)d\lambda \tag{5.102}$$

Similarly,

$$n'_2(t) = \int_{-\infty}^{\infty} c\left(\lambda - \hat{t}_d - \frac{\Delta}{2}T_c\right)n(\lambda)h_{bp}(t - \lambda)d\lambda \tag{5.103}$$

Since the received noise $n(t)$ *is a white Gaussian noise process* and when using linear filters, the process $n'_1(t)$ and $n'_2(t)$ are white Gaussian processes represented by:

$$n'_j(t) = \sqrt{2}\{n'_{jI}(t)\cos(\omega_{IF}t) - n'_{jQ}(t)\sin(\omega_{IF}t)\} \tag{5.104}$$

where $j = 1, 2$. The output of the phase discriminator is given by:

$$\varepsilon(t, \delta) = (\bar{y}_2(t) + n'_2(t))^2 - (\bar{y}_1(t) + n'_1(t))^2 \tag{5.105}$$

It is shown in Appendix B that the tracking jitter, σ_δ^2 is:

$$\sigma_\delta^2 = \frac{PN_0}{2} \left[\left(\left\{ 1 + \left(\delta - \frac{\Delta}{2} \right) \left(1 + \frac{1}{N} \right) \right\}^2 + \left\{ 1 - \left(\delta + \frac{\Delta}{2} \right) \left(1 + \frac{1}{N} \right) \right\}^2 \right) \right.$$

$$\left. + \frac{1}{2} [N_0]^2 B_n \right] \frac{1}{a_d^2} B_\ell \tag{5.106}$$

where

$$a_d = P \left(1 + \frac{1}{N} \right) \left[2 - \left(1 + \frac{1}{N} \right) \Delta \right] \tag{5.107}$$

Now when $N \gg 1$ and $\Delta = 1$, it can be shown that (5.106) and (5.107) reduces to:

$$a_d = P \tag{5.108}$$

$$\sigma_\delta^2 = \left\{ \frac{PN_0}{2} \left(\delta^2 + \frac{1}{2} \right) + \frac{1}{2} [N_0]^2 B_n \right\} \frac{1}{P^2} B_\ell \tag{5.109}$$

Since $\delta^2 \ll \dfrac{1}{2}$ $\hspace{4cm}$ (5.110)

Therefore $\hspace{1cm} \sigma_\delta^2 = \left\{ \frac{PN_0}{4} + \frac{1}{2} [N_0]^2 B_n \right\} \frac{1}{P^2} B_\ell = \frac{N_0}{4P} B_\ell \left[1 + 2 \frac{N_0}{P} B_n \right]$ $\hspace{0.5cm}$ (5.111)

Define the signal-to-noise ratio in the loop bandwidth $= \rho_\ell = \frac{2P}{N_0 B_\ell}$, and the signal-to-noise ratio in the IF filter bandwidth $= \rho_{bp} = \frac{P}{N_0 B_n}$. Therefore the tracking jitter simplified to:

$$\sigma_\delta^2 = \frac{1}{2\rho_\ell} \left[1 + \frac{2}{\rho_{bp}} \right] \tag{5.112}$$

Example 5.10

Consider a spread-spectrum communication system using a noncoherent early–late tracking loop having IF noise bandwidth of 10 kHz and loop filter's bandwidth 0.4 kHz. Assume $N \gg 1$ and $\Delta = 1$, calculate the variance of the tracking jitter, assuming a received power of 0.1 mW and a one-sided noise spectral density at the input of the tracking loop of 10^{-7} W/Hz.

Solution

$$\rho_\ell = \frac{2P}{N_0 B_\ell} = \frac{2 \times 0.1 \times 10^{-3}}{10^{-7} \times 0.4 \times 10^3} = 5$$

$$\rho_{bp} = \frac{P}{N_0 B_n} = \frac{0.1 \times 10^{-3}}{10^{-7} \times 10 \times 10^3} = 0.1$$

$$\sigma_\delta^2 = \frac{1}{2\rho_\ell} \left[1 + \frac{2}{\rho_{bp}} \right] = \frac{1}{2 \times 5} \left[1 + \frac{2}{0.1} \right] = 2.1V^2$$

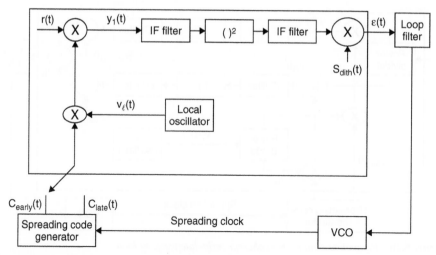

Figure 5.19 *τ-Dither noncoherent early–late tracking loop.*

5.8 τ-Dither early–late noncoherent tracking loop

The noncoherent early–late tracking loop, described in the previous section, is widely used to track code phase changes in spread-spectrum systems. However, it requires restricted signal amplitude balance in the two channels of the discriminator. Unequal gains that may exist in the two channels, could cause a discriminator *output offset* such that the output is not zero when the loop generates zero code phases tracking error. Furthermore, since the two channels are conceptually similar, it may be cost effective to *time share a single channel*. Indeed, these are the reasons for inventing the τ-dither tracking loop depicted in Figure 5.19.

The single correlator channel in Figure 5.19 is used alternately as the early correlator and late correlator using signal $s_{dith}(t)$ at a dithering (switching) frequency. The dithering frequency f_{dith} is too low, relative to the IF filter bandwidth, to cause no filter transience, but is significantly high, compared with the bandwidth of the loop filter, to ignore f_{dith} harmonics. The drawback of this refined solution is that there is a slight degradation in loop noise performance and is a very complex system to analyse.

We proceed with the analysis of the τ-dither loop using a similar method to that described in the previous section. We start the analysis by considering the desired signal only to define the S-curve of the tracking loop. We then introduce the noise into the loop and find the tracking jitter influence on the loop performance.

5.8.1 *Noncoherent τ-dither tracking loop in a noiseless channel*

The 2-channel discriminator loop equivalent to the τ-dither tracking loop is shown in Figure 5.20.

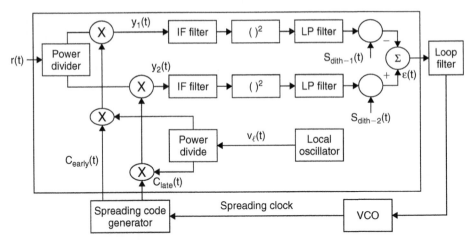

Figure 5.20 *Equivalent τ-dither noncoherent early–late tracking loop.*

The switching functions $s_{dith-1}(t)$ and $s_{dith-2}(t)$ can be expressed in terms of $s_{dith}(t)$ as follows:

$$s_{dith-1}(t) = 0.5 * [1 + s_{dith}(t)] \tag{5.113}$$

$$s_{dith-2}(t) = 0.5 * [1 - s_{dith}(t)] \tag{5.114}$$

The output of the correlators $\bar{y}_1(t)$ and $\bar{y}_2(t)$ given by equations (5.94) and (5.95) which are repeated here for convenience:

$$\bar{y}_1(t) = \sqrt{P}R_c\left(\delta + \frac{\Delta}{2}\right) T_c \cos[\omega_{IF}t + \theta_d(t - t_d) + (\phi - \phi')] \tag{5.94}$$

$$\bar{y}_2(t) = \sqrt{P}R_c\left(\delta - \frac{\Delta}{2}\right) T_c\cos[\omega_{IF}t + \theta_d(t - t_d) + (\phi - \phi')] \tag{5.95}$$

Assuming the IF filters introduce no distortion, the discriminator output signal is:

$$\varepsilon(t, \delta) = [\bar{y}_2^2(t) * s_{dith-2}(t) - \bar{y}_1^2(t) * s_{dith-1}(t)]_{LPF} \tag{5.113}$$

Substituting for $\bar{y}_1(t)$ and $\bar{y}_2(t)$ from (5.94) and (5.95) and eliminating all harmonics of IF frequency by lowpass filtering, it can be shown that:

$$\varepsilon(t, \delta) = \frac{P}{4}\left\{\left[R_c^2\left(\delta - \frac{\Delta}{2}\right) T_c - R_c^2\left(\delta + \frac{\Delta}{2}\right) T_c\right] \right.$$
$$\left. - s_{dith}(t)\left[R_c^2\left(\delta - \frac{\Delta}{2}\right) T_c - R_c^2\left(\delta - \frac{\Delta}{2}\right) T_c\right]_{LPF}\right\} \tag{5.114}$$

The first term is the desired tracking error and the second term consists of harmonics of the dithering frequency which are eliminated by the loop filter, so that:

$$\varepsilon(t, \delta) = \frac{P}{4}\left[R_c^2\left(\delta - \frac{\Delta}{2}\right) T_c - R_c^2\left(\delta + \frac{\Delta}{2}\right) T_c\right] \tag{5.115}$$

Substituting for $D_{nc.\Delta}(\delta)$ from (5.98) in (5.115), we get:

$$\varepsilon(t, \delta) = \frac{P}{4} [D_{nc.\Delta}(\delta)] \tag{5.116}$$

5.8.2 Noncoherent τ-dither tracking loop in AWGN channel

In the following analysis, we assume that the Gaussian white noise process with two-sided power density $\frac{N_0}{2}$ is present in the loop. The discriminator output signal is:

$$\varepsilon(t, \delta) = [\bar{y}_2(t) + n_2'(t)]_{LPf}^2 * s_{dith-2} - [\bar{y}_1(t) + n_1'(t)]_{LPF}^2 * s_{dith-1}(t) \tag{5.117}$$

Substituting for $\bar{y}_1(t)$ and $\bar{y}_2(t)$ from (5.94) and (5.95), $n_j'(t)$ from (5.104) (repeated here for convenience), and eliminating all harmonics of the IF frequency by lowpass filtering, $\varepsilon(t,\delta)$ in (5.117) becomes:

$$n_j'(t) = \sqrt{2}\{n_{jI}'(t)\cos(\omega_{IF}t) - n_{jQ}'(t)\sin(\omega_{IF}t)\} \tag{5.104}$$

$$\varepsilon(t, \delta) = [\bar{y}_2^2(t) + 2\bar{y}_2(t)n_2'(t) + n_2'^2(t)]_{LPF} * s_{dith-2}(t)$$
$$- [\bar{y}_1^2(t) + 2\bar{y}_1(t)n_1'(t) + n_1'^2(t)]_{LPF} * s_{dith-1}(t)$$

$$\varepsilon(t, \delta) = \left\{\frac{P}{2}R_c^2\left(\delta - \frac{\Delta}{2}\right)T_c + \sqrt{2P}\,R_c\left(\delta - \frac{\Delta}{2}\right)T_c\,[n_{2I}'(t)\cos(\theta_d(t - t_d) + (\phi - \phi'))\right.$$

$$\left. + n_{2Q}'(t)\sin(\theta_d(t - t_d) + (\phi - \phi'))] + n_{2I}'^2(t) + n_{2Q}'^2(t)\right\}^* s_{dith-2}$$

$$- \left\{\frac{P}{2}R_c^2\left(\delta + \frac{\Delta}{2}\right)T_c + \sqrt{2P}\,R_c\left(\delta + \frac{\Delta}{2}\right)T_c\,[n_{1I}'(t)\cos(\theta_d(t - t_d) + (\phi - \phi'))\right.$$

$$\left. + n_{1Q}'(t)\sin(\theta_d(t - t_d) + (\phi - \phi'))] + \frac{P}{2}\left\{R_c^2\left(\delta - \frac{\Delta}{2}\right)T_c\,{}^* s_{dith-2}(t)\right.\right.$$

$$\left.\left. - R_c^2\left(\delta + \frac{\Delta}{2}\right)T_c\,{}^* s_{dith-1}(t)\right\}n_{1I}'^2(t) + n_{1Q}'^2(t)\right\}^* s_{dith-1} \tag{5.118}$$

$$\varepsilon(t, \delta) = +\sqrt{2P}\left[n_{2I}'(t)R_c\left(\delta - \frac{\Delta}{2}\right)T_c\,{}^* s_{dith-2}(t) - n_{1I}'(t)R_c\left(\delta + \frac{\Delta}{2}\right)T_c\,{}^* s_{dith-1}(t)\right]$$

$$\{\cos((t - t_d) + (\phi - \phi'))\}$$

$$+\sqrt{2P}\left[n_{2Q}'(t)R_c\left(\delta - \frac{\Delta}{2}\right)T_c\,{}^* s_{dith-2}(t)\right.$$

$$\left. - n_{1Q}'(t)R_c\left(\delta + \frac{\Delta}{2}\right)T_c\,{}^* s_{dith-1}(t)\right]\sin((t - t_d) + (\phi - \phi'))\}$$

$$+ \left[n_{2I}'^2(t)\,{}^* s_{dith-2}(t) + n_{2Q}'^2(t)\,{}^* s_{dith-2}(t) - n_{1I}'^2(t)\,{}^* s_{dith-1}(t) - n_{1Q}'^2(t)\,{}^* s_{dith-1}(t)\right] \tag{5.119}$$

The error signal consists of two components, the desired error correction signal for the code tracking plus noise process that is assumed to be a lowpass white Gaussian process with two-sided power spectral density $\frac{\eta}{2}$ W/Hz. Assuming as before that the error code phase is small, then the τ-dither loop linear model is identical to that shown in Figure 5.A2.

Analysis leading to the phase jitter requires the evaluation of $\frac{\eta}{2}$ at the input of the linear model for the τ-dither tracking loop. The autocorrelation function $R_\varepsilon(\tau)$ which is given by:

$$R_\varepsilon(\tau) = E[\varepsilon(t, \delta)\varepsilon(t + \tau, \delta)] \tag{5.120}$$

where $\varepsilon(t, \delta)$ is now given by (5.119). As in the previous section, the Fourier transform of $R_\varepsilon(\tau)$ is the power spectrum of the error signal $\varepsilon(t, \delta)$. The expression of the power spectrum contains a term for the desired error signal for the tracking loop, the dither frequency harmonics plus (signal $\times$ noise) and (noise $\times$ noise) components. Taking into consideration the noise components of the power spectrum and counting for up to the third harmonic of the dither frequency and denoting the noise power spectral density at the input of the linear model be $S_{n''}(f)$, the power spectrum of the tracking jitter $S_\delta(f)$ is:

$$S_\delta(f) = |H(f)|^2\, S_{n''}(f) \tag{5.121}$$

The tracking jitter, σ_δ^2 is:

$$\sigma_\delta^2 = \int\limits_{-\infty}^{\infty} S_{n''}(f)\, |H(f)|^2\, df \tag{5.122}$$

where $n''(t)$ is the input noise to the linear model of the tracking loop is given by the signal $\times$ noise plus noise $\times$ noise terms in the Fourier transform of $R_\varepsilon(\tau)$. The input noise is assumed to be white Gaussian noise, that it is approximately flat over the loop bandwidth having a two-sided power spectral density $\frac{\eta}{2}$.

$$\sigma_\delta^2 = \frac{\eta}{2}\, B_\ell \tag{5.123}$$

It is shown in Peterson et al. (1995) that:

$$\sigma_\delta^2 \approx \frac{N_0 B\ell}{8P\left(1 - \dfrac{\Delta}{2}\right)^2} \left\{ R_c^2\left(\delta - \frac{\Delta}{2}\right) T_c + R_c^2\left(\delta + \frac{\Delta}{2}\right) T_c \right\} \left(1 + \frac{8}{\pi^2}\right)$$

$$+ \frac{N_0^2 B_n B\ell}{8P^2\left(1 - \dfrac{\Delta}{2}\right)^2} \left\{ 1 + \frac{8}{\pi^2}\left(1 - \frac{f_q}{B_n}\right) + \frac{8}{9\pi^2}\left(1 - \frac{3f_q}{B_n}\right) \right\} \tag{5.124}$$

Example 5.11

The τ-dither tracking system, which is equivalent to the noisy noncoherent tracking system in Example 5.10, is considered with dithering frequency of 2 kHz. Calculate the new variance of the tracking jitter.

Solution

When $N \gg 1$ and $\Delta = 1$, Equation (5.124) reduces to:

$$\sigma_\delta^2 \approx \frac{N_0 \, B\ell}{8P \left(1 - \frac{\Delta}{2}\right)^2} \left\{ R_c^2\left(\delta - \frac{\Delta}{2}\right) T_c + R_c^2\left(\delta + \frac{\Delta}{2}\right) T_c \right\} \left(1 + \frac{8}{\pi^2}\right)$$

$$+ \frac{N_0^2 B_n B\ell}{8P^2 \left(1 - \frac{\Delta}{2}\right)^2} \left\{ 1 + \frac{8}{\pi^2}\left(1 - \frac{f_q}{B_n}\right) + \frac{8}{9\pi^2}\left(1 - \frac{3f_q}{B_n}\right) \right\}$$

For $\delta \approx 0$

$$R_c\left(\delta - \frac{\Delta}{2}\right) T_c \approx \frac{1}{2}$$

$$R_c\left(\delta + \frac{\Delta}{2}\right) T_c \approx \frac{1}{2}$$

Substituting for B_n, B_ℓ, f_{dith} and $R_c \left(\delta \pm \frac{\Delta}{2}\right) T_c$, we get

$$\sigma_\delta^2 \approx \frac{10^{-7} \times 0.4 \times 10^3}{2 \times 0.1 \times 0^{-3}} \left\{ \frac{1}{4} + \frac{1}{4} \right\} \left(1 + \frac{8}{\pi^2}\right) + \frac{10^{-14} \times 10 \times 10^3 \times 0.4 \times 10^3}{2 \times (10^{-4})^2}$$

$$\times \left\{ 1 + \frac{8}{\pi^2}\left(1 - \frac{2 \times 10^3}{10 \times 10^3}\right) + \frac{8}{9\pi^2}\left(1 - \frac{3 \times 2 \times 10^3}{10 \times 10^3}\right) \right\}$$

$$\sigma_\delta^2 \approx 3.55 \, \text{V}^2$$

5.9 Time synchronization of spread-spectrum systems in mobile fading channels

In this section we deal with a time synchronization problem of direct sequence spread-spectrum system operating in *mobile multipath fading* channels. In the analysis of the problem we use the same approach as in Sections 5.2 and 5.7 by considering the acquisition loops first and then deal with the tracking issues.

5.9.1 Code acquisition in fading channels

The material presented so far has focused on providing code phase synchronization schemes for stationary spread-spectrum systems operating in AWGN channels. In such an environment, only a *single signal path* exists between transmitter and receiver, and the receiver acquires the correct code phase at a *single correct timing state* (H_D). For applications in mobile communications, multipath Rayleigh fading channels exist, and the fading-induced delay spread could be much higher than chip duration. Since search step is usually less than, or equal to, $\frac{T_c}{2}$, multi-resolvable paths within the uncertainty region exists, and hence *multiple correct timing states* (H_D).

Considering the multipath fading channels, the signal power is disseminated into a number of resolvable paths. As the number of paths increase, the signal-to-interference ratio decreases, making the decision on the received data hard. This is sometimes unreliable compared to a decision on a signal power total contained in a single path.

When the initial offset in the correct timing state (H_D) is equal *exactly to* $\frac{T_c}{2}$ and the multipath delay spread of the channel is T_m, the number of H_D sub-cells are $2\left(\text{integer}\left[\frac{T_m}{T_c}\right]\right) + 2$ and 3 in the *frequency-selective* and *non-selective* cases, respectively. However, when the initial offset in the H_1 region is *less than* $\frac{T_c}{2}$, the number of sub-cells is $2\left(\text{integer}\left[\frac{T_m}{T_c}\right]\right) + 1$ and 2 for the *frequency-selective* and *non-selective* cases, respectively. Consequently, for the L_p-path non-selective propagation channel with offset less than $\frac{T_c}{2}$, the total H_D cells are $2L_p$. A comprehensive treatment of the code acquisition in a multipath signal taking into account the direction (angle) of arrival of the received signal is presented in Katz et al. (2001).

The RAKE receiver concept is commonly utilized in the multipath signal propagation to enhance the signal reception through diversity at the receiver and, to amalgamate with this concept, a number of novel models are proposed for code acquisition.

The mean acquisition time of a parallel matched filter acquisition system in both non-fading and Rayleigh fading channels is evaluated in Sourour and Gupta (1990) and Ibrahim and Aghvami (1994). The salient conclusions are that: channel fading increases the mean acquisition time compared to a non-fading channel so that an inappropriate choice of the search and the verification modes of the acquisition system can increase the mean acquisition time several times its minimum value; the sensitivity of the mean acquisition time to the thresholds decreases when SNR/chip increases, and there is about 2.5 dB savings in SNR if the parallel system is used instead of a serial one in a non-fading channel, and this improvement can increase to about 4 dB in the fading channel.

In a multipath channel, several versions of the spreading code arrive at the receiver at different time delays (multipath) and paths have different complex gains. Consequently, correlating the multipath signal with the locally generated spreading code produces multi-Autocorrelation Functions (ACFs), scaled by the complex gains and separated by different time delays. The output of a matched filter acquisition is the sum of these autocorrelation functions. Ideally, we want the non-zero delay ACFs to be always zero and the AFC at zero delay to be one, i.e. the ACF has no sidelobes. The non-ideal ACF of the spreading code causes interpath interference which is significant when short spreading codes are used. Furthermore, the interpath interference increases the mean acquisition time, especially when a low probability of false alarm is required.

Assuming equal power per path, the mean acquisition time in a fixed L-path channel using matched filter acquisition is (Iinatti, 1998; Iinatti and Kerhue, 1999):

$$\overline{T}_{acq} = \frac{P_m^L[LT_c + (N-L)(T_c + T_{fa})]}{1 - P_m^L} + \frac{[NT_c + (N-L)(T_c + T_{fa})] \sum_{i=0}^{L-1} i P_m^i}{N \sum_{i=0}^{L-1} P_m^i}$$

$$+ \frac{(N-L)(N-L+1)(T_c + T_{fa}P_{fa})}{2N} + (N+1)T_c \qquad (5.125)$$

where P_m^i is the probability of miss for the ith path, P_{fa} is the probability of false alarm and T_{fa} is the penalty time caused by a false alarm.

An important issue to consider when dealing with the acquisition of spread-spectrum signals in multipath channels is the fact that there may exist more than one correct timing H_1 cell in the uncertainty region of the spreading code phase. The mean acquisition time performance of a serial acquisition system operating in multipath mobile channels and using joint two-cell search is considered in Yang and Hanzo (2001). The mean acquisition time for the joint two-cell acquisition system is:

$$\overline{T}_{acq} \approx \frac{[2 - P_{d1} - (1 - P_{d1})P_{d2}](1 + \alpha P_{fa})}{2[P_{d1} + (1 - P_{d1})P_{d2}]} \qquad (5.126)$$

where P_{d1} and P_{d2} represent the detection probabilities of the first and second H_1 cell and α is an integer representing the penalty time associated with a false alarm and re-entering the search mode. The mean acquisition time in bit duration versus SNR/chip performance for the serial search mode using the conventional cell by cell detection and joint two-cell detection for 10 users is shown in Figure 5.21. Clearly, the mean acquisition time based on joint two-cell detection is much lower than that with cell by cell detection under similar conditions. Furthermore, the mean acquisition time is sensitive to variation in the threshold as shown in Figure 5.22.

The code phase acquisition of a spread-spectrum system operating in fading channels using *differential correlation* techniques, performed at the output of matched filter acquisition, is shown in Figure 5.23.

The idea behind differential correlation is first introduced in Chung (1995) where the differential correlation was performed prior to matched filtering. The received signal is first processed at baseband before input to the matched filter using a complex differential detector with one-chip time delay. Then a coherent partial correlation is performed with the product of the local spreading code and its one-chip delayed phase. The mean acquisition time performance of both parallel and serial acquisition systems using differential correlation operating in fast Rayleigh fading has been investigated to show that the serial differential correlation acquisition prevails over its parallel counterpart. This is in contrast

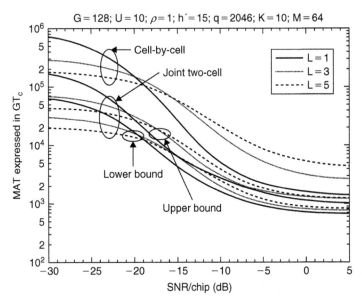

Figure 5.21 *MAT expressed in bit durations. Reproduced with permission from IEEE (Figure 6, Yang, L.-L. and Hanzo, L., 2001, IEEE Transactions on Vehicular Technology, Vol. 50, No. 2, pp. 617–628). G = number of chips/bit; U = number of users; $\rho = \frac{MAI\ from\ each\ interferring\ signal}{received\ power} = \frac{P_I}{P_R}$; h′ = normalized threshold; q = length of uncertainty = 2 × length of code for step size $= \frac{T_c}{2}$; K = number of chips in the penalty time; M = number of chips in the dwell time.*

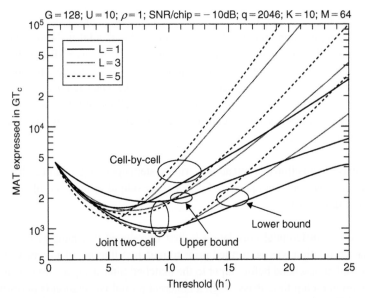

Figure 5.22 *MAT expressed in bit durations. Reproduced with permission from IEEE (Figure 7, Yang, L.-L. and Hanzo, L., 2001, IEEE Transactions on Vehicular Technology, Vol. 50, No. 2, pp. 617–628). G = number of chips/bit; U = number of users; $\rho = \frac{MAI\ from\ each\ interferring\ signal}{received\ power} = \frac{P_I}{P_R}$; h′ = normalized threshold; q = length of uncertainty = 2 × length of code for step size $= \frac{T_c}{2}$; K = number of chips in the penalty time; M = number of chips in the dwell time.*

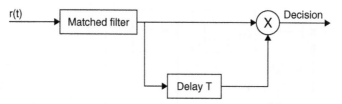

Figure 5.23 *Coherent differential correlation acquisition scheme.*

to the previous conclusions that a parallel scheme outperforms the corresponding serial scheme for a fixed one path system in an AWGN channel.

A similar concept is used in the coherent acquisition system described in Zarrabizadeh and Sousa (1997) which gives a remarkable improvement in SNR over the noncoherent detector of approximately 5 dB. In Ristaniemi and Joutsensalo (2001) the idea behind the differential correlation is exploited as follows: when constant preamble or an unmodulated pilot channel is available to the desired user, the receiver can efficiently filter noise and interferers out by the differential correlations using suitable time lags and consequently it is possible to estimate the desired user's delays even under severe Multiple Access Interference (MAI).

An analysis for accurate evaluation of the code acquisition time performance in Rayleigh fading channels is presented in Sheen and Wang (2001) since direct and flow-chart acquisition approaches are not applicable to fading channels. Code phase acquisition schemes, which use fixed dwell times, are inefficient since they usually require a long fixed correlation duration before a decision on the search process can be taken, especially for long code applications. Removal of the fixed dwell time requirement is therefore very attractive. In addition, multiple dwell time acquisition techniques have shown an improvement over single dwell acquisitions. However, these techniques do not address in detail the frequency offset and data modulation effects. A noncoherent sequential code phase acquisition system using sliding correlation can deal effectively with data modulation, frequency offset and chip-asynchronization simultaneously. Such a system can achieve low error probabilities and outperform their fixed dwell time counterpart by roughly 2 to 4 dB in terms of SNR (Lin, 2002). The effects of the threshold setting on the code acquisition are discussed in Iinatti (2000).

Clearly it is a big task to cover all of the acquisition models for mobile fading channels proposed in the literature. However, we can present an interesting scheme in this section: *the joint triple-cell detection* to code acquisition for spread-spectrum systems operating in such environments (Shin and Lee, 2001).

This joint triple-cell detection acquisition system uses a serial search noncoherent detector with an *active correlator* that has two modes of operation: the *search mode* and the *verification mode*. It is assumed that the signal is received without data modulation. The joint triple-cell detection may be viewed as an extension of the *joint two-cell detection* (Yang and Hanzo, 2001). The detector output corresponds to the cell under test as well

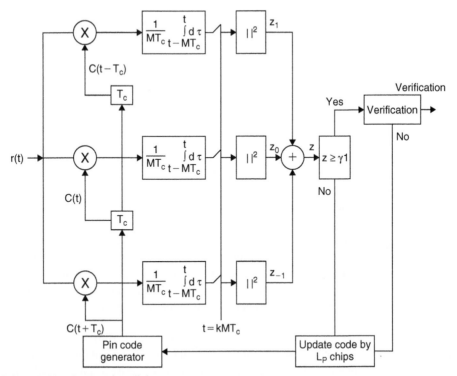

Figure 5.24   *Joint triple-cell detection acquisition receiver.*

as to the previous and next cells combined to form the decision variable. In the search strategy, cells in the uncertainty region are tested in a non-consecutive manner with L_p step size of chips where L_p denotes the number of resolvable paths that are known to the receiver. The non-consecutive search is implemented by advancing the phase of local code generator by L_p chips in the code phase update component.

The joint triple-cell detection, depicted in Figure 5.24, consists of three parallel non-coherent detectors; each incorporates an active correlator followed by a squarer. The outputs z_{-1}, z_0, and z_1 from the three noncoherent detectors are staggered in time, using two delay elements, to correspond to the *previous, current and next* cell, respectively.

The number of cells in the whole uncertainty region (L) is divided into L_p disjoint sub-regions $\Re_\ell(\ell = 1, 2, \ldots\ldots\ldots, L_p)$. Let n_ℓ denotes the number of cells in $\Re_\ell$. Further, the acquisition scheme assumed the H_D cells are uniformly distributed over the whole uncertainty region so that each uncertainty sub-region $\Re_\ell$ contains *one and only one* H_D^ℓ cell corresponding to the ℓth resolvable path at node (ℓ, n_ℓ) and the other nodes denote H_0 cells. The acquisition system avoids testing the same cell again until all cells in the uncertainty region are tested. This requires the *phase of the local code* to be adjusted each time a number of cells have been tested, e.g. n_1, $n_1 + n_2$, $n_1 + n_2 + n_3$.

The decision variables z_{-1}, z_0, and z_1 shown in Figure 5.24, are independent central chi-square variables with two degrees of freedom. The probability of detection for exponentially decaying multipath intensity profile at the ℓth resolvable path for a given decision threshold γ is:

$$P_d^\ell = \sum_{k=\ell-1}^{\ell+1} c_{\ell k} \exp\left(-\frac{\gamma}{2\sigma_k^2}\right) \tag{5.127}$$

where

$$c_{\ell k} = \Pi_{m=\ell-1}^{\ell+1} \frac{1}{\left(1 - \dfrac{\sigma_m^2}{\sigma_k^2}\right)} \tag{5.128}$$

$\sigma_0^2 = \sigma_I^2$ and σ_I^2 denotes the interference power. The probability of false alarm at the ℓth resolvable path is:

$$P_{fa}^\ell = \exp\left(-\frac{\gamma}{2\sigma_I^2}\right) \sum_{k=0}^{2} \frac{1}{k!} \left(\frac{\gamma}{2\sigma_I^2}\right)^k \tag{5.129}$$

The probabilities of detection and false alarm for the ℓth resolvable path for the search and verification modes can be denoted as P_{d1}^ℓ, P_{d2}^ℓ and P_{fa1}^ℓ and P_{fa2}^ℓ, respectively.

The probabilities of detection and false alarm in the search mode can be calculated using (5.126) and (5.129):

$$P_{d1}^\ell = P_d^\ell(\gamma 1) \tag{5.130}$$

$$P_{fa}^\ell = P_{fa}^\ell(\gamma 1) \tag{5.131}$$

Let the decision thresholds in the search and verification modes be γ_1 and γ_2, respectively. Assume that if at least B out A decision variables exceeded the new decision threshold, the acquisition is declared and the tracking system is enabled. The probabilities of detection and false alarm in the verification mode are given by the binomial distribution:

$$P_{d2}^\ell = \sum_{j=B}^{A} \binom{A}{j} (P_d^\ell(\gamma 2))^j (1 - P_d^\ell(\gamma 2))^{A-j} \tag{5.132}$$

$$P_{fa2}^\ell = \sum_{j=B}^{A} \binom{A}{j} (P_{fa2}^\ell(\gamma 2))^j (1 - P_{fa}^2(\gamma 2))^{A-j} \tag{5.133}$$

The dwell time (correlation interval) in the search mode $T = MT_c$. The correlation interval in the verification mode is $2T$.

The effects of the number of resolvable paths L_p on $\overline{T}_{acq}$ is shown in Figure 5.25 for a *uniform multipath intensity profile*, $M = 256$, the signal to interference (SIR/chip) $= -2$ dB, code length $L = 2^{15}$, $A = 4$, $B = 2$, and the penalty factor is 2^5. The *conventional* scheme is based on matched filter acquisition in Rayleigh fading channels (Ibrahim and Aghvami, 1994). The Non-Consecutive Search through Cell by Cell (NCS-CC) and the

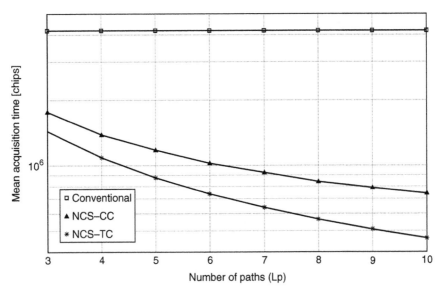

Figure 5.25 *Effects of the number of resolvable paths on mean acquisition time. Reproduced with permission from IEEE (Figure 8, Shin, O.-S. and Lee, K.B., 2001, IEEE Transactions on Communications, Vol, 49, pp. 734–743).*

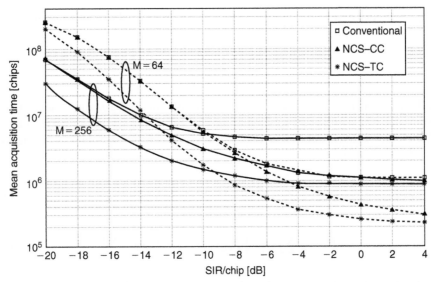

Figure 5.26 *Mean acquisition time versus SIR/CHIP. Reproduced with permission from IEEE (Figure 4, Shin, O.-S. and Lee, K.B., 2001, IEEE Transactions on Communications, Vol, 49, pp. 734–743).*

Non-Consecutive Search through Triple Cell (NCS-TC) detection both deliver a lower $\overline{T}_{acq}$ compared with the matched filter acquisition in fading channels.

The effects of the signal-to-interference power ratio SIR/chip on $\overline{T}_{acq}$ are shown in Figure 5.26 for a *uniform multipath intensity profile* and $L_p = 5$ which outperform the

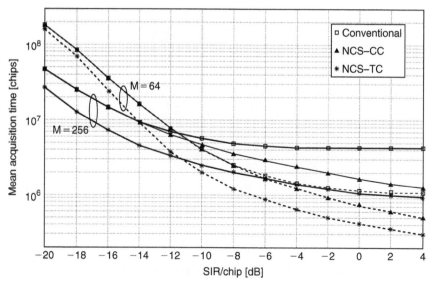

Figure 5.27 *Mean acquisition time versus SIR/CHIP. Reproduced with permission from IEEE (Figure 6, Shin, O.-S. and Lee, K.B., 2001, IEEE Transactions on Communications, Vol, 49, pp. 734–743).*

conventional scheme and Figure 5.27 shows $\overline{T}_{acq}$ for an exponentially decaying multipath intensity profile with the decay rate equal 1 and $L_p = 5$. It is clear from Figure 5.27 that the NCS-TC scheme outperforms both the conventional search and the NCS-CC search schemes.

Example 5.12

An acquisition system using a non-consecutive search with triple cell detection is used in synchronizing a spread-spectrum system. Given the multipath intensity profile is exponential, the interference power $\sigma_I^2 = 0.25$ W, and the threshold $\gamma = 2$ V. Calculate the probabilities of detection and false alarm of the ℓth path given the standard deviations for the ℓth, $\ell - 1^{th}$ and $\ell + 1^{th}$ are 1.34, 2.7 and 2.3, respectively.

Solution

The probability of detection in the search mode is given in (5.127)

$$
\begin{aligned}
P_d^\ell &= \sum_{k=\ell-1}^{\ell+1} c_{\ell k} \exp\left(-\frac{\gamma}{2\sigma_k^2}\right) \\
&= c_{\ell(\ell-1)} \exp\left(-\frac{\gamma}{2\sigma_{\ell-1}^2}\right) + c_{\ell\ell} \exp\left(-\frac{\gamma}{2\sigma_\ell^2}\right) + c_{\ell(\ell+1)} \exp\left(-\frac{\gamma}{2\sigma_{\ell+1}^2}\right) \\
&= c_{\ell(\ell-1)} \exp\left(-\frac{2}{2 \times 2.7}\right) + c_{\ell\ell} \exp\left(-\frac{2}{2 \times 1.34}\right) + c_{\ell(\ell+1)} \exp\left(-\frac{2}{2 \times 2.3}\right) \\
&= 0.6905 c_{\ell(\ell-1)} + 0.4741 c_{\ell\ell} + 0.6474 c_{\ell(\ell+1)}
\end{aligned}
$$

$$c_{\ell k} = \Pi_{m=\ell-1}^{\ell+1} \frac{1}{\left(1 - \frac{\sigma_m^2}{\sigma_k^2}\right)}$$

$$c_{\ell(\ell-1)} = \frac{1}{1 - \frac{\sigma_\ell^2}{\sigma_{\ell-1}^2}} \cdot \frac{1}{1 - \frac{\sigma_{\ell+1}^2}{\sigma_{\ell-1}^2}} = \frac{1}{1 - \frac{1.34^2}{2.7^2}} \cdot \frac{1}{1 - \frac{2.3^2}{2.7^2}} = 4.8370$$

$$c_{\ell(\ell)} = \frac{1}{1 - \frac{\sigma_{\ell-1}^2}{\sigma_\ell^2}} \cdot \frac{1}{1 - \frac{\sigma_{\ell+1}^2}{\sigma_\ell^2}} = \frac{1}{1 - \frac{2.7^2}{1.3^2}} \cdot \frac{1}{1 - \frac{2.3^2}{1.34^2}} = 0.1679$$

$$c_{\ell(\ell+1)} = \frac{1}{1 - \frac{\sigma_{\ell-1}^2}{\sigma_{\ell+1}^2}} \cdot \frac{1}{1 - \frac{\sigma_\ell^2}{\sigma_{\ell+1}^2}} = \frac{1}{1 - \frac{2.7^2}{2.3^2}} \cdot \frac{1}{1 - \frac{1.34^2}{2.3^2}} = -4.0037$$

$$P_d^\ell = 0.6905 \times 4.8370 + 0.4741 \times 0.1679 - 0.6474 \times 4.0037$$

$$= 0.8276$$

The probability of false alarm in the search mode is given in (5.129)

$$P_{fa}^\ell = \exp\left(-\frac{\gamma}{2\sigma_I^2}\right) \sum_{k=0}^{2} \frac{1}{k!} \left(\frac{\gamma}{2\sigma_I^2}\right)^k$$

$$= \exp\left(-\frac{\gamma}{2\sigma_I^2}\right) \left\{1 + \frac{\gamma}{2\sigma_I^2} + \frac{1}{2}\left(\frac{\gamma}{2\sigma_I^2}\right)^2\right\}$$

$$\frac{\gamma}{2\sigma_I^2} = \frac{2}{2 \times 0.25} = 4$$

$$P_{fa}^\ell = \exp(-4)\{1 + 4 + 0.5 * (4)^2\} = 0.2381.$$

5.9.2 Code tracking in fading channels

Tracking the acquired code phase in a real spread-spectrum system is usually confronted with two major problems. These are the effects of multipath interference and the Multi-user Access Interference (MAI) on the tracking loop driving it out of lock.

The probability of the tracking loop to remain in lock is analytically difficult to evaluate. The Mean Time to Lose Lock (MTLL) is an alternative measure. The out-of-lock occurs when a normalized delay error exceeds the range of the discriminator characteristic range.

In the next section we consider the tracking loop performance when a single user accesses the spread-spectrum system operating in multipath fading channel. Once we have formulated this problem we proceed to the problem of multi-user accessing the system in multipath channels.

5.9.2.1   Code tracking of a single user's code in multipath fading channels

The performance of the code tracking loops were derived in previous sections when the channel was corrupted by additive white Gaussian noise only and the system exposed to essentially line-of-sight transmission. However, in cellular wireless applications, the tracking loops performance is severely degraded by the presence of Doppler, Rayleigh multipath fading, and Multiple Access Interference (MAI). The channel also suffers from log-normal distribution fading with shadowing effects. The affects of multipath fading on tracking loop performance, originally discussed in Ward (1965), used open loop to suppress the multipath components prior to delay estimation.

There are two types of signal fading: *slow signal fading and fast signal fading*. In the slow signal fading the multipath structure is not likely to change very often and the paths are not very likely to appear or disappear frequently as in the fast signal fading. On the other hand, *frequency-selective fading* is a linear distortion of the signal that affects a relatively narrow band of the signal spectrum. The fading bandwidth is usually smaller than the loop bandwidth and we can ignore its affect of tracking error on the performance. However, the affects of the tracking errors cannot be ignored when the fading bandwidth is greater than the tracking loop bandwidth. Tracking loops optimized for operation over AWGN channels severely degrade loop performance when operated over frequency-selective channels.

The key parameters in the code tracking loops are the users' delays that maximize the output correlation and phase/amplitude estimation of the signal over the mobile link. Several promising delay estimation algorithms have been developed (Iltis, 1990; 2001). Furthermore, the Rayleigh fading model for the signal amplitude is no longer valid for wideband CDMA when chips duration (T_c) is less than the delay spread of the channel, and the log-normal fading channel is the appropriate model to use.

The analysis of the MTLL for first order tracking loop in the absence of Doppler shift is presented in Holmes and Biederman (1978). The first order tracking loop for spread-spectrum systems operating over frequency-selective and non-selective slowly fading channels is analysed and evaluated in Sheen and Stuber (1994). The effects of multipath fading are evaluated in terms of the MTLL and the root mean square tracking error in Sheen and Stuber (1994). Their conclusions are that the multipath fading has a significant affect on the tracking loop performance when the loop SNR is large since, in such a case, the interference has the dominant affects.

The performance of the first and second order noncoherent *digital* tracking loop operating in mobile environment is considered in Su and Yen (1997) where the mobile channel is characterized by Rayleigh fading and Doppler shift. Their conclusions are that the first order loop has a favourable performance in the absence of Doppler shift since it has a

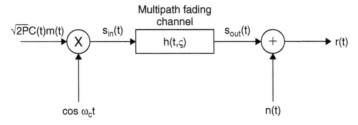

Figure 5.28 *Spread-spectrum transmission on a multipath fading channel.*

lower timing error variance and a higher MTLL than the second order loop under same loop SNR. However, in the presence of severe Doppler, the second order loop has a superior performance compared to the first order loop because the former has a zero mean timing error and large MTLL and the MTLL increases significantly for log-normal fading channel compared to the Rayleigh channel. Furthermore, the performance of the loop enhances as the spread signal bandwidth increases compared to the channel coherent bandwidth. The analysis of the digital tracking loop in the presence of Doppler shift in terms of timing error and MTLL is presented in Yen et al. (1996).

The multipath fading channel is assumed to be a wide-sense stationary uncorrelated scattering (WSSUS) process. The multipath Rayleigh fading channel is represented by zero mean complex Gaussian process in time t, which can be described by a lowpass impulse response $\tilde{h}(t, \varsigma)$ as shown in Figure 5.28 where n(t) is zero mean AWGN with a two-sided power spectral density equal $\frac{N_0}{2}$ W/Hz.

The autocorrelation function of $\tilde{h}(t, \varsigma)$ is

$$R_{\tilde{h}}(\Delta t; \varsigma_1, \varsigma_2) \approx \beta_{\tilde{h}}(\Delta t; \varsigma_1)\delta(\varsigma_1 - \varsigma_2) \tag{5.134}$$

where $\beta_{\tilde{h}}(\Delta t; \varsigma)_{\Delta t=0}$ is the multipath channel intensity profile. The noiseless bandpass received signal is

$$s_{out}(t) = Re\{\sqrt{2P}\tilde{s}(t) \exp(j\omega_c t)\} \tag{5.135}$$

where

$$\tilde{s}(t) = \sum_{k=0}^{\infty} \int_{kT_c}^{(k+1)T_c} m(t - t_d - \varsigma)C(t - t_d - \varsigma), \tilde{h}(t, t_d + \varsigma)\,d\varsigma \tag{5.136}$$

Since C(t) is the code sequence assumed to be m-sequence with a NRZ shaping function so that

$$C(kT_c + \varsigma) = C(kT_c) \tag{5.137}$$

Assuming the time delay is an integer number of chip duration.

$$t - t_d = iT_c \quad \text{for some integer i.} \tag{5.138}$$

Substituting (5.137) and (5.138) in (5.136) we get:

$$\tilde{s}(t) = \sum_{k=0}^{L} m(t - t_d - kT_c)C(t - t_d - kT_c)h_k(t) \tag{5.139}$$

where L is the number of paths representing the multipath fading channel and:

$$h_k(t) = \int_{kT_c}^{(k+1)T_c} \tilde{h}(t, t_d + \varsigma)d\varsigma \tag{5.140}$$

The complex k^{th} lowpass channel can be represented as:

$$h_k(t) = a_k(t)e^{j\theta_k(t)} \tag{5.141}$$

Substituting the above equation in the AWGN channel model we get:

$$s_{out}(t) = \sqrt{2P} \sum_{k=0}^{L} a_k(t)m(t - t_d - kT_c)C(t - t_d - kT_c) \cos(\omega_c t + \theta_k(t)) + n(t) \tag{5.142}$$

When data modulation is removed, i.e. $m(t) = 1$ for all t in (5.142) reduces to:

$$s_{out}(t) = \sqrt{2P} \sum_{k=0}^{L} a_k(t)C(t - t_d - kT_c) \cos(\omega_c t + \theta_k(t)) + n(t) \tag{5.143}$$

The tracking loop for direct sequence spread-spectrum system operating in frequency-selective fading channels is shown in Figure 5.29. The key feature of the loop is the use of a RAKE-like correlator to resolve the received multipath components and using maximum ratio combining reduces the effect of the AWGN. In each branch of the channel model, the received signal is first correlated with the early and late version of the delayed spreading functions.

The tracking loop in Figure 5.29 consists of L branches in the channel parameters estimation unit. Each branch, say branch k, estimates the amplitude $a_k(t)$ and the phase $\theta_k(t)$. The coherent tracking loop correlates the received signal with the function $c_\Delta(t - \hat{t}_d(t) - kT_c)$ where

$$c_\Delta(t - \hat{t}_d(t) - kT_c) \, c(t - \hat{t}_d(t) - kT_c - \Delta T_c) - c(t - \hat{t}_d(t) - kT_c + \Delta T_c) \tag{5.144}$$

The lowpass signal at the kth output $y_k(t)$ is:

$$y_k(t) = \sqrt{\frac{P}{2}} \sum_{m=1}^{L} a_m(t)\hat{a}_k(t) \cdot [R_c(\delta + m - k - \Delta)T_c - R_c(\delta + m - k + \Delta)T_c]$$

$$\cdot \cos(\theta_m(t) - \hat{\theta}_k(t)) + \frac{1}{\sqrt{2}}\hat{a}_k(t)n_k(t) \tag{5.145}$$

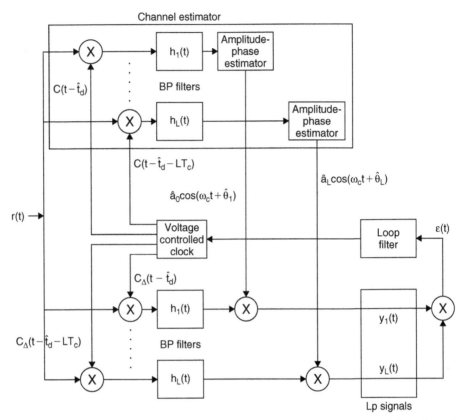

Figure 5.29 *Tracking loop for spread-spectrum operating in frequency-selective fading channels.*

where
$$n_k(t) = \eta_k^c(t)\cos\left(\hat{\theta}(t)\right) + \eta_k^s(t)\sin\left(\hat{\theta}(t)\right) \tag{5.146}$$

and
$$\eta_k^c(t) = n_c(t)c_\Delta(t - \hat{t}_d - kT_c) * \hat{h}_k(t)|_{BPF} \tag{5.147}$$

$$\eta_k^s(t) = n_s(t)c_\Delta(t - \hat{t}_d - kT_c) * \hat{h}_k(t)|_{BPF} \tag{5.148}$$

where $\delta = \frac{t_d - \hat{t}_d}{T_c}$ as before.

The error signal $\varepsilon(t)$ is (Sheen and Stuber, 1995):

$$\varepsilon(t) \approx \sum_{k=1}^{L} y_k(t) = \sqrt{\frac{P}{2}}S(\varepsilon) + n_T(t) \tag{5.149}$$

where
$$n_T(t) = \frac{1}{\sqrt{2}}\sum_{k=1}^{L} \hat{a}_k(t)n_k(t) \tag{5.150}$$

The S-curve of the loop is:

$$S(\varepsilon) = \sum_{k=1}^{L} \sum_{m=1}^{L} a_m(t)\hat{a}_k(t)[R_c(\delta+m-k-\Delta)T_c - R_c(\delta+m-k+\Delta)T_c]\cos(\theta_m(t)-\hat{\theta}_k(t))$$

(5.151)

The error signal is passed onto the loop filter and used to derive the voltage controlled clock (VCC) for updating the code phase. The operation of the VCC is described by (5.A3)

$$\frac{\hat{t}_d}{T_c} = g \int_0^t [\varepsilon(t') * h_\ell(t')]dt'$$

(5.152)

where g is the gain of the VCC and $h_\ell(t)$ is the impulse response of the *loop filter*. Substitute for $\frac{\hat{t}_d(t)}{T_c}$ from (5.152) in (5.73)

$$\delta(t) = \frac{t_d(t)}{T_c} - g \int_0^t [\varepsilon(t') * h_\ell(t')]dt'$$

(5.153)

Differentiating both sides of (5.153), we get the stochastic equation describing the dynamic behaviour of the tracking loop:

$$\frac{d\delta}{dt} = \beta_D - g\left\{\frac{P}{2}S(\delta) + n_T(t)\right\} * h_\ell(t)$$

(5.154)

where the code Doppler shift $\beta_D = \frac{1}{T_c}\frac{dt_d(t)}{dt}$

(5.155)

Figure 5.30 shows the mean square tracking error given by the linear/non-linear model of the tracking loop with respect to averaged received SNR for early–late discriminator offset $\Delta = 0.5$ and zero Doppler shift. The product of the averaged nominal loop bandwidth B_ℓ and the bit duration T_b is taken to be 0.01 and 0.1. The linear model predicts accurately the mean square tracking error for SNR ≥ -10 dB when $B_\ell T_b = 0.01$, but for $B_\ell T_b = 0.1$ it is accurate only when SNR ≥ 0 dB. However, outside these ranges, the linear model estimate of the tracking error deviates substantially from the estimate given by the non-linear model.

Figure 5.31 compares the mean square tracking error of the fading channel tracking loop with the tracking error from conventional non-coherent early–late tracking loops operating in fading environment for maximum Doppler frequency 83 Hz, $\Delta = 0.5$, $B_\ell T_b = 0.01$ and power ratio of the first and second channel taps $R = \frac{E[a_0^2]}{E[a_1^2]} = 0$ dB. Clearly, the tracking loop designed to operate in fading channels outperforms the traditional early–late tracking loop operating in the same fading environment.

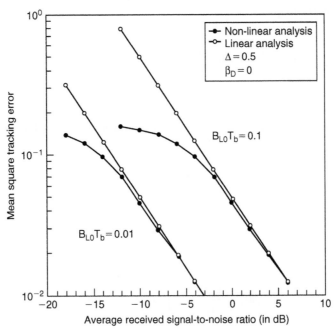

Figure 5.30 *Comparisons of linear and non-linear analyses for the new tracking loop. Reproduced with permission from IEEE (Figure 5, Sheen, Wern-Ho and Stuber, Gordon L., 1995, IEEE Transactions on Communications, Vol. 43, No. 12, pp. 3063–3072).*

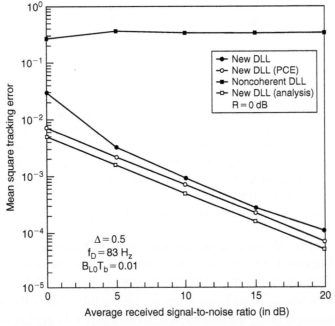

Figure 5.31 *The tracking-error performance of the new tracking loop and the conventional DLL. Reproduced with permission from IEEE (Figure 8, Sheen, Wern-Ho and Stuber, Gordon L., 1995, IEEE Transactions on Communications, Vol. 43, No. 12, pp. 3063–3072).*

5.9.2.2   *Multi-users code tracking in multipath channels*

The noncoherent tracking loops are important because they are relatively insensitive to data modulation. Their performance when used by a single user with a single path received signal in AWGN channels is evaluated in Section 5.7.7 in terms of the tracking jitter σ_δ^2. When a multi-user spread-spectrum system operates in mobile wireless channels, the performance of the noncoherent tracking loops is severely degraded due to the possible presence of Doppler shift, multipath fading and Multiple Access Interference (MAI).

The effects of multipath fading and Doppler shifts on the performance of code tracking loop are discussed above. In this section we now discuss and evaluate the effects of the MAI on the loop's performance.

The MAI can be approximated as *Gaussian process*. However, such an approximation is not always accurate for short codes (Guenach and Vandendrope, 2001), high SNR, when there are few interferers, and when the contribution from a single interferer dominates the MAI. However, an exact computation of the effects of MAI requires knowledge of the statistics of cross-correlation functions of the code sequences and accurate approximation for the aperiodic cross-correlation functions between each of two code sequences. Fading channel estimation is discussed in (Yen and Hanzo, 2001; Bhashyam and Aszhang, 2002).

The computation of the MAI requires the computation of (K-1) cross-correlation functions which is exceptionally expensive if N is large. An alternative approach, which is relatively less costly, is to treat the aperiodic cross-correlation function $C_{k,i}(\ell)$ defined in Section 3.6.2 as a random variable with triangle PDF symmetrical about the origin given by:

$$P[C_{k,i}(\ell) = X] = \frac{1}{\Lambda^2}[\Lambda - |X|] \quad |X| \leq \Lambda$$
$$= 0 \qquad\qquad \text{elsewhere} \tag{5.156}$$

where $\Lambda = \lfloor \sqrt{3N} - 1 \rfloor / N$ and $\lfloor a \rfloor$ is the largest integer less than a. The mean value of $C_{k,i}(\ell)$ is zero and its variance is equal $\frac{\Lambda(\Lambda+2)}{6N^2}$.

Clearly, increasing the number of interferers will increase the MAI, which significantly reduces the MTLL and increases the variance of the tracking error. For a fixed number of users, longer spreading codes can improve the tracking performance (Caffery and Stuber, 2000). While the MAI does not affect the S-curve of the tracking system, it adds noise terms to the error signal at the input to the loop filter of the early–late tracking system.

The multi-user delay tracking loops are of two types: those based on the Extended Kalman Filter (EKF) (Iltis, 1990; Ham et al., 2002) which jointly estimate the code delay, and multipath coefficients for systems operating in fading channels. The EKF is a linear approximation to the MAP estimator but might have convergence problems because of the round-off errors. Furthermore, it has a high computation requirement and is too complex for some applications.

The other tracking loops (Zha and Blostein, 2002) are based on multistage interference cancellation combined with early–late tracking loops that integrate the tracking loop into a multi-user detection system to significantly improve the tracking performance and drive the tracker to be insensitive to a near-far situation.

Let us now analyse the affects of the MAI on the tracking error when multi-users share an asynchronous uplink multipath fading channel in direct sequence spread-spectrum system. Consider K active users, each with spreading factor $N = \frac{T}{T_c}$ where T is the symbol duration and T_c is the chip duration. The number of paths received by each user is the same on the downlink multipath channel as the base station transmitted signal for all users travelling the same paths. This is not necessarily true for the uplink multipath channel, as signals from an individual user may travel on different paths. However, for simplicity, we will assume that the number of paths (L) for the uplink channel under consideration is the same for all users. The equivalent baseband received signal at base-station is:

$$r(t) = \sum_{i=-\infty}^{\infty} \sum_{k=1}^{K} \sum_{\ell=1}^{L} h_{k,\ell}(i)b_k(i)\tilde{C}_k(t - iT - \tau_{k,\ell}) + n(t) \qquad (5.157)$$

where $h_{k,\ell}(i)$ is the complex channel for the kth user receiving signal from the ℓth path in the ith time interval, $b_k(i) \in \{+1, -1\}$ is the kth user's data symbols received in the ith time interval, $\tau_{k,\ell} \in [0, T]$ is the kth user's propagation delay for the ℓth path and $\tilde{C}_k(t)$ is the kth user's spreading waveform is:

$$\tilde{C}_k(t) = \sum_{j=0}^{N-1} C_k(j)p(t - jT) \qquad (5.158)$$

where $\{C_k(t)\}_{j=0}^{N-1} \in [+1, -1]$ is the kth user's spreading code and p(t) is the chip pulse shape. The received signal is chip matched filtered and sampled at the chip rate. For each user, a blank bit interval is inserted every M bit intervals to eliminate the edge effects (Hart et al., 1996). However, M data symbols are detected using an observation window of length (M + 1)T. At the mth data symbol, the received vector r(m) is:

$$r(m) = [r(Nm + 1), r(Nm + 2), \ldots\ldots\ldots\ldots, r(Nm + N)]^T \qquad (5.159)$$

The received vector **r** for an observation window of length (M + 1)T is

$$\mathbf{r} = [\mathbf{r}^T(1), \mathbf{r}^T(2), \ldots\ldots\ldots\ldots\ldots\ldots, \mathbf{r}^T(M + 1)] \qquad (5.160)$$

The discrete version of the received signal given in (5.160) is:

$$\mathbf{r} = \sum_{i=1}^{M} \sum_{k=1}^{K} \sum_{\ell=1}^{L} h_{k,\ell}(i)b_k(i) \quad \mathbf{d}_{k,\ell}(i) + \mathbf{n} \qquad (5.161)$$

where $\mathbf{d}_{k,\ell}(i)$ is the kth user's discrete spreading waveform for the ith symbol and the ℓth path. The time delay of ℓth path of the kth user is:

$$\tau_{k,\ell} = [p_{k,\ell} + \delta_{k,\ell}]T_c \qquad (5.162)$$

where $p_{k,\ell}$, $\delta_{k,\ell}$ are the integer and the fractional parts of the time delay, respectively, such that $p_{k,\ell} \in [0, 1, \ldots\ldots\ldots\ldots, N - 1]$ and $\delta_{k,\ell} \in [0, 1]$. The estimated time delay is:

$$\hat{\tau}_{k,\ell} = [\hat{p}_{k,\ell} + \hat{\delta}_{k,\ell}]T_c \tag{5.163}$$

We will consider the simple case where the *relative position of the true and estimated time delay occur in the same sampling interval* so that the true and estimated integer part of the time delay are equal, i.e. $p_{k,\ell} = \hat{p}_{k,\ell}$.

Now define the kth user's spreading code vector for the $(M + 1)T$ length as:

$$\mathbf{C}_k = [C_k(0), C_k(1), \ldots\ldots\ldots\ldots, C_k(N - 1), 0, 0, \ldots\ldots\ldots, 0]^T \tag{5.164}$$
$$\longleftarrow NM \longrightarrow$$

Denote $\mathbf{C}_k(p_{k,\ell}, i)$ as $\mathbf{C}_k$ right-shifted by $[(i - 1)N + p_{k,\ell}]$. Let the vector $\mathbf{g}_{k,\ell}$ be the ℓth path of the kth user's *matched filter response* at the chip-rate sampling points. The mth element $g_{k,\ell}(m)$ can be expressed in terms of the chip shape p(t) with duration T_c.

$$g_{k,\ell}(m) = \int_{-\infty}^{\infty} p(\tau - \delta_{k,\ell}T_c)p^*(mT_c - \tau)d\tau \tag{5.165}$$

For a rectangular chip shape, we get:

$$\mathbf{g}_{k,\ell} = (1 - \delta_{k,\ell})\delta_{k,\ell} \tag{5.166}$$

The received user's code at the sampling instant is:

$$\mathbf{d}_{k,\ell}(i) = \mathbf{C}_k(p_{k,\ell}, i) * \mathbf{g}_{k,\ell} \tag{5.167}$$

For rectangular chip shape, the discrete spreading waveform is (Zha and Blostein, 2002):

$$\mathbf{d}_{k,\ell}(i) = (1 - \delta_{k,\ell})\mathbf{C}_k(p_{k,\ell}, i) + \delta_{k,\ell}\mathbf{C}_k(p_{k,\ell} + 1, i) \tag{5.168}$$

The time delay error is:

$$\Delta\tau_{k,\ell} = \tau_{k,\ell} - \hat{\tau}_{k,\ell} = [\delta_{k,\ell} - \hat{\delta}_{k,\ell}]T_c \tag{5.169}$$

Generally, $\Delta\tau_{k,\ell}$ is small but has a profound impact on the bit error performance of linear CDMA receivers (Parkvall et al., 1996).

We will now show that the discrete code waveform in (5.168) can be expressed as the weighted sum of two signals: the estimated code vector $\hat{\mathbf{d}}_k(i)$ and the code error vector $\Delta\hat{\mathbf{d}}_k(i)$. Equation (5.168) can be re-written as:

$$\mathbf{d}_{k,\ell}(i) = (1 - \hat{\delta}_{k,\ell})\mathbf{C}_k(p_{k,\ell}, i) + \hat{\delta}_{k,\ell}\mathbf{C}_k(p_{k,\ell} + 1, i)$$
$$+ (\delta_{k,\ell} - \hat{\delta}_{k,\ell})[\mathbf{C}_k(p_{k,\ell} + 1, i) - \mathbf{C}_k(p_{k,\ell}, i)] \tag{5.170}$$

The first two terms in (5.170) represent $\hat{\mathbf{d}}_{k,\ell}$:

$$\hat{\mathbf{d}}_{k,\ell} = (1 - \hat{\delta}_{k,\ell})\mathbf{C}_k(p_{k,\ell}, i) + \hat{\delta}_{k,\ell}\mathbf{C}_k(p_{k,\ell} + 1, i) \qquad (5.171)$$

The last two terms represents the weighted error vector $\Delta\hat{\mathbf{d}}_k$ (i):

$$\Delta\hat{\mathbf{d}}_k(i) = [\mathbf{C}_k(p_{k,\ell} + 1, i) - \mathbf{C}_k(p_{k,\ell}, i)] \qquad (5.172)$$

Consequently,

$$\mathbf{d}_{k,\ell}(i) = \hat{\mathbf{d}}_{k,\ell} + (\delta_{k,\ell} - \hat{\delta}_{k,\ell})\Delta\hat{\mathbf{d}}_k(\mathbf{i}) \qquad (5.173)$$

The above equation implies that *the received code waveform for each user is decomposed into two virtual users, one with the estimated code vector* $\hat{\mathbf{d}}_{k,\ell}$ *(i) and the other with error code vector* $\Delta\hat{\mathbf{d}}_k$ *(i)*.

Now substitute for $\mathbf{d}_{k,\ell}$ (i) from (5.173) in (5.161)

$$\mathbf{r} = \sum_{i=1}^{M}\sum_{k=1}^{K}\sum_{\ell=1}^{L} h_{k,\ell}(i)b_k(i)[\hat{\mathbf{d}}_{k,\ell}(i) + (\delta_{k,\ell} - \hat{\delta}_{k,\ell})\Delta\hat{\mathbf{d}}_k(i)] + \mathbf{n}$$

which simplifies to:

$$\mathbf{r} = \sum_{i=1}^{M}\sum_{k=1}^{K}\sum_{\ell=1}^{L} h_{k,\ell}(i)b_k(i)\hat{\mathbf{d}}_{k,\ell}(i) + \sum_{i=1}^{M}\sum_{k=1}^{K}\sum_{\ell=1}^{L} h_{k,\ell}(i)b_k(i)(\delta_{k,\ell} - \hat{\delta}_{k,\ell})\Delta\hat{\mathbf{d}}_{k,\ell}(i)] + \mathbf{n}$$
$$(5.174)$$

The estimated received signal for the kth user is:

$$\hat{\mathbf{r}}_k = \mathbf{r} - \sum_{i=1}^{M}\sum_{n=1}^{k-1}\sum_{\ell=1}^{L} h_{n,\ell}(i)b_n(i)\hat{\mathbf{d}}_{n,\ell}(i) + \sum_{i=1}^{M}\sum_{n=1}^{k-1}\sum_{\ell=1}^{L}(\delta_{n,\ell} - \hat{\delta}_{n,\ell})h_{n,\ell}(i)b_n(i)\Delta\hat{\mathbf{d}}_{n,\ell}(i)]$$

$$- \sum_{i=1}^{M}\sum_{n=k+1}^{K}\sum_{\ell=1}^{L} h_{n,\ell}(i)b_n(i)\hat{\mathbf{d}}_{n,\ell}(i) - \sum_{i=1}^{M}\sum_{n=k+1}^{K}\sum_{\ell=1}^{L}(\delta_{n,\ell} - \hat{\delta}_{n,\ell})h_{n,\ell}(i)b_n(i)\Delta\hat{\mathbf{d}}_{n,\ell}(i)]$$

$$- (\delta_{k,\ell} - \hat{\delta}_{k,\ell})h_{k,\ell}(i)b_k(i)\Delta\hat{\mathbf{d}}_{k,\ell}(i) \qquad (5.175)$$

Clearly, when the estimated delay $\hat{\delta}_{n,\ell} \cong \delta_{n,\ell}$, then:

$$\hat{\mathbf{r}}_k = \mathbf{r} - \sum_{i=1}^{M}\sum_{n=1}^{k-1}\sum_{\ell=1}^{L} h_{n,\ell}(i)b_n(i)\hat{\mathbf{d}}_{n,\ell}(i) - \sum_{i=1}^{M}\sum_{n=k+1}^{K}\sum_{\ell=1}^{L} h_{n,\ell}(i)b_n(i)\hat{\mathbf{d}}_{n,\ell}(i) \qquad (5.176)$$

That is: $\hat{\mathbf{r}}_k = \mathbf{r} - \text{MAI}$

Where the MAI is:

$$\text{MAI} = \sum_{i=1}^{M}\sum_{n=1}^{k-1}\sum_{\ell=1}^{L} h_{n,\ell}(i)b_n(i)\hat{\mathbf{d}}_{n,\ell}(i) + \sum_{i=1}^{M}\sum_{n=k+1}^{K}\sum_{\ell=1}^{L} h_{n,\ell}(i)b_n(i)\hat{\mathbf{d}}_{n,\ell}(i) \qquad (5.177)$$

When $\hat{\delta}_{n,\ell} \neq \delta_{n,\ell}$ such that there exists interference due to the timing error so that (5.175) becomes:

$$\hat{\mathbf{r}}_k = \mathbf{r} - \text{MAI} - \text{interference caused by the timing error } \delta_{n,\ell} - \hat{\delta}_{n,\ell} \qquad (5.178)$$

Let us investigate then how the timing error $\delta_{n,\ell} - \hat{\delta}_{n,\ell}$ degrades the performance of the tracking system. To simply (5.175) we define the following vectors:

$$\hat{\mathbf{r}}_n = \sum_{i=1}^{M}\sum_{\ell=1}^{L} \hat{b}_n(i)h_{n,\ell}(i)\hat{\mathbf{d}}_{k,\ell}(i) \qquad (5.179)$$

$$\hat{\mathbf{e}}_n = \sum_{i=1}^{M}\sum_{\ell=1}^{L} \hat{b}_n(i)h_{n,\ell}(i)\Delta\hat{\mathbf{d}}_{k,\ell}(i) \qquad (5.180)$$

$$\hat{\Delta}\mathbf{a} = \sum_{i=1}^{M}\sum_{\ell}^{L} \delta_{k,\ell} - \hat{\delta}_{k,\ell} \qquad (5.181)$$

We now substitute the vectors defined in (5.179–5.180) in (5.175) in an iterative implementation and divide the users in two groups for interference computations: interference from the $(k+1)^{\text{th}}$ user to the kth user will be cancelled in the jth iteration and interference from first user to the $(k-1)^{\text{th}}$ user will be cancelled in the $(j+1)^{\text{th}}$ iteration.

$$\hat{\mathbf{r}}_k^{j+1} = \mathbf{r} - \sum_{n=1}^{k-1}(\hat{\mathbf{r}}_n^{j+1} + \hat{\Delta}\mathbf{a}_n^{j+1}\hat{\mathbf{e}}_n^{j+1}) - \sum_{n=k+1}^{K}(\hat{\mathbf{r}}_n^{j} + \hat{\Delta}\mathbf{a}_n^{j}\hat{\mathbf{e}}_n^{j}) - \hat{\Delta}\mathbf{a}_k^{j}\hat{\mathbf{e}}_k^{j} \qquad (5.182)$$

It is assumed that the base station receiver has knowledge of the spreading codes of all users and that the channel changes relatively slowly over the $(M+1)$ symbols.

The delay is updated in small steps (i.e. $\pm 0.05 T_c$) and then used in the next sliding window to estimate and cancel the MAI + timing error interference.

5.10 Summary

Code time synchronization is essential to the delivery of information over spread-spectrum systems. It is required to de-spread the received signal and recover the information. Code synchronous transmission is usually achieved in two steps: *code acquisition* searching for an estimate for the locally generated code phase that is to be aligned with the received code sequence to within one chip duration. When this estimated code phase is verified, the synchronizer transmission is declared *in lock* and the loop tracks any changes in the received code phase in a process known as *code tracking*.

This chapter began by considering the optimum acquisition that achieves the *minimum acquisition time* when cells are tested in parallel; although such a system requires excessive hardware that it is not suitable for many practical systems. A practical acquisition system achieves sub-optimal performance through a compromise between acquisition time and complexity. Sub-optimum acquisition systems using both parallel and serial search were considered in Sections 5.2.2 and 5.2.3. The performance of these systems were theoretically analysed in terms of the average and variance acquisition times for given probability of false alarm and detection and presented in Section 5.3.

The time synchronization of a single user spread-spectrum system operating in an AWGN environment was considered first. The code phase acquisition of such a system was discussed in detail in Sections 5.2 to 5.6. Brief deliberation of the optimum synchronizer was given in Section 5.2.1 followed by the sub-optimum acquisition systems using both parallel and serial search in Sections 5.2.2 and 5.2.3. The performance of these systems were theoretically analysed in terms of the average and variance acquisition times for given probability of false alarm and detection and presented in Section 5.3. Other means of acquisition such as sequential and matched filtering were considered in Sections 5.4 and 5.5. The effects of frequency errors on the code acquisition were presented in Section 5.6.

Two different code tracking loops have been considered in the chapter. An introduction to the optimum tracking system based on the maximum likelihood estimates was presented and its complexity explored. We looked for sub-optimal tracking loop arrangements that are approximation of the optimum loop. There are two different code tracking loops considered in this chapter. The first loops are known as the *early–late tracking loops*, and these were dealt with in Section 5.7 for both coherent and noncoherent systems operating in AWGN channels for a single user and a single path received signal. The second tracking loops are the time-sharing version of the early–late loops and known as the τ-dither loops. Section 5.8 discussed these. The analysis of these tracking loops provides expressions for the *S-curve characteristics of the tracking loop* and the *variance of the tracking error* expressed in terms of the received signal-to-noise power ratio within the loop bandwidth.

The time synchronization of spread-spectrum system operating in *mobile multipath fading* channels was considered in Section 5.9. Here we followed the same approach in our treatment to the synchronization of a spread-spectrum system operating in fixed AWGN channels by first considering the acquisition problem in multipath single-user fading channels in Section 5.9.1. The analysis of code acquisition in fading channels using matched filter, joint two-cell and joint three-cell acquisition system provides expression for the mean acquisition time, detection and the false alarm probabilities.

The performance of code tracking of a single user in multipath channels is considered in Section 5.9.2 where we derived expression for the S-curve of the loop and stochastic equation describing the dynamic behaviour of the tracking loop. The multi-user code tracking in multipath channels was also introduced in Section 5.9.2 to derive the interference contribution due to the timing error.

Problems

5.1 A BPSK direct sequence spread-spectrum system transmits data using a spreading code of length $N = 31$ and a clock of 1.8822 MHz. Assume the carrier frequency is a priori known and the acquisition system uses a single dwell serial search in step of $0.25 T_c$. Given the probability of detection of $p_d = 0.75$, probability of false alarm $P_{fa} = 10^{-3}$ and the average acquisition time not exceeding 60 chip durations. Calculate:
   i. false alarm penalty
   ii. variance of the acquisition time

5.2 Consider the acquisition system in Problem 5.1 when the system is subjected to code Doppler of 2.3 kHz. Calculate:
   i. mean acquisition time
   ii. variance of the acquisition time

5.3 A serial search scheme is used to acquire the code phase of a direct sequence spread-spectrum acquisition. The acquisition scheme integrates the squarer output for interval of 0.78 s and the output is sampled at intervals of 0.078 s. The probability of false alarm is 10^{-2}. Given that the received SNR is 2.5 dB, calculate the probability of detection.

5.4 A matched filter acquisition is used to synchronize a direct sequence spread-spectrum system. The signal-to-noise ratio per chip at the output of the filter is 2.5 dB. Given that the chip rate is 1.2288 Mc/s, the probability of false alarm 10^{-3} and the integration time is 12 μs, calculate:
   i. ratio of the threshold to the noise variance
   ii. probability of detection
   iii. probability of miss

5.5 Consider the matched filter acquisition system in Problem 5.4. If the instability of the carrier frequency of the oscillators in the transmitter and the receiver causes a frequency error of 10 kHz calculate the new:
   i. probability of false alarm
   ii. probability of detection

5.6 A linear model for the early–late tracking loop is considered in the example. The received code sequence has a length of 64 chips and the code rate $R_c = 1.2288$ Mc/s, the received signal power $P = 150$ mW, noise spectral density $N_0 = 0.1$ μW/Hz, the early and late code phase shifts are $\Delta = \pm 0.5$. Calculate the variance of the tracking error σ_δ^2.

5.7 A noncoherent early–late tracking loop is used to synchronize a direct sequence spread-spectrum system. The loop filter bandwidth 350 Hz, $N = 64$, $\Delta = 1$, received power $P = 0.05$ mW, and a one-sided noise spectral density at the input of the tracking loop of 10^{-7} W/Hz. If the variance of the tracking jitter is not to exceed 2, calculate IF noise bandwidth.

5.8   A dithering frequency of 2.4 kHz is introduced to the noisy noncoherent tracking system in Problem 5.7. Calculate the new variance of the tracking jitter.

5.9   A triple cell detection acquisition system is used in synchronizing a direct sequence spread-spectrum system operating in a mobile environment with an exponential multipath intensity profile. Given that the interference power $\sigma_I^2 = 0.25$ W and the decision threshold is $\gamma = 2$, calculate the probability of a false alarm of the system.

References

Bhashyam, S. and Aszhang, B. (2002) Multiuser channel estimation and tracking for long-code CDMA systems, *IEEE Transactions on Communications*, 1081–1090.

Caffery, J. and Stuber, G.L. (2000) Effects of multiple-access on the non-coherent delay lock loop, *IEEE Transactions on Communications*, 48, 2109–2119.

Chung, C.-D. (1995) Differentially coherent detection technique for direct–sequence code acquisition in a Rayleigh fading mobile channel, *IEEE Transactions on Communications*, 43(2/3/4), 1116–1126.

Davenport, W.B. and Root, W.L. (1958) An Introduction to the Theory of Random Signals and Noise. McGraw-Hill.

Guenach, M. and Vandendrope, L. (2001) Tracking performance of DA and DD multiuser timing synchronizers short code DS-CDMA systems, *IEEE Journal on Selected Areas in Communications*, 19, 2452–2461.

Ham, D., Luo, J., Pattipati, K. and Willett, P. (2002) A PDA-Kalman approach to multiuser detection in asynchronous CDMA, *IEEE Communications Letters*, 6, 475–477.

Hart, B.W., Van Nee, R.D.J. and Parsad, R. (1996) Performance degradation due to code tracking errors in spread-spectrum code-division multiple-access systems, *IEEE Transactions on Selected Areas in Communications*, 14, 1669–1679.

Holmes, J.K. and Chen, C.C. (1977) Acquisition time of PN spread-spectrum system, *IEEE Transactions on Communications*, 25, 778–784.

Holmes, J.K. and Biederman, L. (1978) Delay-lock-loop mean time to lose lock, *IEEE Transactions on Communications*, COM-26, 1549–1557.

Ibrahim, B.B. and Hamid Aghvami, A. (19894) Direct sequence spread-spectrum matched filter acquisition in frequency-selective Rayleigh fading channels, *IEEE Journal on Selected Areas in Communications*, 12, 885–890.

Iinatti, J. (1998) Mean acquisition time of DS code synchronization in fixed multipath channel, *Proc. of the ISSSTA '98*, MSun City, South-Africa, Vol. 1, pp. 116–120.

Iinatti, J. and Kerhuel, S. (1999) Code synchronisation in two-path channel: effect of Interpath interference on acquisition, *Electronics Letters*, 35(12), 960–962.

Iinatti, J.H.J. (2000) On the threshold setting principles in code acquisition of DC-SS signals, *IEEE Journal on Selected Areas in Communications*, 18(1), 62–72.

Iltis, R.A. (1990) Joint estimation of PN code delay and multipath using the extended Kalman filter, *IEEE Transactions on Communications*, 38, 1677–1685.

Iltis, R.A. (2001) A DS-CDMA tracking mode receiver with joint channel/delay estimation and MMSE detection, *IEEE Transactions on Communications*, 49, 770–1779.

Katz, M.D., Iinatti, J.H. and Glisic, S. (2001) Two-dimensional code acquisition in time and angular domains, *IEEE Journal on Selected areas in Communications*, 19, 2441–2451.

Lee, J.S. and Miller, L.E. (1998) *CDMA Engineering Handbook*, Artech House Publishers.

Lin, J.-C. (2002) Noncoherent sequential PN code acquisition using sliding correlation for ship-asynchronous direct-sequence spread-spectrum communications, *IEEE Transactions on Communications*, 50, 664–676.

Meyr, H. (1976) Delay-lock tracking of stochastic signals, *IEEE Transactions on Communications*, 24, 331–339.

Parkvall, S., Strom, E. and Ottersten, B. (1996) The impact of time error on the performance of linear DS-CDMA, *IEEE Journal of Selected Areas in Communications*, 14(8), 1660–1668.

Peterson, R.L., Ziemer, R.E. and Borth, D.E. (1995) *Introduction to Spread-spectrum Communications*, Prentice Hall.

Pickholtz, R.L., Schilling, D.L. and Milstein, L.B. (1982) Theory of spread-spectrum communications: A tutorial, *IEEE Transactions on Communications*, 30, 855–884.

Polydoros, A. and Simon, M.K. (1984) Generalized serial search code acquisition: the equivalent circular diagram approach, *IEEE Transactions on Communications*, COM-32(12), 1260–1268.

Polydoros, A. and Weber, C.L. (1984) A unified approach to serial search spread-spectrum code acquisition – Part 1: General theory, *IEEE Transactions on Communications*, 32, 542–549.

Proakis, J.G. (1995) *Digital Communications*. McGraw-Hill.

Ristaniemi, T. and Joutsensalo, J. (2001) Code timing acquisition for DS-CDMA in fading channels by differential correlations, *IEEE Transactions on Communications*, 49, 899–910.

Selin, I. and Tuteur, F. (1963) Synchronization of coherent detectors, *IEEE Transactions on Communication Systems*, 100–109.

Sheen, W.-H. and Stuber, G.L. (1994) Effects of multipath fading on delay-locked loops for spread-spectrum systems, *IEEE Transactions on Communications*, 42, 1947–1956.

Sheen, W.-H. and Stuber, G.L. (1995) A new tracking loop for direct sequence spread systems on frequency-selective fading channels, *IEEE Transactions on Communications*, 43, 3063–3072.

Sheen, W.-H. and Wang, H.-C. (2001) A new analysis of direct-sequence pseudonoise code acquisition on Rayleigh fading channels, *IEEE Transactions on Communications*, 19, 2225–2232.

Shin, O.-S. and Lee, K.B. (2001) Utilization of multipaths for spread-spectrum code acquisition in frequency-selective Rayleigh fading channels, *IEEE Transactions on Communications*, 49, 734–743.

Simon, M.K., Omura, J.K., Scholtz, R.A. and Levit, B.K. (1985) *Spread-spectrum Communications*, Vol. III. Computer Science Press.

Sourour, E.A. and Gupta, S.C. (1990) Direct-sequence spread-spectrum parallel acquisition in fading mobile channels, *IEEE Transactions on Communications*, 38, 992–998.

Spilker, J.J. and Magill, D.T. (1961) The Delay-Lock Discriminator: An Optimum Tracking Device, *Proc. IRE*, 49, 1403–1416.

Su, S.-L. and Yen, N.-Y. (1997) Performance of digital code tracking loops for direct-sequence spread-spectrum signals in mobile radio channels, *IEEE Transactions on Communications*, 45, 596–604.

Ward, R.B. (1965) Acquisition of pseudo noise signals by sequential estimation, *IEEE Transactions on Communication Technology*, 13, 474–483.

Yang, L.-L. and Hanzo, L. (2001) Serial acquisition of DS-CDMA signals in multipath fading mobile channel, *IEEE Transactions on Vehicular Technology*, 50, 617–628.

Yen, K. and Hanzo, L. (2001) Genetic algorithm assisted joint multiuser symbol detection and fading channel estimation for synchronous CDMA systems, *IEEE Journal on Selected Areas in Communications*, 19, 985–998.

Yen, N.-Y., Su, S.-L. and Hsieh, S.-C. (1996) Performance analysis of digital delay lock loops in the presence of Doppler shift, *IEEE Transactions on Communications*, 44, 668–674.

Zarrabizadeh, M.H. and Sousa, E.S. (1997) A differentially coherent PN code acquisition receiver for CDMA systems, *IEEE Transactions on Communications*, 45(11), 1456–1465.

Zha, W. and Blostein, S.D. (2002) Multiuser receivers that are robust to delay mismatch, *IEEE Transactions on Communications*, 50(12), 2072–2080.

Appendix 5.A

Let the impulse response of the loop filter be h(t), then loop filter output voltage is given by:

$$v(\lambda) = \int_{-\infty}^{\lambda} \varepsilon(\alpha)h(\lambda - \alpha)d\alpha \tag{5.A1}$$

The phase at the output of the VCO is:

$$\frac{\hat{t}_d}{T_c} = g \int_0^t v(\lambda)d\lambda \tag{5.A2}$$

where g is the VCO gain in Hz/volt. Substituting for $v(\lambda)$ from (5.A1) in (5.A2), we get:

$$\frac{\hat{t}_d}{T_c} = g \int_0^t \int_{-\infty}^{\lambda} \varepsilon(\alpha)h(\lambda - \alpha)d\alpha \, d\lambda \tag{5.A3}$$

Substituting for $\varepsilon(\alpha)$ in (5.94), we get:

$$\frac{\hat{t}_d}{T_c} = g \int_0^t \int_{-\infty}^{\lambda} \sqrt{\frac{P}{2}} \left[D_\Delta(\delta(\alpha)) + \sqrt{\frac{1}{P}} n'(\alpha) \right] h(\lambda - \alpha)d\alpha \, d\lambda \tag{5.A4}$$

Equation (5.A4) represents the non-linear model for the tracking loop in AWGN channel. The non-linearity is introduced by $D_\Delta(\delta)$. For small tracking error (δ), the non-linear function $D_\Delta(\delta)$ can be expressed linearly in terms of δ as shown in Figure 5.A1.

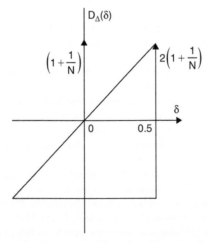

Figure 5.A1   *Linear relationship between $D_\Delta(\delta)$ and tracking error δ for small values of δ.*

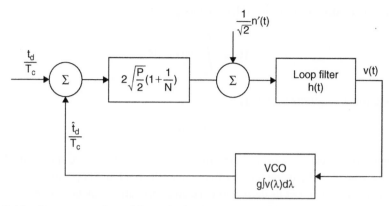

Figure 5.A2 *Linear equivalent of the early–late loop in noisy channel.*

From Figure 5.A1, the slope of the S-curve at $\delta = 0$ is $2\left(1 + \frac{1}{N}\right)$ and $D_\Delta(\delta)$ is given by:

$$D_\Delta(\delta) = 2\left(1 + \frac{1}{N}\right)\delta \quad 0 < \Delta < 1 \tag{5.A5}$$

The linear equivalent of (5.A4) is:

$$\frac{\hat{t}_d}{T_c} = g \int_0^d \int_{-\infty}^\lambda \sqrt{\frac{P}{2}}\left[2\left(1 + \frac{1}{N}\right) + \sqrt{\frac{1}{P}}n'(\alpha)\right]h(\lambda - \alpha)d\alpha\, d\lambda \tag{5.A6}$$

The simplified linear model is shown in Figure 5.A2.

We now derive the expression for the variance of the tracking error (σ_δ^2) in terms of the received signal-to-noise ratio in the loop bandwidth and start by defining σ_δ^2 as:

$$\sigma_\delta^2 = \int_{-\infty}^\infty S_{n''}(f)|H(f)|^2 df \tag{5.A7}$$

where $S_{n''}(f)$ and $H(f)$ represent noise spectral density at the input of the linear equivalent model and the tracking loop transfer function, respectively. Since we assume white noise input, $S_{n''}(f)$ is constant over all frequencies:

$$S_{n''}(f) = S_{n''}(0) \tag{5.A8}$$

Thus equation (5.A7) is simplified to:

$$\sigma_\delta^2 = S_{n''}(0) \int_{-\infty}^\infty |H(f)|^2 df \tag{5.A9}$$

Define the two-sided noise bandwidth of the loop B_ℓ as:

$$B_\ell = \int\limits_{-\infty}^{\infty} |H(f)|^2 df \qquad (5.A10)$$

Therefore, we can re-write (5.A9) as:

$$\sigma_\delta^2 = S_{n''}(0)B_\ell \qquad (5.A11)$$

To compute the tracking error variance from equation (5.A11), we must calculate the loop noise bandwidth B_ℓ using transfer function of the tracking loop and the noise power spectral density $S_{n''}(0)$. Let us derive now the autocorrelation function of the input noise $R_{n'}(\tau)$.

$$R_{n'}(\tau) = E\left\{ n(t)n(t+\tau)\left[C\left(t-\hat{t}_d - \frac{\Delta}{2}T_c\right) - \left(t-\hat{t}_d + \frac{\Delta}{2}T_c\right) \right] \right.$$
$$\left. \times \left[C\left(t+\tau-\hat{t}_d - \frac{\Delta}{2}T_c\right) - C\left(t+\tau-\hat{t}_d + \frac{\Delta}{2}T_c\right) \right] \right\} \quad (5.A12)$$

Since noise $n(t)$ is independent of the code waveform, so the expected values of the noise and the code waveform are also independent.

$$R_{n'}(\tau) = E\{n(t)n(t+\tau)\}E\left\{ \left[C\left(t-\hat{t}_d - \frac{\Delta}{2}T_c\right) - C\left(t-\hat{t}_d + \frac{\Delta}{2}T_c\right) \right] \right.$$
$$\left. \times \left[C\left(t+\tau-\hat{t}_d - \frac{\Delta}{2}T_c\right) - C\left(t+\tau-\hat{t}_d + \frac{\Delta}{2}T_c\right) \right] \right\} \quad (5.A13)$$

Now
$$E\{n(t)n(t+\tau)\} = \frac{N_0}{2}\delta(\tau) \qquad (5.A14)$$

where $\frac{N_0}{2}$ is the two-sided noise density in W/Hz.

Substituting (5.A14) for the expected value of the noise into (5.A13) simplifies the expression for $R_{n'}(\tau)$ to:

$$R_{n'}(\tau) = \frac{N_0}{2}\delta(\tau)E\left\{ \left[C\left(t-\hat{t}_d - \frac{\Delta}{2}T_c\right) - C\left(t-\hat{t}_d + \frac{\Delta}{2}T_c\right) \right] \right.$$
$$\left. \times \left[C\left(t+\tau-\hat{t}_d - \frac{\Delta}{2}T_c\right) - C\left(t+\tau-\hat{t}_d + \frac{\Delta}{2}T_c\right) \right] \right\} \quad (5.A15)$$

Now
$$\frac{N_0}{2}\delta(\tau) = 0 \qquad \text{for } \tau \neq 0$$
$$= \frac{N_0}{2} \qquad \text{for } \tau = 0$$

Thus we can simplify (5.A13) to:

$$R_{n'}(\tau) = \frac{N_0}{2}\delta(\tau)E\left\{\left[C\left(t - \hat{t}_d - \frac{\Delta}{2}T_c\right) - C\left(t - \hat{t}_d + \frac{\Delta}{2}T_c\right)\right]^2\right\} \quad (5.A16)$$

$$= \frac{N_0}{2}\delta(\tau)\left\{E\left[C^2\left(t - \hat{t}_d - \frac{\Delta}{2}T_c\right)\right]\right. \quad (5.A17)$$

$$\left. - 2E\left[C\left(t - \hat{t}_d - \frac{\Delta}{2}T_c\right) * C\left(t - \hat{t}_d + \frac{\Delta}{2}T_c\right)\right]\right.$$

$$\left. + E\left[C^2\left(t - \hat{t}_d + \frac{\Delta}{2}T_c\right)\right]\right\}$$

Since the chips in C(t) take on only values of ± 1, therefore $C^2(t) = 1.0$ and (5.A17) can be simplified to:

$$R_{n'}(\tau) = N_0\delta(\tau)\left[1 - E\left[C\left(t - \hat{t}_d - \frac{\Delta}{2}T_c\right) * C\left(t - \hat{t}_d + \frac{\Delta}{2}T_c\right)\right]\right]$$

Thus, $\qquad R_{n'}(\tau) = N_0\delta(\tau)[1 - R_c(\Delta T_c)]$ \qquad (5.A18)

Now substitute for the autocorrelation for m-sequence autocorrelation:

$$R_{n'}(\tau) = \Delta N_0\delta(\tau)\left[1 + \frac{1}{N}\right] \quad \text{for } 0 < \Delta < 1 \quad (5.A19)$$

$$R_{n'}(\tau) = N_0\delta(\tau)\left[1 + \frac{1}{N}\right] \quad \text{for } 1 \leq \Delta \leq (N - 1)$$

The two-sided power spectral density $S_{n'}(f)$ of $n'(t)$ is the Fourier transform of $R_{n'}(\tau)$ such that:

$$S_{n'}(f) = F[R_{n'}(\tau)] \quad (5.A20)$$

$$= \Delta N_0\left[1 + \frac{1}{N}\right] \quad \text{for } 0 < \Delta < 1$$

$$= N_0\left[1 + \frac{1}{N}\right] \quad \text{for } 1 \leq \Delta \leq (N - 1)$$

The noise spectral density $S_{n''}(f)$ at the input of the linear model is related to $S_{n'}(f)$ as:

$$S_{n''}(f) = \left[\frac{1}{\sqrt{\dfrac{P}{2}}\dfrac{d}{d\delta}(D_\Delta(\delta))|_{\delta=0}}\right]^2 S_{n'}(f) \quad (5.A21)$$

Now $\frac{d}{d\delta}(D_\Delta(\delta))|_{\delta=0}$ is the slope of the S-curve at $\delta = 0$ which is $2(1 + \frac{1}{N})$, so that:

$$S_{n''}(f) = \left[\frac{1}{2\sqrt{\frac{P}{2}\left(1 + \frac{1}{N}\right)}} \right]^2 S_{n'}(f) \tag{5.A22}$$

where N is the code sequence period.

Substituting for $S_{n'}(f)$ in the above equation, we get:

$$S_{n''}(f) = \left[\frac{1}{2\sqrt{P}\left(1 + \frac{1}{N}\right)} \right]^2 \Delta N_0 \left[1 + \frac{1}{N}\right] = \frac{\Delta N_0}{4P\left[1 + \frac{1}{N}\right]} \quad \text{for } 0 < \Delta < 1$$

$$S_{n''}(f) = \left[\frac{1}{2\sqrt{P}\left(1 + \frac{1}{N}\right)} \right]^2 N_0 \left[1 + \frac{1}{N}\right] = \frac{N_0}{4P\left[1 + \frac{1}{N}\right]} \quad \text{for } 1 \leq \Delta \leq (N-1)$$

Equation (5.A11) becomes:

$$\sigma_\delta^2 = \frac{\Delta N_0}{4P\left[1 + \frac{1}{N}\right]} B_\ell \quad \text{for } 0 < \Delta < 1 \tag{5.A23}$$

$$\sigma_\delta^2 = \frac{N_0}{4P\left[1 + \frac{1}{N}\right]} B_\ell \quad \text{for } 1 \leq \Delta \leq (N-1) \tag{5.A24}$$

Now define the loop signal-to-noise ratio $(SNR)_\ell$ as:

$$(SNR)_\ell = \frac{P}{\frac{N_0}{2}B_\ell} \tag{5.A25}$$

Substituting for loop signal-to-noise ratio in (5.A23) and (5.A24) we get:

$$\sigma_\delta^2 = \frac{\Delta}{2(SNR)_\ell\left[1 + \frac{1}{N}\right]} \quad \text{for } 0 < \Delta < 1 \tag{5.A26}$$

$$\sigma_\delta^2 = \frac{1}{2(SNR)_\ell\left[1 + \frac{1}{N}\right]} \quad \text{for } 1 \leq \Delta \leq (N-1) \tag{5.A27}$$

Appendix 5.B

Using $\bar{y}_1(t)$ and $\bar{y}_2(t)$ in (5.94), (5.95) and noise in (5.104), which are repeated here, we get:

$$\bar{y}_1(t) = \sqrt{P}R_c\left(\delta + \frac{\Delta}{2}T_c\right)\cos\left[\omega_{IF}t + \theta_d(t - t_d) + (\phi - \phi')\right] \qquad (5.94)$$

$$\bar{y}_2(t) = \sqrt{P}R_c\left(\delta - \frac{\Delta}{2}T_c\right)\cos\left[\omega_{IF}t + \theta_d(t - t_d) + (\phi - \phi')\right] \qquad (5.95)$$

$$n_j'(t) = \sqrt{2}\{n_{jI}'(t)\cos(\omega_{IF}t) - n_{jQ}'(t)\sin(\omega_{IF}t)\} \qquad (5.104)$$

where $j = 1, 2$. The output of the phase discriminator is given by:

$$\varepsilon(t) = (\bar{y}_2(t) + n_2'(t))^2 - (\bar{y}_1(t) + n_1'(t))^2$$

$$\varepsilon(t) = \left\{\sqrt{P}R_c\left(\delta - \frac{\Delta}{2}\right)T_c\cos(\omega_{IF}t + \theta_d(t - t_d) + (\phi - \phi')) + n_2'(t)\right\}^2$$

$$- \left\{\sqrt{P}R_c\left(\delta + \frac{\Delta}{2}\right)T_c\cos(\omega_{IF}t + \theta_d(t - t_d) + (\phi - \phi')) + n_1'(t)\right\}^2$$

$$\varepsilon(t) = \frac{P}{2}\left\{R_c^2\left(\delta - \frac{\Delta}{2}\right)T_c - R_c^2\left(\delta + \frac{\Delta}{2}\right)T_c\right\} + n_2'^2 - n_1'^2$$

$$+ 2(\sqrt{P})\left\{R_c\left(\delta - \frac{\Delta}{2}\right)T_c n_2'(t) - R_c\left(\delta + \frac{\Delta}{2}\right)T_c n_1'(t)\right\}$$

$$\cos(\omega_{IF}t + \theta_d(t - t_d) + (\phi - \phi'))$$

$$(5.B1)$$

Substituting for $n_1'(t)$ and $n_2'(t)$ from (5.104), we get:

$$\varepsilon(t, \delta) = \frac{P}{2}\left\{R_c^2\left(\delta - \frac{\Delta}{2}\right)T_c - R_c^2\left(\delta + \frac{\Delta}{2}\right)T_c\right\} + 2\{n_{2I}(t)\cos(\omega_{IF}(t))$$

$$- n_{2Q}(t)\sin(\omega_{IF}(t)\}^2 - 2\{n_{1I}(t)\cos(\omega_{IF}(t)) - n_{1Q}(t)\sin(\omega_{IF}(t)\}^2$$

$$+ 2(\sqrt{P})\left\{R_c\left(\delta - \frac{\Delta}{2}\right)T_c\sqrt{2}[n_{2I}(t)\cos(\omega_{IF}(t)) - n_{2Q}(t)\sin(\omega_{IF}(t))]\right.$$

$$\left. - R_c\left(\delta + \frac{\Delta}{2}\right)T_c\sqrt{2}[n_{1I}(t)\cos(\omega_{IF}(t)) - n_{1Q}(t)\sin(\omega_{IF}(t))]\right\}$$

$$\cos(\omega_{IF}t + \theta(t - t_d) + (\phi - \phi'))$$

$$(5.B2)$$

Simplifying the above equation, we get:

$$\epsilon(t, \delta) = \frac{P}{2} \left\{ R_c^2 \left(\delta - \frac{\Delta}{2} \right) T_c - R_c^2 \left(\delta + \frac{\Delta}{2} \right) T_c \right\} + \{ n_{2I}^2(t) + n_{2Q}^2(t) - n_{1I}^2(t) - n_{1Q}^2(t) \}$$

$$+ 2(\sqrt{P}) \left[R_c \left(\delta - \frac{\Delta}{2} \right) T_c \sqrt{2} \left[n_{2I}(t) \frac{\cos(\theta_d(t - T_d) + (\phi - \phi'))}{2} \right. \right.$$

$$+ n_{2Q}(t) \frac{\sin(\theta_d(t - T_d) + (\phi - \phi'))}{2} \right]$$

$$- R_c \left(\delta + \frac{\Delta}{2} \right) T_c \sqrt{2} \left[n_{1I}(t) \frac{\cos(\theta_d(t - T_d) + (\phi - \phi'))}{2} \right.$$

$$\left. \left. - n_{1Q} \frac{\sin(\theta_d(t - T_d) + (\phi - \phi'))}{2} \right] \right] \tag{5.B3}$$

$$\epsilon(t, \delta) = \frac{P}{2} \left\{ R_c^2 \left(\delta - \frac{\Delta}{2} \right) T_c - R_c^2 \left(\delta + \frac{\Delta}{2} \right) T_c \right\} + \sqrt{2P} \left\{ R_c \left(\delta - \frac{\Delta}{2} \right) T_c \, n_{2I}(t) \right.$$

$$\left. - R_c \left(\delta + \frac{\Delta}{2} \right) T_c \, n_{1I}(t) \right\} \cos(\theta_d(t - T_d) + (\phi - \phi'))$$

$$+ \sqrt{2P} \left\{ R_c \left(\delta - \frac{\Delta}{2} \right) T_c \, n_{2Q}(t) - R_c \left(\delta + \frac{\Delta}{2} \right) T_c \, n_{1Q}(t) \right\}$$

$$\sin(\theta_d(t - T_d) + (\phi - \phi')) + \{ n_{2I}^2(t) + n_{2Q}^2(t) - n_{1I}^2(t) - n_{1Q}^2(t) \} \tag{5.B4}$$

$$R_\epsilon(\tau) = E[\epsilon(t)\epsilon(t + \tau)] \tag{5.B5}$$

$$E[\cos(\phi(t)] = E[\cos(\phi(t + \tau)] = E[\sin(\phi(t)] = E[\sin(\phi(t + \tau)] = 0 \tag{5.B6}$$

Using (5.B6) we can simplify (5.B5):

$$R_\epsilon(\tau) = \frac{(P)^2}{4} \left\{ R_c^2 \left(\delta - \frac{\Delta}{2} \right) T_c - R_c^2 \left(\delta + \frac{\Delta}{2} \right) T_c \right\}^2$$

$$+ 2P \, E \left\{ R_c^2 \left(\delta - \frac{\Delta}{2} \right) T_c \, n_{2I}(t) n_{2I}(t + \tau) + R_c^2 \left(\delta + \frac{\Delta}{2} \right) T_c n_{1I}(t) n_{1I}(t + \tau) \right\}$$

$$\times E[\cos(\theta_d(t + \tau - t_d) + (\phi - \phi')) \cos(\theta(t - t_d) + (\phi - \phi'))]$$

$$+ 2P \, E \left\{ R_c^2 \left(\delta - \frac{\Delta}{2} \right) T_c \, n_{2Q}(t) n_{2Q}(t + \tau) + R_c^2 \left(\delta + \frac{\Delta}{2} \right) T_c \, n_{1Q}(t) n_{1Q}(t + \tau) \right\}$$

$$\times E[\sin(\theta_d(t + \tau - t_d) + (\phi - \phi'))\sin(\theta(t - t_d) + (\phi - \phi'))]$$

$$+ E\left\{ n_{2I}^2(t)n_{2I}^2(t + \tau) + n_{2Q}^2(t)n_{2Q}^2(t + \tau) + n_{1I}^2(t)n_{1I}^2(t + \tau) + n_{1Q}^2(t)n_{1Q}^2(t + \tau) \right.$$

$$\left. - n_{2I}^2(t)n_{1I}^2(t + \tau) - n_{2Q}^2(t)n_{1Q}^2(t + \tau) - n_{1I}^2(t)n_{2I}^2(t + \tau) - n_{1Q}^2(t)n_{2Q}^2(t + \tau) \right\}$$

$$(5.B7)$$

Let $n_b(t)$ denote any one of the four baseband noise processes $n_{1I}(t), n_{1Q}(t), n_{2I}(t), n(t)$.

Therefore, (5.B7) becomes:

$$R_\varepsilon(\tau) = \left(\frac{P}{2}\right)^2 \left\{ R_c^2\left(\delta - \frac{\Delta}{2}\right)T_c - R_c^2\left(\delta + \frac{\Delta}{2}\right)T_c \right\}^2$$

$$+ 2P\left\{ R_c^2\left(\delta - \frac{\Delta}{2}\right)T_c + R_c^2\left(\delta + \frac{\Delta}{2}\right)T \right\} E[n_b(t)n_b(t + \tau)]E[\cos(\theta_d(t - t_d)$$

$$- \theta_d(t + \tau - t_d)] + 4E[n_b(t)n_b(t + \tau)]^2 - 4E^2[n_b^2(t)] \qquad (5.B8)$$

$$E\{\cos^2(\theta_d(t - T_d) - \theta_d(t + \tau - T_d)\} = R_\theta(\tau) = \frac{1}{2} \qquad (5.B9)$$

$$S_{\theta_d}(f) = 2\int\limits_{-\infty}^{\infty} R_\theta(\tau)e^{-j\omega\tau}d\tau = 2\int\limits_{-\infty}^{\infty} \frac{1}{2}e^{-j\omega\tau}d\tau = \delta(f) \qquad (5.B10)$$

Fourier Transform of the first term in $R\varepsilon(\tau)$ is:

$$\left(\frac{P}{2}\right)^2 \left\{ R_c^2\left(\delta - \frac{\Delta}{2}\right)T_c - R_c^2\left(\delta + \frac{\Delta}{2}\right)T_c \right\}^2 \delta(f) \qquad (5.B11)$$

Fourier Transform of the second term in $R\varepsilon(\tau)$ is:

$$E[n_b(t)n_b(t + \tau)] = R_{n_b}(\tau) \qquad (5.B12)$$

Fourier Transform of $R_{n_b}(\tau) = S_{n_b}(f)$

Fourier Transform of $E[n_b(t)n_b(t+\tau)]E[\cos[\theta_d(t-T_d) - \theta_d(t+\tau-T_d)] = S_{n_b}(f) * S_{\theta_d}(f)$

$$(5.B13)$$

Therefore, Fourier Transform of the second term in $R\varepsilon(\tau)$ is:

$$2P\left\{ R_c^2\left(\delta - \frac{\Delta}{2}\right)T_c + R_c^2\left(\delta + \frac{\Delta}{2}\right)T_c \right\} [S_{n_b}(f) * S_{\theta_d}(f)] \qquad (5.B14)$$

The third term is $4 \times$ autocorrelation of a square law device with $n_b(t)$ as its input:

$$4\{E[n_b^2(t)n_b^2(t + \tau)] - (E[n_b^2(t)])^2\} = 4E\{[n_b(t)n_b(t + \tau)]^2 - (E[n_b^2(t)])^2\} \qquad (5.B15)$$

It was shown in Davenport and Root (1958, pp. 168) that if x_1, x_2, x_3, x_4 are random variables with Gaussian distribution and zero mean then:

$$E[x_1 x_2 x_3 x_4] = E[x_1 x_2]E[x_3 x_4] + E[x_1 x_3]E[x_2 x_4] + E[x_1 x_4]E[x_2 x_3]$$

Thus:

$$
\begin{aligned}
E[n_b^2(t)n_b^2(t+\tau)] &= E[n_b(t)n_b(t)n_b(t+\tau)n_b(t+\tau)] \\
&= E[n_b(t)n_b(t)]E[n_b(t+\tau)n_b(t+\tau)] \\
&\quad + E[n_b(t)n_b(t+\tau)]E[n_b(t)n_b(t+\tau)] \\
&\quad + E[n_b(t)n_b(t+\tau)]E[n_b(t)n_b(t+\tau)] \\
&= E[n_b^2(t)]E[n_b^2(t+\tau)] \\
&\quad + R_{nb}(\tau)\,R_{nb}(\tau) \\
&\quad + R_{nb}(\tau)\,R_{nb}(\tau) \\
&= E^2[n_b^2(t)] + 2R_{n_b}^2(\tau) \quad\quad\quad (5.B16)
\end{aligned}
$$

Now $E[n_b^2(t)] = \sigma_{n_b}^2$

Thus
$$E[n_b^2(t)n_b^2(t+\tau)] = [\sigma_{n_b}^2]^2 + 2R_{n_b}^2(\tau) \quad\quad\quad (5.B17)$$

The third term is:

$$4\{\sigma_{nb}^4 + 2R_{nb}^2 - \sigma_{nb}^4\} = 8R_{nb}^2 \quad\quad\quad (5.B18)$$

The output power spectrum of third term is:

$$8S_{n_b}(f) * S_{n_b}(f)] \quad\quad\quad (5.B19)$$

The power spectrum $S_\varepsilon(f)$ of error signal $\varepsilon(t)$ is:

$$
\begin{aligned}
S_\varepsilon(f) &= \left(\frac{P}{2}\right)^2 \left\{R_c^2\left(\delta - \frac{\Delta}{2}\right)T_c - R_c^2\left(\delta + \frac{\Delta}{2}\right)T_c\right\}^2 \delta(f) \\
&\quad + 2P\left\{R_c^2\left(\delta - \frac{\Delta}{2}\right)T_c + R_c^2\left(\delta + \frac{\Delta}{2}\right)T_c\right\} S_{nb}(f) * S_{\theta d}(f) \\
&\quad + 8S_{nb}(f) * S_{nb}(f) \quad\quad\quad (5.B20)
\end{aligned}
$$

The noise component $S_{n_\varepsilon}(f)$ in $S_\varepsilon(f)$ is due to:

$$S_{n_\varepsilon}(f) = 2P\left\{R_c^2\left(\delta - \frac{\Delta}{2}\right)T_c + R_c^2\left(\delta + \frac{\Delta}{2}\right)T_c\right\} S_{nb}(f)^* S_{\theta d}(f) + 8S_{nb}(f)^* S_{nb}(f)$$
$$(5.B21)$$

The noise outputs of the correlators are given by (5.102) and (5.103) and since noise and the spreading waveform are independent, the power spectrum of noise $n_b(t)$ can be calculated by convolving the spectrum on $n(t)$ and spectrum of the spreading waveform. That is:

$$S_{nb}(f) = S_c(f) * S_n(f) = \frac{N_0}{4} \quad\quad\quad (5.B22)$$

Assume the IF bandpass filters are perfect brick wall filters with one-sided noise bandwidth B_n then at the output of the filter:

$$S_{nb}(f) * S_{nb}(f) = \int_{-\frac{B_n}{2}}^{\frac{B_n}{2}} \left[\frac{N_0}{4}\right]^2 df = \left[\frac{N_0}{4}\right]^2 B_n \qquad (5.B23)$$

$$S_{n_\varepsilon}(f) = \frac{\eta}{2} = \frac{2PN_0}{4} \left\{ R_c^2\left(\delta - \frac{\Delta}{2}\right) T_c + R_c^2\left(\delta + \frac{\Delta}{2}\right) T_c \right\} \delta(f) + 8\left[\frac{N_0}{4}\right]^2 B_n \quad (5.B24)$$

$$\frac{\eta}{2} = \frac{PN_0}{2} \left\{ R_c^2\left(\delta - \frac{\Delta}{2}\right) T_c + R_c^2\left(\delta + \frac{\Delta}{2}\right) T_c \right\} + \frac{1}{2}[N_0]^2 B_n \qquad (5.B25)$$

$$D_{nc\cdot\Delta}(\delta) = \left[R_c^2\left(\delta - \frac{\Delta}{2}\right) T_c - R_c^2\left(\delta + \frac{\Delta}{2}\right) T_c \right] \qquad (5.B26)$$

$$R_c\left(\delta - \frac{\Delta}{2}\right) = 1 + \left(\delta - \frac{\Delta}{2}\right)\left(1 + \frac{1}{N}\right)$$

$$R_c\left(\delta + \frac{\Delta}{2}\right) = 1 - \left(\delta + \frac{\Delta}{2}\right)\left(1 + \frac{1}{N}\right) \quad \text{for } -\frac{\Delta}{2} \le \delta \le \frac{\Delta}{2} \qquad (5.B27)$$

$$
\begin{aligned}
D_{nc\cdot\Delta}(\delta) &= \left[1 + \left(\delta - \frac{\Delta}{2}\right)\left(1 + \frac{1}{N}\right)\right]^2 - \left[1 - \left(\delta + \frac{\Delta}{2}\right)\left(1 + \frac{1}{N}\right)\right]^2 \\
&= 1 + 2\left(\delta - \frac{\Delta}{2}\right)\left(1 + \frac{1}{N}\right) + \left(\delta - \frac{\Delta}{2}\right)^2\left(1 + \frac{1}{N}\right)^2 \\
&\quad - \left[1 - 2\left(\delta + \frac{\Delta}{2}\right)\left(1 + \frac{1}{N}\right) + \left(\delta + \frac{\Delta}{2}\right)^2\left(1 + \frac{1}{N}\right)^2 \right] \\
&= 1 - 1 + 2\left(1 + \frac{1}{N}\right)\left(\delta - \frac{\Delta}{2} + \delta + \frac{\Delta}{2}\right) \\
&\quad + \left(1 + \frac{1}{N}\right)^2\left[\delta^2 - \delta\Delta + \left(\frac{\Delta}{2}\right)^2 - \delta^2 - \delta\Delta - \left(\frac{\Delta}{2}\right)^2 \right] \\
&= 4\delta\left(1 + \frac{1}{N}\right) - 2\delta\Delta\left(1 + \frac{1}{N}\right)^2 \\
&= 2\left(1 + \frac{1}{N}\right)\left[2 - \left(1 + \frac{1}{N}\right)\Delta \right]\delta \qquad (5.B28)
\end{aligned}
$$

$$\frac{d}{d\delta}(D_{nc\cdot\Delta}(\delta))_{\delta=0} = 2\left(1 + \frac{1}{N}\right)\left[2 - (1 + \frac{1}{N})\Delta \right]$$

$$a_d = \frac{P}{2}\frac{d}{d\delta}(D_{nc\cdot\Delta}(\delta))_{\delta=0} = P\left(1 + \frac{1}{N}\right)\left[2 - \left(1 + \frac{1}{N}\right)\Delta\right] \tag{5.B29}$$

The tracking jitter, σ_δ^2, is given by:

$$\sigma_\delta^2 = \frac{\eta}{2a_d^2}B_\ell \tag{5.B30}$$

Therefore, the tracking jitter becomes:

$$\sigma_\delta^2 = \frac{PN_0}{2}\left\{R_c^2\left(\delta - \frac{\Delta}{2}\right)T_c + R_c^2\left(\delta + \frac{\Delta}{2}\right)T_c\right\} + \left\{\frac{1}{2}[N_0]^2B_n\right\}\frac{1}{K_d^2}B_L \tag{5.B31}$$

where a_d is given by (5.B29) and the first term in (5.B31) is given by (5.B27).

$$\left\{R_c^2\left(\delta - \frac{\Delta}{2}\right)T_c + R_c^2\left(\delta + \frac{\Delta}{2}\right)T_c\right\} = \left\{1 + \left(\delta - \frac{\Delta}{2}\right)\left(1 + \frac{1}{N}\right)\right\}^2$$

$$+ \left\{1 - \left(\delta + \frac{\Delta}{2}\right)\left(1 + \frac{1}{N}\right)\right\}^2 \tag{5.B32}$$

$$\sigma_\delta^2 = \frac{PN_0}{2}\left(\left\{1 + \left(\delta - \frac{\Delta}{2}\right)\left(1 + \frac{1}{N}\right)\right\}^2\right.$$

$$\left. + \left\{1 - \left(\delta + \frac{\Delta}{2}\right)\left(1 + \frac{1}{N}\right)\right\}^2 + \frac{1}{2}[N_0]^2B_n\right)\frac{1}{a_d^2}B_\ell \tag{5.B33}$$

6

Cellular Code Division Multiple Access (CDMA) Principles

By the end of the first decade of the new millennium, the population served by wireless mobile phones is expected to exceed that served by wire line. Clearly, to ensure fair use of the common radio resource, users must conform to a specific access technology. The implementation of efficient 'multiple access' schemes permit numerous users with diverse requirements to obtain network services simultaneously from the shared radio spectrum.

The origin of multiple access techniques has been around for some time (Viterbi, 1999), when Marconi's 'Tuned Circuit' was the enabling technology for Frequency Division Multiplexing (FDM). This was eventually developed into the Frequency Division Multiple Access (FDMA) technique where multiple modulated carriers provided separate channels.

Traditionally, this form of multiple access is based on the division of the available frequency band into a number of sub-bands, each containing the transmission from an individual user. Channels can be assigned on demand or on a permanent basis. The number of available channels limits the ultimate number of users accessing the system simultaneously. This scheme was used in first generation wireless networks such as the Total Access Connection System (TACS) in the UK.

The designers of satellite communications systems in the 1960s were faced with a dilemma when they used FDMA techniques. To make best use of limited satellite transmitter power, it was necessary to drive the on-board power amplifier into saturation. However, when multiple carriers arrived at the satellite simultaneously, the non-linearity of the amplifier generated inter-modulation products that cause distortion. Consequently, FDMA was not well-suited to this application.

The search for an alternative scheme contributed to the invention of Time Division Multiple Access (TDMA). Just as FDMA is related to FDM, TDMA is related to Time Division Multiplexing (TDM), a communications technique developed from the parallel to serial conversion methods used from the introduction of the computer technology in the 1950s.

In TDMA, only one signal is received at any instant, which avoids the generation of cross-modulation products. Each signal is transmitted within a particular time slot in a time frame, and a particular signal utilizes the entire available spectrum for the duration of the allocated time slot. TDMA superseded FDMA for satellite communications in the 1970s. It is the multiple access technique used in the second generation (GSM) mobile terrestrial wireless communication (Macario, 1997; Grag and Wilkes, 1996). In this application, analogue modulation (FDMA) can be seen as the technique of the first generation wireless networks, and digital modulation (TDMA) as the technique of the second generation wireless networks.

In more recent developments, the separation of the received signal into its constituent elements is done on a code sequence basis. This scheme is called Code Division Multiple Access (CDMA). Its origin dates back as early as World War II when it was used with great sophistication in order to protect military communications from hostile jamming, interference or interception. CDMA became more important to military satellite communications in the 1970s and 1980s and is used by the current low orbiting satellite networks.

In the CDMA scheme, each user is assigned a unique code sequence (signature sequence). Transmission from active users is not coordinated in time or frequency, resulting in Multiple Access Interference (MAI). This effect limits the total number of users accessing the network simultaneously. CDMA was standardized in 1993 for cellular telephony (IS-95) with the aim of supporting millions of subscribers globally (Glisic and Vucetic, 1997). By the end of the last century, the standards for the third generation (3G) high-speed wireless data networks based on using CDMA technology were universally endorsed (Steele and Hanzo, 1999).

In this chapter, we introduce CDMA communications technology in cellular systems and explore single-user detection techniques. Multi-user detection is dealt with in the next chapter. In Section 6.1, we consider the characteristics of the radio channel carrying the CDMA signals, including the scattering effects of the terrestrial environment on radio wave propagation, statistics of the signal parameters, and path loss predictions. The parameters affecting the performance of the cellular CDMA network such as Near–Far effects, power control schemes are considered in Section 6.2, and interference in Section 6.3, while Sections 6.4, 6.5 and 6.6 present the single-user receiver. The system capacity is discussed in Section 6.5, while uplink capacity is considered in Section 6.9. Section 6.10 presents the effects of power control errors on link capacity, while call blocking on the uplink is covered in Section 6.11. The salient points of the chapter are summarized in Section 6.12.

6.1 Wideband mobile channel

The high data rate service provisions of current and future wireless systems substantiate our consideration for the wideband characteristics of the mobile channel.

In wireless communications, transverse Electromagnetic (EM) waves propagate from/to a fixed base-station to/from a mobile receiver. The electric and magnetic fields of the waves are orthogonal to each other and to the direction of the energy propagation. The two fields are closely related so that there is a continuous flow of energy from one to the other. The range of EM wave frequencies, which are used in mobile radios, are subject to reflection and scattering by solid objects, and when they penetrate into buildings their propagation is altered. There is seldom direct line-of-sight transmission between base station and mobile transceivers in modern wireless systems. These physical phenomena cause both short-term and long-term signal fading with specific fading statistical characteristics. Additionally, the radio wave path losses depend on the nature of the terrain between the base station and the mobile unit.

6.1.1 Propagation of radio waves

We start this section by reminding the reader of the basic phenomena that take place during transmission over a mobile wireless channel.

Radio waves emitted from transmitting antenna are reflected and scattered by surrounding buildings, hills, and other obstacles in the radio path so that they arrive at the mobile receiver from different directions and at different times. A single radio wave pursues multiple paths and, consequently, a number of copies of the original wave with different amplitudes and phases reach the receiving antenna. This phenomenon is called multipath radio wave propagation. The superposition of these waves sometimes leads to a destructive addition when the waves are out of phase causing the radio signal to fade. Radio wave reflections in built-up areas and the effects of terrain undulation over long enough distances between base station and mobile unit cause long-term variation in the average signal level called *slow (log-normal) fading*. The multipath propagation (Lee, 1974) caused by reflection from local scatterers, causes the signal level to vary quite substantially with location. These rapid short-term fluctuations are called *fast (Rayleigh) fading*.

Bello (1963) has characterized the scattering process in mobile radio channels as 'uncorrelated-scattering', since the reflected echoes have different path lengths and are uncorrelated in amplitude. The mainly non-stationary character of multipath propagation has a long-term constant average power and can be described as a *wide-sense stationary process*. Consequently, mobile radio channels are described by Bello as *wide-sense stationary uncorrelated scattering (WSSUS)* channels.

The propagation of the radio wave over different paths means that the arrival of each component is spread out in time. This variation of the radio wave arrival time is known as '*delay spread*' τ_d, and is the cause of inter-symbol interference in digital communications. In fact, $1/\tau_d$ defines the maximum symbol rate which can be transmitted over the multipath radio channel with moderate distortion. The coherent bandwidth of a signal, B_c, is the bandwidth for which the frequency components of the signal are correlated and is related to the delay spread of the channel as:

$$B_c \approx \frac{1}{\tau_d} \tag{6.1}$$

When the bandwidth of the transmitted radio signal is greater than B_c, the channel is called *'frequency-selective'*, causing severe distortion to the signal. In such a case, frequency components greater than B_c are subjected to uncorrelated fading while components less than B_c are affected by flat fading. Therefore, the channel is *'frequency non-selective'* when the bandwidth of the signal is less or equal to B_c.

Any relative motion of the receiver with respect to the transmitter changes the frequency of the transmitted radio wave f_c by Doppler shift f_d such that the frequency of the received signal $f_i(\alpha)$ is given by:

$$f_i(\alpha) = f_c + \frac{v}{c}f_c \cos \alpha = f_c\left(1 + \frac{v}{c}\cos \alpha\right) \tag{6.2}$$

where v is the speed of the receiver (in a vehicle for example) and c is the speed of light $(3 \times 10^8$ m/s) and α is the angle between direction of signal received and direction of the vehicle. Since $-1 \le \cos \alpha_i \le 1$, the received energy will be spread into the range $f_c \pm \frac{v}{c}f_c$. The maximum Doppler frequency shift, f_d, is also known as *maximum fade rate* and is given by:

$$f_d = \frac{v}{c}f_c \tag{6.3}$$

To appreciate the effects of receiver motion on the signal reception, consider a vehicle travelling at speed v and receiving a single unmodulated carrier transmitted by a vertically polarized omni-directional antenna. In an urban environment, the receiver acquires a large number of independent scattered waves (N) as shown in Figure 6.1 where $I = 1, 2, \ldots, N$. Each of these waves has random amplitude A_i, uniformly distributed phase θ_i and an incident angle α_i to the direction of vehicle travel.

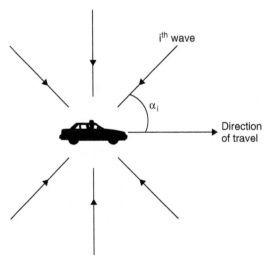

Figure 6.1   *Reception of scattered waves by a mobile station.*

Thus, the ith received wave, $r_i(t)$, can be expressed as:

$$r_i(t) = A_i \cos(\omega_i(\alpha_i)t + \theta_i) \tag{6.4}$$

The resulting received signal, $r(t)$, is the sum on N waves as:

$$r(t) = \sum_{i=0}^{N-1} A_i \cos(\omega_i(\alpha_i)t + \theta_i) \tag{6.5}$$

Let us denote the statistical distribution of α by $p(\alpha)$. The power received in a differential angle $d\alpha$ is then given by $P \cdot p(\alpha) \cdot d\alpha$ where P is the average power received by an isotropic antenna.

Consider an ideal transmission scenario where variation of received power is mainly due to the Doppler effect, so that the received power within differential frequency df, is given by $S(f)^* df$ where $S(f)$ is the Doppler Power Spectral Density (PSD). Consequently, we have:

$$P \cdot [g(\alpha)p(\alpha) + g(-\alpha)p(-\alpha)]|d\alpha| = S(f) \cdot |df|$$

where $g(\alpha)$ is the antenna gain at angle α. Now $f = f_C + f_d \cos\alpha$ so that:

$$|df| = f_d \sin\alpha|d\alpha| = f_d\sqrt{1 - \left(\frac{f - f_c}{f_d}\right)^2}|d\alpha|$$

Therefore,
$$S(f) = \frac{P \cdot [g(\alpha)p(\alpha) + g(-\alpha)p(-\alpha)]}{f_d\sqrt{1 - \left(\frac{f-f_c}{f_d}\right)^2}} \quad \text{for } |f - f_c| < f_d$$
$$= 0 \quad \text{otherwise} \tag{6.6}$$

Consider an omni-directional receiving antenna with constant gain so that $g(\alpha) = g(d - \alpha) = g$ and assuming that the incident angle is uniformly distributed between $0 \leq \alpha \leq \pi$, so that $p(\alpha) = \frac{1}{\pi}$. Substituting for $p(\alpha)$ in (6.6), we get Doppler PSD as:

$$S(f) = \frac{P.g}{\pi f_d\sqrt{1 - \left(\frac{f-f_c}{f_d}\right)^2}} \quad \text{for } |f - f_c| < f_d$$
$$= 0 \quad \text{otherwise} \tag{6.7}$$

The $S(f)$ given by (6.7) is plotted in Figure 6.2 which shows the classical Doppler power spectral density of Jakes (1994).

Measurements by Bug et al. (2001) on broadband channels with a bandwidth of 8 MHz using the WSSUS theory of Bello (1963) have shown that Doppler PSD for the wideband channels are very similar to the *Clarke-Jakes model*. Narasimhan and Cox (1999) have derived a general model for the Doppler power spectrum in a 3D wireless environment that reduces to the Clarke-Jakes model in a 2D environment.

In CDMA systems, the individual paths can be resolved if their relative delay is larger than chip duration T_c. Experiments in Japan have shown that in urban areas, the delay spread of a wideband fading channel ranges from 1 to 10 μs which is larger than the chip duration specified in IS-95 which is 0.8 μs.

Multipath fading in wideband channels has been studied by a number of researchers in the last two decades. A multipath wave propagation model for wideband mobile radio channels based on time delay and Doppler frequency characterization is described in Parson and Bajwa (1982). The channel is represented by a two-port system (filter) with time-varying coefficients that are related to the average delay and delay spread. In Iwai and Karasawa (1993) a proposal for a wideband fading model which creates random scattered waves according to Poisson's distribution is presented. The concept of the model is that waves are generated and disappear continuously in a random process, as shown in Figure 6.3. The progress of each scattered wave from generation to disappearance is independent of the other scattered waves.

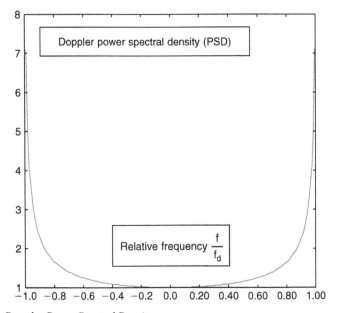

Figure 6.2   *Doppler Power Spectral Density.*

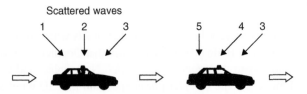

Figure 6.3   *Generation and disappearance of scattered waves.*

Example 6.1

Consider the Doppler PSD given in (6.7). Derive an expression for the Doppler received power at the following frequencies:

i. $f = f_c$

ii. $f - \dfrac{f_d}{2} < f < f + \dfrac{f_d}{2}$

Solution

i. At $f = f_c$, $S(f_c) = \dfrac{P \cdot g}{\pi f_d} =$ Doppler received power at f_c.

ii. The received Doppler power (P_D) is given by:

$$P_D = \int_{f_c - \frac{f_d}{2}}^{f_c + \frac{f_d}{2}} S(f)df = \int_{f_c - \frac{f_d}{2}}^{f_c + \frac{f_d}{2}} \frac{Pg}{\pi f_d \sqrt{1 - \left(\frac{f - f_c}{f_d}\right)^2}} df$$

Let $x = \dfrac{f - f_c}{f_d}$, then we can write the above expression as:

$$P_D = \int_{-0.5}^{0.5} \frac{Pg}{\pi \sqrt{1 - x^2}} dx = \frac{Pg}{\pi} [\sin^{-1}(0.5) + \sin^{-1}(0.5)]$$

$$P_D = \frac{Pg}{3}$$

6.1.2 Statistics of mobile radio channel

The radio signal transported over a mobile channel can be described by a random process with certain statistics that are closely related to the terrain configuration. While the base-station antenna is commonly clear of its neighbour's coverage, the mobile unit's antenna may be within the coverage of more than one base station. The mobile will receive a number of independent reflected waves, each with random phase and amplitude and combined at the receiver. The total received signal undergoes short-term fading, due to multipath propagation. The received signal can be described by quadrature components, which are uncorrelated Gaussian processes with a zero mean and variance σ^2. The envelope of the received signal, R, can be described by a Rayleigh Probability Density Function (PDF) p(R) as:

$$p(R) = \frac{R}{\sigma^2} e^{-\frac{R^2}{2\sigma^2}} \quad \text{for } R \geq 0$$

$$= 0 \text{ otherwise} \tag{6.8}$$

The phase of the received signal has a uniform distribution between $-\frac{\pi}{2}$ and $\frac{\pi}{2}$. The mean power of the multipath signal is σ^2.

When the received signal is made up of a line-of-sight component of amplitude A, plus multiple reflected waves, the envelope of the received signal has Rician PDF expressed as:

$$p(R) = \frac{R}{\sigma^2} \exp\left[-\frac{1}{2\sigma^2}(R^2 + A^2)\right] I_0\left(\frac{AR}{\sigma^2}\right) \quad \text{for } R \geq 0$$
$$= 0 \quad \text{otherwise} \tag{6.9}$$

where $I_0(\cdot)$ represents Bessel's function of zero order.

It is worth noting that if the multipath propagation is eradicated, the Rician random envelope transmitted in a Gaussian channel is modified to a Gaussian envelope describing the envelope of the direct (unfaded) wave.

Large obstacles – like hills and buildings, walls and furniture – often block the propagation of radio waves. The received signal variation due to these obstacles is called *shadow fading* (Lee, 1991). Measurements of the received power taken at a specified location between transmitter and receiver have shown the received power to be a random variable with certain distributions. Let the received power be p_r watts, then received power in dBm $= p_r' = 10 \log 10 \left(\frac{p_r}{1000}\right)$. The random variable p_r' is known to have *log-normal distribution*. Let $\hat{p}_r$ be the average power in dBm taken at the same location and let the standard deviation of the shadowing process be σ_{shadow} in dB, and then the PDF of the shadow fading is given by:

$$f_{\text{shadow}}(p_r') = \frac{1}{\sigma_{\text{shadow}}\sqrt{2\pi}} \exp\left[-\frac{(p_r' - \hat{p}_r)^2}{2\sigma_{\text{shadow}}^2}\right] \tag{6.10}$$

The values for σ_{shadow} in dB depend on the type of the obstruction and typically range between 6 and 8 dB. For fixed transmitted power, the received power is attenuated due to shadowing effects and the random attenuation of the shadow fading changes as the mobile station moves around the obstacle's area.

6.1.3 Path losses

When there is a clear line-of-sight between a transmitter antenna of power gain g_t and a mobile receiver antenna of power gain g_r located at a distance (r) apart, and the transmitted and received powers are P_t and P_r, respectively, then the path loss is given by Rapport (2001):

$$\frac{P_r}{P_t} = \frac{c^2 g_t g_r}{(4r\pi f)^2} \tag{6.11}$$

where f is the frequency of transmission and c is the speed of light. Thus, free space power loss is proportional to r^{-2} and can be expressed in dB as:

$$L_{\text{Los}} = -32.44 + G_t + G_r - 20\log_{10}(r) - 20\log_{10}(f)\text{dB} \tag{6.12}$$

where $G_t = 10 \log_{10}(g_t)$ dB, $G_r = 10 \log_{10}(g_r)$ dB, r is the distance in Km and f is the frequency in MHz. However, there is seldom a clear radio path between base station and the mobile unit and the path loss of radio waves in a real mobile environment is proportional to r^{-n} where the exponent n typically ranges from 3 to 4.

The propagation of radio waves is strongly dependent on the nature of the transmission environment. Therefore, for similar distances between transmitter and receiver, the power loss in urban areas is different from the power loss in suburban and rural areas. Consequently, closed-form mathematical expressions for calculating path loss are complicated. However, there are several empirical models that can be employed for predicting the path losses.

The prediction models proposed for path loss estimation were derived initially for narrowband radio signals. In the rest of this section we consider if these models provide accurate enough estimates of path loss for wideband signal such as CDMA.

Let us consider a narrowband signal transmitted at frequency f_c, from a base station to a mobile unit, when there is no clear line of sight. The path loss is given by Lee (1991):

$$\frac{P_r}{P_t} = \frac{c^2 g_t g_r}{(4r^2\pi)^2 f_c^3} \tag{6.13}$$

The path loss in dB (L) is:

$$L = -152.44 + G_t + G_r - 40 \log_{10}(r) - 30 \log_{10}(f) \tag{6.14}$$

where G_t and G_r are as defined for (6.12).

We now estimate the wideband path loss by considering the radio propagation of a signal with a bandwidth B Hz and assume the power spectral density, $S(f)$ is uniform over the bandwidth B such that:

$$S(f) = \frac{P_t}{B}$$

The received power, dP_r, within infinitesimal spectrum (df) is:

$$dP_r = \frac{c^2 g_t g_r}{(4r^2\pi)^2 f^3} S(f) \cdot df$$

The total received power is given by:

$$P_r = \int_{f_c - \frac{B}{2}}^{f_c + \frac{B}{2}} dP_r \cdot df$$

Substituting for dP_r, we get:

$$P_r = \int\limits_{f_c-\frac{B}{2}}^{f_c+\frac{B}{2}} \frac{c^2 g_t g_r}{(4\pi r^2)^2 f^3} S(f) \cdot df$$

which simplifies to:

$$\frac{P_r}{P_t} = \frac{c^2 g_t g_r}{(4\pi r^2)^2} \cdot \frac{1}{f_c^3 \left[1 - \left(\dfrac{B}{2f_c}\right)^2\right]^2} \qquad (6.15)$$

We are now in a position to consider two transmitters sending equal power to a mobile receiver at distance (r) from each of them. One transmitter is sending a carrier frequency, f_c, and the other is sending a signal on the same carrier but a wider spectrum of B (e.g. CDMA signal). The path loss difference in dB between the received signals, assuming the reception is carried out by the same receiver and under the same environment, is given by:

$$\frac{\text{Path loss (Narrowband)}}{\text{Path loss (wideband)}} = \left[1 - \left(\frac{x}{2}\right)^2\right]^2 \qquad (6.16)$$

where x represents the ratio of the spectrum bandwidth to the carrier frequency. Generally, wireless communications using CDMA systems are designed such that $x \ll 1$.

Clearly, equation (6.16) is too simple to accurately predict path losses. However, it does show that loss models used for narrowband transmission are capable of estimating path losses occurring during radio wave propagation in a wideband media. The difference in dBs of path loss of a narrowband signal to that of a wideband signal at equal distance from the two transmitters using (6.16) is plotted in Figure 6.4. It can be seen from the graph that the difference is -0.5 dB when $B = 0.5\,f_c$ and becomes -1 dB when $B = 0.65 f_c$. Since most wideband signals have bandwidth $B < 0.5 f_c$, we can conclude that the path loss of a wideband signal can, in most practical cases, be estimated using path loss models of narrowband signal at the same distance from transmitter.

6.1.4 Prediction of path loss

In the previous section, we established that path loss prediction formulas used so far for narrowband signals can also be used to estimate the path loss due to wideband signals. Obstacles such as high buildings, bridges and trees sometimes obstruct the radio path over the total signal bandwidth causing the shadowing effects as mentioned before. However, these objects are much larger than the signal wavelength and can cause diffractions and scattering. The shadowing effects produce very slow variations in the local mean received power and can be described by a lognormal distribution in the long-term variation of the signal as discussed previously.

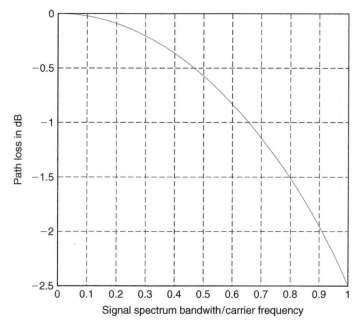

Figure 6.4 *Difference in path losses applying narrowband equations to wideband signal.*

There are many other factors that can change the level of the received signal and the most important among these are foliage effects, street orientation effects, and tunnel effects (Lee, 1993). Furthermore, it is an accepted fact that there is no single prediction model that fits all propagation environments. Several empirical models have been proposed as a result of intensive research carried out in Japan, UK and the USA (Hata, 1980; Ibrahim and Parson, 1983; Okumura et al., 1968). In this section, we review these models which are adopted by the ITU standards.

Okumura et al. (1968) carried out his measurement in the 1960s in and around Tokyo. He used an antenna height of 200 m at the transmitter and 3 m at the mobile receiver, and a frequency range of 100–3000 MHz. He proposed an empirical model for predicting path loss based on data presented in graphical form. Hata (1980) generalized the Okumura model in the 1980s for applications in urban, suburban, and rural radio transmission, and proposed empirical expressions for predicting path loss. The CCIRs model for path loss prediction in urban areas is given by the following expressions (Macario, 1997):

$$L(dB) = 69 + 26 \log(f_c) - 14 \log(h_t) + [45 - 6.5 \log(h_t)] \log(r) - A(h_r) \quad (6.17)$$

where $A(h_r) = [\log(f_c) - 0.7]h_r - [1.6 \log(f_c) - 0.8]dB$
where h_t, h_r are in metres, r in Km, and f_c is in MHz.

Ibrahim and Parson (1983) carried out measurements in the London area in 1983 and proposed the 'Ibrahim–Parson' model as expressed by the following equation:

$$L(dB) = -20.\log(0.7h_t) - 8\log(h_r) + \frac{f_c}{40} + 26\log\left(\frac{f_c}{40}\right) - 86\log\left(\frac{f_c + 100}{156}\right)$$

$$\left\{40 - 14.15\log\left(\frac{f_c + 100}{156}\right)\right\}\log(r) + 0.256L_1 - 0.37H + K \qquad (6.18)$$

where K = 0.087 U − 5.54 for city centre, otherwise K = 0

h_t, h_r = transmitting and receiving antennas heights (m), $h_r \leq 3$ m

L_1 = Land usage factor

U = degree of urbanization factor

H = difference in height between transmitter and receiver

r = range in metres, r ≤ 10 Km

f_c = Transmission frequency in MHz.

A simplified model for the 'Ibrahim–Parson' model can be expressed as:

$$L(dB) = 40 \cdot \log(r) - 20\log(h_t \cdot h_r) + \beta \qquad (6.19)$$

Where β is the clutter factor in dB given by:

$$\beta = 20 + \frac{f_c}{40} + 0.18L_1 - 0.34H + K \qquad (6.20)$$

Parameters in (6.20) are defined in the previous model except that K = 0.09 U − 5.9 for city centre, otherwise K = 0.

Example 6.2

Estimate the path loss at a distance 10 Km from the transmitter in a large metropolitan city using:

i. Ibrahim–Parson's model
ii. Ibrahim–Parson's simplified model
iii. CCIR standard model

The radio system has the following parameters:

Base station antenna height = 30 m
Mobile station antenna height = 3 m
Carrier frequency = 850 MHz
Land usage factor (L) = 30%
Difference in heights between the ground contains the transmitter and the receiver = 10 m

Solution

i. Substituting in the Ibrahim–Parson's equation, we get:

$$L(dB) = -20\log(0.7 \times 30) - 8\log(3) + \frac{850}{40} + 26\log\left(\frac{850}{40}\right) - 86\log\left(\frac{950}{156}\right)$$
$$+ \left[40 + 14.15\log\left(\frac{950}{156}\right)\right]\log(r) + 0.256 \times 30 - 0.37 \times 10$$

$$L(dB) = 166.4\,dB$$

$$= 158.72\,dB \text{ if we ignore land usage factor contribution.}$$

ii. Substituting in the Ibrahim–Parson's simplified equation, we get:

$$L(dB) = 40\log(10{,}000) - 20\log(30 \times 3) + 20 + \frac{850}{40} + 0.18 \times 30 - 0.37 \times 10$$

$$L(dB) = 164.16\,dB$$

$$= 158.76\,dB \text{ if we ignore contribution from land usage factor.}$$

iii. Substituting in CCIR standard model, we get:

$$L(dB) = 69 + 26\log(850) - 14\log(30) + [45 - 6.5\log(30)]\log(10) - A(h_r)$$
$$A(h_r) = [\log 850 - 0.7] \times 3 - [1.6\log 850 - 0.8] = 2.8$$

Thus $L(dB) = 157.08\,dB$

The calculations presented above show that the path loss in dB obtained from the CCIR standard model and from Ibrahim–Parson's model are in fairly close agreement if we ignore the loss contribution from the land usage factor. This difference is reasonable and not unexpected as land usage is not considered in the CCIR model.

Example 6.3

Consider a mobile unit at distance r_1 from a transmitter, transmitting a wideband signal with a bandwidth $B = 0.3\,f_c$. The same mobile receives a tone signal at frequency f_c and at a distance r_2 from a transmitter. The tone signal level is at $-1\,dB$ below the level of the received wideband signal. Assuming that the transmitting and receiving system are identical in both cases, compute the ratio r_1/r_2.

Solution

From the previous analysis, we have:

$$P_r\,(\text{wideband}) = \frac{K}{r_1^4} \cdot \frac{1}{\left[1 - \left(\frac{B}{2f_c}\right)^2\right]^2}$$

$$P_r \text{ (narrowband)} = \frac{K}{r_2^4}$$

where K is a constant defined by the transmitting-receiving system parameters. Substituting the values given in the example in the above equations, we get:

$$10 \cdot \log \left\{ \left(\frac{r_1}{r_2}\right)^4 \left[1 - \left(\frac{0.3}{2}\right)^2\right]^2 \right\} = -1$$

Thus, simplifying this expression gives:

$$\frac{r_1}{r_2} = 0.955$$

This result endorses our previous conclusion that propagation path loss for a wideband signal can be computed using expressions derived for narrowband empirical formulas.

6.2   The Cellular CDMA system

Having discussed the characteristics of the wideband channel and predictions of its path loss, we now turn our attention to the cellular concept of the CDMA system.

There are two basic strategies which can be followed in the design of wireless systems intended to cover a large service area. Many commercial radio and television broadcast systems operate at the maximum power and with the highest possible antenna. In such systems, the frequency used by the transmitter cannot be used by any other transmitter until there is enough geographical separation to ensure that radio interference is avoided. The separation required between transmitters using the same frequency defines the frequency re-use factor. A cellular radio takes the opposite approach by dividing the coverage area into small cells which vary in size and use low power transmitters so that radio resources can be used efficiently. We shall cover this in Section 6.5. In CDMA communication systems, all transmitters in the network use the same radio channels, making the frequency re-use factor 100%. It is worth noting that the wireless cellular concept would not function without the use of a fixed telecommunications infrastructure. Base station transceivers in the cellular network are connected by microwave links or fibre optical cable and they are also connected to the local telephone exchanges.

6.2.1   The cellular concept

The cellular structure of a wireless network consists of clusters of cells. Each cell consists of a transceiver at the fixed base station with suitable transmitting power and a number of mobile units scattered within the coverage area of the cell. Each mobile unit operates a transceiver comprising of a sensitive receiver and a low power transmitter operating in a wireless environment characterized by multipath propagation, fading, and access interference. The subscriber is able to move within the network with seamless communications

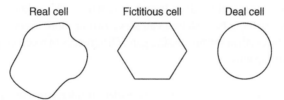

Figure 6.5 *Representation of a single cell.*

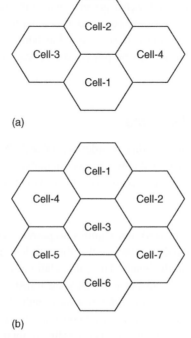

Figure 6.6 *(a) 4-cell cluster; (b) 7-cell cluster.*

while being unaware of the cell structure. In real wireless networks, the radiation pattern of the transmitter antenna defines the shape of the cell and, in most cases, it has a non-uniform shape. Ideally, the cell is represented by a circle with the base station at the centre so that network cells usually overlap, but it is common practice to represent the radio cell as a hexagon that would fit into the corresponding circle. The single cell system is shown in Figure 6.5.

The cellular structure formation usually exists as a cluster of 4, 7, or 13 cells. A cellular cluster of 4 and 7 cells are shown in Figure 6.6.

6.2.2 The Near–Far effect

A phenomenon that degrades the service quality of a wireless network is the Near–Far effect which occurs when the interference from a user transmitting near the base station overpowers the weaker signal received from a distant user. This is known as the Near–Far (N-F) effect. Essentially, a base station can serve two or more users if the difference in their power level is less than the processing gain. Differences in received power levels tend to degrade system capacity.

To examine the N-F effect in more detail, consider an interfering signal with a received power (n) times higher than the power received from the target user. The interfering signal will have approximately the same affect on link capacity and system performance as (n) separate interferers, each with signal power equal to the power of the target signal. Consequently, power control algorithm is essential in combating the N-F effects to optimize system capacity, as we will see in the next section. Furthermore, to account for the rapid changes in path loss and user movement, power control schemes have to be adaptive but fast and accurate.

6.2.3 Power control schemes

In the previous section we indicated that, in order to combat the N-F effect, cellular wireless systems must control the transmitted power from each mobile unit as well as from the base station. Furthermore, an efficient power control scheme will also reduce the multiple access interference.

In a single cell, the forward (down) link requires no power control mechanism since the base station transmits all signals together so that the path loss experienced on passage to each receiver is the same for all signals. Thus, the relationship between interference and the user's required signal level remains as it was at the transmitter. However, where there are multiple cells, the user receives interference from multiple users in the serving base station's area as well as interference from neighbouring cells. Therefore, control methods are used to allocate power to individual users according to their exposure to interference.

On the reverse link or the uplink, users' signals are exposed to different path losses defined by the distance from the mobile unit to its operating base station, and the variations in radio propagation path. Thus, even if all mobile units transmit at the same power, their signals have different powers when they reach the base station. Power control in the reverse link not only combats interference from shadowing and the N-F problem, but also keeps user transmitter power at the minimum necessary level for acceptable performance, thus prolonging battery life in portable handsets.

There are two methods to manage user transmit power in the reverse link. These methods are based on *open-loop* and *closed-loop control schemes*. In the open-loop power control method, the mobile unit uses the strength of the received pilot on the forward link to determine what adjustment of the transmitter power on the reverse link is needed to achieve the required radio power at the base station receiver. The mobile receiver controls its transmitter power by applying Automatic Gain Control (AGC) measurements. The weakness of this design is that the forward and reverse links may be subject to different levels of path loss.

In the closed-loop control method, the base station provides each mobile unit with continuous information about the power of their signal to enable frequent adjustments of reverse link power. The power received on the reverse link at the base station is compared to a desired level and the power difference is hard-quantized to obtain a power command bit that is transmitted to the user on the forward link. The power command changes the user transmitter power by a fixed step. A variant of this method is to use a soft-quantizer rather than a hard-quantizer. The soft-quantizer requires more bits to code the transmit power command. To reduce delay in the power control, the command bit is transmitted on the forward link unprotected, so it is exposed to channel errors.

Power control can be used to follow only the slow log-normal power variation caused by shadowing effects where the signal remains correlated, or to track a rapidly fading signal. Power control systems designed to mitigate fast multiple path fading have power updates sent at a much higher rate than the maximum fade rate of the received signal. Measurement on a real system with theoretically ideal power control showed that the power controlled received at the base station exhibits log-normal variation with a standard deviation of $\frac{E_b}{N_0}$ is typically between 1 and 2 dB (Glisic and Vucetic, 1997).

6.3 Interference considerations

Multiple access interference is the limiting factor to system capacity of CDMA systems. Therefore, interference in cellular CDMA transmissions deserve detailed treatment in this section. Considering the cellular wireless system, there are two types of interferences: interference generated from mobile users in the same cell, called the Intra cell interference ($I_{intracell}$), and interference coming from users in surrounding cells, called the Inter cell interference ($I_{intercell}$). Furthermore, interference on the forward link is different from interference on the reverse link. We will deal with intra and inter cell interference on the two links separately.

6.3.1 Interference on the reverse link

The received signal at the base station is affected by Intracell and Intercell, probably at dissimilar levels. The individual user's power within a given cell is controlled and is continually adjusted by the serving base station to a pre-defined level (S). Clearly, the maximum intra cell interference level is:

$$I_{intracell} = \rho \cdot (K - 1) \cdot S$$

where K is number of active users and ρ is voice activity factor. For large K, we have:

$$I_{intracell} = \rho \cdot (K - 1) \cdot E_b \cdot R_b \approx \rho K \, E_b R_b. \tag{6.21}$$

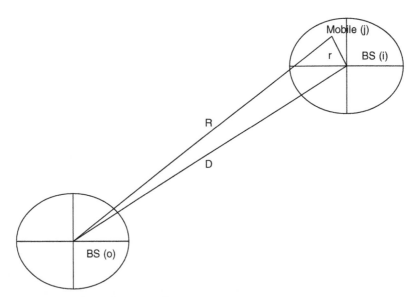

Figure 6.7   *Inter-cell interference analysis.*

6.3.2   The inter cell interference (Heath and Newson, 1992; Newson, 1992; Viterbi et al., 1994)

The path loss between base station (i) and mobile user (j), distance $r_{j,i}$ apart is composed of two components. One component is due to multipath propagation where signal power decays with index α such that $2 \le \alpha \le 6$. The other component is due to the log normally distributed shadow effects represented by $\gamma_{j,i}$ dB, which is a Gaussian random variable with zero mean and standard deviation σ_S of 8 dB (Viterbi, 1995). The total path loss, defined as $PL_{j,i}$, is given by:

$$PL_{j,i} = r_{j,i}^{-\alpha} \cdot 10^{\frac{\gamma_{j,i}}{10}} \qquad (6.22)$$

Let mobile user (j) be at distance $R_{j,O}$ from the base station (o) as shown in Figure 6.7. For an ideal power control, the power received from an active user at its serving base station is limited to S. Thus, user (j) transmitted power, $W_{j,I}(r)$ is:

$$W_{j,i}(r) = S \cdot r_{j,i}^{\alpha} \cdot 10^{-\frac{\gamma_{j,i}}{10}} \qquad (6.23)$$

The presence of shadowing means that users may not necessarily operate to the nearest base station but to the base station for which path loss is minimal.

Let the mobile user (j) communicate within the coverage area of base station (i). So, in order to keep the received power at the serving base station fixed at S, we get the following from (6.23).

User (j) transmitted power to base station (i) < user (j) transmitted power to base station (O), thus:

$$r_{j,i}^{\alpha} \cdot 10^{-\frac{\gamma_{j,i}}{10}} < R_{j,o}^{\alpha} \cdot 10^{-\frac{\delta_{j,o}}{10}}$$

which becomes:

$$\left(\frac{r_{j,i}}{R_{j,o}}\right)^{\alpha} \cdot 10^{\frac{\delta_{ji,o}-\gamma_{j,i}}{10}} < 1 \qquad (6.24)$$

where $\delta_{j,o}$ = Shadowing component between user (j) and base station (O)

$\gamma_{j,i}$ = Shadowing component between user (j) and base station (i)

Let us define the function $f_0(.)$ such that:

$$f_0\left[(\delta_0 - \gamma_i), \frac{r}{R}\right] = 1 \quad \text{if} \quad \left(\frac{r_{j,i}}{R_{j,o}}\right)^{\alpha} \cdot 10^{\frac{\delta_{j,o}-\gamma_{j,i}}{10}} < 1 \qquad (6.25)$$

$$= 0 \quad \text{otherwise}$$

The inter cell interference from user (j) to the coverage area of base station (O) is given by:

$$\Delta(R_{j,o}/r_{j,i}) = \psi \cdot S \cdot \frac{r_{j,i}^{\alpha}}{R_{j,o}^{\alpha}} \cdot 10^{\frac{(\delta_{j,o}-\gamma_{j,i})}{10}} \cdot f_0\left[(\delta_0 - \gamma_i), \frac{r}{R}\right] \qquad (6.26)$$

where ψ is a random variable defines voice activity such that:

$$\psi = 1 \quad \text{with probability } \rho$$
$$= 0 \quad \text{with probability } 1 - \rho$$

Assume that cell i is circular with a radius R_d and there are K users in the cell. Active users are distributed uniformly inside the cell so that there are $\frac{K}{\pi \cdot R_d^2}$ users per unit area (Cooper and Nettleton, 1978).

Consider the element shaded area dA shown in Figure 6.8 due to radius width dr and element angle dθ, then

$$dA \approx \text{arc length} \cdot dr$$

But arc length $\approx r \cdot d\theta$

Therefore, $dA \approx r \cdot d\theta \cdot dr$

The number of users in the shaded area $= \dfrac{K}{\pi \cdot R_d^2} \cdot r \cdot d\theta \cdot dr$

The inter cell interference contribution from shaded area $= \Delta(R_{j,o}/r_{j,i}) \dfrac{K}{\pi \cdot R_d^2} \cdot r \cdot d\theta \cdot dr$

The inter cell interference contribution from all users in cell (i) is given by:

$$\int_0^{2\pi} \int_0^{R_d} \Delta(R_{j,o}/r_{j,i}) \cdot \frac{K}{\pi \cdot R_d^2} \cdot r \cdot f_0\left[(\delta_i - \gamma_i), \frac{r}{R}\right] \cdot dr \cdot d\theta \qquad (6.27)$$

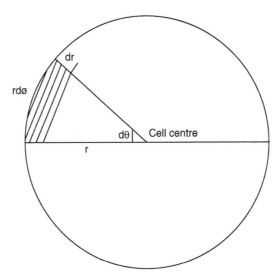

Figure 6.8   *Interference element of area dA.*

Consider the triangle ijo in Figure 6.7, the distance between mobile (j) and base station (o) is given by:

$$R_{j,o}^2 = D_i^2 - 2D_i r \cos\theta + r^2$$

Total interference from N_c cells is given by:

$$I_{intercell} = \sum_{i=1}^{N_c-1} \left[\int_0^{2\pi} \int_0^{R_d} \Delta I(R_{j,o}/r_{j,i}) \cdot \frac{K}{\pi \cdot R_d^2} \cdot r \cdot f_0\left[(\delta_o - \gamma_i), \frac{r}{R}\right] \cdot dr \cdot d\theta \right]$$

$$= \sum_{i=1}^{N_c-1} \left[\int_0^{2\pi} \int_0^{R_d} S \cdot \psi \cdot \frac{r^\alpha \cdot 10^{\frac{(\delta_o - \gamma_i)}{10}}}{(D_i^2 - 2D_i \cdot r \cdot \cos\theta + r^2)^{\frac{\alpha}{2}}} \cdot \frac{K}{\pi \cdot R_d^2} \cdot r \right.$$

$$\left. \cdot f_0\left[(\delta_{oi} - \gamma_i), \frac{r}{R}\right] \cdot dr \cdot d\theta \right]$$

$$I_{intercell} = \frac{S \cdot K\psi}{\pi \cdot R_d} \sum_{i=1}^{N_c-1} \left[\int_0^{2\pi} \int_0^{R_d} \frac{r^{\alpha+1} \cdot 10^{\frac{(\delta_o - \gamma_i)}{10}}}{(D_i^2 - 2D_i \cdot r \cdot \cos\theta + r^2)^{\frac{\alpha}{2}}} \cdot f_0\left[(\delta_o - \gamma_i), \frac{r}{R}\right] \cdot dr \cdot d\theta \right]$$

$$\frac{I_{intercell}}{S} = \frac{K\psi}{\pi \cdot R_d} \sum_{i=1}^{N_c-1} \left[\int_0^{2\pi} \int_0^{R_d} \frac{r^{\alpha+1} \cdot 10^{\frac{(\delta_o - \gamma_i)}{10}}}{(D_i^2 - 2D_i \cdot r \cdot \cos\theta + r^2)^{\frac{\alpha}{2}}} \cdot f_0\left[(\delta_o - \gamma_i), \frac{r}{R}\right] \cdot dr \cdot d\theta \right]$$

$$(6.28)$$

The above expression can only be evaluated by numerical integration. Consider a cell of three sectors, and cell radius is normalized to unity, then average inter cell interference is shown in Gilhousen et al. (1991) as:

$$E\left(\frac{I_{intercell}}{S}\right) \leq 0.24 \cdot N_s \tag{6.29}$$

The variance of inter cell interference is:

$$Var\left(\frac{I_{intercell}}{S}\right) \leq 0.078 \cdot N_s \tag{6.30}$$

where N_s = number of users per sector.

6.3.3 Interference on the forward link

We have already stated that the forward link in a single cell system requires no power control arrangement. However, in a multi-cell system, users near the cell boundary receive a significant amount of interference, which necessitates the use of a power control scheme in the forward link of the cellular CDMA system. The power control scheme entails power allocation by the serving base station to individual users. This is achieved when the user acquires the highest power pilot and measures its energy. The base station from which maximum power $(S_{T1})_i$ is received by the ith user is designated as the user's serving base station. The user also measures total energy received from surrounding base stations and grouped the received power according to the power level as:

$$(S_{T1})_i > (S_{T2})_i > (S_{T3})_i > - - - - - (S_{TN_c})_i > 0$$

where N_c denotes the number of surrounding cells that contribute significant inter cell interference, assuming the base stations beyond the second ring contribute negligible interference.

6.4   Single-user receiver in a multi-user channel

6.4.1   The multi-user channel

In this section we consider the design of a receiver for a single user sharing the CDMA wireless channel with a group of interfering users. This means that the single-user transmission will be degraded by the multiple users accessing the CDMA channel simultaneously (Poor and Verdu, 1988). On the other hand, the group detection schemes that reduce or eliminate the multiple access interference are dealt with in Chapter 7.

We start with the total signal received by the single-user receiver. The cellular CDMA system allows simultaneous two-way communications and the transmission and reception is carried out within the same spectrum. The logical channel used for transmission of information from the base station to the mobile unit is called the forward (down) link. The reverse (up) link is used for transmission of information from the mobile unit to the base station.

Consider a group of K active users sharing a CDMA radio channel operating in a single cell. Each user is transmitting a block of data containing N_b bits. The kth user data $b_k(t)$ and code sequence $C_k(t)$ have elements of $\{+1, -1\}$ and can be expressed by:

$$b_k(t) = \sum_i b_k^i P_b(t - iT)$$

$$C_k(t) = \sum_j c_k^j P_c(t - iT_c)$$

where T and T_c are symbol and code chip durations, respectively; $P_b(t)$ and $P_c(t)$ are the data and code chip pulse shapes, respectively. When data and code sequence consists of unit amplitude rectangular pulses, the pulse shape is given by:

$$P(t) = 1 \text{ inside pulse duration and } P(t) = 0 \text{ otherwise.}$$

Consider the link of the CDMA system shown in Figure 6.9.

The kth user transmitted signal generated at the base station can be expressed as:

$$s_k(t) = C_k(t) \cdot b_k(t) \cdot \cos(\omega_0 + \theta_0) \tag{6.31}$$

where ω_0 represents the centre radian frequency of a unit amplitude carrier. The total signal, s(t), transmitted by the base station to the K users is given by:

$$s(t) = \sum_{k=1}^{K} C_k(t) \cdot b_k(t) \cdot \cos(\omega_0 t + \phi_0) \tag{6.32}$$

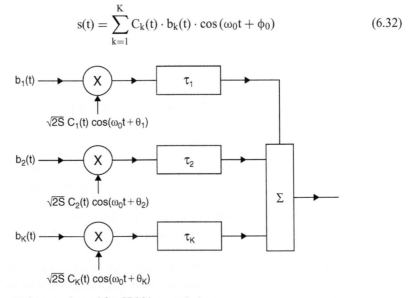

Figure 6.9 *Multi-user channel for CDMA transmission.*

The total signal received in AWGN channel by the kth user (indeed by any of the K users) is:

$$r(t) = \sum_{k=1}^{K} C_k(t) \cdot b_k(t) \cos(\omega_0 t + \phi_0) + n(t) \tag{6.33}$$

Clearly, here we have assumed that the transmission over the forward link of the CDMA system is to be completely synchronous.

The transmission on the reverse link is uncoordinated so the signal transmitted by the kth user transmitted can be expressed as:

$$s_k(t) = C_k(t - \tau_k) \cdot b_k(t - \tau_k) \cdot \cos(\omega_0 t + \phi_k) \tag{6.34}$$

The total signal received by the base station is:

$$r(t) = \sum_{k=1}^{K} C_k(t - \tau_k) \cdot b_k(t - \tau_k) \cos(\omega_0 t + \phi_k) + n(t) \tag{6.35}$$

where τ_k represents the total delay of signal received from the kth user referenced to a target user, ϕ_k represents the initial phase shift of carrier used by the kth user plus shift originated by delay τ_k. The channel noise $n(t)$ is assumed to be a white additive process. Equation (6.35) models an asynchronous channel in the reverse link CDMA system.

6.4.2 The conventional receiver

The conventional receiver is the classical detection method used in a spread spectrum transmission technique. It is simple and cheap to implement but its bit error rate performance is degraded by the multiple access interference since no further processing is used in the detection method to reduce or eliminate the interference.

The conventional receiver consists of a filter matched to the pulse shape of the target user's code sequence. The output of the matched filter is correlated with the user code sequence to maximize the signal-to-noise power ratio at the receiver output. The single-user receiver makes no use of information regarding the interference created by other users. Consequently, the performance of such receivers is limited by the interference contribution from users sharing the CDMA channel rather than by Gaussian channel noise. The received signal is given by:

$$r(t) = \left[\sum_{k=1}^{K} C_k(t - \ell T - \tau_k) \cdot b_k(t - \ell T - \tau_k) \cdot \cos(\omega_0 t + \phi_1) \right] + n(t) \tag{6.36}$$

We will assume the code sequence of the target user to be $C_1(t)$. The block diagram of the single-user receiver is shown in Figure 6.10.

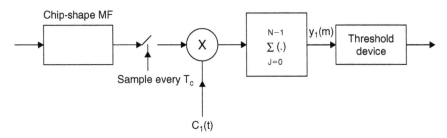

Figure 6.10 *CDMA conventional receiver.*

Furthermore, an ideal carrier phase tracking is assumed for clarity and a matched filter gain of 2 is used to easily simplify the expression as we will show later. At the end of the mth symbol interval, the output signal is:

$$y_1(m) = \frac{1}{T} \cdot \int_{mT+\tau_1}^{(m+1)T+\tau_1} r(t) \cdot \cos(\omega_0 t + \phi_1) \cdot C_1(t - mT - \tau_1) \cdot dt \tag{6.37}$$

Substituting r(t) from (6.36) in (6.37), we get:

$$y_1(m) = \frac{1}{T} \cdot \int_{mT+\tau_1}^{(m+1)T+\tau_1} n(t) \cdot \cos(\omega_0 t + \phi_1) \cdot C_1(t - mT - \tau_1) \cdot dt$$

$$+ \frac{1}{T} \cdot \int_{mT+\tau_1}^{(m+1)T+\tau_1} \sum_{k=1}^{K} C_k(t - \ell T - \tau_k) \cdot b_k(t - \ell T - \tau_k)$$

$$\cdot \cos^2(\omega_0 t + \phi_1) \cdot C_1(t - mT - \tau_1) \cdot dt$$

Now $\cos^2(\omega_0 t + \phi_1) = \dfrac{1 + \cos 2(\omega_0 t + \phi_1)}{2}$

The gain of the matched filter is set to 2 and the double frequency component will be blocked by the lowpass filtering so that at the correlator input $\cos 2(\omega_0 t + \phi_1) \approx 1$. The noise component that appears at the output of the first user detector at the end of symbol m is $n_1(m)$ given by:

$$n_1(m) = \frac{1}{T} \cdot \int_{mT+\tau_1}^{(m+1)T+\tau_1} n(t) \cdot \cos(\omega_0 t + \phi_1) \cdot C_1(t - mT - \tau_1) \cdot dt \tag{6.38}$$

Thus, $y_1(m)$ is given by:

$$y_1(m) = \frac{1}{T} \cdot \int_{mT+\tau_1}^{(m+1)T+\tau_1} \sum_{k=1}^{K} C_k(t - \ell T - \tau_k) \cdot b_k(t - \ell T - \tau_k) \cdot C_1(t - mT - \tau_1) \cdot dt + n_1(m)$$

The receiver acquires target user code and successfully achieves code tracking so that:

$$y_1(m) = b_1(m) \cdot \frac{1}{T} \cdot \int_{mT+\tau_1}^{(m+1)T+\tau_1} C_1(t - mT - \tau_1) \cdot C_1(t - mT - \tau_1) \cdot dt$$

$$+ \frac{1}{T} \cdot \int_{mT+\tau_1}^{(m+1)T+\tau_1} \sum_{k=2}^{K} C_k(t - \ell T - \tau_k) \cdot b_k(t - \ell T - \tau_k) \cdot C_1(t - mT - \tau_1) \cdot dt$$

$$+ n_1(m)$$

Now $\frac{1}{T} \cdot \int_{mT+\tau_1}^{(m+1)T+\tau_1} C_1(t - mT - \tau_1) \cdot C_1(t - mT - \tau_1) \cdot dt$ represents the autocorrelation $R_1(0)$ of the target user code sequence, that is:

$$R_1(0) = \frac{1}{T} \cdot \int_{mT+\tau_1}^{(m+1)T+\tau_1} C_1(t - mT - \tau_1) \cdot C_1(t - mT - \tau_1) \cdot dt \qquad (6.39)$$

The cross-correlation of code sequences $R_{k1}(\tau)$ is defined as:

$$R_{k1}(\tau) = \frac{1}{T} \cdot \int_{mT+\tau_1}^{(m+1)T+\tau_1} C_k(t - \ell T - \tau_k) \cdot C_1(t - mT - \tau_1) \cdot dt$$

Therefore the decoded signal at the receiver output is given by:

$$y_1(m) = b_1(m) + \sum_{k=2}^{K} R_{k1}(\tau) \cdot b_k(t - \ell T - \tau_k) + n_1(m) \qquad (6.40)$$

The first term in the above expression is the transmitted symbol; the second term is the interference resulting from $(K - 1)$ users sharing the CDMA channel. The last term represents the noise sample at sampling instant m.

In deriving (6.40) we assumed that the received amplitude for all users is normalized to unity. We now consider the general case where the received energy at chip level of the kth user be E_{ck} so that all users signal energies can be written as **A** where:

$$\mathbf{A} = \text{diag}[\sqrt{E_{c1}}, \sqrt{E_{c2}}, \ldots, \sqrt{E_{cK}}]$$

We can also write the output from the kth user receiver as $y_k(m)$ where $k = 1, 2, \ldots, K$ and all outputs can be written as column **y** where:

$$\mathbf{y} = [y_1, \ldots, y_K]^T$$

The data transmitted by the users can be represented by a column **b** where:

$$\mathbf{b} = [b_1, b_2, \ldots, b_K]^T$$

The matrix form of the output of all receivers is:

$$\mathbf{y} = \mathbf{RAb} + \mathbf{n} \tag{6.41}$$

where $\mathbf{R}$ is $K \times K$ normalized cross-correlation matrix for K-user system.

6.5   Improved single-user receivers

6.5.1   Introduction

The performance of the conventional receiver discussed in the previous section was degraded by the Multiple Access Interference (MAI) even though matched filter detection commonly achieves minimum Bit Error Rate (BER) in an Additive White Gaussian Noise (AWGN) channel by maximizing the SNR at its output (Proakis, 1995).

In the conventional receiver considered, the filter impulse response is matched to the code sequence of the target user disregarding the interference generated by other users sharing the channel. However, the noise is dominated by MAI so that even though the total interference acquires Gaussian statistical distribution, the noise power spectral density is not white. The degradation in the conventional receiver's performance is due to the fact that the multiple users' channels alter the code sequence of the target user, making the filter matching inaccurate. Furthermore, the conventional receiver's performance is disposed to the N-F effect and, therefore, there is a need for the CDMA system to use a strict power control scheme.

When code sequences of the active users are orthogonal to each other, the conventional receiver is optimal since the multi-user CDMA channel will be mainly corrupted by Gaussian noise. However, users' channels orthogonality cannot be maintained all of the time in a mobile environment. The conventional receiver works well avoiding the N-F problem if users' channels are almost orthogonal and the received powers at the base station from all active users are kept equal.

What is needed at the mobile unit is a robust receiver that detects the single signal belonging to its user and requires only the knowledge of the spreading code and timing for that particular signal. The receiver also needs to be relatively modest in its computation requirements to be cost effective. This is in contrast with the multi-user receiver selecting a single-user signal while it carries out simultaneous detection of signals belonging to a group of users using knowledge of all users' spreading codes, timings, and received powers. The techniques used by the multi-user (fully centralized) receiver for optimum detection have computation requirements that are currently unacceptably costly for many applications. This is discussed in the next chapter. Furthermore, network security restrictions may not allow the distribution of all users' code sequences to all receiving terminals. Consequently, specialized single-user signal detection techniques, also known in the literature as 'decentralized detector', where the receiver has no knowledge of the interferers' code sequences, delays, and tracking phases, is an important and valuable area of development in the design of single-user receivers.

The N-F resistance of a receiver is defined by its performance in the presence of the worst-case MAI compared to its performance in the absence of MAI. The N-F resistance is given by the ratio between the signal energy due to the component of the target signal vector that is orthogonal to the interference vector space and the energy of the target signal vector. Clearly, the N-F resistance is zero only if the target signal vector is contained within the interference space. The N-F problem can be alleviated somewhat by using power control. However, such schemes are relatively costly to implement in wireless mobile communications. The ideal way to handle the N-F problem is to use N-F resistant receivers.

The optimum single-user receiver in multi-user CDMA channel described by Verdu (1986) and Poor and Verdu (1988) requires excessive computation. The complexity of the computation grows exponentially with the number of users (K) and is as large as that required for multi-user detection with fully locked users. An optimum single-user receiver can be based on the conventional receiver strategy but with target user's signature modified to eradicate MAI.

Consideration has been given to the design of single-user receivers which deliver sub-optimum (slightly degraded) performance but offer the advantage of lower computation costs. Such receivers represent a compromise between the conventional (decentralized) receivers that neglect the presence of the MAI, and the optimum multi-user (centralized) receivers that need multiple users' parameters to mitigate MAI. Most of the sub-optimum single-user receiver designs exploit the cyclostationary nature of the CDMA signal to remove the interference. It is well-known that cyclostationary signals exhibit correlation with a frequency-shifted version of themselves and the frequency shift corresponds to the baud rate which is also the period of cyclostationarity.

The CDMA signal exhibits three periodicities based on data rate, chip rate, and code rate. When these three rates are integer multiples of each other, the signal exhibits a significant amount of spectral correlation, and the signal spectrum can be viewed as the summation of several versions of the baseband spectrum of the data signal translated by different carriers which are spaced apart by the code rate.

The optimum adaptive linear filter response for cyclostationary signals such as CDMA is periodically time-dependent and is known as Frequency-Shift (FRESH) filter or Time-Dependent Adaptive Filter (TDAF). The TDAF shifts the frequency, weighs and combines the input samples to yield the filter response. The TDAF filter combines the correlated replica of the data spectrum to give a very good data estimate by rejecting the interference (Gardner, 1986, 1993).

The sub-optimum single-user receiver designs described in the literature can be grouped into two categories. One category is based on non-adaptive modified conventional receivers that are capable of combating the effects of the MAI. The other category applies the adaptive concept on the received signal based on the minimum mean square error criterion. In the following sections, we will consider various schemes that are available in the recent literature to enhance the performance of single-user receivers.

6.5.2 Modified conventional receivers schemes

A straightforward approach to improving the performance of the conventional receiver is to use the 'whitened matched filter' solution (Monk et al., 1994). The concept of this solution is similar to the 'noise whitening' concept used in sequence detection in additive coloured Gaussian noise.

The noise generated from the MAI usually has a non-flat amplitude spectrum. The noise whitening solution equalizes the spectrum of the signal, making it relatively flat, similar to the white noise spectrum, by enhancing the low level spectral components and attenuating the high level ones.

Traditionally, noise whitening involves the processing of the conventional receiver outputs by a filter derived from the users' correlation matrix $\mathbf{R}$. In CDMA channels, a white noise model can be obtained by factoring the positive definite matrix of cross-correlation as (Duel-Hallen, 1993):

$$\mathbf{R} = \mathbf{F}^T \mathbf{F} \tag{6.42}$$

where $\mathbf{F}$ is a lower triangular matrix. If a filter with response $(\mathbf{F}^T)^{-1}$ is applied to the sampled output of the conventional receiver (6.41), the output vector $\tilde{y}$ is given by:

$$\tilde{y} = \mathbf{F}^{-T} \mathbf{y} \tag{6.43}$$

Substituting for $\mathbf{y}$ from (6.41), we get:

$$\tilde{y} = (\mathbf{F}^T)^{-1} \mathbf{R} \mathbf{A} \mathbf{b} + (\mathbf{F}^T)^{-1} \mathbf{n}$$

The output signal $= (\mathbf{F}^T)^{-1} \mathbf{R} \mathbf{A} \mathbf{b} = (\mathbf{F}^T)^{-1} \mathbf{F}^T \mathbf{F} \mathbf{A} \mathbf{b} = \mathbf{F} \mathbf{A} \mathbf{b} \tag{6.44}$

The output noise $\tilde{\mathbf{n}} = (\mathbf{F}^T)^{-1} \mathbf{n}$

Noise covariance matrix $= \mathbf{E}[\tilde{\mathbf{n}}\tilde{\mathbf{n}}^T] = (\mathbf{F}^T)^{-1} \mathbf{E}(\mathbf{n}\mathbf{n}^T)\mathbf{F}^{-1}$

But
$$\mathbf{E}(\mathbf{n}\mathbf{n}^T) = \frac{N_0}{2}\mathbf{R} \tag{6.45}$$

Thus, noise covariance matrix $= (\mathbf{F}^T)^{-1}\frac{N_0}{2}\mathbf{R}\,\mathbf{F}^{-1}$

$$= \frac{N_0}{2}(\mathbf{F}^T)^{-1}\mathbf{F}^T\mathbf{F}\,\mathbf{F}^{-1} = \frac{N_0}{2}\mathbf{I} \tag{6.46}$$

where $\mathbf{I}$ is the $K \times K$ identity matrix. Therefore, the noise at the output of the whitening filter is the Gaussian white noise with variance $\frac{N_0}{2}$. It is clear from equation (6.45) that the kth user's signal is corrupted by interference from users $1, 2, \ldots k - 1$, and the kth user's signal contains the same level of interference as the output of the kth user's conventional receiver.

Example 6.4

Let us examine the basics of the noise whitening analysis by an example. Consider three users sharing the CDMA channel and the code spreading sequences assigned to the users are:

$$\mathbf{C}_1 = [1 \ -1 \ -1 \ -1 \ 1 \ -1 \ \ 1 \ -1]$$
$$\mathbf{C}_2 = [1 \ \ \ 1 \ \ \ 1 \ \ \ 1 \ 1 \ -1 \ -1 \ \ 1]$$
$$\mathbf{C}_3 = [1 \ -1 \ -1 \ \ \ 1 \ 1 \ -1 \ -1 \ \ 1]$$

Let each user transmits the following four symbols with unit energy per bit:

$$\mathbf{b}_1 = [1 \ \ \ 1 \ \ \ 1 \ -1]$$
$$\mathbf{b}_2 = [1 \ \ \ 1 \ -1 \ \ \ 1]$$
$$\mathbf{b}_3 = [1 \ -1 \ \ \ 1 \ \ \ 1]$$

Calculate the output signal and the MAI at the output of the conventional receiver without and with noise whitening. Assume zero channel background noise.

Solution

We can write the matrix of the normalized $\mathbf{R}$ as:

$$\mathbf{R} = \begin{array}{|c|c|c|} \hline 1 & -0.25 & 0 \\ \hline -0.25 & 1 & 0.25 \\ \hline 0 & 0.25 & 1 \\ \hline \end{array}$$

From (6.42) $\mathbf{R} = \mathbf{F}^T \mathbf{F}$

Where

$$\mathbf{F} = \begin{array}{|c|c|c|} \hline F_{11} & 0 & 0 \\ \hline F_{21} & F_{22} & 0 \\ \hline F_{31} & F_{32} & F_{33} \\ \hline \end{array}$$

And

$$\mathbf{F}^T = \begin{array}{|c|c|c|} \hline F_{11} & F_{21} & F_{31} \\ \hline 0 & F_{22} & F_{32} \\ \hline 0 & 0 & F_{33} \\ \hline \end{array} \tag{6.47}$$

$$\mathbf{R} = \begin{array}{|c|c|c|} \hline F_{11} & 0 & 0 \\ \hline F_{21} & F_{22} & 0 \\ \hline F_{31} & F_{32} & F_{33} \\ \hline \end{array} \times \begin{array}{|c|c|c|} \hline F_{11} & F_{21} & F_{31} \\ \hline 0 & F_{22} & F_{32} \\ \hline 0 & 0 & F_{33} \\ \hline \end{array}$$

Thus
$$\mathbf{R} = \begin{array}{|c|c|c|} \hline F_{11}^2 & F_{11}F_{21} & F_{11}F_{31} \\ \hline F_{11}F_{21} & F_{21}^2 + F_{22}^2 & F_{21}F_{31} + F_{22}F_{32} \\ \hline F_{11}F_{31} & F_{31}F_{21} + F_{22}F_{32} & F_{31}^2 + F_{32}^2 + F_{33}^2 \\ \hline \end{array}$$

Comparing this **R** matrix with the given **R** matrix, we get:

$$\mathbf{F} = \begin{array}{|c|c|c|} \hline 1 & 0 & 0 \\ \hline -0.25 & 0.968 & 0 \\ \hline 0 & 0.258 & 0.966 \\ \hline \end{array}$$

And
$$\mathbf{F}^{\mathrm{T}} = \begin{array}{|c|c|c|} \hline 1 & -0.25 & 0 \\ \hline 0 & 0.968 & 0.258 \\ \hline 0 & 0 & 0.966 \\ \hline \end{array}$$

From (6.44), the output signal $= \tilde{\mathbf{y}} = \mathbf{F A b} = \begin{array}{|c|c|c|} \hline 1 & 0 & 0 \\ \hline -0.25 & 0.968 & 0 \\ \hline 0 & 0.258 & 0.966 \\ \hline \end{array} \times \begin{array}{|c|c|c|c|} \hline 1 & 1 & 1 & -1 \\ \hline 1 & 1 & -1 & 1 \\ \hline 1 & -1 & 1 & 1 \\ \hline \end{array}$

$$\tilde{\mathbf{y}} = \begin{array}{|c|c|c|c|} \hline 1 & 1 & 1 & -1 \\ \hline 0.718 & 0.718 & -1.218 & 1.218 \\ \hline 1.224 & -0.708 & 0.708 & 1.224 \\ \hline \end{array}$$

From (6.41), the output of the MF is:

$$\mathbf{y} = \mathbf{R A b}$$

$$\mathbf{y} = \begin{array}{|c|c|c|} \hline 1 & -0.25 & 0 \\ \hline -0.25 & 1 & 0.25 \\ \hline 0 & 0.25 & 1 \\ \hline \end{array} \times \begin{array}{|c|c|c|c|} \hline 1 & 1 & 1 & -1 \\ \hline 1 & 1 & -1 & 1 \\ \hline 1 & -1 & 1 & 1 \\ \hline \end{array}$$

$$= \begin{array}{|c|c|c|c|} \hline 0.75 & 0.75 & 1.25 & -1.25 \\ \hline 1 & 0.5 & -1 & 1.5 \\ \hline 1.25 & -0.75 & 0.75 & 1.25 \\ \hline \end{array}$$

The MAI at the output of the whitened MF is $= \tilde{\mathbf{y}} - \mathbf{b} = \begin{array}{|c|c|c|c|} \hline 0 & 0 & 0 & 0 \\ \hline -0.282 & -0.282 & -0.218 & 0.218 \\ \hline 0.224 & -0.292 & -0.292 & 0.224 \\ \hline \end{array}$

The MAI at the output of the MF is $= \mathbf{y} - \mathbf{b} = \begin{array}{|c|c|c|c|} \hline -0.25 & -0.25 & 0.25 & -0.25 \\ \hline 0 & -0.5 & 0 & 0.5 \\ \hline 0.25 & 0.25 & -0.25 & 0.25 \\ \hline \end{array}$

So far we have seen how the whitening filter is able to alter MAI distribution into relatively flat power spectral density which makes the BER performance optimal in the conventional receiver if it is preceded by the whitening filter. Our attention should be focused on methods to implement the whitening filter. The optimum filter Q(w) for the detection of a given signal in presence of arbitrary noise is (Pawelec, 2002):

$$Q(\omega) = \frac{S^*(\omega)}{S_n(\omega)} \tag{6.48}$$

where $S(\omega)$ is the PSD of the desired signal, and $S_n(\omega)$ represents the PSD of the noise which may contain AWGN and access interference. The noise-whitening filter $Q(\omega)$ can be placed in front of the conventional (matched filter) receiver to maximize the output SNR. This approach, as we will see in the sequel, does not require knowledge of other users' code spreading sequences, chip timing, and carrier phase offset so that the detector does not lock on and de-spread other users' signals. The baseband noise-whitening system is shown in Figure 6.11.

For simplicity, we will assume the desired signal acquired flat-shape and unit-valued spectrum, then equation (6.48) becomes:

$$Q(\omega) = \frac{1}{S_n(\omega)} \tag{6.49}$$

Further, we assume a perfect power control, the multiple access PSD is given in Monk et al. (1994):

$$I_0 = (K - 1)\frac{E_b}{G_p}|R_c\,P_c(\omega)|^2 \tag{6.50}$$

The AWGN plus access interference PSD $= S_n(\omega) = N_0 + (K - 1)\frac{E_b}{G_p}|R_cP_c(\omega)|^2$

The noise-whitening filter transfer function is:

$$Q(\omega) = \frac{1}{N_0 + \dfrac{E_b(K - 1)}{G_p}|R_cP_c(\omega)|^2} \tag{6.51}$$

Where $P_c(\omega)$ is the chip pulse shape, E_b is the energy per data bit, K is the number of active users sharing the CDMA channel, R_c is the chip rate, and G_p is the processing gain. When

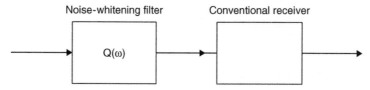

Figure 6.11 *Noise-whitening receiver.*

the additive white noise becomes much smaller than MAI, the filter $Q(\omega)$ becomes nearly proportional to $\frac{1}{|P_c(\omega)|^2}$ which splits into two parts, $\frac{1}{P_c(\omega)}$ and $\frac{1}{P_c^*(\omega)}$, and filter $Q(\omega)$ is then equivalent to the part coded at the transmitter; the other part is decoded at the receiver.

Research on noise whitening in the 1990s was aimed at the strongest interference from certain locked users (Yoon and Leib, 1996). The proposed scheme utilizes the chip delays and signal power of users with the strongest interference to further maximize the output SNR and can only operate in the centralized form since it requires information that is only available at the base station. Consider a group of K users transmitting data asynchronously with the target user receiver locked on to K_L users and K_U is the number of unlocked users so that $K = K_L + K_U$.

The key merit of this scheme is its ability to suppress interference from the dominant locked users and is achieved at the expense of a limited increase in the complexity of the filter impulse response computation over that required for a conventional receiver. This modified conventional receiver has shown to be N-F resistant but its resistance depends on the number of locked users and their chip delay distribution. When all the interferers are unlocked, the scheme approaches the conventional noise-whitening solution but if all the interferers are locked, the N-F resistance of the receiver is maximized and there is a significant reduction in computing time. The proposed receiver in Yoon and Leib (1996) is as shown Figure 6.12.

The modified impulse response h(t) is made up of three terms as shown in Figure 6.12. The first term is matched to the code sequence of the target user ($C_0(t)$), the second term comprises K_L components representing the interference estimates generated by each of the locked interferers, and the last term gives the estimate of the interference from the unlocked users.

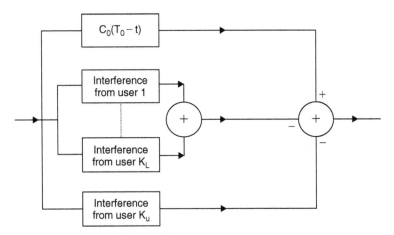

Figure 6.12   *Improved noise-whitening filter.*

The interference contributions are subtracted from the matched filter response making the output signal interference free. The way the interference is removed may resemble the parallel interference cancellation methods. However, in this case the multiple access interference is removed from the impulse response of the filter to maximize the output SNR. This solution does not account for the ISI since it assumes ideal chip pulses that are non-zero only over $0 \leq t \leq T_c$.

The single-user receiver described in Patel and O'Farrell (1998), can completely cancel the MAI at considerably reduced complexity compared to the multi-user detection while maintaining an improvement in BER performance over the conventional receiver. Consider a received signal from a synchronous CDMA channel shared by K with $T_k = 0$ for $k = 1, 2, \ldots, K$, and A_k is the received amplitude of the kth user. With no loss of generality we assume $\ell = 0$ in the received signal in (6.36) so that the received signal is simplified to:

$$r(t) = \left[\sum_{k=1}^{K} C_k(t) \cdot A_k \cdot b_k(t) \cdot \right] + n(t) \tag{6.52}$$

Further, we will assume the code sequence of the target user to be $C_1(t)$ so that for N samples of the received signal, the received vector r(N) is given by:

$$r(N) = C_1(N)A_1b_1 + \sum_{k=2}^{K} C_k(N) \cdot A_k b_k + n(N) \tag{6.53}$$

The received vector given by (6.53) consists mainly of the wanted signal, $r_S(N)$, given by the first term; and the corrupting signal $r_n(N)$ given by the last two terms where the second term represents the MAI, and the third term, the AWGN signal. That is:

$$r(N) = r_s(N) + r_n(N) \tag{6.54}$$

Using (6.53), we have:

$$r_n(N) = r_{MAI}(N) + r_{AWGN}(N) \tag{6.55}$$

From (6.53), we have:

$$r_{MAI}(N) = \sum_{k=2}^{K} C_k(N) \cdot A_k b_k \tag{6.56}$$

$$r_{AWGN}(N) = n(N) \tag{6.57}$$

The $N \times N$ matrix autocorrelation function of $r_n(N)$ is given by the summation of the autocorrelation matrix of the MAI and the autocorrelation matrix of the AWGN:

$$R_n = E[r_n(N)r_n^T(N)] = R_{MAI} + R_{AWGN} \tag{6.58}$$

From (4.46), we have:

$$R_{AWGN} = \frac{N_0}{2}I = \sigma^2 I \tag{6.59}$$

The autocorrelation of the MAI is given by:

$$\mathbf{R}_{MAI} = E[\mathbf{r}_{MAI}(N) \cdot \mathbf{r}_{MAI}^T(N)] \tag{6.60}$$

Where $R_{MAI}(i,j) = \frac{1}{K-1} \left[\sum_{k=2}^{K} C_{k,i}(N) \cdot C_{k,j}(N) \right]$

The filter coefficients that maximize the SNR at the out of the single-user receiver are represented by h which is given by:

$$\mathbf{h} = \mathbf{R}_n^{-1} \cdot C_1(N) \tag{6.61}$$

There are two extreme cases:

i. $\mathbf{r}_{MAI}(N) << \mathbf{r}_{AWGN}(N)$ (6.62)

In this case, the interference is mainly due to AWGN and the linear filter behaves like a conventional matched filter where the filter impulse response (h) is given by:

$$\mathbf{h} = C_1(N) = \text{ target user code sequence} \tag{6.63}$$

ii. $\mathbf{r}_{MAI}(N) >> \mathbf{r}_{AWGN}(N)$

The interference is mainly due to MAI and:

$$\mathbf{R}_n = \mathbf{R}_{MAI} \tag{6.64}$$

Thus the impulse response of the filter is:

$$\mathbf{h} = \mathbf{R}_{MAI}^{-1} C_1(N) \tag{6.65}$$

The single-user receiver described in (6.52)–(6.58) is not N-F resistant and requires tight power control in order to achieve an acceptable BER performance. Also the matrix R_n has to be computed for each user and the receiver needs prior knowledge of the code sequences of all active users at the detection instant.

Concluding the discussion on this subject, it is worth mentioning that in certain applications the CDMA channels are corrupted by multiple access interference and additive impulsive noise that is non-Gaussian. The strong impulses are generated by atmospheric phenomena, switching processors in networks, and discharge in electronic appliances. Several methods are described in the literature for treating the effects of the impulsive noise, among them the adaptive filtering (Kim and Efron, 1995), and non-linear correlators (Aazhang and Poor, 1989; Delich and Hocanin, 2000).

6.6 Adaptive single-user receivers

The time-varying channel characteristics and the changing number of users means that the single-user receiver has to be adaptive to the transmission environment to attain optimum performance. Adaptive receivers use digital filtering with adjustable coefficients to equalize the received signal. The coefficients can be optimized according to certain criterion in order to suppress received signal corruption generated from multiple access, intersymbol and multipath interference.

Adaptive receivers can be broadly classified into chip-rate linear receivers and fractionally spaced receivers. The fractionally spaced receivers have a structure similar to fractionally spaced equalizers and are of two types: either linear receivers using Time Dependent Adaptive Filters (TDAF) (Reed and Hsia, 1990) or non-linear receivers using decision feedback filters. In addition, the fractionally spaced linear filters can either use complex coefficients or real coefficients. The various structures for adaptive single-user receivers are shown in Figure 6.13.

Let the received sequence be $r(m)$ with $m = 1, 2, \ldots, M$ so that the estimated auto-correlation $\hat{\mathbf{R}}$ is given by:

$$\hat{\mathbf{R}} = (1/M) \cdot \sum_{m=1}^{M} \mathbf{r}(m) \cdot \mathbf{r}^*(m) \tag{6.66}$$

Several schemes can be used to optimize the filter coefficients $\mathbf{W}$. The most prominent of these schemes is based on the Minimum Mean Square Error (*MMSE*) criterion so that the estimated $\hat{\mathbf{W}}$ is:

$$\hat{\mathbf{W}} = \hat{\mathbf{R}}^{-1} \cdot \hat{\mathbf{P}} \tag{6.67}$$

where the correlation matrix $\hat{\mathbf{R}}$ is as defined above and

$$\hat{\mathbf{P}} = E[d_1(m)r(m)] \tag{6.68}$$

$d_1(m)$ is the data symbols transmitted by the target user.

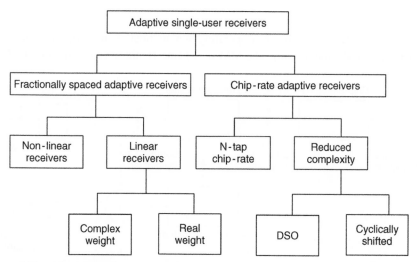

Figure 6.13 *Adaptive single-user receivers.*

There are two adaptive algorithms commonly used to optimize the filter coefficients: the Least Mean Squares (LMS) algorithm and the standard Recursive Least Squares (RLS) algorithm (Proakis, 1995) are discussed in Chapter 2.

A delay estimator is presented in Strom et al. (1996), Parkvall et al. (1999) to evaluate the propagation delay in an asynchronous link and correct the received signal for the target user to synchronize the data transmission without the need for a training sequence.

We now present a survey for a number of adaptive single-user receivers found in the literature (Madhow and Honig, 1994; Majmunder et al., 2000). A variant of these schemes is also considered in Madhow and Honig (1994) where only a few strong interferers exist in the system because of strict power control.

6.6.1 The chip-rate linear adaptive receivers

In these receivers, adaptive processing is carried out at the chip-rate. The linear adaptive receiver is presented in Madhow and Honig (1944) and its block diagram is shown in Figure 6.14. The adaptive N-tap linear transversal filter exploits the cyclostationary nature of the CDMA signal to mitigate interference. The adaptive algorithm is based on the MMSE criterion and the LMS adaptive filter needs to span only one symbol interval to suppress MAI, but for suppressing the ISI generated by multipath signal propagation, the filter has to span more than one symbol interval. The output of the adaptive filter is sampled at symbol rate to estimate the data symbol.

The receiver is N-F resistant and its BER performance is much better than that of the conventional receiver. However, its computation cost is higher compared to that of the conventional receiver as a result of the adaptive nature of the receiver (Strom and Miller, 1999). The receiver is able to reject up to $(N-1)$ synchronous interferers or $(N-1)/2$ asynchronous interferers.

An N-tap chip-rate adaptive receiver operating in a multipath fading context is presented in Foerster and Milstein (2000). The MMSE scheme is still used to suppress the MAI,

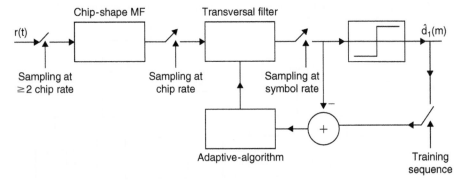

Figure 6.14 *N-tap chip-rate adaptive receiver.*

but convolutional coding and interleaving are employed to combat the effects of the flat Rayleigh fading. In a rapidly time-varying channel, fast RLS algorithm is suggested for fast filter coefficients adaptation and algorithm fast convergence.

These receivers can be modified to reduce the computation cost but with, in most cases, worse BER performance. The modified receivers can be classified into Cyclically Shifted Filter Bank (CSFB) linear adaptive receivers and Data Symbol Oversampling (DSO) linear adaptive receivers. In these receivers, described in Madhow and Honig (1994), the despread of the received signal is carried out prior to the adaptive processing to reject the MAI. The block diagram of the (CSFB) linear adaptive receiver is shown in Figure 6.15.

The sampled received signal is fed to a bank of D non-adaptive N-tap linear transversal filters to reduce the computation cost and D is chosen to be less than N. The first filter in the bank is matched to the code sequence of the target user. The following (D-1) filters are a cyclically shifted version of the first filter. The outputs of the filter bank are sampled at the data rate and are adaptively combined to obtain an estimate of the data. The weights of the combiner are optimized according to the MMSE criterion. Assuming the target user's spreading sequence enjoys a near impulse autocorrelation function, then the weight vectors of the D filters are approximately orthogonal to each other and up to (D-1) synchronous interferers can be rejected. Since two consecutive symbols of each interferer overlap with the target user's symbol, the receiver rejects a maximum of only (D-1)/2 asynchronous interferers.

A block diagram of the other modified linear adaptive receiver, i.e. the DSO, is shown in Figure 6.16.

The received signal is first despread using the desired user spreading code C, and the output of the matched filter is sampled D times per symbol interval to feed the adaptive transversal filter. Unlike the CSFB and the N-tap MMSE schemes, where the interfering signal contributes segments from two successive symbol intervals, the interfering signal in DSO

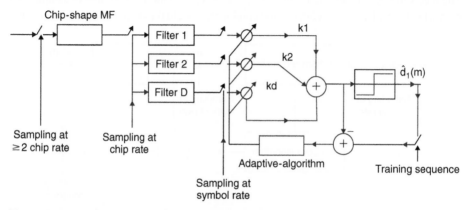

Figure 6.15 *Cyclically shifted filter bank adaptive receiver.*

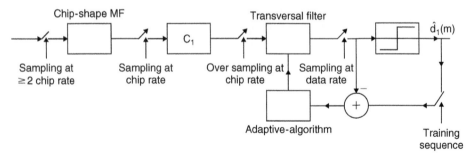

Figure 6.16   *Oversampling adaptive receiver.*

linear adaptive receivers contains segments from more than two different symbol intervals. Consequently, the performance of the DOS linear adaptive receivers is worse than that of the CSFB and N-tap receivers but significantly better than the performance of conventional receivers particularly when the number of strong asynchronous interferers is small.

6.6.2   Fractionally spaced adaptive receivers

The fractionally spaced adaptive receivers have a similar structure to that of fractionally spaced equalizers. But, while the conventional equalizers reject ISI, these adaptive detectors de-spread the target user signal, reject MAI and can combine multipath signals as in RAKE receivers (Monogioudis et al., 1994). These receivers are generally N-F resistant. Since fractionally spaced equalizers perform relatively better than symbol spaced equalizers, it is expected that fractionally spaced adaptive receivers perform better than chip spaced adaptive receivers discussed earlier in the section.

A block diagram of the fractionally spaced linear adaptive receivers (Majmunder et al., 2000) is shown in Figure 6.17.

The received signal is sampled at the rate $\frac{1}{T_s}$ ($T_s \leq T_c$) generating a number of samples per chip which are adaptively weighted, and combined to yield the output of the adaptive filter.

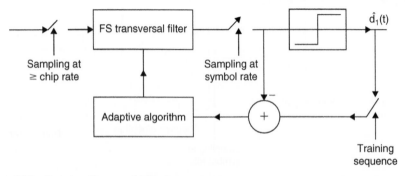

Figure 6.17   *Fractionally spaced adaptive receivers.*

Filter weights are updated at the symbol rate. Figure 6.17 shows this adaptive receiver mitigating the MAI.

The adaptive single-user receivers operating in asynchronous CDMA system are considered in Papajic and Vucetic (1994). The proposed structures are classified on the basis of the number of detected symbols available for processing. The simplest structure is capable of removing MAI to a great extent without requiring the knowledge of the timing, signature, and carrier phases of other users. The receivers are near–far resistant and thus do not need accurate power control. They achieve significant improvements in system capacity (the number of users assessing the network simultaneously), compared to conventional single-user receivers with computation costs, although slightly higher than that of conventional receivers, are independent of the number of users. The adaptive scheme is capable of removing the effect of multipath distortion if the fading rate is slower than the convergence speed of the adaptation algorithm. The SNR at the output of the adaptive linear receiver is:

$$\text{SNR}_{\text{output}} = 10 \log\left(\frac{\sigma_a^2}{\varepsilon_{\text{opt}}}\right) \tag{6.69}$$

where ε_{opt} is *minimum mean square error* and σ_a^2 is the average symbol power of target user. The SNR of the conventional receiver is given by:

$$\text{SNR}_{\text{output MF}} = 10 \log\left(\frac{\sigma_a^2}{\sigma_n^2 + \dfrac{(K-1)\sigma_a^2}{G_p}}\right) \tag{6.70}$$

where K = number of active users, G_p is the processing gain, and σ_n^2 is the Gaussian noise variance. Simulation results presented in Papajic and Vucetic (1994) show that the maximum number of users accessing the system is about 77% of the processing gain compared to 10% of the processing gain for the conventional matched filter receiver.

The block diagram of the fractionally spaced decision-feedback adaptive receivers is shown in Figure 6.18.

Fractionally spaced decision-directed adaptive receivers have a structure and properties similar to the decision-feedback equalizer but with a fractionally spaced forward filter and a symbol spaced feedback filter which uses feedback of previous symbol decisions. Since the symbol decisions are fed back into the filter, erroneous symbol decisions propagate back and cause a burst of errors resulting in poor convergence characteristics.

The adaptive algorithm, when applied to a fast fading CDMA channel, is unable to track the fading on any of the interfering users (Foerster and Milstein, 2000) but some interfering users can be tracked in a slow fading channel. However, the adaptive receiver will often lose

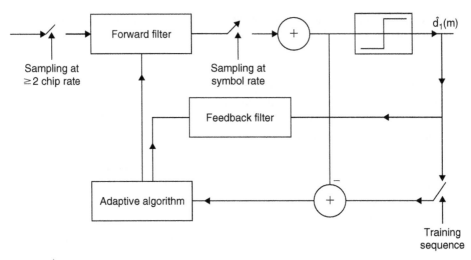

Figure 6.18   *Fractionally spaced decision feedback adaptive receiver.*

phase lock on the target signal when that signal dips into a deep fade. This phenomenon is caused by incorrect decisions being fed to the adaptation algorithm. The adaptive filter may emerge from the deep fade in any of three possible cases: it may be correctly locked on the target signal phase; it may be locked 180 degrees out of phase to the correct phase of the user's signal and this phase reversal can be combated using differential encoding; or it may be locked either in phase or 180 degrees out of phase to any of the interfering signals, which is the most common behaviour in a N-F environment and thus renders the MMSE receiver ineffective. Modifying the MMSE receiver to track channel phase variations and operate adequately in Rayleigh fading channels can alleviate the problem. A way of achieving this is to predict the channel phase and remove the predicted phase from the received signal before entering the MMSE filter.

Consider an Lth order linear prediction of the channel. The phase $\hat{\theta}_1$ and amplitude, $\hat{\beta}(m)$, of the channel during the current symbol interval is estimated in terms of the channel phase during previous symbol intervals as:

$$\hat{\beta}(m) = \sum_{i=1}^{L} a_i \cdot \beta(m - i) \tag{6.71}$$

The phase estimate of the channel during mth symbol interval $= \hat{\theta}_1(m) = <\hat{\beta}(m)$.

The coefficients of the linear predictor $\mathbf{a} = [a_1, a_2, a_3, \ldots, a_L]^T$ are chosen to provide the MMSE solution. It is found that (Barbosa and Miller, 1998):

$$\mathbf{a} = \mathbf{C}^{-1} \cdot \mathbf{v} \tag{6.72}$$

where $[v]_i = R_c(i.T_b)$ and $R_c(\tau)$ is the autocorrelation of the fading signal. A simplified expression for vector C is:

$$\mathbf{C} = \mathbf{B} + \frac{1}{\dfrac{E_b}{N_0}} \cdot \mathbf{I} \qquad (6.73)$$

I is identity matrix, $[B]_{i,j} = R_C((i-j)T_b)$. For a fading channel, $R_c(\tau)$ can be expressed as an approximation to the Bessel autocorrelation function over the region $|\tau| < (L+1) \cdot T_b$:

$$R_c(\tau) = 1 - (\pi \cdot f_d \cdot \tau)^2 \qquad (6.74)$$

The BER of the MMSE receiver over a flat-fading channel can be approximated by the following closed form expression:

$$P_e \approx \left(1 - \frac{1}{\sqrt{1 + \dfrac{1}{\dfrac{E_b}{\hat{N}_0}}}} \right) \cdot 10^{\xi \frac{K-1}{G_p}} \qquad (6.75)$$

where G_p = processing gain, K = the number of active users the system can support and ξ is constant given in Table 6.1 (Barbosa and Miller, 1998):

Table 6.1 *Values of ξ for various values of processing gain N and signal-to-noise ratio $\frac{E_b}{N_0}$*

G_p in dB	$\frac{E_b}{N_0}$ in dB		
	10	20	30
15	0.46	1.17	2.17
31	0.529	1.26	2.28
63	0.508	1.22	2.12
127	0.468	1.12	2.0

The MMSE receiver offers a substantial improvement in BER and capacity, with only a small increase in computation cost over the conventional receivers.

Example 6.5

Consider the fractionally spaced decision-feedback adaptive receivers shown in Figure 6.18 with 2nd order predictor (L = 2) and $\frac{E_b}{N_0} = 10$ dB for slow fading CDMA channel at a maximum Doppler frequency $f_d = 8$ Hz, and fast fading channel at $f_d = 100$ Hz. If the data rate $R_b = 9.6$ kb/sec. Find the coefficients of the predictor.

Solution

The coefficients of the linear predictor are given by:

$$\mathbf{a} = [a_1, a_2]^T$$

Define
$$\mathbf{v} = [v_1, v_2]^T$$

From (6.73), $\mathbf{C}$ is given by:

$$\mathbf{C} = \begin{array}{|c|c|} \hline R_c(0) & R_c(-T_b) \\ \hline R_c(T_b) & R_c(0) \\ \hline \end{array} + \left(\frac{E_b}{N_0}\right)^{-1} \times \begin{array}{|c|c|} \hline 1 & 0 \\ \hline 0 & 1 \\ \hline \end{array}$$

$$\left(\frac{E_b}{N_0}\right)^{-1} = 0.1$$

Slow fading: $f_d = 8\,Hz$

From (6.74), we have:

$$R_c(0) = 1$$
$$R_c(T_b) = 1 - (\pi \cdot f_d \cdot T_b)^2 \approx 1 = R_c(-T_b)$$

Therefore substituting in the expression for $\mathbf{C}$, we get:

$$\mathbf{C} = \begin{array}{|c|c|} \hline 1 & 1 \\ \hline 1 & 1 \\ \hline \end{array} + 0.1 \begin{array}{|c|c|} \hline 1 & 0 \\ \hline 0 & 1 \\ \hline \end{array}$$

Simplifying the above equation, we get:

$$\mathbf{C} = \begin{array}{|c|c|} \hline 1.1 & 1 \\ \hline 1 & 1.1 \\ \hline \end{array}$$

Det. $\mathbf{C} = (1.1)^2 - (1)^2 = 0.21$

$$\mathbf{C}^{-1} = 1/Det.\mathbf{C} * \begin{array}{|c|c|} \hline 1.1 & -1 \\ \hline -1 & 1.1 \\ \hline \end{array}$$

$$\mathbf{C}^{-1} = \begin{array}{|c|c|} \hline 5.2381 & -4.7619 \\ \hline -4.7619 & 5.2381 \\ \hline \end{array}$$

Now we have:

$$[\mathbf{v}]_i = R_{c(iT_b)}$$
$$[\mathbf{v}]_1 = R_{c(T_b)} = 1$$
$$[\mathbf{v}]_2 = R_{c(2T_b)} = 1 - 4(\pi \cdot f_d \cdot T_b)^2 \approx 1$$

Putting these results together, we have:

$$
\begin{vmatrix} a_1 \\ a_2 \end{vmatrix} = \begin{vmatrix} 5.2381 & -4.7619 \\ -4.7619 & 5.2381 \end{vmatrix} \begin{vmatrix} 1 \\ 1 \end{vmatrix}
$$

Therefore
$$a_1 = 0.4762$$
$$a_2 = 0.4762$$

Fast fading: $f_d = 100\,\text{Hz}$

$$R_c(0) = 1 \quad R_c(T_b) = 1 - (\pi \cdot f_d \cdot T_b)^2 = 0.9989$$
$$R_c(2T_b) = 1 - 4(\pi \cdot f_D \cdot T_b)^2 = 0.9957$$

$$
\mathbf{C} = \begin{vmatrix} 1 & 0.9989 \\ 0.9989 & 1 \end{vmatrix} + 0.1 \begin{vmatrix} 1 & 0 \\ 0 & 1 \end{vmatrix}
$$

Simplifying the above equation, we get:

$$
\mathbf{C} = \begin{vmatrix} 1.1 & 0.9989 \\ 0.9989 & 1.1 \end{vmatrix}
$$

Det. $\mathbf{C} = (1.1)^2 - (0.9989)^2 = 0.2122$

$$
\mathbf{C}^{-1} = 1/0.2122 \begin{vmatrix} 1.1 & -0.9989 \\ -0.9989 & 1.1 \end{vmatrix}
$$

$$
\mathbf{C}^{-1} = \begin{vmatrix} 5.1838 & -4.7073 \\ -4.7073 & 5.1838 \end{vmatrix}
$$

Now we have:

$$[\mathbf{v}]_i = R_{c(iT_b)}$$
$$[\mathbf{v}]1 = R_{c(T_b)} = 0.9989$$

$$[\mathbf{v}]2 = R_{c(2T_b)} = 1 - 4(\pi \cdot f_D \cdot T_b)^2 = 0.9957$$

Putting these results together, we have:

$$
\begin{vmatrix} a_1 \\ a_2 \end{vmatrix} = \begin{vmatrix} 5.1838 & -4.7073 \\ -4.7073 & 5.1838 \end{vmatrix} \begin{matrix} \times \\ \times \end{matrix} \begin{vmatrix} 0.9989 \\ 0.9957 \end{vmatrix}
$$

Therefore
$$a_1 = 0.4910$$
$$a_2 = 0.4594$$

Example 6.6

Consider the fractionally spaced decision-feedback adaptive receivers shown in Figure 6.18 when users are sharing a fading CDMA channel with processing gain equal to 31 and a probability of error not exceeding 3%. Calculate the number of users the system can support with $\frac{E_b}{N_0} = 20\,dB$ and 30 dB.

Solution

From (6.75), we have:

$$P_e \approx \left(1 - \frac{1}{\sqrt{1 + \dfrac{1}{\dfrac{E_b}{\hat{N}_0}}}} \right) \cdot 10^{\xi \frac{K-1}{G_p}} = 0.03$$

$\frac{E_b}{N_0} = 20\,dB$, $G_p = 31$, and $\xi = 1.26$ from Table 6.1. Substitute in the equation for probability of error:

$$0.03 = \left(1 - \frac{1}{\sqrt{1 + \dfrac{1}{100}}} \right) \cdot 10^{1.26 \cdot \frac{K-1}{31}}$$

Hence $K \approx 45$

For $\frac{E_b}{N_0} = 30\,dB$, $N = 31$, and $\xi = 2.28$ from Table 6.1,

$$0.03 = \left(1 - \frac{1}{\sqrt{1 + \dfrac{1}{1000}}} \right) \cdot 10^{2.28 \frac{K-1}{31}}$$

Hence $K \approx 56$.

6.6.3 Adaptive receiver for multipath fading channel

In a multipath environment, there are two approaches to suppress multiple access interference (Latva-aho and Juntti, 2000): (1) it can be suppressed after combining the multipath signals and this is known as the post-combining scheme; and (2) the second approach is to suppress the interference before combining the multipath signals and is known as the pre-combining scheme.

The post-combining MMSE receiver requires the channel coefficients used by all users to be adapted as the channel changes. Thus, in relatively fast fading, the adaptive post-combining receivers have severe convergence problem. A pre-combining receiver is shown in Figure 6.19. The computation used for the pre-combining receiver is increased only moderately over the computation used in the conventional receiver. Furthermore, the parameters of the CDMA channel used by the target user must be known or accurately estimated. The adaptive pre-combining receiver does not necessarily need a training sequence as the decisions made by proceeding conventional RAKE receiver are used to train the adaptive filter. Referring to Figure 6.19, each receiver branch consists of three blocks, channel estimator, adaptive FIR filter, and LMS algorithm so that it is adapted independently to suppress the MAI from each path.

The channel estimator consists of a pilot channel correlator followed by a moving average filter. The length of averaging depends on the rate of fading. The decisions made by the conventional RAKE can be used to obtain reference signal in the adaptive method. The adaptive FIR filter employs the LMS algorithm and the MMSE criterion to optimize the coefficients. The disadvantage of this receiver is that it increases the gradient noise that degrades receiver performance.

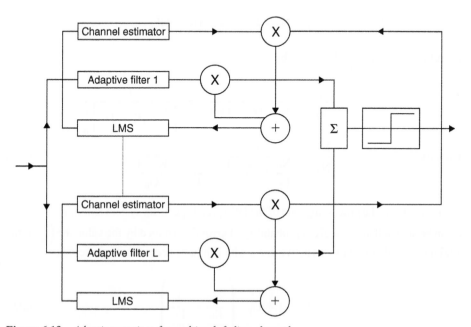

Figure 6.19   *Adaptive receiver for multipath fading channel.*

6.7   CDMA system capacity

6.7.1   *Single cell capacity with ideal power control*

A CDMA system compromises a single cell shared by a group of K active users. On the down link, signals are sent together to the users by the serving base station and attenuated equally during signal propagation. Thus, there is no requirement for power control.

However, on the uplink, the power received at the base station from each individual user is kept at the same level (S) by the power control scheme whatever the physical position of the particular user. Thus, the total interfering power (I) at the base station is:

$$I = (K - 1) \cdot S \qquad (6.76)$$

The interfering power density (I_0) is given by:

$$I_0 = \frac{I}{W} = (K - 1)\frac{S}{W} = (K - 1)E_c \qquad (6.77)$$

where E_c is the energy per chip respectively. The energy per bit E_b is given by:

$$E_b = N \cdot E_c \qquad (6.78)$$

where N is number of chips per symbol period. The total noise density ($\hat{N}_0$) is given by sum of a Gaussian noise density (N_0) and interference power density (I_0). Thus:

$$\hat{N}_0 = N_0 + I_0 = N_0 + (K - 1)E_c$$

Therefore, the ratio $\left(\frac{E_b}{\hat{N}_0}\right)$ is given by:

$$\frac{E_b}{\hat{N}_0} = \frac{E_b}{N_0 + (K - 1) \cdot E_c} = \frac{1}{\dfrac{N_0}{E_b} + (K - 1) \cdot \dfrac{E_c}{E_b}}$$

Therefore,
$$\frac{E_b}{\hat{N}_0} = \frac{1}{\dfrac{N_0}{E_b} + \dfrac{K - 1}{N}} = \frac{1}{\dfrac{N_0}{E_b} + \dfrac{K - 1}{G_p}} \qquad (6.79)$$

where $N = G_p$ is the processing gain of the DS-spread spectrum system assuming $T = NT_c$. It can be shown that, for a given probability of error determined by the value for $\left(\frac{E_b}{\hat{N}_0}\right)$, the total number of simultaneous active users who can have access to the system is given by:

$$K \le 1 + G_p \cdot \left(\frac{1}{\dfrac{E_b}{\hat{N}_0}} - \frac{1}{\dfrac{E_b}{N_0}} \right) \qquad (6.80)$$

When the Gaussian noise is much smaller than multiple access interference and $K \gg 1$, then:

$$K \approx G_p \cdot \left(\frac{1}{\frac{E_b}{\hat{N}_0}} \right) \tag{6.81}$$

We now consider the maximum data rate that the system achieves in a CDMA channel shared by K users. Each user transmits a signal of bandwidth B and an average power S. Assume that the two-sided noise density is $\frac{N_0}{2}$ and total noise power is N_{noise}. Let us consider the asynchronous CDMA link first where signals from other $(K-1)$ users appear as interference at the kth user receiver. The kth user's maximum data rate R_k is given by the Shannon equation:

$$R_k = B \cdot \log_2 \left[1 + \frac{S}{N_{noise}} \right] = B \cdot \log_2 \left[1 + \frac{S}{N_0 \cdot B + (K-1)S} \right] \tag{6.82}$$

Substitute for received power S in terms of energy per bit E_b and (6.82) becomes:

$$R_k = B \cdot \log_2 \left[1 + \frac{R_k \cdot E_b}{N_0 \cdot B + (K-1)R_k \cdot E_b} \right]$$

$$= B \cdot \log_2 \left[1 + \frac{R_k}{B} \frac{\frac{E_b}{N_0}}{1 + (K-1) \cdot \frac{R_k}{B} \cdot \frac{E_b}{N_0}} \right]$$

Simplifying the above expression to become:

$$\frac{R_k}{B} = \log_2 \left[1 + \frac{R_k}{B} \frac{\frac{E_b}{N_0}}{1 + (K-1) \cdot \frac{R_k}{B} \cdot \frac{E_b}{N_0}} \right]$$

Thus

$$2^{\frac{R_k}{B}} = \left[1 + \frac{R_k}{B} \frac{\frac{E_b}{N_0}}{1 + (K-1) \cdot \frac{R_k}{B} \cdot \frac{E_b}{N_0}} \right].$$

Now take $\log_e (\cdot)$ to both sides, we get:

$$\frac{R_k}{B} \cdot \log_e (2) = 0.69 \cdot \frac{R_k}{B} = \log_e \left[1 + \frac{R_k}{B} \frac{\frac{E_b}{N_0}}{1 + (K-1) \cdot \frac{R_k}{B} \cdot \frac{E_b}{N_0}} \right]$$

Now in most practical cases, $\dfrac{R_k}{B} \dfrac{\dfrac{E_b}{N_0}}{1+(K-1)\cdot\dfrac{R_k}{B}\cdot\dfrac{E_b}{N_0}} \ll 1$, so that:

$$\log_e\left[1+\dfrac{R_k}{B}\dfrac{\dfrac{E_b}{N_0}}{1+(K-1)\cdot\dfrac{R_k}{B}\cdot\dfrac{E_b}{N_0}}\right] \approx \dfrac{R_k}{B}\dfrac{\dfrac{E_b}{N_0}}{1+(K-1)\cdot\dfrac{R_k}{B}\cdot\dfrac{E_b}{N_0}}$$

Thus, $0.69\cdot\dfrac{R_k}{B} = \dfrac{R_k}{B}\dfrac{\dfrac{E_b}{N_0}}{1+(K-1)\cdot\dfrac{R_k}{B}\cdot\dfrac{E_b}{N_0}}$

That is $0.69 = \dfrac{\dfrac{E_b}{N_0}}{1+(K-1)\cdot\dfrac{R_k}{B}\cdot\dfrac{E_b}{N_0}}$ which is simplified to:

$$\dfrac{R_k}{B} = \dfrac{1.44}{K} - \dfrac{1}{K\cdot\dfrac{E_b}{N_0}}$$

For a transmission channel with $K \cdot \frac{E_b}{N_0} \gg 1$, then the kth user's Shannon capacity is given by:

$$R_k \approx \dfrac{1.44}{K}\cdot B \tag{6.83}$$

Therefore, the total Shannon capacity of the CDMA system (R_{cdma}) when K active users are transmitting simultaneously is:

$$R_{cdma} = K \cdot R_k \approx 1.44 \cdot B \tag{6.84}$$

It is clear from (6.84) that the total Shannon capacity of the CDMA system does not increase with K as in FDMA and TDMA. This is because, during the detection process, individual receivers consider other users as an extra source of additive noise.

Substitute the approximate expression for $\frac{E_b}{N_0}$ given by (6.79) in the probability of error at the output of kth user receiver:

$$P_e(\text{approx}) = Q\left(\sqrt{2 \cdot \frac{E_b}{\hat{N}_0}}\right) = Q\left(\sqrt{\frac{2}{\frac{N_0}{E_b} + \frac{K-1}{G_p}}}\right)$$

$$P_e(\text{approx}) = Q\left(\sqrt{\frac{1}{\frac{N_0}{2 \cdot E_b} + \frac{K-1}{2 \cdot G_p}}}\right) = Q\left[\frac{N_0}{2 \cdot E_b} + \frac{K-1}{2 \cdot G_p}\right]^{-\frac{1}{2}} \qquad (6.85)$$

It is interesting to compare this approximate expression for probability of error with the more rigorous expression derived by Pursley (1977) as:

$$P_e = Q\left[\frac{N_0}{2 \cdot E_b} + \frac{K-1}{3 \cdot G_p}\right]^{-\frac{1}{2}} \qquad (6.86)$$

Let us now consider the synchronous link with the individual code sequences being orthogonal to each other. Assuming the received power of individual users is kept equal, we have (Proakis, 1995):

$$\sum_{k=1}^{K} R_k \leq B \log_2\left(1 + \frac{K \cdot S}{BN_0}\right) \qquad (6.87)$$

In this case, we see that the total Shannon capacity of the CDMA system increases with K as is the case in FDMA and TDMA.

Example 6.7

A group of K users access a single cell CDMA system and each user transmits data at the rate of 19.2 k bit/sec employing a code sequence at chip rate of 1.2288 MHz. The data is used to modulate a BPSK carrier of amplitude one volt. The effective signal-to-noise ratio $\left(\frac{E_b}{N_0}\right)$ and the density of white noise at the input of user matched filter receiver (N_0) are 5 dB and 1 μW/Hz respectively. Calculate:

i. Maximum number of users (K) that can access the system at the specified probability of error.
ii. Ratio of bit energy to interference density $\frac{E_b}{I_0}$
iii. Total capacity of the system
iv. Probability of error using (6.85) and (8.86).

Solution

i. The maximum number of users is given by

$$K = 1 + G_p \cdot \left(\frac{1}{\dfrac{E_b}{\hat{N}_0}} - \frac{1}{\dfrac{E_b}{N_0}} \right)$$

The processing gain G_p is given by:

$$G_p = \frac{1228.8}{19.2} = 64$$

$$\frac{E_b}{\hat{N}_0} = 5 \, dB = 3.162$$

The normalized average power of the carrier $= 0.5$

$$\frac{E_b}{N_0} = \frac{0.5}{10^{-6} \times 19.2 \times 10^3} = 26.04 = 14.157 \, dB$$

Thus, $K = 1 + 64 \cdot \left(\dfrac{1}{3.162} - \dfrac{1}{26.04} \right) = 18.78 \approx 18$ users

ii. $\dfrac{E_b}{\hat{N}_0} = \dfrac{E_b}{N_0 + \text{interference}} = \dfrac{1}{\dfrac{1}{\dfrac{E_b}{N_0}} + \dfrac{1}{\dfrac{E_b}{I_0}}}$

Thus $\dfrac{1}{\dfrac{E_b}{I_0}} = \dfrac{1}{\dfrac{E_b}{\hat{N}_0}} - \dfrac{1}{\dfrac{E_b}{N_0}} = \dfrac{1}{3.162} - \dfrac{1}{26.04} = 0.2778$

Therefore $\dfrac{E_b}{I_0} = 3.599 = 5.56 \, dB$

iii. Total Shannon capacity of the system per Hz of channel bandwidth is given by:

$$\frac{R_k}{B} = \frac{1.44}{K} - \frac{1}{K \cdot \dfrac{E_b}{N_0}}$$

that is $K \cdot \dfrac{R_k}{B} = 1.44 - \dfrac{1}{\dfrac{E_b}{N_0}} = 1.4 \, b/sec/Hz$

iv. The approximate probability of error is given by:

$$P_e(\text{approx}) = Q\left[\frac{N_0}{2 \cdot E_b} + \frac{K-1}{2 \cdot G_p}\right]^{-\frac{1}{2}}$$

Substituting for $G_p = 64$, $K = 18$ and $\dfrac{E_b}{N_0} = 26.04$, we get:

$$P_e \text{ (approx)} = Q[2.5648] = 0.0052$$

The Pursley's expression for probability of error is given as:

$$P_e = Q\left[\frac{N_0}{2 \cdot E_b} + \frac{K-1}{3 \cdot G_p}\right]^{-\frac{1}{2}} = Q[3.0469] = 0.0012$$

6.7.2 Single cell capacity improvement methods (Gilhousen et al., 1991)

Multiple access interference can be reduced by only transmitting using cell sectorization. The voice activity factor is the proportion of the time during which the user's signal is present. Research on telephone conversations in the late 1960s established that a voice is active for only about three-eights (3/8) of the call duration. Cell sectorization is achieved by using a directional antenna arrangement and, therefore, is capable of reducing the number of interfering signals reaching that sector in the cell. For example, in a three-sector cell, each sector will have an antenna with 120° effective beamwidth and the number of active users $K = 3\, N_S$ where N_S is the number of active users per sector.

Further reduction of the access interference can be achieved by switching off the user's transmitter when there is no voice activity. Let us denote the voice activity factor by (ρ). Therefore, the effective $\frac{E_b}{\hat{N}_0}$ becomes:

$$\frac{E_b}{\hat{N}_0} = \frac{1}{\dfrac{N_0}{E_b} + \dfrac{(N_s - 1) \cdot \rho}{G_p}} \tag{6.88}$$

When the additive white Gaussian noise power density is much smaller than the multiple access interference power density, we get:

$$\frac{E_b}{\hat{N}_0} \Rightarrow \frac{E_b}{I_0} = \frac{1}{\dfrac{(N_s - 1) \cdot \rho}{G_p}}$$

Assuming $\rho = 3/8$, equation (6.81) becomes:

$$K_s = 1 + \frac{1}{\rho} \cdot \frac{G_p}{\frac{E_b}{I_0}} \approx \frac{1}{\rho} \cdot \frac{G_p}{\frac{E_b}{I_0}} \tag{6.89}$$

and

$$K = 3N_s = 3 \cdot \frac{1}{\rho} \cdot \frac{G_p}{\frac{E_b}{I_0}} = 3 \cdot \frac{8}{3} \cdot \frac{G_p}{\frac{E_b}{I_0}} = 8 \frac{G_p}{\frac{E_b}{I_0}} \tag{6.90}$$

Thus the total number of active users (K) is increased by a factor of 8 compared with the omni-directional antenna arrangement without voice activity monitoring.

Example 6.8

Consider a single cell supporting 300 active users uniformly distributed within the cell and make simultaneous access to the system. The cell is divided into three sectors and each user employs a directional antenna and voice activity provision. The effective $\frac{E_b}{I_0}$ at the input of a conventional receiver is 4.5 dB and the noise contributions from the channel are negligible. Calculate:

i.   Maximum number of users in each sector
ii. Processing gain in dB
iii. The processing gain in dB that is required to keep the probability of error for the same total number of active users the same as when the cell sectorization and voice activity monitoring techniques are not used.

Solution

i. $N_s = \dfrac{300}{3} = 100$

ii. $\dfrac{E_b}{\hat{N}_0} = 4.5\,\text{dB} = 2.818$

$$K = 300 = 8\frac{G_p}{\frac{E_b}{I_0}} = 8 \cdot \frac{G_p}{2.818}$$

Thus $G_p = 105.675 = 20.2397\,\text{dB}$.

iii. $K = \dfrac{G_p}{\frac{E_b}{I_0}}$

Substitute in the above expression, we get:

$$300 = \frac{G_p}{2.818}$$

Thus $G_p = 845.4 = 29.27\,\text{dB}$

6.8 Capacity of cellular CDMA system

The cellular CDMA network consists of multiple cells, each cell with a base station. Base stations from different cells belonging to the service network are interconnected to a gateway to the Public Switched Telephone Network (PSTN).

So far we have focused our attention on the MAI generated by users uniformly distributed within a single cell. However, CDMA systems consist of a number of these cells in clusters of 4, 7 or 13 as we have seen in Section 6.2.1. The impact of multiple cells on our previous analysis of a single cell is mostly due to the interference generated by adjacent cells which degrade the capacity of each individual cell.

6.8.1 Capacity of the uplink (reverse link)

The capacity of a given CDMA system for a specified BER performance is limited by the number of users sharing the reverse link. To achieve the required performance, an upper limit is set on the total interference within a given cell, and user transmitted power must be tightly controlled by the cell base station.

Consider a group of K active users accessing the CDMA network that operates a perfect power control to deliver $\left(\frac{E_b}{\hat{N}_0}\right)_{req.}$ for the specific performance. The access interference on the reverse link is given by (6.21) in Section 6.3.1 and (6.28) to (6.30) in Section 6.3.2.

In order for the CDMA system to deliver the required SNR, the interference plus background noise must be limited to a certain value $\hat{N}_0 \cdot W$ (Viterbi and Viterbi, 1993). Thus,

intra cell interference + inter cell interference + channel white noise $\leq \hat{N}_0 \cdot W$.

Substituting for each term in this equation, we get:

$$\rho \cdot K \cdot E_b \cdot R_b + I_{intercell} + N_0 \cdot W \leq \hat{N}_0 W \qquad (6.91)$$

Define $\eta = \dfrac{N_0}{\hat{N}_0}$ and $f = \dfrac{I_{intercell}}{I_{intracell}}$

Substituting in (6.91), we get:

$\rho \cdot K \cdot E_b \cdot R_b(1+f) \leq \hat{N}_0 \cdot W - N_0 \cdot W$, which simplifies to:

$$K \leq \frac{\frac{W}{R_b}}{\left(\frac{E_b}{\hat{N}_0}\right)_{req}} \cdot \frac{(1-\eta)}{\rho \cdot (1+f)} \qquad (6.92)$$

6.9   System Link Outage

6.9.1   Uplink outage

In practice, the number of users accessing the network at any time is a random variable that has Poisson's distribution and this number depends on other random variables; such as inter cell interference and voice activity. We can re-write (6.91) to express the lower bound of the required bit energy to interference ratio to deliver the service quality, but let us express the $I_{intracell}$ as:

$$I_{intracell} = S \sum_{i=2}^{K} \psi_i \tag{6.93}$$

where ψ_i represents the ith user accessing the CDMA system and defined in Section 6.3.2. Equation (6.91) can be written as:

$$S \sum_{i=2}^{K} \psi_i + I_{intercell} + N_0 \cdot W \leq \hat{N}_0 W \text{ which is the same as:}$$

$$\sum_{i=2}^{K} \psi_i + \frac{I_{intercell}}{S} + \frac{N_0 \cdot W}{S} \leq \frac{\hat{N}_0 W}{S}$$

Thus

$$\frac{\hat{N}_0 W}{S} \geq \sum_{i=2}^{K} \psi_i + \frac{I_{intercell}}{S} + \frac{N_0 \cdot W}{S} \tag{6.94}$$

$$\frac{\hat{N}_0 W}{E_b R_b} \geq \sum_{i=2}^{K} \psi_i + \frac{I_{intercell}}{S} + \frac{\sigma_n^2}{S}$$

$$\frac{\hat{N}_0}{E_b} \geq \frac{\sum\limits_{i=2}^{K} \psi_i + \frac{I_{intercell}}{S} + \frac{\sigma_n^2}{S}}{\frac{W}{R_b}} \tag{6.95}$$

Generally we define $\frac{E_b}{\hat{N}_0}$ such that:

$$\left(\frac{E_b}{\hat{N}_0} \right)_{req} = \frac{\frac{W}{R_b}}{\sum\limits_{i=2}^{K} \psi_i + \frac{I_{intercell}}{S} + \frac{\sigma_n^2}{S}} \tag{6.96}$$

where $\sigma_n^2 =$ channel white Gaussian noise power. When the user's $\frac{E_b}{N_0}$ falls below $\left(\frac{E_b}{N_0} \right)_{req}$, the uplink performance is degraded below the accepted performance since the bit error rate exceeded the specified level. The *outage probability* P_0 is the probability that the link is degraded so it is not operable.

The probability that the uplink is available (P_a) with $\frac{E_b}{\hat{N}_0}$ at or higher than the required level is given by:

$$P_a = \Pr\left\{\left(\frac{E_b}{\hat{N}_0}\right) > \left(\frac{E_b}{\hat{N}_0}\right)_{req}\right\} \tag{6.97}$$

The probability this link is in outage:

$$P_0 = 1 - P_a = \Pr\left\{\left(\frac{E_b}{\hat{N}_0}\right) < \left(\frac{E_b}{\hat{N}_0}\right)_{req}\right\} \tag{6.98}$$

Considering (6.96), $\frac{W}{R_b}$ and $\frac{\sigma_n^2}{S}$ are constants so that $\frac{E_b}{\hat{N}_0}$ falls below the required level only when interference increases so that:

$$\sum_{i=2}^{K} \psi_i + \frac{I_{intercell}}{S} > \frac{\frac{W}{R_b}}{\frac{E_b}{\hat{N}_0}} - \frac{\sigma_n^2}{S} \tag{6.99}$$

$$P_0 = 1 - P_a = \Pr\left\{\left(\frac{E_b}{\hat{N}_0}\right) < \left(\frac{E_b}{\hat{N}_0}\right)_{req}\right\} = \Pr\left\{\sum_{i=2}^{K} \psi_i + \frac{I_{intercell}}{S} > \frac{\frac{W}{R_b}}{\frac{E_b}{\hat{N}_0}} - \frac{\sigma_n^2}{S}\right\}$$

$$P_0 = \Pr\left\{\sum_{i=2}^{K} \psi_i + \frac{I_{intercell}}{S} > \phi\right\} \tag{6.100}$$

where

$$\phi = \frac{\frac{W}{R_b}}{\frac{E_b}{I_0}} - \frac{\sigma_n^2}{S} \tag{6.101}$$

Assuming binomial distribution for the random variables ψ_i, Gaussian distribution for the inter cell interference with mean $E\left(\frac{intercell}{S}\right)$ and variance $Var\left(\frac{intercell}{S}\right)$, and that all the variables are mutually independent, the uplink probability of outage in (6.101) is developed to (6.102) by Glisic and Vucetic (1997):

$$P_0 = \sum_{k=0}^{K-1} \binom{K-1}{k} \cdot \rho^k \cdot (1-\rho)^{K-1-k} \cdot Q\left[\frac{\phi - k - E\left(\frac{intercell}{S}\right)}{\sqrt{Var\left(\frac{intercell}{S}\right)}}\right] \tag{6.102}$$

6.9.2 Downlink (forward link) outage

As noted earlier, a single cell CDMA system requires no power control on the downlink while, in the cellular system, downlink power control takes the form of power allocation according to the individual user needs as specified by its signal-to-noise ratio defined as the ratio of power received from its serving base station to the total power received.

In any cell of the cellular CDMA system, the ith user monitors the received power from neighbouring base stations and declares the base station from which the received power $(S_{T1})_i$ is the largest as its serving base station by acquiring (correlating to) the highest power pilot. The other neighbouring base stations are denoted as $(S_{Tj})_i$ where $j = 2, \ldots,$ N_c and N_c denotes the number of surrounding cells that contribute significant intercell interference where $S_{T1} > S_{T2} > \cdots > S_{TN_c}$. The serving base station allocates fraction (β) of its power $(S_{T1})_I$ to all users in the cell and fraction $(1 - \beta)$ of $(S_{T1})_i$ to the pilot (normally 20% of base station transmitter power) to aid pilot acquisition and tracking. The base station allocates fraction ϕ_i of $\beta \cdot (S_{T1})_I$ to the ith user. Therefore, the received $\left(\frac{E_b}{\hat{N}_0}\right)_i$ is given by:

$$\left(\frac{E_b}{\hat{N}_0}\right)_i \geq \frac{\dfrac{\phi_i \cdot \beta \cdot (S_{T1})_i}{R_b}}{\displaystyle\sum_{j=1}^{N_c} \dfrac{(S_{Tj})_i}{W} + N_0} \tag{6.103}$$

Clearly

$$\sum_{i=1}^{N_c} \phi_i \leq 1$$

Simplify (6.103) to get an expression for ϕ_i:

$$\left(\frac{E_b}{\hat{N}_0}\right)_i \geq \frac{\dfrac{\phi_i \cdot \beta \cdot W}{R_b}}{\left[1 + \left(\dfrac{\displaystyle\sum_{j=2}^{N_c} S_{Tj}}{S_{T1}}\right)_i + \dfrac{N_0 \cdot W}{(S_{T1})_i}\right]} \tag{6.104}$$

Define thermal noise power $= N = N_0 \cdot W$

and $f_i \equiv \left(1 + \dfrac{\displaystyle\sum_{j=2}^{N_c} S_{Tj}}{S_{T1}}\right)_i$ where $i = 1, 2, \ldots K$; so that:

$$\phi_i \leq \frac{\left(\dfrac{E_b}{\hat{N}_0}\right)_i}{\beta \cdot \dfrac{W}{R_b}}\left[f_i + \frac{N}{(S_{T1})_i}\right] \tag{6.105}$$

With summation of (6.105) for all users, we get:

$$\sum_{i=1}^{K} \phi_i \le \frac{\left(\frac{E_b}{\hat{N}_0}\right)_i}{\beta \cdot \frac{W}{R_b}} \left[\sum_{i=1}^{K} f_i + \sum_{i=1}^{K} \frac{N}{(S_{T1})_i}\right] \tag{6.106}$$

Using equations (6.104–6.106), we get:

$$\left[\sum_{i=1}^{K} f_i + \sum_{i=1}^{K} \frac{N}{(S_{T1})_i}\right] \le \frac{\beta \frac{W}{R_b}}{\left(\frac{E_b}{\hat{N}_0}\right)_i} $$

Thus

$$\sum_{i=1}^{K} f_i \le \frac{\beta \frac{W}{R_b}}{(\frac{E_b}{\hat{N}_0})_i} - \sum_{i=1}^{K} \frac{N}{(S_{T1})_i} \tag{6.107}$$

Define

$$\Delta_f = \frac{\beta \cdot \frac{W}{R_b}}{\frac{E_b}{I_0}} - \sum_{i=1}^{K} \frac{N}{(S_{T1})_i} \tag{6.108}$$

$$\Delta_f \approx \frac{\beta \cdot \frac{W}{R_b}}{\frac{E_b}{\hat{N}_0}} \qquad \text{since } \sum_{i=1}^{K} (S_{T1})_i \gg N$$

and

$$\sum_{i=1}^{K} f_i \le \Delta_f \tag{6.109}$$

The specified BER is given by the required bit energy to noise power density $\left(\frac{E_b}{\hat{N}_0}\right)_{req}$ which gives Δ_f':

$$\Delta_f' \approx \frac{\beta \cdot \frac{W}{R_b}}{\left(\frac{E_b}{\hat{N}_0}\right)_{req}} \tag{6.110}$$

The BER increased higher than the specified level as $\left(\frac{E_b}{\hat{N}_0}\right) < \left(\frac{E_b}{\hat{N}_0}\right)_{req}$ so that outage probability is given by:

$$P_{outage} = Pr\left[\sum_{i=1}^{K} f_i > \Delta_f'\right] \tag{6.111}$$

6.10   Effects of power control errors on link capacity (Leung, 1996; Viterbi et al., 1993; Wang and Yu, 2001)

In our previous analysis of the capacity of the cellular CDMA system, we have assumed perfect power control. We now focus our attention on the effects of power control errors on system performance.

6.10.1   Power control errors in the uplink

The steady-state power control errors will be identical for signals received at the base station since these signals are received by common antenna and are further processed by common RF circuits and DSP. The main power control errors arise from power measurement errors, finite step sizes within the control process, and delay due to fast fading of received signals. The received power is a log-normally distributed random variable so that we may expect the power control errors to be modelled as log-normally distributed variables as well.

Clearly, the power control errors modify the interference characteristics and, therefore, change user's $\left(\frac{E_b}{\hat{N}_0}\right)$ to fall below the required level and impairs system capacity. The reverse link capacity is extremely sensitive to power control errors. For example, for an RMS power control error of 2 dB, the number of users is reduced by 50% for 90% link availability and to less than 40% for 98% link availability (Heath and Newson, 1992). Therefore, in order to be able to yield the capacity advantages of CDMA systems compared to the more conventional multiple access techniques, extremely accurate power control is essential. However, realistically a certain level of RMS power control error may exist, which requires an increase of the system $\left(\frac{E_b}{N_0}\right)$ in order to obtain the required level of link availability. This can be achieved in several ways, one of which is to reduce the number of users. Define factor (P) as:

$$p = \frac{\left(\frac{E_b}{\hat{N}_0}\right)_{inc}}{\left(\frac{E_b}{\hat{N}_0}\right)_{req}} \tag{6.112}$$

where $\left(\frac{E_b}{N_0}\right)_{inc}$ is the increment in system $\frac{E_b}{N_0}$ above the required level in perfect power control. The number of users of a cellular CDMA is reduced by factor $\left(\frac{1}{P}\right)$ due to errors within power control process.

6.10.2   Pilot signal interference on the down link

The base station continuously transmits pilot signals on the down link so that users may initiate system acquisition. This enables the coherent demodulation, which reduces the level of $\left\{\frac{E_b}{\hat{N}_0}\right\}_{req}$ and improves system performance. However, the pilot signal is an additional source of interference, so the increase in system capacity due to coherent demodulation is offset by the reduction in capacity due to pilot signal interference. Practically, the pilot power must be much larger than the signal power in order to ensure successful system operation. Furthermore, capacity enhancing features, such as voice activity and sectorization, do not reduce the effects of pilot signal interference on the system capacity.

6.11 Call blocking probability on the uplink

The probability of call blocking ($P_{blocking}$) is a measure of network quality of service and is defined as the probability that a user accessing the CDMA network finds all channels busy. The reader may recall from previous sections that a user call is not blocked if (Viterbi, 1995; Viterbi and Viterbi, 1993):

Intra cell interference + inter cell interference + channel white noise $\leq \hat{N}_0 \cdot W$

which can be expressed mathematically as:

$$(1 + f) \cdot S \cdot \sum_{i=2}^{K} \psi_i \leq W \cdot (\hat{N}_0 - N_0) \tag{6.113}$$

Simplifying (6.113), we get:

$$\sum_{i=2}^{K} \psi_i \leq \frac{(1 - \eta)}{(1 + f)} \cdot \frac{\dfrac{W}{R_b}}{\dfrac{E_b}{\hat{N}_0}}$$

Define

$$\frac{(1 - \eta)}{(1 + f)} \cdot \frac{\dfrac{W}{R_b}}{\dfrac{E_b}{\hat{N}_0}} = \Delta' \tag{6.114}$$

Thus

$$\sum_{i=2}^{K} \psi_i \leq \Delta' \tag{6.115}$$

A call blockage occurs when $\sum\limits_{i=2}^{K} \psi_i > \Delta'$ and

$$P_{blocking} = Pr\left[\sum_{i=2}^{K} \psi_i > \Delta'\right] \tag{6.116}$$

Equation (6.116) is valid if it is written as:

$$P_{blocking} < Pr\left[\left(\sum_{i=1}^{K} \psi_i\right) > \Delta'\right] \tag{6.117}$$

As mentioned previously, the number of users accessing the CDMA system at any time can be modelled by Poisson's distribution, but for large number of users, a Gaussian process can approximate the distribution and the probability of call blocking under such environment is given in Glisic and Vucetic (1997):

$$P_{blocking} \approx Q\left\{ \frac{\Delta' - (1 + f) \cdot \rho \cdot \left(\dfrac{\lambda}{\mu}\right)}{\sqrt{(1 + f) \cdot \rho \cdot \left(\dfrac{\lambda}{\mu}\right)}} \right\} \tag{6.118}$$

and $\frac{\lambda}{\mu}$ is the average traffic in Erlangs. The same equation can be applied for a single cell with $f = 0$, and ρ is the voice activity.

6.12 Summary

CDMA is a modulation and a multiple access technique employed for contemporary wireless communications. Multiple user transmissions using other techniques, such as FDMA, and the TDMA required tight synchronization in time and frequency. The CDMA scheme is based on unique signatures assigned to the users, and transmissions from different users require no coordination in time or frequency by the serving network.

The wave propagation in a mobile environment can be modelled by the WSSUS theory of Bello, and the Doppler PSD of the CDMA signal is very similar to the Doppler PSD of the narrowband signal as described by the Clark-Jakes model. In this chapter, the statistics of the mobile wireless channel such as Rayleigh fading and shadow fading were briefly outlined and several models for predicting the path loss described. It was stressed that narrowband path loss formulas can be used to predict CDMA signal path loss with small error (less than 1 dB).

Cellular wireless communication suffers degraded performance due to the N-F effect. This can be combated by using a tight power control scheme or by using N-F resistant receiver designs. The requirements for tight power control methods on both the down link and the uplink were discussed and the performance degradation due to errors in the power control was considered. The other main source of performance degradation is the interference generated by multiple users' access from within the user cell (Intracell interference) or from adjacent cells (Intercell interference). Multiple access interference was, therefore, considered in some detail in Section 6.3.

Single-user receivers were covered in Sections 6.4 and 6.5. The conventional receiver consisting of a bank of matched filters was described first and its performance in a radio environment with multiple access interference and the N-F effect was outlined. The principle of noise whitening to modify the conventional receiver was discussed and several examples were given.

Single-user receivers operating in a mobile wireless channel have to be adaptive to the time-varying channel states to optimize the performance as discussed in Section 6.6. The adaptive receiver consists of a digital FIR filter whose coefficients are computed according to the MMSE criterion and an adaptation algorithm (such as LMS) is used to continually adjust these coefficients. These MMSE adaptive receivers have reduced complexity when compared to optimal multi-user designs and can be implemented with only knowledge of the target user parameters. The adaptive receivers considered in this chapter can have adaptive filters either chip rate spaced or fractionally spaced. The chip rate spaced can either be N-tap or reduced complexity linear receivers. The fractionally spaced can either be linear receivers or non-linear including decision-feedback filters.

System capacity was considered in Section 6.7, starting with a single cell and ideal power control, and then capacity optimization, such as cell sectorization and voice activity consideration, was presented. The capacity of CDMA system of multiple cells with intra and inter cell interference was presented in Section 6.8. The CDMA system capacity operating in the AWGN channel is limited by the uplink multiuser access interference while the capacity of the same system operating in the usual multipath fading channel is limited by the down link. The outage of the down link and uplink was covered in Section 6.9, while the effects of power control errors on system capacity was presented in Section 6.10. The call blocking probability on the uplink was discussed in Section 6.11. Finally, the chapter was summarized in Section 6.12.

Problems

6.1 A radio signal transmitted at carrier frequency of 850 MHz. The average power received by an isotropic antenna is 120 mW. The receive antenna gain is 3 dBi and the receiving station is travelling at 120 km/hour. Calculate:
   i.   maximum Doppler frequency shift (f_d)
   ii. power received at carrier frequency (f_C)
   iii. power received at frequency $= f_c \pm \frac{f_d}{2}$

6.2 In an experiment, a transmitter is operating on frequency 850 MHz with a transmit antenna gain $G_t = 1.2$ dBi and is transmitting narrowband signal to two receiving stations. The fixed station is at distance $r_1 = 1$ Km with receive antenna gain $G_{r1} = 2.5$ dBi and is experiencing line-of-sight transmission path. The mobile station at distance $r_2 = 0.5$ Km with a receive antenna gain $G_{r2} = 3.5$ dBi and experiencing multipath propagation transmission. Calculate the path loss in each case.

6.3 Three users sharing a CDMA channel and the code spreading sequences assigned to the users are:

   $C_1 = \begin{bmatrix} 1 & 1 & -1 & -1 & 1 & 1 & -1 & -1 \end{bmatrix}$
   $C_2 = \begin{bmatrix} 1 & -1 & 1 & -1 & 1 & -1 & -1 & 1 \end{bmatrix}$
   $C_3 = \begin{bmatrix} 1 & -1 & -1 & 1 & 1 & 1 & -1 & 1 \end{bmatrix}$

   Let each user transmit the following four symbols with unit energy per bit:

   $b_1 = \begin{bmatrix} 1 & -1 & 1 & -1 \end{bmatrix}$
   $b_2 = \begin{bmatrix} 1 & 1 & -1 & 1 \end{bmatrix}$
   $b_3 = \begin{bmatrix} 1 & 1 & 1 & -1 \end{bmatrix}$

   Calculate the output signal and the MAI at the output of the conventional receiver without and with noise whitening. Assume zero channel background noise.

6.4 A CDMA link is shared by 64 active users to transmit data at the rate of 9.6 kb/s and a spreading rate of 1.2288 Mc/s. A fractionally spaced adaptive receiver is used with minimum mean square error 0.1 V^2. Assume the average symbol power to be unity and channel Gaussian noise variance 0.1 W. Calculate the SNR at the output of the receiver and compare with that from a conventional matched filter.

6.5 A CDMA link is shared by 64 active users to transmit data at the rate of 9.6 kb/s and a spreading rate of 1.2288 Mc/s. A fractionally spaced decision feedback adaptive receiver is used with probability of error not exceeding 2.5%. Calculate the number of users the system can support with $\frac{E_b}{N_0} = 10$ dB. You may assume $\xi = 0.468$.

6.6 A group of K users access a single cell CDMA system and each user transmits data at the rate of 9.6 kb/s employing a code sequence at chip rate of 1.2288 MHz. The data is used to modulate a BPSK carrier of amplitude one volt. The effective signal-to-noise ratio $\left(\frac{E_b}{N_0}\right)$ and the density of white noise at the input of user matched filter receiver (N_0) are 4.7 dB and 2.1 μW/Hz, respectively. Calculate:

   i. maximum number of active users (K) that can access the system at the specified probability of error.

   ii. approximate probability of bit error.

6.7 Consider a single cell supporting 300 active users uniformly distributed within the cell and make simultaneous access to the system. The cell is divided into three sectors and each user employs a directional antenna and voice activity provision. The effective $\frac{E_b}{I_0}$ at the input of a conventional receiver is 3.5 dB and the noise contributions from the channel are negligible. Calculate:

   i. processing gain in dB

   ii. processing gain in dB when the link does not use cell sectorization and voice activity monitoring techniques but keeps the same total number of active users.

6.8 A group of K users access a multi-cell CDMA system and each user transmits data at the rate of 19.2 k bit/sec employing a code sequence at chip rate of 1.2288 MHz. The data is used to modulate a BPSK carrier of amplitude one volt. The effective signal-to-noise ratio $\frac{E_b}{N_0}$ and the density of white noise at the input of user matched filter receiver N_0 are 4.7 dB and 1 μW/Hz respectively. Assume a voice activity factor of 3/8 and the inter cell interference = 0.65 Intra cell interference, calculate the maximum number of users K that can access the system.

6.9 A group of K = 6 users per sector access a multi-cell CDMA system and each user transmits data at the rate of 9.6 kb/s employing a code sequence at chip rate of 1.2288 MHz. The data is used to modulate a BPSK carrier of amplitude one volt. The effective signal-to-noise ratio $\frac{E_b}{N_0}$ and the variance of white noise at the input of user matched filter receiver are 5 dB and 10^{-6} W, respectively. Assume a voice activity factor of 3/8, the average and the variance on the inter cell interference normalize to received power to be 0.24 * K and 0.078* K, calculate the uplink probability of outage.

6.10 A group of users access a multi-cell CDMA system and each user transmits data at the rate of 19.2 k bit/sec employing a code sequence at chip rate of 1.2288 MHz. The data is used to modulate a BPSK carrier of amplitude one volt. The effective signal-to-noise ratio $\frac{E_b}{N_0}$ and the density of white noise at the input of user matched filter receiver are 3.5 dB and 10^{-6} W/Hz respectively. Assume a voice activity factor of 3/8, the average traffic $\rho\left(\frac{\lambda}{\mu}\right)$ = 190 Erlangs and the inter cell interference = 0.55 of the intra cell interference, calculate the probability of blocking a call on the uplink.

References

Aazhang, B. and Poor, H.V. (1989) An analysis of non-linear direct sequence correlators, *IEEE Transactions on Communications*, 37(7), 723–731.

Barbosa, A.N. and Miller, S.L. (1998) Adaptive detection of DS/CDMA signals in fading channels, *IEEE Transactions on Communications*, 46(1), 115–124.

Bello, P.A. (1963) Characterization of randomly time-variant linear channels, *IEEE Transactions on Communications Systems*, CS-11(4), 360–393.

Bug, St., Wengerter, Ch., Gaspard, I. and Jakoby, R. (2001) Channel modeling based on comprehensive measurements for DVB-T mobile applications, *Proceedings of the 18th IEEE Instrumentation and Measurement Technology Conference*, Budapest.

Cooper, G.R. and Nettleton, R.W. (1978) A spread–spectrum technique for high-capacity mobile communications, *IEEE Transactions on Vehicular Technology*, VT-27(4), 264–275.

Delich, H. and Hocanin, A. (2000) Performance of robust single-user detection in DS-CDMA systems, *IEEE Wireless Communications and Networking Conference 2000*, Sept 2000, Chicago, USA.

Duel-Hallen, A. (1993) Decorrelating decision-feedback multiuser detector for synchronous code-division multiple-access channel, *IEEE Transactions on Communications*, 41(2), 285–290.

Foerster, J.R. and Milstein, L.B. (2000) Coding for a coherent DS-CDMA system employing an MMSE receiver in a Rayleigh fading channel, *IEEE Transactions on Communications*, 48(6), 1012–1021.

Gardner, W.A. (1990) *Introduction to Random Processes with Applications to Signals and Systems*, 2nd edn, Macmillan Publishing.

Gardner, W.A. (1993) Cyclic wiener filtering: theory and method, *IEEE Transactions on Communications*, 41(1), 151–163.

Gilhousen, K.S., Jacobs, I.M., Padovani, R., Viterbi, A.J., Weaver, L.A. and Wheatley, C.E. (1991) On the capacity of a cellular CDMA system, *IEEE Transactions on Vehicular Technology*, 40(2), 303–312.

Glisic, S. and Vucetic, B. (1997) *Spread Spectrum CDMA Systems for Wireless Communications*, Artech House.

Grag, V.K. and Wilkes, J.E. (1996) *Wireless and Personal Communications Systems*, Prentice Hall.

Hata, M. (1980) Empirical formula for propagation loss in land mobile radio, *IEEE Transactions on Vehicular Technology*, VT-29, 317–325.

Heath, M.R. and Newson, P. (1992) On the capacity of spread spectrum CDMA for mobile radio, *Proceedings IEEE Vehicular Technology Conference*, Denver, Colorado, pp. 985–988.

Ibrahim, M.F. and Parson, J.D. (1983) Signal strength prediction in built up area, *IEE Proceedings*, Part I, 130(pt F), 377–384.

Iwai, H. and Karasawa, Y. (1993) Wideband propagation model for the analysis of the effect of the multipath fading on the near–far problem in CDMA mobile radio systems, *IEICE Transactions on Communications*, E76-B(2), 103–112.

Jakes, W.-C. (1994) *Microwave Mobile Communications*, Wiley.

Kim, S.R. and Efron, A. (1995) Adaptive robust impulse noise filtering, *IEEE Transactions on Signal Processing*, 43(8), 1855–1866.

Latva-aho, M. and Juntti, M.J. (2000) LMMSE detection for DS-CDMA systems in fading channels, *IEEE Transactions on Communications*, 48(2).

Lee, W.C.Y. (1974) *Mobile Communications Design Fundamentals*, Wiley.

Lee, W.C.Y. (1991) Overview of cellular CDMA, *IEEE Transactions on Vehicular Technology*, 40(2), 291–302.

Lee, W.C.Y. (1993) *Mobile Communications Design Fundamentals*, Wiley.

Leung, Y.W. (1996) Power control in cellular networks subject to measurement error, *IEEE Transactions on Communications*, 44(7), 772–775.

Macario, R.C.V. (1997) *Cellular Radio*, Macmillan Press.

Madhow, U. and Honig, M.L. (1994) MMSE Interference suppression for Direct-Sequence spread spectrum CDMA, *IEEE Transactions on Communications*, 42(12), 3178–3188.

Majmunder, M., Sanhu N. and Reed, J.H. (2000) Adaptive single-user receivers for direct-sequence spread-spectrum CDMA systems, *IEEE Transactions on Vehicular Technology*, 49(2), 379–389.

Monk, A.M., Davis, M., Milstein, B. and Helstrom, C.W. (1994) A noise-whitening approach to multiple access noise rejection. Part 1: Theory and background, *IEEE Journal on Selected Areas in Communications*, 12(5), 817–827.

Monogioudis, P., Tafazoli, R. and Evans, B.G. (1994) LFSE interference cancellation in CDMA systems, *Proceedings of IEEE International Conference on Communications*, 2, 1160–1163.

Narasimhan, R. and Cox, D.C. (1999) A generalized Doppler power spectrum for wireless environments, *IEEE Communications Letter*, 3(6), 164–165.

Newson, P. (1992) The effect of power control on the capacity of a direct sequence CDMA for mobile radio, *IEE Colloquium on Spread Spectrum Techniques for Radio Communications*.

Okumura, Y., Ohmori, E., Kawano, T. and Fukuda, K. (1968) Field strength and its variability in VHF and UHF land mobile service, *Review of the Electrical Communication Laboratory*, 16, 825–875.

Papajic, P.B. and Vucetic, B.S. (1994) Adaptive receivers structures for asynchronous CDMA systems, *IEEE Journal on Selected Areas in Communications*, 12(4), 685–697.

Parkvall, S., Strom, E.G., Milstein, L.B. and Ottersten, B.E. (1999) Asynchronous near–far resistance DS-CDMA receivers without a priori synchronization, *IEEE Transactions on Communications*, 47(1), 78–88.

Parson, J.D. and Bajwa, A.S. (1982) Wideband characterisation of fading mobile radio channels, *IEE Proceedings*, Vol. 129(pt. F, o. 2), pp. 95–101.

Patel, N.R. and O'Farrell, T. (1998) Optimum single-user detection for DS/CDMA systems achieving complete MAI cancellation, *Proceedings of the IEEE ISSSTA '98*, South Africa, 3, 941–945.

Pawelec, J.J. (2002) An adaptive non-AWGN SSMA receiver, *IEEE Communications Magazine*, 40(8), 126–127.

Poor, H.V. and Verdu, S. (1988) Single-user detectors for multiuser channels, *IEEE Transactions on Communications*, 36(1), 50–60.

Proakis, J.G. (1995) *Digital Communications*, 3rd Edn, McGraw Hill.

Pursley, M. (1977) Performance evaluation for phase coded spread spectrum multiple access communications. Part I, *IEEE Transactions on Communications*, 25(8), 797–799.

Rapport, T.S. (2001) *Wireless Communications: Principles and Practice*, Prentice Hall.

Reed, J.H. and Hsia, T.C. (1990) The performance of time-dependent adaptive filters for interference rejection, *IEEE Trans Acoust., Speech Signal Processing*, 38, 1373–1385.

Steele, R. and Hanzo, L. (1999) *Mobile Radio Communications*, Wiley.

Strom, E.G. and Miller, S.L. (1999) Properties of the single-bit single-user MMSE receiver for DS-CDMA systems, *IEEE Transactions on Communications*, 47(3), 416–427.

Strom, E.G., Parkvall, S., Miller, S.L. and Ottersten, B.E. (1996) Propagation delay estimation in asynchronous direct-sequence code division multiple access systems, *IEEE Transactions on Communications*, 44(1), 84–93.

Verdu, S. (1986) Minimum probability of error for asynchronous Gaussian multiple-access channels, *IEEE Transactions on Information Theory*, 32(Issue 1), 85–96.

Viterbi, A.J. (1995) CDMA: *Principles of Spread Spectrum Communication*, Addison-Wesley.

Viterbi, A.J. (1999) *The History of Multiple Access and the Future of Multiple Services Through Wireless Communication*, QUALCOMM.

Viterbi, A.J., Viterbi, A.M. and Zehavi, E. (1993) Performance of power- controlled wideband terrestrial digital communication, *IEEE Transactions on Communications*, 41(4), 559–569.

Viterbi, A.J., Viterbi, A.M. and Zehavi, E. (1994) Other-cell interference in cellular power-controlled CDMA, *IEEE Transactions on Communications*, 42(No. 2/3/4), 58–61.

Viterbi, A.M. and Viterbi, A.J. (1993) Erlang capacity of power-controlled CDMA system, *IEEE Journal on Selected Areas in Communications*, 11(6), 892–900.

Wang, J. and Yu, A. (2001) Open-loop power control error in cellular CDMA overlay systems, *IEEE Journal on Selected Areas in Communication*, 19(7), 1246–1254.

Yoon, Y.C. and Leib, H. (1996) Matched filters with interference suppression capabilities for DS-CDMA, *IEEE Journal on Selected Areas in Communications*, 14(8), 1510–1521.

Multi-User Detection in CDMA Cellular Radio

7.1 Introduction

The superposition of the signals transmitted by the users in a multiple access spread spectrum, also known as Code Division Multiple Access (CDMA), system cause considerable interference to the desired signal if the users' signature waveforms are not orthogonal to each other all the time, a situation which is unlikely to occur in mobile-originated calls.

A lot of research has focused on reducing or cancelling the multiple access interference in order to improve the CDMA receivers' performance. The initial approach was to *design an improved single-user detector operating efficiently in multi-user channel* by applying advanced adaptive signal processing algorithms. An overview of the work in this area is presented in Section 6.6 of the last chapter. It is worth noting that these detectors are preferred by individual mobile users because knowledge of the parameters (signature waveforms, timing, amplitude and phase) of the interfering users is not desired.

The second approach considers the detection of signals associated with a group of users where spreading codes, timing information and possibly signals amplitude and phase are known and used jointly to better detect each user. These devices are called *multi-user detectors* and are suitably used at the base station since the parameters of the group are known or can be estimated.

Multi-user detection comes under titles such as *group* or *joint detection* and related topics are *interference cancellation* and *co-channel interference suppression*. These topics will be treated in more detail in this chapter. The theoretical work of the optimum multi-user detector was published by Verdu (1998). Unfortunately, Verdu's optimum detectors are far too complex for practical systems. In most cases, Verdu's detectors are used as a performance benchmark for the sub-optimal solutions. Multi-user detection is still the most active research area in digital communications.

We begin this chapter by considering the optimum detection strategies for CDMA systems operating in *synchronous* and *asynchronous* channels in Section 7.2. We have considered the probability of error in these two types of channels in the additive Gaussian environment.

Having characterized the complexity and feasibility of implementing the optimal detectors, and found that such detectors are too expensive to operate, we take the performance of the optimum detectors as a benchmark for any other non-optimal detectors. We focus our attention on finding sub-optimal detector solutions. We have classified the sub-optimal detectors into two categories: *linear detectors* and *interference cancellation detectors*. The linear detectors employ linear mapping (transformation) at the output of the conventional detector to reduce the access interference and provide better performance. The linear detectors are considered in Section 7.3.

In Section 7.3.1, we have presented the linear interference suppression scheme known as the *decorrelator* operating in synchronous and asynchronous channels. We have also considered the Near–Far (N-F) resistance of this detection scheme. In Section 7.3.2, we consider another linear scheme known as the *Minimum Mean Square Error (MMSE) detection*.

Section 7.4 presents techniques that combat the multiple access interference such as *smart antennas; space diversity algorithms; beam forming techniques* and *Bell labs layered space-time (BLAST) architectures*.

In Section 7.5, we introduce various techniques to remove multi-access interference such as *Successive Interference Cancellation* (SIC), *Parallel Interference Cancellation* (PIC) methods and *hybrid of SIC and PIC* schemes. This section concludes with a consideration of the *Iterative (Turbo) interference cancellation* techniques. The chapter as a whole is concluded with a summary in Section 7.6.

7.2   Optimal multi-user CDMA detection

Optimum receivers for multiple access CDMA systems are designed according to two different strategies: the *individually optimum strategy* used to minimize the probability of error for each individual user in the group; that is, for the ith user, select the estimated data $\hat{b}_i$ that minimizes $P[\hat{b}_i \neq b_i]$. The other strategy, called the *jointly optimum detection*, maximizes the *a posteriori* probability $P[b_i|\{y(t)\}]$ for $i = 1, 2, \ldots, K$ where K is the number of active users sharing the CDMA channel. In the latter scheme we maximize the likelihood decisions for the group of users. However, this strategy may not achieve minimum probability of error for each individual user in the group.

The individual optimum and the joint optimum don't usually result in the same decisions under high levels of Gaussian noise, as we will demonstrate in the following example:

Example 7.1

Consider a 2-user CDMA system operating in an additive Gaussian noise environment and assume that a priori probability takes the following values within the symbol duration T such that for a received $y(t)$ at any time t where $0 \leq t \leq T$:

$$P[(+1,+1)|\{y(t)\} = 0.10 \quad P[(+1,-1)|\{y(t)\} = 0.25$$
$$P[(-1,+1)|\{y(t)\} = 0.37 \quad P[(-1,-1)|\{y(t)\} = 0.28$$

Determine the data estimates under individually optimum decisions and jointly optimum detection.

Solution

Since there is only one pair that has the largest *a posteriori* probability among the four possible pairs, the jointly optimum decisions are $(\hat{b}_1, \hat{b}_2) = (-1 +1)$.

The individually optimum decisions use the following expression:

$$P[b_1|\{y(t)\}] = P[((b_1,+1)\{y(t)\}] + P[((b_1,-1)|\{y(t)\}]$$

For $b_1 = +1$, substitute in the above expression to get

$$P[b_1|\{y(t)\}] = P[(b_1,+1)|\{y(t)\}] + P[(b_1,-1)\{y(t)\}] = 0.10 + 0.25 = 0.35$$

Similarly if we consider $b_1 = -1$ to substitute in (7.1), we get

$$P[b_1|\{y(t)\}] = P[(b_1,+1)|\{y(t)\}] + P[(b_1,-1)|\{y(t)\}] = 0.37 + 0.28 = 0.65$$

Thus it is very likely that $\hat{b}_1 = -1$

Considering all other possible values for b_1 and b_2, we can construct the following:

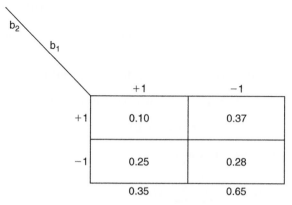

It is clear, therefore, that the individually optimum receiver estimates are:

$$(\hat{b}_1, \hat{b}_2) = (-1, -1)$$

In the above example, we considered a special case where the noise level is moderately high and, consequently, the two decisions differ with high probability. However, as signal-to-noise power ratio increases, the two decisions conform to the same data estimates $(\hat{b}_1, \hat{b}_2)$.

In Chapter 6, we showed that the output of the single-user matched filter receiver, given by (6.40), consists of three terms: the first term is proportional to the desired symbol of the intended user, and the third term represents channel noise samples. The second term, however, represents the multiple access interference generated from a large number of equal power users. The interference term can be envisaged as describing a random process consisting of a large number of random variables. Applying the central limit theorem to this random process one finds that it can be *approximated by Gaussian distribution*. One may conclude that the conventional matched filter detector can achieve optimum decisions for channels with a large number of equal power users.

However, the *Gaussian distribution* justification is false since matched filter estimates are based on one shot received sample from the desired user signal y_1 in the interval corresponding to desired symbol b_1 and ignores the interference generated by other users sharing the CDMA channel. Clearly, this interference makes the matched filter non-optimal. On the other hand, the outputs $(y_1, y_2, \ldots, y_K)$ produce sufficient statistics for $(b_1, b_2, \ldots, b_K)$ to be jointly detected.

In synchronous CDMA transmission, each user transmits one symbol in a given time interval which interferes with other users' symbols in the same time interval. However, in asynchronous transmission system, two symbols from each interfering user overlaps the desire user symbol in any given time interval. Let us now consider the K-user basic CDMA channel where each user is transmitting a block of data of length $(2M+1)$. Let the symbols transmitted at the ith time interval be $\mathbf{b}(i)$ given by:

$$\mathbf{b}(i) = [b_1(i), \ldots \ldots, b_k(i), \ldots \ldots \ldots, b_K(i)]^T \qquad (7.1)$$

where $i = -M, -M+1, \ldots, 0, \ldots, M-1, M$. Then all the symbols transmitted by all users, $\mathbf{B}$, are given by:

$$\mathbf{B} = [\mathbf{b}(-M), \ldots \ldots \ldots \mathbf{b}(0) \ldots \ldots, \mathbf{b}(M)] \qquad (7.2)$$

The receiver input signal from additive Gaussian noise channel is:

$$r(t) = s(t, \mathbf{B}) + n(t) \qquad (7.3)$$

where $n(t)$ is white Gaussian noise and:

$$s(t, \mathbf{B}) = \sum_{i=-M}^{M} \sum_{k=1}^{K} b_k(i)\sqrt{E_k(i)}\, C_k(t - iT - \tau_k) \qquad (7.4)$$

For the kth user in the ith time interval, E_k is the received energy, τ_k the delay and $C_k(t)$ is the normalized signature waveform user that is zero outside the interval $[0, T]$.

The optimum detection that maximizes the decision on the transmitted symbols based on the observation of received signal r(t) in each time interval is to maximize the *a posteriori* probability, MAP criterion. Using Bayes' rule, the *a posteriori* probability is given by Proakis (1995):

$$P(\mathbf{B}|r(t)) = P(r(t)|\mathbf{B})P[\mathbf{B}]/P(r(t)) \qquad (7.5)$$

For equal probability symbols, $P[\mathbf{B}]$ is constant equal $\frac{1}{2M+1}$. Furthermore, $P(r(t))$ is independent of which symbols are transmitted. Consequently, the detector designed using MAP criterion and the detector designed using Maximum Likelihood (ML) criterion make the same decisions. Therefore, we have:

$$P(\mathbf{B}|r(t)) = P(r(t)|\mathbf{B}) \qquad (7.6)$$

The optimum detector will search for symbol sequence, **B**, that maximizes,

$$P(r(t)|\mathbf{B}) = C \exp\left[-\Lambda(\mathbf{B})/2\sigma^2\right] \qquad (7.7)$$

Where C is constant independent of **B**, σ^2 is noise variance and $\Lambda(\mathbf{B})$ is the log likelihood function.

7.2.1 Optimum synchronous detector

In the synchronous case, all the delays, τ_k for $k = 1, \ldots, K$ in (7.4) are zero and it is sufficient to separately consider each symbol time interval (Figure 7.1) so that the summation $(\sum_{-M}^{M})$ can be dropped and hence **B** will be replaced by **b**.

Consequently, the log likelihood function in the synchronous channel can be obtained separately for each symbol interval time, and the detector is called a *one-shot detector*. So without loss of generality, we can derive the criterion for interval time $i = 0$. The log likelihood function, $\Lambda(\mathbf{b})$, is therefore given by:

$$\Lambda(\mathbf{b}) = \int_0^T \left[r(t) - \sum_{k=1}^{K} \sqrt{E_k(0)}\, b_k(0)\, C_k(t)\right]^2 dt \qquad (7.8)$$

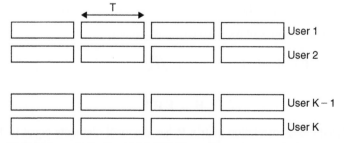

Figure 7.1 *Multi-user symbols aligned for CDMA synchronous transmission.*

The optimum detector will choose **b** that minimizes $\Lambda(\mathbf{b})$ in (7.8). Expand the right-hand side of (7.8) gives:

$$\Lambda(\mathbf{b}) = \int_0^T r(t)^2 dt - 2 \sum_{k=1}^K \sqrt{E_k(0)} \, b_k(0) \int_0^T r(t) C_k(t) dt$$

$$+ \sum_{k=1}^K \sum_{j=1}^K \sqrt{E_k(0)} \sqrt{E_j(0)} \, b_k(0) b_j(0) \int_0^T C_k(t) C_j(t) dt \qquad (7.9)$$

The term including $r(t)^2$ in (7.9) is common to all symbols **b** and can be neglected without making any difference to the minimization of (7.8). In addition, $\int_0^T r(t) C_k(t) \, dt] = y_k(0)$ with $[k = 1, \ldots\ldots\ldots, K]$ represents the matched filter output for the kth user and $\int_0^T C_k(t) C_j(t) dt] = \rho_{kj}(0)$ with $[k = 1, \ldots\ldots\ldots\ldots, K]$ is the cross-correlation of signature waveform of users k and j.

Thus, equation (7.9) can be simplified to:

$$\Lambda(\mathbf{b}) = \int_0^T r(t)^2 dt - 2 \sum_{k=1}^K \sqrt{E_k(0)} \, b_k(0) y_k(0) + \sum_{k=1}^K \sum_{j=1}^K \sqrt{E_k(0)} \sqrt{E_j(0)} \, b_k(0) b_j(0) \rho_{kj}(0)$$

$$(7.10)$$

Neglecting the first term in (7.10) and taking the negative of second and last terms in (7.10) we have:

$$\Omega(\mathbf{b}) = 2 \sum_{k=1}^K \sqrt{E_k(0)} \, b_k(0) y_k(0) - \sum_{k=1}^K \sum_{j=1}^K \sqrt{E_k(0)} \sqrt{E_j(0)} \, b_k(0) b_j(0) \rho_{kj}(0) \qquad (7.11)$$

Thus minimizing the log likelihood function in (7.10) is equivalent to maximizing $\Omega(\mathbf{b})$ in (7.11). We now write (7.11) in matrix form (Verdu 1998) as a correlation metrics:

$$\Omega(\mathbf{b}) = 2\mathbf{b}^T \mathbf{E} \mathbf{y} - \mathbf{b}^T \mathbf{H} \mathbf{b} \qquad (7.12)$$

where $$\mathbf{y} = [y_1, \ldots\ldots\ldots\ldots, y_k, \ldots\ldots\ldots\ldots y_K]^T \qquad (7.13)$$

E is $K \times K$ diagonal matrix of received energies given by:

$$\mathbf{E} = \text{diag}[\sqrt{E_1}, \ldots\ldots\ldots, \sqrt{E_k}, \ldots\ldots\ldots\ldots, \sqrt{E_K}] \qquad (7.14)$$

$$\mathbf{H} = \mathbf{E} \mathbf{R} \mathbf{E} \qquad (7.15)$$

The normalized cross-correlation matrix **R** has diagonal elements equal to 1 and (k, j) element equal to $\rho_{kj}(0)$. Therefore, the criterion for the optimum multi-user detection in

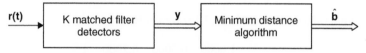

Figure 7.2 *Optimum detector for synchronous DS-CDMA.*

a synchronous CDMA Gaussian channel, which maximizes the correlation metrics, is given by:

$$\hat{b} = \text{Arg.} \max_{b \in \{+1, -1\}^K} \{2b^T E y - b^T H b\} \tag{7.16}$$

It is clear that in order to compute (7.16), the optimum detector has to have knowledge of the received signal energies **E** and user signatures. The synchronous optimum detector computes the 2^K values of the correlation metrics in (7.15) for the 2^K possible choices of the transmitted symbols from K users and selects the symbol sequence corresponding to the largest correlation metrics. In practice, the synchronous optimum detector consists of a bank of K matched filter detectors followed by an algorithm that selects the sequence that maximizes the correlation metrics or chooses the sequences corresponding to minimum Edulidean distance. A block diagram of the synchronous optimum receiver is shown in Figure 7.2.

Example 7.2

Consider a CDMA system accessed by three users: each user is transmitting one symbol. The received normalized energy for each user is unity so that the energy matrix is:

$$\begin{matrix} 1 & 0 & 0 \\ E = 0 & 1 & 0 \\ 0 & 0 & 1 \end{matrix}$$

There are eight possible combinations of the possible symbols that are transmitted given by:

$$\begin{matrix} 0 & 0 & 0 \\ 0 & 0 & 1 \\ b = - & - & - \\ - & - & - \\ 1 & 1 & 1 \end{matrix}$$

Assume the samples of the received signals y_1, y_2, y_3 at the matched filter output are given by:

$$y = \{0.45, 0.32, 0.75\}^T$$

Furthermore, let the normalized cross-correlation matrix **R** be:

$$\begin{matrix} & 1.0000 & 0.5000 & 0.2000 \\ R = & 0.5000 & 1.0000 & 0.6700 \\ & 0.2000 & 0.6700 & 1.0000 \end{matrix}$$

Substituting the values for **E** and **R** in (7.15) we get **H**:

$$\mathbf{H} = \begin{array}{ccc} 1.0000 & 0.5000 & 0.2000 \\ 0.5000 & 1.0000 & 0.6700 \\ 0.2000 & 0.6700 & 1.0000 \end{array}$$

Now substituting for **E**, **y**, **H** in (7.12) for all possible symbols **b**, we get Table 7.1.

Thus the maximum value for $\Omega(b)$ is 0.7 and **b** which maximizes $\Omega(b)$ is $\mathbf{b} = [1, -1, 1]$, which is the most likely symbols transmitted by the three users.

The upper bound for the minimum BER of the kth user is given by Verdu (1998) as:

$$P_k(\sigma) \leq \sum_{\epsilon \in F_k} 2^{-\omega(\epsilon)} Q\left(\frac{\sqrt{\epsilon^T H \epsilon}}{\sigma} \right) \tag{7.17}$$

where F_k represents the indecomposable error vectors (IEV). An error vector ϵ is said to be decomposable if there exists an ϵ' and ϵ'' such that (Luo and Ephremides, 2001; Ma et al., 2001):

(i) $\epsilon = \epsilon' + \epsilon''$
(ii) if $\epsilon_i = 0$ then $\epsilon'_i = 0$ and $\epsilon''_i = 0$ $\qquad\qquad$ (7.18)
(iii) $\epsilon'^T H \epsilon'' \geq 0$

Otherwise, ϵ is said to be indecomposable.

$$\omega(\epsilon) = \sum_{k=1}^{K} |\epsilon_k| \tag{7.19}$$

Table 7.1 *Correlation metrics vs. possible symbol sequences*

$\Omega(b)$	b		
−8.7800	−1	−1	−1
−2.3000	−1	−1	1
−2.8200	−1	1	−1
−1.7000	−1	1	1
−4.1800	1	−1	−1
0.7000	**1**	**−1**	**1**
−2.2200	1	1	−1
−2.7000	1	1	1

Let the transmitted symbols from the three users be **b**, the estimated symbols be $\hat{\mathbf{b}}$ and the error vector $\in$ where:

$$\mathbf{b} = [+1 \quad +1 \quad -1]^T$$
$$\hat{\mathbf{b}} = [-1 \quad +1 \quad +1]^T$$

Thus
$$\in = [+1 \quad 0 \quad -1]^T$$

For this particular error vector

$$\omega(\in) = 2$$

For the value of **H** specified in Example 7.2, we have:

$$\in^T \mathbf{H} \in= 1.6 \text{ so } \sqrt{\in^T \mathbf{H} \in} = 1.265$$

Now in general, the transmitted symbols **b** are not known at the receiver. Considering all the possibilities for **b** (eight possibilities, each user is transmitting a single symbol), we can generate a set of error vectors $\in$ as follows:

Align the transmitted and estimated symbols and put the vector component to zero if no error occurs at that component or equal to the transmitted component otherwise. For example, if the transmitted vector is $(-1, -1, -1)$ and assume the estimated vector is $(-1, \mathbf{1}, \mathbf{1})$ then:

$$\mathbf{b} = [-1 \quad -1 \quad -1]^T$$
$$\hat{\mathbf{b}} = [-1 \quad 1 \quad 1]^T$$

Thus
$$\in = [\ 0 \quad -1 \quad -1]^T$$

Using this method we can generate Table 7.2.

Table 7.2 *Possible symbol sequences vs. error vectors*

$\mathbf{b}^T$			$\in^T$		
−1	−1	−1	0	−1	−1
−1	−1	1	0	−1	0
−1	1	−1	0	0	−1
−1	1	1	0	0	0
1	−1	−1	+1	−1	−1
1	−1	1	+1	−1	0
1	1	−1	+1	0	−1
1	1	1	+1	0	0

The set of error vectors that affects the kth user are given by the kth column in the error table whose components are not zero. For example, for user 1, considering the first column in Table 7.2, the shaded error vectors are those used in computing user 1 BER. That is, the error vectors which affect user 1 are rows *5, 6, 7 and 8*. Similarly, the error vectors affecting user 2 are rows *1, 2, 5, 6* and for user 3 they are rows *1, 3, 5, and 7*.

Consider BER for user 1 then. The set of error vectors which have to be considered are:

+1	−1	−1
+1	−1	0
+1	0	−1
+1	0	0

Now we will test the set of error vectors above to search for decomposable vectors using (7.18). In most cases, an error vector, decomposable or not, will satisfy (i) and (ii). Therefore, our task is to test the vectors to satisfy (iii). In this simple example, we could carry out the test easily but for a more complex problem, the IEVs are obtained using an exhaustive search computer program. We need to identify the IEVs to compute the upper bound of BER. Clearly, only the first error vector $[+1 \ -1 \ -1]$ is decomposable and the IEV that contributes to first user error are $[+1 \ 0 \ 0]$, $[1 \ -1 \ 0]$ and $[1 \ 0 \ -1]$. Assuming the values of **H** in Example 7.2, and computing the values of $\boldsymbol{\varepsilon}^T \mathbf{H} \boldsymbol{\varepsilon}$, we get:

$$
\mathbf{H} = \begin{matrix}
1.0000 & 0.5000 & 0.2000 \\
0.5000 & 1.0000 & 0.6700 \\
0.2000 & 0.6700 & 1.0000
\end{matrix}
$$

$\boldsymbol{\varepsilon}^T$			$\boldsymbol{\varepsilon}^T \mathbf{H} \boldsymbol{\varepsilon}$
[+1	0	0]	1
[1	−1	0]	1
[1	0	−1]	1.6

Using (7.17), the upper bound for the minimum BER is:

$$
P_1(\sigma) \leq 2^{-1} Q\left(\frac{\sqrt{1}}{\sigma}\right) + 2^{-2} Q\left(\frac{\sqrt{1}}{\sigma}\right) + 2^{-2} Q\left(\frac{\sqrt{1.6}}{\sigma}\right)
$$

Let us assume $\sigma = 1$ then:

$$
P_1(\sigma) \leq \frac{1}{2} Q(1) + \frac{1}{4} Q(1) + \frac{1}{4} Q(1.265)
$$

$$
P_1(\sigma) \leq \mathbf{0.1447}
$$

The lower bound for the BER for the kth user is given by Verdu (1998) as:

$$
P_k(\sigma) \geq 2^{1-\omega_{k,min}} Q\left(\frac{d_{k,min}}{\sigma}\right) \tag{7.20}
$$

where

$$\omega_{k,min} = \min_{\substack{||S(\in)|| = d_{k,min} \\ \in \in F_k}} \omega(\in)$$

Now from tables above, for user 1 we have:

$$\omega_{k,min} = 2$$

$$d_{k,min} = 1$$

Thus

$$P_1(\sigma) \geq \mathbf{0.0793}$$

Let us check Verdu's BER limits with the work of Proakis using the probability of error for a matched filter single-user detector as an upper limit for the case in this example. Using expressions in Proakis (1995, p. 860) the average probability of error is:

$$P_1 < \left(\frac{1}{2}\right)^{K-1} (K-1)Q(\sqrt{2(\mathrm{SNR})_{min}}) \tag{7.21}$$

where

$$(\mathrm{SNR})_{min} = \frac{1}{N_0} \left[\sqrt{E_k} - \sum_{\substack{j=1 \\ j \neq k}}^{K} \sqrt{E_j} |\rho_{jk}(0)| \right]^2 \tag{7.22}$$

Now we assumed that $\sigma = 1$, therefore:

$$\sigma^2 = 1 = 0.5\,N_0$$

$$N_0 = 2$$

For user $k = 1$

$$(\mathrm{SNR})_{min} = \frac{1}{N_0} \left[\sqrt{E_1} - \sum_{j=2}^{K} \sqrt{E_j} |\rho_{jk}(0)| \right]^2$$

$$(\mathrm{SNR})_{min} = \frac{1}{N_0}[1 - |\rho_{21}(0)| - |\rho_{31}(0)|]^2 = \frac{1}{2}[0.3]^2 = 0.045$$

Hence

$$P_1 < \left(\frac{1}{2}\right)^2 {}^* 2^* Q(\sqrt{0.09}) = \mathbf{0.191}$$

7.2.2   Optimum asynchronous detector

In the asynchronous CDMA channel, a possible arrangement for the multi-user symbols is shown in Figure 7.3.

It is clear from Figure 7.3 that each desired symbol is overlapped by exactly two consecutive symbols from each undesired user (interferer). Consider K users sharing the asynchronous CDMA channel; each is transmitting $(2M + 1)$ symbols. The multi-user optimum decisions of the detector are obtained by a maximum likelihood sequence detector that selects the most likely sequence of transmitted symbols for the given observations.

The log likelihood function $\Lambda(\mathbf{b})$ can be computed using the following integral:

$$\Lambda(\mathbf{b}) = \int_{-MT}^{MT+2T} \left[r(t) - \sum_{k=1}^{K} \sum_{i=-M}^{K} \sqrt{E_k(i)}\, b_k(i) C_k(t - ik - \tau_k) \right]^2 dt \qquad (7.23)$$

$$= \int_{-MT}^{MT+2T} \left[r(t)^2 \, dt - 2 \sum_{k=1}^{K} \sum_{i=-M}^{K} \sqrt{E_k(i)}\, b_k(i) \int_{-MT}^{MT+2T} r(t) C_k(t - iT - \tau_k)\, dt \right.$$

$$+ \sum_{k=1}^{K} \sum_{j=1}^{K} \sum_{i=-M}^{M} \sum_{m=-M}^{M} \sqrt{E_k(i) E_j(m)}\, b_k(i) b_j(m) \int_{-MT}^{MT+2T} C_k(t - iT - \tau_k)$$

$$C_j(t - mT - \tau_j)\, dt \qquad (7.24)$$

where $\mathbf{b}$ is $K(2M + 1)$ vector representing the data sequences from the K users where:

$$\mathbf{b} = \begin{array}{ccccc} b_1(-M) & b_1(-M+1) & b_1(0) & b_1(M-1) & b_1(M) \\ \cdots\cdots & \cdots\cdots & \cdots\cdots & \cdots\cdots & \cdots\cdots \\ \cdots\cdots & \cdots\cdots & \cdots\cdots & \cdots\cdots & \cdots\cdots \\ \cdots\cdots & \cdots\cdots & \cdots\cdots & \cdots\cdots & \cdots\cdots \\ \cdots\cdots & \cdots\cdots & \cdots\cdots & \cdots\cdots & \cdots\cdots \\ b_k(-M) & b_k(-M+1) & b_k(0) & b_k(M-1) & b_k(M) \\ \cdots\cdots & \cdots\cdots & & \cdots\cdots & \cdots\cdots \\ \cdots\cdots & \cdots\cdots & & \cdots\cdots & \cdots\cdots \\ \cdots\cdots & \cdots\cdots & & \cdots\cdots & \cdots\cdots \\ b_K(-M) & b_K(-M+1) & b_K(0) & b_K(M-1) & b_K(M) \end{array} \qquad (7.25)$$

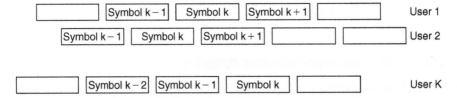

Figure 7.3 *Multi-user symbols aligned for asynchronous CDMA transmission.*

and
$$\mathbf{b}(i) = [b_1(i)\, b_2(2)\ldots\ldots\ldots\ldots\ldots b_K(i)]^T \qquad (7.26)$$

As in the synchronous case, $r(t)^2$ is common to all users and the integral involving $r(t)^2$ can be ignored. Let us define the signal energy by a $(2M+1)K * (2M+1)K$ diagonal matrix $\mathbf{E}$ given by:

$$\mathbf{E} = \text{diag}([\sqrt{E_1(-M)},\ldots\ldots,\sqrt{E_K(-M)},\ldots\ldots,\sqrt{E_1(M)},\ldots\ldots,\sqrt{E_K(M)}]) \quad (7.27)$$

The normalized cross-correlation $K*K$ matrix $\mathbf{R}$ (m) with elements:

$$R_{kj}(m) = \int\limits_{-\infty}^{\infty} C_k(t - \tau_k)C_j(t - mT - \tau_j)dt \qquad (7.28)$$

Now since the users' signature waveforms are zeros outside the time interval [0,T], then:

$$\mathbf{R}(m) = 0 \quad |m| > 1$$
$$\mathbf{R}(-m) = \mathbf{R}^T(m) \qquad (7.29)$$

Let us define a $(2M+1)K * (2M+1)K$ matrix $\mathfrak{R}$ to represent the $\mathbf{R}(m)$ matrices such that:

$$\mathfrak{R} = \begin{matrix}
\mathbf{R}(0) & \mathbf{R}^T(1) & 0\ldots & \ldots\ldots & \ldots\ldots & 0 \\
\mathbf{R}(1) & \mathbf{R}(0) & \mathbf{R}^T(1) & 0 & \ldots\ldots & 0 \\
\cdot\cdot\cdot\cdot & \cdot\cdot\cdot\cdot & \cdot\cdot\cdot\cdot\cdot & \cdot\cdot\cdot & \cdot\cdot\cdot\cdot\cdot & \cdot\cdot\cdot\cdot \\
\cdot\cdot\cdot\cdot & \cdot\cdot\cdot\cdot & \cdot\cdot\cdot\cdot\cdot & \cdot\cdot\cdot & \cdot\cdot\cdot\cdot\cdot & \cdot\cdot\cdot\cdot \\
0 & 0 & 0 & \mathbf{R}(1) & \mathbf{R}(0) & \mathbf{R}^T(1) \\
0 & 0 & 0 & 0 & \mathbf{R}(1) & \mathbf{R}(0)
\end{matrix} \qquad (7.30)$$

Matrix $\mathfrak{R}$ is defined as the *symmetric block-Toeplitz matrix*. The log likelihood function $\Lambda(\mathbf{b})$ in (7.23) can be expressed in terms of correlation metric that involves K matched filter detectors, one for each user. The output is defined by $(2M+1)$ K vector $\mathbf{y}$.

$$\mathbf{y} = \mathfrak{R}\mathbf{Eb} + \mathbf{n} \qquad (7.31)$$

$$\mathbf{n} = [\mathbf{n}^T(-M)\mathbf{n}^T(-M+1)\ldots\ldots\ldots\ldots\ldots\ldots\ldots\ldots\ldots.\mathbf{n}^T(M)]^T \quad (7.32)$$

where
$$\mathbf{n}(i) = [n_1(i)\, n_2(i)\ldots\ldots\ldots\ldots\ldots\ldots\ldots\ldots\ldots n_K(i)]^T \qquad (7.33)$$

The Gaussian noise vectors $\mathbf{n}(i)$ have zero mean and autocorrelation matrix which depends on the cross-correlation matrix $\mathbf{R}(i)$:

$$E[\mathbf{n}(i)\mathbf{n}^T(k)] = \sigma^2 \mathbf{R}(i - k) \qquad (7.34)$$

Where $\mathbf{R}(i - k)$ is defined in (7.28) and thus is non-zero only when $|i - k| \leq 1$.

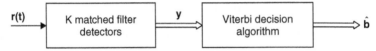

Figure 7.4   *Optimum detector for asynchronous CDMA systems.*

Following a similar procedure used for the synchronous case, the criterion for the optimum multi-user detection in an asynchronous Gaussian channel is to maximize the correlation metrics given by:

$$\hat{b} = \arg \max_{b \in \{+1,-1\}^K} \{2\mathbf{b}^T \mathbf{E} \mathbf{y} - \mathbf{b}^T \mathbf{E} \Re \mathbf{E} \mathbf{b}\} \qquad (7.35)$$

A direct approach for the solution of the optimization will select the sequences **b** represented by vector of size $2(M+1)K$ from all possible $2^{(2M+1)K}$ vectors which indicates the cost of computation being exponentially dependent on both K and M. In contemporary CDMA systems both K and M are large so the direct approach for the optimization problem is impractical.

An efficient solution to the optimization of (7.35) is proposed by Verdu using the Viterbi algorithm (VA) which selects a sequence **b** that maximizes the correlation metric in (7.35) which makes the cost of computation exponentially increase with K and M.

The optimum decision algorithm proposed by Verdu (1995) consists of K matched filter detectors followed by the maximum-likelihood Viterbi decision algorithm as shown in Figure 7.4.

Example 7.3

Let us now consider two users on the system, as shown in Figure 7.4. Each user is transmitting three symbols at a time $i = -1, 0, 1$. Let the two matched filter detectors' hard decision output **y** be given by:

$$\mathbf{y} = 01 \ \ 10 \ \ 11 \qquad (7.36)$$

We now focus our attention on the output of the maximum-likelihood Viterbi decision algorithm $\hat{b}$. To simplify the example, assuming the energies of the two users are equal $(E_1 = E_2)$. Using the simplified expression for the correlation metric (Verdu, 1995) for this example:

$$\Omega(\mathbf{b}) = \sum_{j=-1}^{4} \lambda_j(b_{j-1}, b_j)$$

$$\lambda_j(b_{j-1}, b_j) = 2b_j y_j - 2\rho_{12} b_j b_{j-1} - 1 \quad \text{if j is even}$$

where

$$= 2b_j y_j - 2\rho_{21} b_j b_{j-1} - 1 \quad \text{if j is odd}$$

and $b_{-2} = 0$, $\rho_{12} = 0.35$, $\rho_{21} = -0.35$.

Now we draw the trellis diagram for the two users:

For $j = -1$ $\lambda_{-1}(b_{-2}, b_{-1}) = 2b_{-1} y_{-1} - 2\rho_{21} b_{-1} b_{-2} - 1$
                    $\lambda_{-1}(0, b_{-1}) = 2b_{-1} y_{-1} - 1$

now b_{-1} could take two possible values ($+1$ or -1). Therefore:

For $b_{-1} = +1$, $\lambda_{-1}(0, +) = 2b_{-1} y_{-1} - 1 = 2 * (+1) * (-1) = -2$
For $b_{-1} = -1$, $\lambda_{-1}(0, -) = 2b_{-1} y_{-1} - 1 = 2 * (-1) * (-1) = 2$

The Survivor path is $\lambda_{-1}(0, -)$.

For $j = 0$ $\lambda_0(b_{-1}, b_0) = 2b_0 y_0 - 2\rho_{12} b_0 b_{-1} - 1$
            $\lambda_0(-, -) = 2 * (-1) * (+1) - 2 * (0.35) * (-1) * (-1) - 1 = \mathbf{-3.7}$
            $\lambda_0(-, +) = 2 * (+1) * (+1) - 2 * (0.35) * (+1) * (-1) - 1 = \mathbf{1.7}$

The Survivor path is $\lambda_{-1}(0, -)$, $\lambda_0(-, +)$.

For $j = 1$ $\lambda_1(+, +) = 2*(+1) * (+1) - 2 * (-0.35) * (+1) * (+1) - 1 = \mathbf{1.7}$
            $\lambda_1(+, -) = 2 * (-1) * (+1) - 2 * (-0.35) * (-1) * (+1) - 1 = \mathbf{-3.7}$

The Survivor path is $\lambda_{-1}(0, -)$, $\lambda_0(-, +)$, $\lambda_1(+, +)$.

For $j = 2$ $\lambda_2(+, +) = 2 * (+1) * (-1) - 2 * (0.35) * (+1) * (+1) - 1 = \mathbf{-3.7}$
            $\lambda_2(+, -) = 2 * (-1) * (-1) - 2 * (0.35) * (-1) * (+1) - 1 = \mathbf{1.7}$

The Survivor path is $\lambda_{-1}(0, -)$, $\lambda_0(-, +)$, $\lambda_1(+, +)$, $\lambda_2(+, -)$.

For $j = 3$ $\lambda_3(-, -) = 2 * (-1) * (+1) - 2 * (-0.35) * (-1) * (-1) - 1 = \mathbf{-2.3}$
            $\lambda_3(-, +) = 2 * (+1) * (+1) - 2 * (-0.35) * (+1) * (-1) - 1 = \mathbf{0.3}$

The Survivor path is $\lambda_{-1}(0, -)$, $\lambda_0(-, +)$, $\lambda_1(+, +)$, $\lambda_2(+, -)$, $\lambda_3(-, +)$.

For $j = 4$ $\lambda_4(+, +) = 2 * (+1) * (+1) - 2 * (0.35) * (+1) * (+1) - 1 = \mathbf{0.3}$
            $\lambda_4(+, -) = 2 * (-1) * (+1) - 2 * (0.35) * (-1) * (+1) - 1 = \mathbf{-2.3}$

The Survivor path is $\lambda_{-1}(0,-)$, $\lambda_0(-,+)$, $\lambda_1(+,+)$, $\lambda_2(+,-)$, $\lambda_3(-,+)$, $\lambda_4(+,+)$.

Therefore $\hat{\mathbf{b}} = [01 \quad 10 \quad 10]$

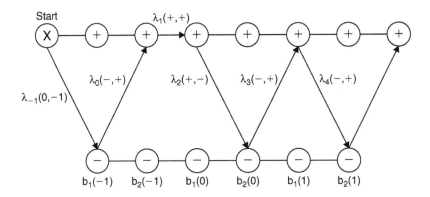

7.3   Linear sub-optimal detectors

7.3.1   Decorrelator detector

The matched filter receiver output $\mathbf{y}$ for a K-user synchronous CDMA system can be expressed in a matrix form as in (6.41) in Chapter 6 which is repeated here for convenience:

$$\mathbf{y} = \mathbf{RA}\,\mathbf{b} + \mathbf{n} \qquad (6.41)$$

where $\mathbf{R}$ is $K \times K$ normalized cross correlation matrix for K-user signature waveforms and vector $\mathbf{n}$ represents AWG noise samples. Let the received energy of the kth user be E_k so that users signal energies can be written as $\mathbf{A}$ where:

$$\mathbf{A} = \text{diag}[\sqrt{E_1}, \sqrt{E_2}, \dots\dots\dots\dots\dots\dots\dots\dots, \sqrt{E_K}] \qquad (7.36)$$

Where diag[ ] is a diagonal matrix. Now let us multiply both sides of (6.41) with the inverse of $\mathbf{R}$ (denoted by $\mathbf{R}^{-1}$) assuming $\mathbf{R}^{-1}$ exists, that is $\mathbf{R}$ is a square matrix, i.e. number of rows equal number of columns.

$$\mathbf{R}^{-1}\mathbf{y} = \mathbf{R}^{-1}\mathbf{RA}\,\mathbf{b} + \mathbf{R}^{-1}\mathbf{n} \qquad (7.36)$$

Since $\mathbf{R}^{-1}\,\mathbf{R} = \mathbf{I}$ where $\mathbf{I}$ is the identity matrix and $\mathbf{I}\,\mathbf{Ab} \equiv \mathbf{Ab}$, we can simplify (7.36) as:

$$\mathbf{R}^{-1}\mathbf{y} = \mathbf{A}\,\mathbf{b} + \mathbf{R}^{-1}\mathbf{n} \qquad (7.37)$$

The decorrelator described in (7.37) linearly transforms the outputs of the conventional matched filter receiver into scaled transmitted sequence $\mathbf{A}\,\mathbf{b}$ plus noise. A block diagram of the decorrelating receiver is given in Figure 7.5.

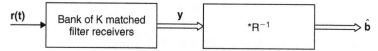

Figure 7.5 *Decorrelating receiver for synchronous CDMA system.*

Example 7.4

A synchronous CDMA system shared by three users, each transmitting three symbols defined by vector **b** is using a decorrelator receiver. Assume:

$$\mathbf{b} = \begin{matrix} 1 & -1 & 1 \\ -1 & 1 & 1 \\ 1 & 1 & 1 \end{matrix}$$

Further, for simplicity we assume the received energy from each user is set to unity, i.e. $A_1 = A_2 = A_3 = 1$. The signature sequences are of one byte for each user:

$$C_1 = [-1 \quad 1 \quad 1 \quad 1 \quad 1 \quad -1 \quad 1 \quad 1]$$
$$C_2 = [1 \quad 1 \quad -1 \quad 1 \quad -1 \quad 1 \quad 1 \quad -1]$$
$$C_3 = [1 \quad -1 \quad 1 \quad 1 \quad -1 \quad 1 \quad 1 \quad 1]$$

Assume the noise vector **n** at the matched filter detectors is given by:

$$\mathbf{n} = \begin{matrix} 0.3 & 0 & 0.5 \\ 0 & 0.5 & 0.3 \\ 0.6 & 0.1 & -0.2 \end{matrix}$$

Now we analyse the decorrelator to estimate its output $\hat{\mathbf{b}}$.

Solution

The normalized correlation matrix **R** is:

$$\mathbf{R} = \begin{matrix} 1 & -0.25 & 0 \\ -0.25 & 1 & 0.25 \\ 0 & 0.25 & 1 \end{matrix}$$

The inverse **R** is computed using Matlab:

$$\mathbf{R}^{-1} = \begin{matrix} 1.0714 & 0.2857 & -0.0714 \\ 0.2857 & 1.1429 & -0.2857 \\ -0.0714 & -0.2857 & 1.0714 \end{matrix}$$

The noise vector at the output of the decorrelator is given by:

$$\mathbf{R}^{-1}\mathbf{n} = \begin{matrix} 0.2786 & 0.1357 & 0.6357 \\ -0.0857 & 0.5429 & 0.5429 \\ 0.6214 & -0.0357 & -0.3357 \end{matrix}$$

Now $$\mathbf{y} = \mathbf{R}\mathbf{A}\mathbf{b} + \mathbf{n}$$

Since we assumed $\mathbf{A} = \mathbf{I}$, therefore $\mathbf{y} = \mathbf{R}\mathbf{b} + \mathbf{n}$ which gives:

$$\mathbf{y} = \begin{matrix} 1.55 & -1.25 & 1.25 \\ -1.0 & 2 & 1.30 \\ 1.35 & 1.35 & 1.05 \end{matrix}$$

The estimated symbols $\hat{\mathbf{b}}$ are given by

$$\mathbf{R}^{-1}\mathbf{y} - \mathbf{R}^{-1}\mathbf{n} = \begin{matrix} 1 & -1 & 1 \\ -1 & 1 & 1 \\ 1 & 1 & 1 \end{matrix}$$

Thus, the decorrelator receiver eliminates the multiple access interference associated with the matched filter detection but changes (generally increases) the level of background noise. The significant advantage of the decorrelator receiver is its low complexity (increases linearly with K) compared to the optimum receiver (complexity increases exponentially with K). However, it needs the knowledge of the signature waveforms, timing information and energies of all users.

The probability of error of quasi synchronous CDMA system of three users applying decorrelator detection, compared with probability of error using matched filter receiver and ideal BPSK receiver, is shown in Figure 7.6 with user code sequences employing 31 Gold sequences, and perfect power control for small time uncertainty ($\pm 1, \pm 2, \pm 3$ chips). It is clear that even if exact power control is used, the use of decorrelator is still preferred over the matched filter CDMA receiver. For uncertainty of ± 1 chip, the probability of error at 10^{-4} is less than 1 dB from the ideal BPSK result.

The decorrelating receiver for the asynchronous CDMA system can be derived using a similar approach but the correlation $\Re$ is time dependent as given by (7.30) and $\mathbf{A}$ is replaced by $\mathbf{E}$ so that multiplying both sides of (7.31) by $\Re^{-1}$ we get:

$$\Re^{-1}\mathbf{y} = \mathbf{E}\,\mathbf{b} + \Re^{-1}\mathbf{n} \tag{7.38}$$

Both (7.37) and (7.38) show the output of the decorrelator receivers and consist of two terms: one due to the users' useful signals which are $\mathbf{Ab}$ and $\mathbf{Eb}$ and the other due to the background noise $\mathbf{R}^{-1}\mathbf{n}$ and $\Re^{-1}\mathbf{n}$ with zero mean. The noise component of the receiver in a synchronous CDMA system has variance given by:

$$E[(\mathbf{R}^{-1}\mathbf{n})(\mathbf{R}^{-1}\mathbf{n})^{\mathrm{T}}] = E[\mathbf{R}^{-1}\mathbf{n}\,\mathbf{n}^{\mathrm{T}}\mathbf{R}^{-1}] \tag{7.39}$$

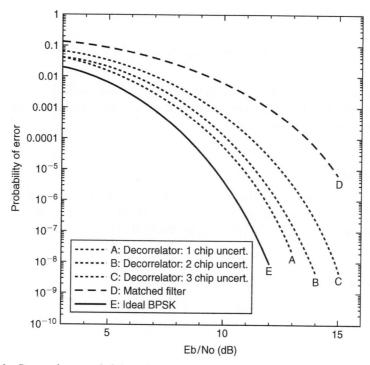

Figure 7.6 *Decorrelator probability of error compared with MF and ideal BPSK. Reproduced with permission from IEEE (Figure 4, Iltis, R.A. and Mailaender, L., 1996, Multiuser detection of quasi-synchronous CDMA signals using linear decorrelators, IEEE Transactions on Communications, 44(11), 1561–1571).*

Clearly the kth user useful signal is orthogonal to the interfering signals meaning the decorrelator receiver is Near–Far resistant. Recall from (6.45) in Chapter 6 that:

$$\mathbf{E}(\mathbf{n}\,\mathbf{n}^T) = \frac{N_0}{2}\mathbf{R} = \sigma^2 \mathbf{R} \qquad (7.40)$$

Substituting (7.40) in (7.39), the variance becomes:

$$E[(\mathbf{R}^{-1}\mathbf{n})(\mathbf{R}^{-1}\mathbf{n})^T] = \sigma^2 E[\mathbf{R}^{-1}\mathbf{R}\,\mathbf{R}^{-1}] = \sigma^2\mathbf{R}^{-1} \qquad (7.41)$$

We now return to the synchronous decorrelator detector to compute its BER performance. The kth user BER P_k is given in Verdu (1995):

$$P_k = Q\left(\frac{A_k}{\sigma\sqrt{(\mathbf{R}^{-1})_{kk}}}\right) \qquad (7.42)$$

where $\frac{1}{(R^{-1})_{kk}}$ is shown to be:

$$\frac{1}{(R^{-1})_{kk}} = 1 - \mathbf{a}_k^T \mathbf{R}_k^{-1} \mathbf{a}_k \qquad (7.43)$$

Substituting (7.43) in (7.40), the kth BER becomes:

$$P_k = Q((A_k/\sigma)\sqrt{1 - \mathbf{a}_k^T \mathbf{R}_k^{-1} \mathbf{a}_k}) \tag{7.44}$$

where $\mathbf{R}_k$ is the $(K-1) \times (K-1)$ matrix that results by removing the kth row and column from matrix $\mathbf{R}$, $\mathbf{a}_k$ is the kth column of matrix $\mathbf{R}$ without the diagonal element and A_k is the received energy from the kth user.

Example 7.5

Consider the synchronous CDMA channel shared by three users with $\mathbf{R}$ given in Example 7.4. We assume that all users are received with equal power and the Gaussian noiser variance $\sigma^2 = 1$. Use (7.44) to compute the BER of user 1, 2 and 3.

Solution
For simplicity we assume that $A_k = 1$ for $k = 1, 2, 3$.

We start with BER for the 1st user.

Recall
$$\mathbf{R} = \begin{matrix} 1 & -0.25 & 0 \\ -0.25 & 1 & 0.25 \\ 0 & 0.25 & 1 \end{matrix}$$

Consequently
$$\mathbf{R}_1 = \begin{matrix} 1 & 0.25 \\ 0.25 & 1 \end{matrix}$$

$$\mathbf{a}_1 = \begin{matrix} -0.25 \\ 0 \end{matrix}$$

Therefore
$$1 - \mathbf{a}_1^T \mathbf{R}_1^{-1} \mathbf{a}_1 = 0.9333$$

And
$$P_1 = Q(1.035) = \mathbf{0.1670}$$

Using a similar method, we compute the BER for the other uses:

$$P_2 = Q(0.9354) = \mathbf{0.1748}$$

$$P_3 = Q(1.1547) = \mathbf{0.1670}$$

7.3.2   Minimum Mean Square Error (MMSE) detection

An efficient approach for estimating a random variable $\mathbf{b}$ based on the observation of $\mathbf{y}$ is to find the function $\mathbf{b(y)}$ which Minimum Mean Square Error (MMSE):

$$E([\mathbf{b} - \mathbf{b(y)}]^2) \tag{7.45}$$

The MMSE detector applies a linear transformation to the output ($\mathbf{y}$) of the conventional detector of a matched filter bank to minimize the difference between the transmitted symbol sequence $\mathbf{b}$ and the estimated symbol sequence $\hat{\mathbf{b}}$ defined by the function $\mathbf{b}(\mathbf{y})$. The optimization of the function $\mathbf{b}(\mathbf{y})$ is accomplished by choosing the $K \times K$ matrix $\mathbf{W}$ that achieves:

$$\min_{\mathbf{W} \in \mathbf{R}} E(\|b - \mathbf{Wy}\|^2) \tag{7.46}$$

where $\mathbf{y}$ is given by (6.41). Now, since $\|X\|^2 = \text{trace}\{\mathbf{X}\mathbf{X}^T\} = \text{trace}\{\text{cov}\{\mathbf{X}\}\}$, expression (7.46) can be written as:

$$\min_{\mathbf{W} \in \mathbf{R}} E(\|b - \mathbf{Wy}\|^2) = \min_{\mathbf{W} \in \mathbf{R}} \text{trace}\{\text{cov}\{b - \mathbf{Wy}\}\} \tag{7.47}$$

The covariance of the error vector $[\text{cov}\{b - \mathbf{Wy}\}]$ is given by:

$$\text{cov}\{b - \mathbf{Wy}\} = E[(b - \mathbf{Wy})(b - \mathbf{Wy})^T]$$

It is shown in Appendix (A) that:

$$\text{cov}\{b - \mathbf{Wy}\} = [\mathbf{I} + \sigma^{-2}\mathbf{ARA}]^{-1} + (\mathbf{W} - \overline{\mathbf{W}})(\mathbf{R}\,\mathbf{A}^2\mathbf{R} + \sigma^2\mathbf{R})(\mathbf{W} - \overline{\mathbf{W}})^T \tag{7.48}$$

Where we assumed $\mathbf{A}$ is non-singular and:

$$\overline{\mathbf{W}} \equiv \mathbf{A}^{-1}[\mathbf{R} + \sigma^2\mathbf{A}^{-2}]^{-1} \tag{7.49}$$

Since the 2nd term of (7.48) is always positive, thus (7.48) becomes:

$$\min_{\mathbf{W} \in \mathbf{R}} E(\|b - \mathbf{Wy}\|^2) = \text{trace}\{[\mathbf{I} + \sigma^{-2}\mathbf{ARA}]^{-1}\} \tag{7.50}$$

Consequently, (7.49) achieves the minimum sum of mean square errors so that the estimated symbol sequence of the kth is given by:

$$\hat{b}_k = \text{sgn}(\overline{\mathbf{W}}\mathbf{y})_k$$
$$= \text{sgn}(A_k^{-1}([\mathbf{R} + \sigma^2\mathbf{A}^{-2}]^{-1}\mathbf{y})_k) \tag{7.51}$$

Since we assumed A_k^{-1} to be unity, (7.51) can be simplified as:

$$\hat{b}_k = \text{sgn}(([\mathbf{R} + \sigma^2\mathbf{A}^{-2}]^{-1}\mathbf{y})_k) \tag{7.52}$$

From the previous analysis, we conclude that the MMSE receiver substitutes the linear transformation $\mathbf{R}^{-1}$ in the decorrelator receiver by:

$$[\mathbf{R} + \sigma^2\mathbf{A}^{-2}]^{-1} \tag{7.53}$$

A block diagram of the MMSE receiver for synchronous CDMA system is shown in Figure 7.7.

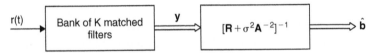

Figure 7.7   *MMSE receiver for synchronous CDMA system.*

Example 7.6

Consider the CDMA link defined in Example 7.4 but with the decorrelator receiver replaced with an MMSE receiver so that we can compare their performances. Using system parameters in Example (7.4), we have:

$$A = \begin{matrix} 1 & 0 & 0 \\ 0 & 1 & 0 \\ 0 & 0 & 1 \end{matrix}$$

Therefore

$$A^{-2} = \begin{matrix} 1 & 0 & 0 \\ 0 & 1 & 0 \\ 0 & 0 & 1 \end{matrix}$$

Since

$$\sigma^2 = 1$$

$$y = \begin{pmatrix} 1.55 & -1.25 & 1.25 \\ -1.0 & 2 & 1.3 \\ 1.35 & 1.35 & 1.05 \end{pmatrix}$$

$$n = \begin{pmatrix} 0.3 & 0 & 0.5 \\ 0 & 0.5 & 0.3 \\ 0.6 & 0.1 & -0.2 \end{pmatrix}$$

Therefore

$$[\mathbf{R} + \sigma^2\mathbf{A}^{-2}]^{-1} = \begin{matrix} 0.5081 & 0.0645 & -0.0081 \\ 0.0645 & 0.5161 & -0.0645 \\ -0.0081 & -0.0645 & 0.5081 \end{matrix}$$

Now

$$[\mathbf{R} + \sigma^2\mathbf{A}^{-2}]^{-1}\mathbf{y} = \begin{matrix} 0.7121 & -0.5169 & 0.7105 \\ -0.5032 & 0.8645 & 0.6839 \\ 0.7379 & 0.5669 & 0.4395 \end{matrix}$$

and

$$[\mathbf{R} + \sigma^2\mathbf{A}^{-2}]^{-1}\mathbf{n} = \begin{matrix} 0.1476 & 0.0315 & 0.2750 \\ -0.0194 & 0.2516 & 0.2000 \\ 0.3024 & 0.0185 & -0.1250 \end{matrix}$$

$$[\mathbf{R} + \sigma^2\mathbf{A}^{-2}]^{-1}\mathbf{y} - [\mathbf{R} + \sigma^2\mathbf{A}^{-2}]^{-1}\mathbf{n} = \begin{matrix} 0.8597 & -0.4855 & 0.9855 \\ -0.5226 & 1.1161 & 0.8839 \\ 1.0403 & 0.5855 & 0.3145 \end{matrix}$$

When elements of MMSE detector are compared with threshold of zero yield the estimated symbols

Therefore $\qquad$ $[\mathbf{R} + \sigma^2\mathbf{A}^{-2}]^{-1}\mathbf{y} - [\mathbf{R} + \sigma^2\mathbf{A}^{-2}]^{-1}\mathbf{n} = \begin{matrix} 1 & -1 & 1 \\ -1 & 1 & 1 \\ 1 & 1 & 1 \end{matrix}$

7.4   Interference combat schemes

Recall from Chapter 3 that the capacity of multi-user spread-spectrum system is limited by the multiple access interference. The most effective technique to mitigate multipath fading and multiple access interference is a useful scheme for transmitter power control. However, there are practical problems with this approach. For example, for the transmitter to implement a power control scheme, it has to *know* the channel state information at the receiver. This information has to be fed back from the receiver to the transmitter, which requires a feedback channel to be set up (but which may not be available all the time). In addition, the use of the feedback channel, when available, causes degradation in bandwidth efficiency. Furthermore, the rate of the transmitter power adjustment has to be fast enough to match the channel fading rate so that the increase in the transmitter power level has to match the signal fading level, but which may be outside the dynamic range of the transmitter. These problems are of considerable complexity in practical systems.

We have considered, in Section 7.3, a number of sub-optimal multi-user detection schemes to mitigate the affects of interference such as the decorrelator and the MMSE receivers. However, these new strategies bring about additional computation costs compared to the conventional matched filter receivers and compromise on the quality of service. For example, the decorrelator receivers combat the interference but enhance background Gaussian noise and, therefore, provide the best performance and exhibit optimum system capacity in a wireless environment of relatively high interference accompanied with low level background noise. On the other hand, the MMSE receivers reduce the impact of background noise at the expense of a reduction in interference mitigation.

An alternative affective and widely applied technique for reducing fading is *antenna diversity*. The classical approach is to use multiple antennas at the receiver and perform signal combining, or to select an antenna switching in order to improve the BER of the received signal.

The main problem with the diversity scheme at the mobile receivers is that it makes the receivers large and more expensive. So far, such techniques have almost exclusively been applied to the base stations, since each base station often serves hundreds to thousands of mobile units. It is, therefore, more economical to add equipment to the base stations rather than to the mobile units. For this reason, transmit diversity schemes for the down links and receive diversity schemes for the uplinks are very attractive for mobile communication systems.

We now focus our attention on the techniques for using multiple antennas at the transmitter and/or receiver. This can reduce interference and consequently increase system capacity and network coverage. An important scheme in such techniques is the *'smart' antenna* that intelligently controls the radiation towards the desired user and significantly reduce the interference to other users.

7.4.1 Smart antennas

An antenna is a port where electromagnetic energy from the radio transmitter is coupled (radiated) into the space through the transmitting antenna. This energy is acquired from the space using a suitable receiving antenna. Traditionally, antennas are classified according to their radiation pattern as omni-directional or directional. The omni-directional antennas have equal gain in all directions (direction here being referred to azimuth) and are known as isotropic antennas. Omni-directional antennas are commonly used in both base stations and mobile systems. The omni-directional radiation reaches the desired user with only a small percentage of the overall energy radiated into the space. Given this limitation, improving desired user performance can only be achieved by simply boosting the transmitted power level which, in return, makes the situation worse, since it increases the interference for other users in the same or adjoining cells.

Directional antennas, on the other hand, have certain fixed preferential transmission and reception directions, meaning that they have more gain in these directions and less in others and are mainly employed at the base stations. A sectorized antenna system takes a traditional cellular area and subdivides it into sectors that are covered using directional antennas looking out from the same base station location. Operationally, each sector can be treated as a different cell in the system – the range of which can be greater than in the omni-directional case, since power can be focused onto a smaller area. The increases in the system capacity due to the use of sectorized antennas are discussed in Chapter 6.

The gain of a directional antenna in the desired direction is specified relative to the gain of an isotropic antenna. For example, a given directional antenna with 6 dBi means that it has a 6 dB radiated signal gain higher than an isotropic antenna with the same dimensions. Both the omni-directional, and to a certain level the directional antennas, radiate power (wastefully) in other directions than toward the desired user, which causes interference to other users.

The need to transmit to a large group of users in a given cell more efficiently, without worsening the interference to other users, led to the evolution of the 'smart antennas'. A clear difference between a *'smart' antenna* and a *'dumb' antenna* is the amount of intelligence provided using Digital Signal Processing (DSP). Smart antenna systems consist of multiple antenna elements whose outputs are processed at the transmitter/receiver side to mitigate the fading and interference (Alexiou and Haardt, 2004; Thompson et al., 1996; Winters, 1998).

The theory behind the smart antenna technique is not new and has been used extensively in military radar and communication systems since World War II as a counter measure to electronic jamming (Lehne and Pettersen, 1999). The basic principle behind this theory is that interferers (i.e. other users accessing the cellular system) are rarely located at the same locations as the desired user. Consequently, by maximizing the antenna gain in the desired direction and simultaneously placing minimal radiation in the direction of the interferers, significant improvement to the quality of the desired users' communications can be achieved. This theory has developed to beam forming smart antennas (Pedersen, 2003; Godara, 1997).

To make antennas 'smart' or intelligent, novel signal processing algorithms are used. Indeed, antennas smartness is gauged by their digital signal processing capabilities.

The antenna is the wireless communications system element that is the input or the output port of the communication channel. In the conventional wireless communication systems, a single omni-directional transmitting antenna at the input of the channel and a single omni-directional receiving antenna at the output of the channel are used. This antenna arrangement is called single input, single output or simply SISO and is shown in Figure 7.8a. Two or more antennas can be used at the transmitter (channel input) or at the receiver (channel output) or both. These antenna systems are called Single In Multiple Out (SIMO); Multiple In, Single Out (MISO) and Multiple In Multiple Out (MIMO) as shown in Figure 7.8b, c and d, respectively.

An early version of the multiple antenna usage consists of two identical antennas separated by several wavelengths, and connected to a conventional receiver. Assuming that there is a sufficient scattering of radio energy to decorrelate the received signals, the probability that both received signals will become extremely weak at the same time is very small. Therefore, to improve the quality of the reception, the strongest received signal is selected to the input of the receiver. This technique is called '*switched diversity*'. Clearly, such a simple scheme improves the system performance by *avoiding signal fading* but will not eliminate or reduce inter symbol interference or co-channel interference so it is incapable of increasing the system capacity or the radio coverage (Figure 7.8).

The geometry of the multiple antenna system can take various forms (Lehne and Pettersen, 1999). For example, the antenna elements can be placed in a simple one-dimensional (x-axis) with uniform spacing of Δx as shown in Figure 7.9a. Or the antenna elements be placed in a circular array of two dimensions (x and y axis) with a uniform angular element spacing of $\frac{2\pi}{N}$ where N is the number of elements in the system as shown in Figure 7.9b. This structure is commonly used for two-dimensional *beam forming* in the horizontal (azimuth) plane for outdoors environments in large cells. The third possible elements position arrangements are when the elements are placed in a two-dimensional plane with a horizontal element spacing of Δx and vertical element spacing of Δy as shown in Figure 7.9c. This structure can perform beam forming within a solid angle. In the final elements arrangement, the elements are placed in a three-dimensional plane to form a cubic or spherical structure where elements are spaced at Δx, Δy, and Δz as shown

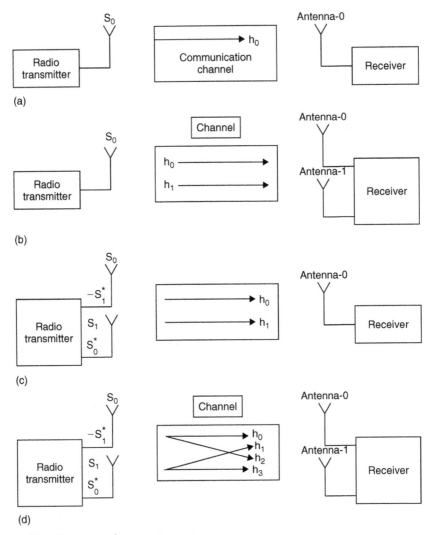

Figure 7.8 *The input and out antenna system.*

in Figure 7.9d. This last array can perform beam forming in the entire space within all the solid angles. On a broader perspective, the elements can take any geometric shape but regular shapes simplify the mathematical analysis of the antenna system.

The transmitter power is commonly divided among the transmit antennas. On the other hand, the outputs of all antennas at the receiver are combined in a technique called '*diversity combining*' to maximize the combined received signal-to-noise power ratio.

In the following sections we will explore the smart antenna characteristics by considering first the antenna diversity techniques and derive the optimum algorithms based on the maximum-likelihood decision rule at the receiver and then we examine the beam forming schemes.

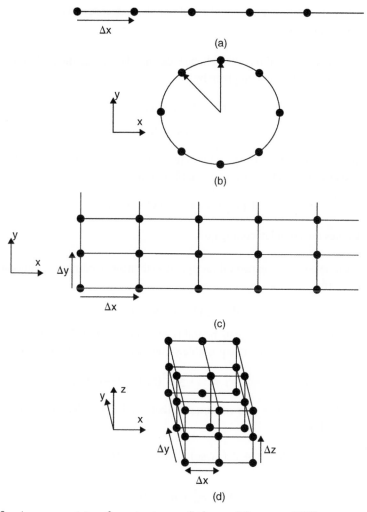

Figure 7.9 *Array geometries of smart antennas (Lehne and Pettersen, 1999).*

7.4.2 Space diversity algorithms (Alamouti, 1998)

7.4.2.1 Single antenna at transmitter – two antennas at the receiver

Consider the system depicted in Figure 7.8b with a single antenna at the transmitter and two antennas at the receiver. At a given instant of time, let the baseband signal sent from the single transmitter antenna be s_0. Denote the complex baseband channel impulse response between the transmitter antenna and receiver antenna-0 by h_0; and the complex baseband channel impulse response between the transmitter antenna and receiver antenna-1 by h_1 where:

$$h_0 = \alpha_0 e^{j\beta_0} \tag{7.54}$$

$$h_1 = \alpha_1 e^{j\beta_1} \tag{7.55}$$

where α_0 and α_1 are the magnitudes of channel impulse response and β_0 and β_1 are channel impulse response phases.

Assuming the channel exhibits little change during the signal interval, the baseband received signals r_0 and r_1 are then given by:

$$r_0 = h_0 s_0 + n_0 \qquad\qquad (7.56)$$

$$r_1 = h_1 s_0 + n_1 \qquad\qquad (7.57)$$

where n_0 and n_1 are complex samples for Gaussian noise plus interference. The squared Euclidean distance $d^2(x,y)$ between two complex signal points x and y is given by:

$$d^2(x, y) = (x - y)(x^* - y^*) \qquad\qquad (7.58)$$

where '*' denotes the complex conjugate.

Suppose the variable y takes on two values y_1 and y_2 then the maximum likelihood decision rule chooses y_1 if:

$$d^2(x, y_1) \leq d^2(x, y_2) \qquad\qquad (7.59)$$

We now combine the received signals r_0 and r_1 by applying rule (7.58) and assuming the transmitted signal s_0 can take on two possible values s_i and s_k. We choose s_k if and only if:

$$d^2(r_0, h_0 s_k) + d^2(r_1, h_1 s_k) \leq d^2(r_0, h_0 s_i) + d^2(r_1, h_1 s_i) \qquad (7.60)$$

Substituting (7.58) in (7.60), the maximum likelihood (ML) decoder estimates s_0 by choosing s_k if and only if:

$$A|s_k|^2 + d^2(\tilde{s}_0, s_k) \leq A|s_i|^2 + d^2(\tilde{s}_0, s_i) \qquad\qquad (7.61)$$

Where $\qquad\qquad A = \alpha_0^2 + \alpha_1^2 - 1$

The combiner output $\qquad \tilde{s}_0 = (\alpha_0^2 + \alpha_1^2)s_0 + h_0^* n_0 + h_1^* n_1 \qquad (7.62)$

$$\alpha_0^2 = |h_0|^2$$
$$\alpha_1^2 = |h_1|^2$$

Clearly the combining algorithm given by (7.61) requires the channels h_0 and h_0 to be estimated at the receiver.

7.4.2.2 *Alamouti's algorithm for two antennas at transmitter – single antenna at the receiver*

We now extend the combining analysis from a single antenna to two antennas at the transmitter and a single antenna at the receiver as shown in Figure 7.8c. As before, our aim is to derive the optimum combining algorithm. At a given instant of time, symbol s_0 and symbol s_1 are transmitted simultaneously from antenna-0 and antenna-1, respectively. During the next symbol interval, symbol $-s_1^*$ and symbol s_0^* are transmitted simultaneously from antenna-0 and antenna-1, respectively. Therefore, Alamouti space–time block code **S** is given by (Haykin and Moher, 2005):

$$\mathbf{S} = \begin{array}{cc} s_0 & s_1 \\ & \xrightarrow{\hspace{1cm}} \text{ Space} \\ -s_1^* \quad \downarrow & s_0^* \\ & \text{time} \end{array} \tag{7.63}$$

We now show that Alamouti's transmission matrix **S** is complex-orthogonal matrix in both space and time. To authenticate the spatial orthogonality we formulate the product:

$$\mathbf{S\,S}^{\mathrm{T}} = \begin{array}{cc} s_0 & s_1 \\ -s_1^* & s_0^* \end{array} \times \begin{array}{cc} s_0^* & -s_1 \\ s_1^* & s_0 \end{array}$$

$$= [|s_0|^2 + |s_1|^2] \times \begin{array}{cc} 1 & 0 \\ 0 & 1 \end{array}$$

$$= |s_0|^2 + |s_1|^2 \times \mathbf{I} \tag{7.64}$$

The product $\mathbf{S\,S}^{\mathrm{T}}$ in (7.64) is equal an identity matrix **I** multiplied by a constant $|s_0|^2 + |s_1|^2$. Similarly, we can show the temporal orthogonality of **S** by formulating the product:

$$\mathbf{S}^{\mathrm{T}}\mathbf{S} = \begin{array}{cc} s_0^* & -s_1 \\ s_1^* & s_0 \end{array} \times \begin{array}{cc} s_0 & s_1 \\ -s_1^* & s_0^* \end{array}$$

$$= \mathbf{S\,S}^{\mathrm{T}} \tag{7.65}$$

where T stands for matrix transpose. The signal received at time t is:

$$r_0 = h_0 s_0 + h_1 s_1 + n_0 \tag{7.66}$$

The received signal at time $t + T$, where T is symbol duration, is:

$$r_1 = -h_0 s_1^* + h_1 s_0^* + n_1 \tag{7.67}$$

The squared Euclidean distance at time t is:

$$d^2(r_0, h_0 s_0 + h_1 s_1) = [r_0 - (h_0 s_0 + h_1 s_1)][r_0^* - (h_0^* s_0^* + h_1^* s_1^*)]$$

$$= |r_0|^2 - r_0 h_0^* s_0^* - r_0 h_1^* s_1^* - r_0^* h_0 s_0 - r_0^* h_1 s_1 + \alpha_0^2 |s_0|^2 + \alpha_1^2 |s_1|^2 \tag{7.68}$$

The squared Euclidean distance at time $T + t$ is:

$$d^2(r_1, -h_0 s_1^* + h_1 s_0^*) = [r_1 - (-h_0 s_1^* + h_1 s_0^*)][r_1^* - (-h_0^* s_1 + h_1^* s_0)]$$

$$= |r_1|^2 + r_1 h_0^* s_1 - r_1 h_1^* s_0 + r_1^* h_1 s_1^* - r_1^* h_1 s_0^* + \alpha_0^2 |s_1|^2 + \alpha_1^2 |s_0|^2 \tag{7.69}$$

We now group the terms in (7.68) and (7.69) associated with s_0 and get:

$$-r_0 h_0^* s_0^* - r_0^* h_0 s_0 + \alpha_0^2 |s_0|^2 - r_1 h_1^* s_0 - r_1^* h_1 s_0^* + \alpha_1^2 |s_0|^2$$

$$= -(r_0 h_0^* + r_1^* h_1) s_0^* - r_0^* h_0 s_0 + \alpha_0^2 |s_0|^2 - r_1 h_1^* s_0 + \alpha_1^2 |s_0|^2$$

$$= -\tilde{s}_0 s_0^* - \tilde{s}_0^* s_0 + (\alpha_0^2 + \alpha_1^2)|s_0|^2 \tag{7.70}$$

where

$$\tilde{s}_0 = r_0 h_0^* + r_1^* h_1 \tag{7.71}$$

Substituting for r_0 and r_1, the combiner output signal $\tilde{s}_0$ simplifies to:

$$\tilde{s}_0 = (\alpha_0^2 + \alpha_1^2)s_0 + h_0^* n_0 + h_1 n_1^* \tag{7.72}$$

Similarly the combiner output signal $\tilde{s}_1$ is given as:

$$\tilde{s}_1 = r_0 h_1^* - r_1^* h_0$$

$$\tilde{s}_1 = (\alpha_0^2 + \alpha_1^2)s_1 - h_0 n_1^* + h_1^* n_0 \tag{7.73}$$

Now

$$d^2(\tilde{s}_0, s_0) = (\tilde{s}_0 - s_0)(\tilde{s}_0^* - s_0^*)$$

$$= |\tilde{s}_0|^2 - \tilde{s}_0 s_0 - \tilde{s}_0 s_0^* + |s_0|^2 \tag{7.74}$$

Use (7.72) to simplify (7.70) to:

$$(\alpha_0^2 + \alpha_1^2 - 1)|s_0|^2 + d^2(\tilde{s}_0, s_0) - |\tilde{s}_0|^2 \tag{7.75}$$

The ML decoder estimates the signal s_0 by choosing s_k if and only if:

$$(\alpha_0^2 + \alpha_1^2 - 1)|s_k|^2 + d^2(\tilde{s}_0, s_k) \leq (\alpha_0^2 + \alpha_1^2 - 1)|s_i|^2 + d^2(\tilde{s}_0, s_i) \tag{7.76}$$

A similar method can be used to estimate s_1. Comparing (7.72) and (7.73) with (7.62) there are phase rotations on the noise components in $\tilde{s}_1$ which have no effects on the combined SNR. Therefore, we conclude the two-branch transmit diversity is equal to the two-branch receiver diversity.

7.4.2.3 *Alamouti's algorithm for two antennas at transmitter – two antennas at the receiver*

This antennas arrangement scheme is the most commonly referenced in the literature and is shown in Figure 7.8d. At a given time t signals s_0 and s_1 are sent from transmitter antenna-0 and antenna-1 simultaneously, and signals r_0 and r_2 are received by receiver antenna-0 and antenna-1. At time $T + t$, signals $-s_1^*$ and s_0^* are sent by antenna-0 and antenna-1 simultaneously, and signal r_1 is received by antenna-0 and signal r_3 is received by antenna-1. Alamouti's transmission code, channels, and the received signals are given in Table 7.3.

Table 7.3 *Transmission code, channels and received signals of Alamouti's MIMO system*

(a) Transmission code

	Tx Antenna-0	Tx Antenna-1
Time t	s_0	s_1
Time t + T	$-s_1^*$	s_0^*

(b) Channels between TX and Rx antennas

	RX Antenna-0	RX Antenna-1
TX Antenna-0	h_0	h_2
Tx Antenna-1	h_1	h_3

(c) Received signals

	Rx Antenna-0	Rx Antenna-1
Time t	r_0	r_2
Time t + T	r_1	r_3

The received signals defined in Table 7.3c are given by:

$$r_0 = h_0 s_0 + h_1 s_1 + n_0$$
$$r_2 = h_2 s_0 + h_3 s_1 + n_2$$
$$r_1 = -h_0 s_1^* + h_1 s_0^* + n_1$$
$$r_3 = -h_2 s_1^* + h_3 s_0^* + n_3 \tag{7.77}$$

We formulate the squared Euclidean distance at time t:

$$d^2[r_0, (h_0 s_0 + h_1 s_1)] + d^2[r_2, (h_2 s_0 + h_3 s_1)] \tag{7.78}$$

and at time t + T:

$$d^2[r_1, (-h_0 s_1^* + h_1 s_0^*)] + d^2[r_3, (-h_2 s_1^* + h_3 s_0^*)] \tag{7.79}$$

Separating the signals s_0 and s_1 in (7.78) and (7.79), it can be shown that the combiner output signals:

$$\widetilde{s}_0 = s_0 \sum_{i=0}^{3} \alpha_i^2 + h_0^* n_0 + h_1 n_1^* + h_2^* n_2 + h_3 n_3^* \tag{7.80}$$

$$\widetilde{s}_1 = s_1 \sum_{i=0}^{3} \alpha_i^2 - h_0 n_1^* + h_1^* n_0 - h_2 n_3^* + h_3^* n_2 \tag{7.81}$$

The ML decoder estimates s_0 and s_i by choosing s_k if and only if:

$$\left[\sum_{i=0}^{3} \alpha_i^2 - 1\right] |s_k|^2 + d^2(\tilde{s}_0, s_k) \leq (\alpha_0^2 + \alpha_1^2 - 1)|s_i|^2 + d^2(\tilde{s}_0, s_i) \qquad (7.82)$$

$$\left[\sum_{i=0}^{3} \alpha_i^2 - 1\right] |s_k|^2 + d^2(\tilde{s}_1, s_k) \leq \left[\sum_{i=0}^{3} \alpha_i^2 - 1\right] |s_i|^2 + d^2(\tilde{s}_1, s_i) \qquad (7.83)$$

The BER performance of uncoded coherent BPSK for Maximum Ratio Receive Combining (MRRC) diversity schemes employing one transmitter and multiple receivers and Alamouti transmit diversity scheme in Rayleigh fading are shown in Figure 7.10. The performance curves are obtained when the total transmit power from Alamouti's two antennas is equal to the single transmit antenna for the MRRC and the receiver has perfect knowledge of the channel. Because of the transmit power assumption, the BER performance of Alamouti's transmit diversity is 3 dB worse than a two-branch MRRC. However, if the transmit power is the same for the two schemes, the performance would be identical. That is Alamouti's scheme provides the same diversity order as MRRC with one antenna and two receive antennas. Alamouti's scheme can be generalized to two transmit antennas and M receive antennas to provide a diversity of order 2M. Alamouti's scheme does not require any feedback from the receiver to the transmitter and its computation complexity is similar to MRRC.

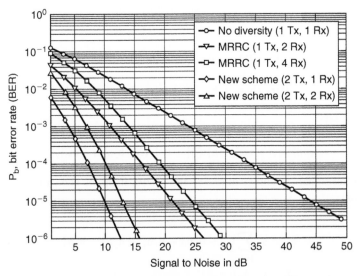

Figure 7.10 *BER performance comparison of Alamouti's algorithms (Figure 4 reproduced with permission from IEEE: Alamouti, S.M., 1998, A simple transmit diversity technique for wireless communications, IEEE Journal on Select Areas in Communications, 16(8), 1451–1458).*

7.4.2.4 The analysis of baseband MIMO channel

Consider a Multiple Transmit–Multiple Receive (MIMO) channel with n_t transmit antennas and n_r receive antennas. The complex transmitted signal **s** at a given discrete time is represented by vector of n_t elements:

$$\mathbf{s} = [s_1, s_2, \ldots\ldots\ldots\ldots\ldots\ldots\ldots, s_{n_t}]^T \tag{7.84}$$

The elements of **s** have zero mean and a variance σ_s^2 so that the transmitted power P is given by:

$$P = n_t \sigma_s^2 \tag{7.85}$$

The correlation matrix $\mathbf{R_s}$ of the transmitted vector **s** is n_t by n_t matrix is given by:

$$\mathbf{R_s} = E[\mathbf{s}\,\mathbf{s}^T] = \sigma_s^2 \mathbf{I}_{n_t} \tag{7.86}$$

where $\mathbf{I}_{n_t}$ is n_t by n_t identity matrix.

The complex received signal **r** at the given time is expressed by vector, given by:

$$\mathbf{r} = [r_1, r_2, \ldots\ldots\ldots\ldots\ldots\ldots\ldots, r_{n_r}]^T \tag{7.87}$$

Denote the flat-fading Gaussian channel between transmit antenna i to receive antenna k at the given discrete time as h_{ki} where $i = 1, 2, \ldots n_t$ and $k = 1, 2, \ldots, n_r$. The complex channel n_t by n_r can be represented by matrix **H**:

$$\mathbf{H} = \begin{matrix} h_{1,1} & h_{1,2} & \ldots\ldots\ldots\ldots & h_{1,n_t} \\ h_{2,1} & h_{2,2} & \ldots\ldots\ldots\ldots & h_{2,n_t} \\ \ldots\ldots & & \ldots\ldots\ldots\ldots & \ldots\ldots \\ \ldots\ldots & & \ldots\ldots\ldots\ldots & \ldots\ldots \\ \ldots\ldots & & \ldots\ldots\ldots\ldots & \ldots\ldots \\ h_{n_r,1} & h_{n_r,2} & \ldots\ldots\ldots\ldots & h_{n_r,n_t} \end{matrix} \quad n_r \text{ receive antennas} \tag{7.88}$$

$$\underleftarrow{\qquad} n_t \text{ transmit antennas} \underrightarrow{\qquad}$$

Each complex element of **H** is expressed by two real components (one for the real part and the other for the imaginary), each with zero mean and standard deviation $\frac{1}{\sqrt{2}}$ as shown:

$$h_{ik} : n\left(0, \frac{1}{\sqrt{2}}\right) + jn\left(0, \frac{1}{\sqrt{2}}\right) \tag{7.89}$$

where n(x,x) represents a real Gaussian distribution so that the amplitude of each element is Rayleigh distributed. Elements of **H** have zero mean and unit variance. The complex channel noise can be expressed by n_r by 1 vector:

$$\mathbf{n} = [n_1, n_2, \ldots\ldots\ldots\ldots\ldots.n_{n_r}]^T \tag{7.90}$$

where elements of **n** are complex Gaussian variables with zero mean and variance σ_n^2. The correlation matrix $\mathbf{R}_n$ of the channel noise is given by:

$$\mathbf{R}_n = \mathrm{E}[\mathbf{n}\,\mathbf{n}^T] = \sigma_n^2 \mathbf{I}_{n_r} \tag{7.91}$$

where $\mathbf{I}_{n_r}$ is n_r by n_r identity matrix.

The complex received signal r_ℓ at the ℓth antenna due to transmitted symbol s_k transmitted by the kth antenna:

$$r_\ell = \sum_{m=1}^{n_t} h_{\ell k} s_k + n_\ell \tag{7.92}$$

Therefore, the complex MIMO channel model can be expressed in matrix form as in (Haykin and Moher, 2005):

$$\mathbf{r} = \mathbf{H}\,\mathbf{s} + \mathbf{n} \tag{7.93}$$

7.4.2.5 The capacity of MIMO channel

In Chapter 2, we stated that the information capacity C bits/s of a real AWGN channel of bandwidth B is given by Shannon expression:

$$C = B \log_2\left(1 + \frac{S}{\sigma_n^2}\right) \mathrm{b/s} \tag{7.94}$$

S and σ_n^2 are the average signal power and noise power ($\sigma_n^2 = N_0 B$), respectively. Since, according to the sampling theory, we need 2 samples per Hz, the AWGN channel capacity per Hz is given by:

$$C_{real} = \frac{1}{2}\log_2\left(1 + \frac{S}{\sigma_n^2}\right) \mathrm{b/s/Hz} \tag{7.95}$$

The capacity for a complex flat-fading SISO channel is given in Thompson et al. (1996):

$$C_{siso} = \mathbf{E}\left[\log_2\left(1 + \frac{|h|^2 \sigma_s^2}{\sigma_n^2}\right)\right] \mathrm{b/s/Hz} \tag{7.96}$$

The scaling factor 1/2 is missing in (7.94) since the complex channel is equivalent in capacity to two real channels, one is carrying the in-phase signal and the other is carrying the quadrature signal. The expectation in (7.96) is taken over the channel gain h(n) at discrete time n. The capacity of complex MIMO flat-fading channel is given in Thompson et al. (1996):

$$C_{mimo} = \mathbf{E}\left[\log_2\left\{\det\left(\mathbf{I}_{Nr} + \frac{\sigma_s^2}{\sigma_n^2}\mathbf{H}\,\mathbf{H}^H\right)\right\}\right] \mathrm{b/s/Hz} \tag{7.97}$$

We now work through two examples showing how to calculate the b/s/H for SISO and MIMO channel systems. We use expression (7.96) in Example 7.7, and expression (7.97), in Example 7.8.

Example 7.7

Consider a complex flat Rayleigh fading SISO channel as a media for data transmission with AWGN of zero mean and unity variance. The samples of the channel complex gain h(n) at discrete time n = 1, 2, 3, ..., 8 are:

$$
\begin{matrix}
0.1477 + j1.619 & 0.633 - j0.0706 & -0.8365 - j0.2766 \\
-0.6501 + j0.4143 & 0.2878 - j0.2447 & -0.4792 + j0.9239 \\
0.9914 + j0.6154 & -0.3533 - j0.0809 &
\end{matrix}
$$

Assuming $\frac{\sigma_s^2}{\sigma_n^2} = 2$, calculate the capacity of the channel in b/s/H.

Solution

We notice that the mean value of the given noise samples is not zero. This is because the number of samples is small and, for the mean to tend to zero, requires a Gaussian process with sizable statistics.

$$
\frac{\sigma_s^2}{\sigma_n^2} = 2
$$

The values of $|h|^2$ are:

$$
|h|^2 = [1.6257, 0.6369, 0.8810, 0.7709, 0.3778, 1.0408, 1.1669, 0.3624]
$$

The channel capacities can be calculated using (7.96):

$$
C_{siso} = E\left[\log_2 \left(1 + \frac{|h|^2 \sigma_s^2}{\sigma_n^2} \right) \right] = E[\log_2 (1 + 2x|h|^2)]
$$

These capacities at the time instants are tabulated below:

Time instant	Channel capacity in b/s/H
1	2.0880
2	1.1851
3	1.4658
4	1.3458
5	0.8119
6	1.6237
7	1.7372
8	0.7865

Thus the expected channel capacity is **1.38** b/s/H.

A quicker method to calculate an estimate of the channel capacity is find the expected value for $|h|^2$:

$$\sum_{i=1}^{8} |h(n)|^2 = 6.8624$$

Thus

$$E(|h|^2) = \frac{6.8624}{8} = 0.8578$$

Using (7.96) the SISO channel capacity in b/s/H is:

$$C_{siso} = \log_2 (1 + 2 * 0.8578) = \mathbf{1.44\,b/s/H}$$

Example 7.8

Consider a complex flat Rayleigh fading MIMO channel as a transmission media with $n_t = n_r = 2$. The AWGN has zero mean and unity variance. The sampled complex gain of the channel h(n) at a given discrete time are:

$$h_0 = 0.4412 + j0.7662$$
$$h_1 = 0.6661 + j0.2760$$
$$h_2 = -0.8365 + j0.2766$$
$$h_3 = 0.7134 + j0.3141$$

Assuming $\frac{\sigma_s^2}{\sigma_n^2} = 2$, calculate the capacity of the channel in b/s/H.

Solution

For 2×2 MIMO channel, the identity matrix is:

$$\mathbf{I}_2 = \begin{matrix} 1 & 0 \\ 0 & 1 \end{matrix}$$

The 2×2 complex channel matrix $\mathbf{H}$ is:

$$\mathbf{H} = \begin{pmatrix} h_{1,1} & h_{1,2} \\ h_{2,1} & h_{2,2} \end{pmatrix} = \begin{pmatrix} h_0 & h_1 \\ h_2 & h_3 \end{pmatrix}$$

$$\mathbf{H} = \begin{pmatrix} 0.4412 + j0.7662 & 0.6661 + j0.2760 \\ -0.8365 + j0.2766 & 0.7134 + j0.3141 \end{pmatrix}$$

The Hermitian Transpose of the channel $\mathbf{H}$ is:

$$\mathbf{H}^H = \begin{matrix} 0.4412 - j0.7662 & -0.8365 - j0.2766 \\ 0.6661 - j0.2760 & 0.7134 - j0.3141 \end{matrix}$$

Therefore

$$\mathbf{H}^* \mathbf{H}^H = \begin{array}{cc} 1.3016 & 0.4048 - j0.7753 \\ 0.4048 + j0.7753 & 1.3838 \end{array}$$

The channel capacity C_{mimo} calculated using (7.97) is:

$$C_{mimo} = \log 2 \left(\det(I_2 + 2^* \mathbf{H}^* \mathbf{H}^H) \right) = 3.4 \, b/s/H$$

7.4.3 Beam forming techniques

The beam forming technology offers a significantly improved solution to reduce inter-ference levels. The principle behind the technology is that each user's signal is transmit-ted/received by the base station only in the direction of the target user (Razavilar et al., 1999). Consequently, the overall interference in the system is reduced. A beam forming transmission system is shown in Figure 7.11. Each individual user's signal is multiplied with complex weights which adjust the magnitude and phase of the signal to and from each antenna so that the output from the transmit/receive antennas form a beam in the desired direction and reduce the interference in other directions.

7.4.3.1 Basic beam forming system

Consider an array of L omni-directional receive and M directional transmit antennas and let the total complex received signal from all M transmitting antennas be $x_\ell(t)$ which consists

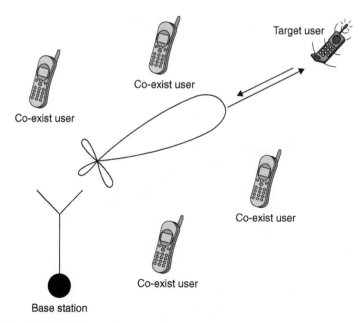

Figure 7.11 *Principles of beam forming transmission.*

of the received signal m(t) plus background noise and interference $n_\ell(t)$:

$$x_\ell(t) = m(t) + n_\ell(t) \tag{7.98}$$

A simple beam former is shown in Figure 7.12 where received signal from each antenna is multiplied by a complex weight ω_ℓ and summed to form the array output y(t) (Godara, 1997):

$$y(t) = \sum_{\ell=1}^{L} \omega_\ell^* x_\ell(t) \tag{7.99}$$

where '*' denotes complex conjugate.

The weights vector of the beam former can be expressed as:

$$\mathbf{W} = [\omega_1, \omega_2, \ldots \ldots \ldots \ldots \ldots \ldots \ldots \ldots \ldots \ldots, \omega_L]^T \tag{7.100}$$

The complex signals at the output of the receiving antennas $\mathbf{x}(t)$ can be expressed as:

$$\mathbf{X} = [x_1(t), x_2(t), \ldots \ldots \ldots \ldots \ldots \ldots \ldots \ldots, x_L(t)]^T \tag{7.101}$$

Therefore, (7.99) can be expressed in vector form as:

$$\mathbf{y} = \mathbf{W}^H \mathbf{X} \tag{7.102}$$

For a given weights vector $\mathbf{W}$, the received power $P(\mathbf{W})$ is given by:

$$P(\mathbf{W}) = E[y(t)y^*(t)] = \mathbf{W}^H \mathbf{R} \mathbf{W} \tag{7.103}$$

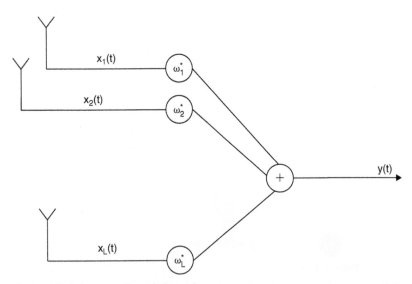

Figure 7.12 *Block diagram of simple beam former structure.*

Where **R** is L by L correlation matrix of the receive signals:

$$\mathbf{R} = E[\mathbf{X}\mathbf{X}^H] \tag{7.104}$$

The elements of matrix **R** denote the correlation between various received signals **X**. For example, R_{ij} denotes the correlation between the signals received by the ith and the jth antennas.

7.4.3.2 Beam forming schemes

The *delay and sum* beam former   In this scheme, elements of **W** are of equal magnitudes and the phases are selected to steer the receiving array in the direction of signal arrival from the transmitting antennas. A two-antenna 'delay and sum' beam former is shown in Figure 7.13. Each received signal is multiplied by a factor 0.5 so that the total gain in the signal direction θ is equal unity.

The two antennas are separated by a distance d. Let us assume that the transmitted signal is arriving from direction θ and that the received signal at the output of the first antenna is s(t). The signal at the second antenna arrives after $\tilde{T}$ seconds where $\tilde{T}$ is given by:

$$\tilde{T} = \frac{d}{c}\cos\theta \tag{7.105}$$

where c is speed of light. Therefore, the signal at the output of the second antenna is $s(t - \tilde{T})$. We delay the signal from the first antenna to produce $s(t - \tilde{T})$ to be in phase with the signal from the second antenna, multiply both signals with a scaling factor 0.5 and add them up to produce an output signal $s(t - \tilde{T})$.

This beam former reduces the power of the uncorrelated noise at the array output and produces a gain in the SNR equal L. Furthermore, it provides maximum output SNR when

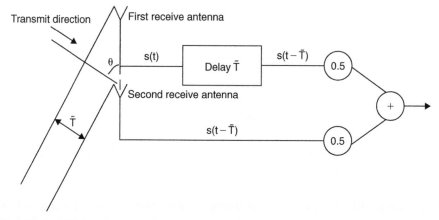

Figure 7.13   *Delay and sum beam former.*

there are no directional interferers operating at the same frequency. However, it is not effective in the presence of directional interferers.

Null steering beam former   This beam former produces null in the response pattern in the direction of arrival of the interfering signal, cancelling the interference arriving from that direction. An early version of this scheme subtracts the estimated interference using a delay and sum beam former, which is very effective for cancelling strong interference from a small number of interferers. However, it becomes cumbersome as the number of interferers grows in applications like mobile cellular communications.

A beam in the desired direction and null in the interference directions can be formed by proper choice of the weights in beam former shown in Figure 7.12. Define the complex steering vectors such that associated with the ith transmit antenna direction by a complex vector s_i:

$$\mathbf{s}_i = [e^{j\omega_0\tau_1}, e^{j\omega_0\tau_2}, \ldots\ldots\ldots\ldots\ldots\ldots e^{j\omega_0\tau_L}] \tag{7.106}$$

where ω_0 is the radian frequency of the desired signal, and τ_i is the time delay measured from the reference receive antenna to the origin for the signal arriving from the ith transmit antenna. *The principle behind the null beam former is to choose the weights that make the receiver's response in the desired direction (steering vector s_1) unity and the response in the other (L − 1) directions (steering vectors $s_2, s_3 \ldots, s_{L-1}$) nulls.* We can express these constraints by the following expressions:

$$\mathbf{W}^H\mathbf{s}_1 = 1 \tag{7.107}$$

$$\mathbf{W}^H\mathbf{s}_i = 0 \quad i = 2, 3, \ldots\ldots\ldots\ldots, L - 1 \tag{7.108}$$

Combine (7.107) and (7.108) in a single matrix expression as:

$$\mathbf{W}^H\mathbf{A} = \varsigma^T \tag{7.109}$$

Where

$$\varsigma = [1, 0, 0, \ldots\ldots\ldots\ldots\ldots\ldots]^T \tag{7.110}$$

$$\mathbf{A} = [\mathbf{s}_0, \mathbf{s}_1, \mathbf{s}_2, \mathbf{s}_3, \ldots\ldots\ldots\ldots\ldots\ldots\ldots, \mathbf{s}_{L-1}] \tag{7.111}$$

Consequently

$$\mathbf{W}^H = \varsigma^T\mathbf{A}^{-1} \tag{7.112}$$

Clearly, $\mathbf{A}$ has to be invertible, that is $\mathbf{A}$ has to be square matrix. If $\mathbf{A}$ is not invertible, its Moore-Penrose pseudo inverse can be used in (7.112).

Optimization of the beam former using a reference signal   A simple scheme for optimizing the antennas' weights is shown in Figure 7.14 which is based on the basic

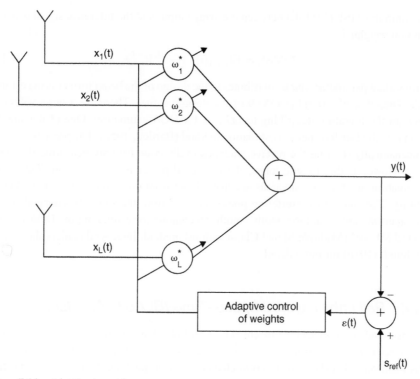

Figure 7.14 *Adaptive beam former.*

beam former in Figure 7.12 with a closed loop added for adaptive optimization of the weights using reference signal $s_{ref}(t)$. The error signal $\varepsilon(t)$ is:

$$\varepsilon(t) = s_{ref}(t) - \mathbf{W}^H \mathbf{X} \tag{7.113}$$

The mean square error (MSE) used to adjust the weights is given by:

$$\begin{aligned} \text{MSE} &= E[|\varepsilon(t)|^2] \\ &= E[|s_{ref}(t)|^2] - 2\mathbf{W}^H E[\mathbf{X}(t)s_r(t)] + \mathbf{W}\mathbf{X}\mathbf{X}^H\mathbf{W}^H \end{aligned}$$

Therefore

$$\text{MSE} = E[|s_{ref}(t)|^2] - 2\mathbf{W}^H \mathbf{z} + \mathbf{W}^H\mathbf{R}\mathbf{W}$$

where

$$\mathbf{z} = E[\mathbf{X}\, s_{ref}(t)] \tag{7.114}$$

$\mathbf{R} = [\mathbf{X}\,\mathbf{X}^H]$ as given by (7.104) and $\mathbf{z}$ is the correlation between $\mathbf{X}$ and the reference signal $s_{ref(t)}$. Minimization of the MSE is given by the Wiener-Hoff equation for the optimal weight vector:

$$\tilde{\mathbf{W}} = \mathbf{R}^{-1}\mathbf{z} \tag{7.115}$$

The minimum MSE (MMSE) between the array output and the reference signal using the optimum weights is:

$$\text{MMSE} = E[|s_{ref}(t)|^2] - \mathbf{z}^H \mathbf{R}^{-1} \mathbf{z} \tag{7.116}$$

An important parameter which contributes to the design of the beam former is the estimate of the Direction of Arrival (DOA) θ of the desired signal. There are a large volume of papers in the literature describing techniques for DOA estimation. One of the simplest schemes is the Bartlett spectral estimation method (Bartlett, 1956). The process is similar to mechanically steering the receive antennas in direction (θ) and measuring the output power. However, due to side lobes, the measured power is not only contributed from the direction in which the antennas are steered, but from directions where side lobes are pointing. The amount of interfering power received from the side lobes depends on the array aperture and main lobe width. Another technique for estimating the DOA is called Spectral MUSIC (**Mu**ltiple **Si**gnal **C**lassification) method. Interested readers should refer to Schmidt (1986) for more details.

7.4.4   Bell Labs Layered Space–Time (BLAST) Architectures

Blast is a transmission scheme invented by Bell Labs that uses multiple antennas at both transmitter and receiver to achieve very high spectral efficiency. Tests carried out by the Bell Labs team on a 30 kHz point to point channel realized payload data rates ranging from 0.5 Mb/s to 1 Mb/s. Traditional transmission methods over the same bandwidth would achieve a maximum rate of 50 Kb/s. Therefore, the BLAST data rates could be up to 20 times higher in comparison with data rates that could be achieved using traditional techniques.

BLAST accomplishes the high spectral efficiency by exploiting the multipath propagation effects for communications with the use of appropriate coding schemes and low complexity interference cancellation techniques. In conventional systems the scattering phenomena that takes place through propagation of radio signals generates multiple copies of the transmitted signal. These signal copies arrive at the receiver at different times to interfere with each other constructively or destructively, resulting in either enhancing or fading of the received signal and degrading the wireless communications. In a rich Rayleigh scattering environment, a significant decorrelation between the signals sampled by the receive array elements (Foschini, 1999) is developed. This decorrelation is exploited to permit the transmission of many parallel subchannels carrying parallel equal rate data streams over the same frequency band.

The number of effective subchannels is related to the number of antennas and the degree of decorrelation. The transmission channel, assumed unknown to the transmitter, is estimated at the receiver by measuring the response to a known training sequence. The channel is assumed to be *flat across the frequency band with a characteristic that is static during each burst of data but may change from burst to burst.* Consequently, channel capacity can change from burst to burst as we have seen in Example 7.7. The main system parameters are the number of transmit antennas n_t, the number of receive antennas n_r and the average signal-to-noise ratio (*averaged over a large number of bursts from a single transmitter to a single receiver*).

There are three BLAST systems proposed in the literature: the *diagonal or D-BLAST* system, the *vertical or V-BLAST* system and the *Turbo BLAST* system. The transmission schemes and signal processing algorithms for these systems are considered in the next section for single-user point to point wireless communications.

7.4.4.1 *The D-BLAST transmission algorithm*

Consider the BLAST system in Foschini (1996) where the user's data is converted into n_t parallel substreams of equal data rate; each substream is mapped to M-PSK or M-QAM complex symbols. Additionally, a Forward Error Correction (FEC) code can be used to encode the data before mapping. The symbols of each substream are dispersed '*diagonally*' across the antennas-time plane and transmitted in the same frequency band of the channel using transmitter antenna of n_t elements. The association of the n_t signals – antenna elements is periodically cycled so that none of the parallel data streams are permanently committed to a specific transmit antenna element (Foschini et al., 2003). The dwelling time on any association is τ seconds so the full cycle takes $n_t\tau$ seconds. *The theoretical rates grow linearly with the number of transmit antennas which approach the Shannon capacity.*

The radiated signals are captured by banks of n_r receiving antenna elements. The antenna elements at each end of the communication link are half wave dipoles separated by about half-wavelength. The received signals are processed by first extracting the strongest signals then proceeding with the remaining weaker signals. The total transmitted power p_{tot} is fixed regardless of the number of transmit elements n_t. The transmission bandwidth is generally chosen to be narrow enough so that the channel frequency response can be treated as flat across the band. A conventional QAM modulation with diagonally layered coding in which code blocks for each layer are lined across diagonals in space time are usually used. A block diagram of the D-BLAST transmitter is shown in Figure 7.15.

The D-BLAST receiver consists of antenna array of n_r elements connected to n_r QAM receivers where $n_t \leq n_r$. The noise vector at the receiver is complex with components statistically independent and of identical noise power σ_n^2 at each of the receive antenna outputs. Furthermore, the noise process is Additive White Gaussian Noise (AWGN) with n_r dimensions where the dimension is referred to space with flat Rayleigh fading. Let p_{avg} be the average signal power at the output of each receiving antenna so that the average SNR at the output of each receive antenna ρ is given by:

$$\rho = \frac{P_{avg}}{\sigma_n^2} \tag{7.117}$$

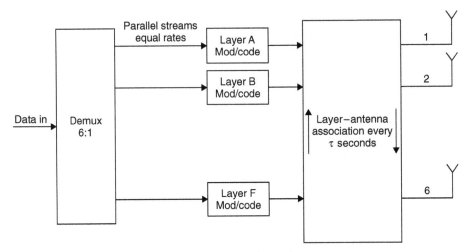

Figure 7.15 *Principle of D-BLAST transmission with $n_t = 6$.*

Let **a** denote the complex vector of the transmitted QAM symbols where:

$$\mathbf{a} = [a_1, a_2, \ldots\ldots\ldots\ldots\ldots\ldots\ldots\ldots, a_{n_t}]^T \tag{7.118}$$

The complex noise vector **n** is given by:

$$\mathbf{n} = [n_1, n_2, \ldots\ldots\ldots\ldots\ldots\ldots\ldots\ldots, n_{n_r}]^T \tag{7.119}$$

The elements of **n** are isolated independently distributed (iid) Gaussian with noise power (variance) σ_n^2. The n_r receivers operate in the same frequency band. The channel frequency transfer function is represented by $n_r \times n_t$ complex matrix **H** with complex elements h_{ij} describing the transfer function from j^{th} transmitter to K^{th} receiver given in (7.88). The received discrete time complex baseband n_r-vector **r** is:

$$\mathbf{r} = \mathbf{Ha} + \mathbf{n} \tag{7.120}$$

The block diagram of D-BLAST receiver is shown in Figure 7.16.

The Blast receiver structure comprises an array of n_r receive antenna elements, each connected to a QAM receiver, the out of which is processed using the D-BLAST detection algorithm. Before data transmission began, the BLAST system runs through a start-up procedure using known training signals to accurately estimate the channel at the receiver.

The D-BLAST signal processing algorithm proceeds along the *diagonals* of the *space–time layers* and moves left to right. After completing the training phase, the system is ready for data detection. Consider each received vector of duration τ seconds and represented by a rectangle and the first 10 of these rectangles during the first 10τ is shown in Figure 7.17. Let us consider the detection of repetitive layers labelled A, B, C, D, E, and F.

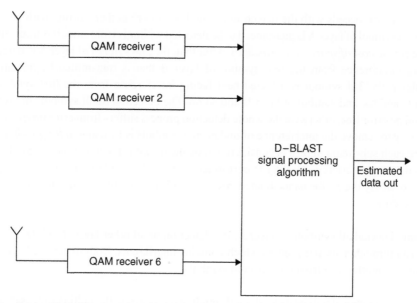

Figure 7.16 *Block diagram of D-BLAST receiver.*

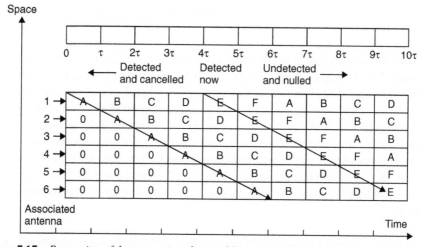

Figure 7.17 *Processing of the space–time diagonal layers in D-BLAST structure.*

During time $0 \leq t \leq \tau$ the first symbol from layer A is transmitter through antenna-1, and during time $\tau \leq t \leq 2\tau$ the 2nd symbol from layer A is transmitted from antenna-2 and the first symbol from layer B is transmitted from antenna-1. Similarly, during time $2\tau \leq t \leq 3\tau$ the 3rd symbol from layer A is transmitted from antenna-3, the 2nd symbol from layer B is transmitted from antenna-2 and the first symbol from layer C is transmitted from antenna-1 and so on. The space–time transmission diagram of the D-Blast system is shown in Figure 7.17.

The receiver proceeds with the detection of one layer after another starting with layer A. The first symbol of layer A is guaranteed to be detected without errors since it is transmitted alone (*at the cost of space–time wastage*). After that, the 2nd symbol of layer A is detected facing interference from the first symbol of layer B that is transmitted by antenna-1. Similarly, the 3rd symbol is detected next facing interference from the first symbol of layer C and the 2nd symbol of layer B and so on. The detection of all symbols of layer A should be error free, otherwise the whole detection process suffers from error propagation. One way to remove the interference of undetected symbols is by using *nulling techniques* where each substream in turn is considered to be the desired signal and the remainder are considered as interferers. Nulling is performed by linearly weighting the received signal so as to satisfy the performance related criterion such as minimum mean squared error or zero forcing.

Having detected all symbols of layer A, the detection of all other layers B, C, D, E and F will run through a similar process which contains interference from undetected symbols as well as interference from the detected symbols.

Interference removal will be carried out through two processes: the nulling process for the undetected symbols (*symbols positioned above the desired layer diagonal*) as described above and the interference cancellation for the detected symbols (*symbols positioned below the desired layer diagonal*). The interference cancellation process entails the reconstruction and subtraction of the interfering signal from the received signal before detection.

Consider the detection of symbols from a typical layer, say layer E symbols as shown in Figure 7.17. During time $4\tau \leq t \leq 5\tau$, the first symbol from layer E transmitted from antenna-1 is subjected to interference from the correctly detected symbols from layers A, B, C, and D. The interfering signals can be reconstructed and cancelled from the received signal enabling the first symbol from layer E to be correctly detected. During time $5\tau \leq t \leq 6\tau$, the 2nd symbol from layer E transmitted from antenna-2 faces interference from the first undetected symbol from layer F and interference from the correctly detected symbols from layers A, B, C, and D. The interference from layer F is nulled and the interference from detected symbols is cancelled. This process is repeated until all symbols of layer E are detected.

The D-BLAST capacity (C_D) in b/s/H per available dimension using optimally selected n_t transmitters when n_r receives are available ($n_t \leq n_r$) is given in Foschini et al. (1999) as:

$$C_D = \underset{n_t}{\text{Max}} \, \frac{1}{n_r} \sum_{k=1}^{n_t} \log_2\left[1 + \left(\frac{\rho}{n_t}\right)(n_r - k + 1)\right] \quad \text{b/s/H/dimension} \qquad (7.121)$$

The capacity in b/s/Hz of the D-BLAST system over fading wireless channel versus number of antennas for various SNR at each receive antenna is shown in Figure 7.18. The figure shows the capacity in b/s/Hz increasing linearly with number of antennas for given average

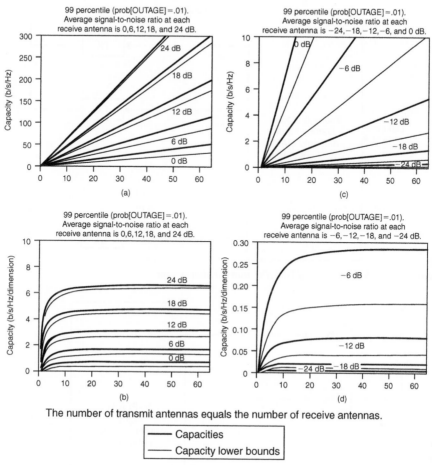

The number of transmit antennas equals the number of receive antennas.

| —— Capacities |
| —— Capacity lower bounds |

Figure 7.18 *Capacity in b/s/Hz of D-BLAST system in Rayleigh fading channel versus number of antennas. Reproduced with permission by Bell Labs Technical Journal (Figure 3, Foschini, G.J. (1996) Layered space-time architecture for wireless communication in fading environment when using multi-element antennas, Bell Labs Technical Journal, 1(2), 41–59).*

SNR at each receive antenna. The capacity per dimension for a given average SNR at each receive antenna is approaching a limit as the number of antennas increases.

7.4.4.2 V-BLAST detection algorithm

The Vertical or V-BLAST proposed in Golden et al. (1999) is a simplified version of D-BLAST in order to reduce the computation complexity, but by doing so, the transmit diversity is lost. A block diagram of the V-BLAST transmit system is depicted in Figure 7.19.

The input data is separated into substreams. The QAM symbols of a certain substream are transmitted through the same antenna (i.e. *each substream is tied up to its antenna*). The

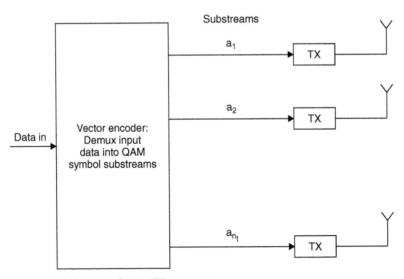

Figure 7.19   *Block diagram of V-BLAST transmission system.*

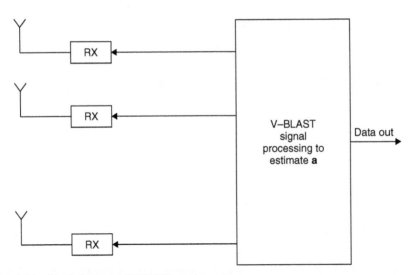

Figure 7.20   *Block diagram of V-BLAST receive system.*

QAM transmitters operate in co-channels at equal symbol rate with synchronized symbol timing. The total radiated power is constant and independent of the number of transmit antennas. A V-BLAST high level receive system is shown in Figure 7.20.

In the following analysis, we assume the V-BLAST is operating in quasi static flat Rayleigh fading channel that is perfectly estimated over every burst (*number of symbols*) and the system experiences ideal symbol timing. The detection is carried out along the vertical alignment of the received substreams which are re-ordered according to the level of their *SNRs*.

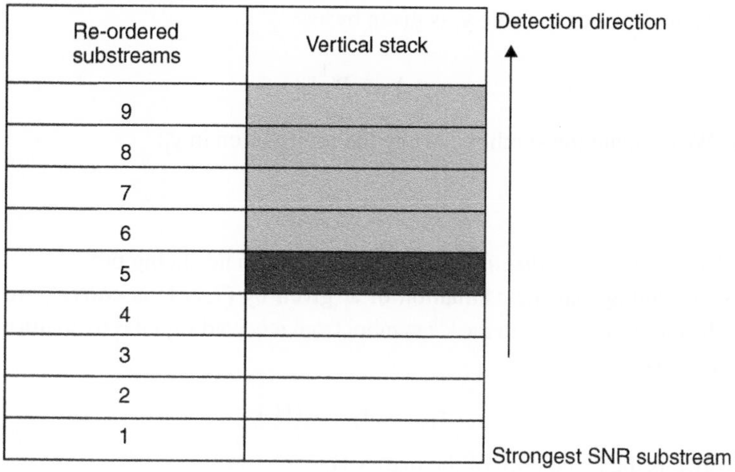

Figure 7.21 *Substreams re-ordering for detection.*

The detector estimates the substream with the strongest SNR and then subtracts the estimated substream from the received vector. After that, the next substream with the next highest SNR is estimated and subtracted and so on. This procedure is outlined in Figure 7.21.

When we consider estimating symbols from substream 5, we should have already successfully estimated substreams 1 to 4 so that we can cancel their interference from the received vector. However, substreams 6 to 9 have to be *nulled* using either a minimum mean squared error or zero-forcing criterion.

The received discrete–time complex baseband signal vector $\mathbf{r}$ is given by (7.120) where the channel frequency transfer function is represented by the $n_r \times n_t$ complex matrix $\mathbf{H}$ and the complex elements $\mathbf{h_{ki}}$ are describing the transfer function from the ith transmitter to the kth receiver. The V-BLAST algorithm operates on the received vector $\mathbf{r}$ to compute the decision statistics y_j. The algorithm, described in Golden *et al.* (1999), sets the initialization values for the first iteration (j = 1):

$$\mathbf{r}_1 = \mathbf{r}$$

$$\mathbf{H}_1 = \mathbf{H}$$

For the jth iteration, we run through the following procedure:

Step 1: We compute the inverse of $\mathbf{H}_j$. If $n_t \neq n_t$ or $\mathbf{H}_j$ is singular, then the inverted matrix $\mathbf{H}_j$ (i.e. $\mathbf{H}_j^{-1}$) does not exist. Under such conditions, we compute the *Moore-Penrose pseudo inverse matrix* $\mathbf{H}_j^+$. We set the nulling vector $\mathbf{w}_j$ equal to the jth column of the Moore-Penrose pseudo inverse matrix: $(\mathbf{H}_j^+)_j$:

$$\mathbf{w}_j = (\mathbf{H}_j^+)_j \tag{7.122}$$

The decision statistics variable y_j is given by:

$$y_j = \mathbf{w}_j^T \mathbf{r}_j \qquad (7.123)$$

Step 2: We estimate the symbol $\hat{a}_j$ using the information in y_j:

$$\hat{a}_j = Q(y_j) \qquad (7.124)$$

where $Q(.)$ denotes *quantization* or any other appropriate slicing process.

Step 3: Assuming that the estimation of a_j given by (7.124) is correct, we can now cancel the interference due to the kth symbol from received signal $\mathbf{r}_j$ to obtain a modified received signal $\mathbf{r}_{j+1}$:

$$\mathbf{r}_{j+1} = \mathbf{r}_j - \hat{a}_j(\mathbf{H}_j)_j \qquad (7.125)$$

where $(\mathbf{H}_j)_j$ is the jth column of the matrix $\mathbf{H}_j$.

Step 4: Zero the kth column of the matrix $\mathbf{H}_j$ to obtain $\mathbf{H}_{j+1}$ and we set k to $j + 1$ and repeat steps 1–4 until all the substreams are detected.

The V-BLAST capacity (C_V) in b/s/H per available dimension is lower than (7.121) since it is given by the worst of all the n_t transmitted substreams that limits capacity. C_V is given in Foschini et al. (1999) as:

$$C_V = \underset{\{0 < \alpha < 1\}}{\text{Max}} \{\alpha \log 2[1 + \rho(\alpha^{-1} - 1)]\} \quad \text{b/s/H/dimension} \qquad (7.126)$$

where ρ is the signal-to-noise seen to be a single receive antenna averaged over a large number of bursts from the same transmitter and $\alpha = n_t/n_r$. For each value of SNR (ρ), the parameter α is varied within its limits to optimize the capacity C_V.

Figure 7.22 shows the block error rate against SNR in dB for $n_t = 8$, $n_r = 12$ V-BLAST system operating at 25.9 b/s/H using nulling with and without interference cancellation. The figure shows a gain of almost 4 dB due to the interference cancellation at block error rate of 10^{-1}. The block is considered in error if a single of its bits is in error.

The capacity in b/s/H/available dimension for D-BLAST and V-BLAST versus average SNR at each receiving antenna is compared in Figure 7.23. For example for a capacity of 3 b/s/H/dimension, the D-BLAST system is better than the V-BLAST by about 4 dB in SNR. The V-BLAST technique has now been extended to the multi-user detection over wireless uplink system (Sfar et al., 2003).

7.4.4.3 Turbo-BLAST system

Another BLAST architecture which combines the original BLAST idea with the Turbo principles is proposed in Sellathurai and Haykin (2002). This scheme randomly layered the space–time coded symbols before transmission. The input data is demultiplexed into substreams that are encoded independently by similar block codes. The coded substreams

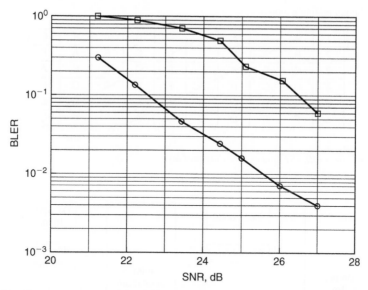

Figure 7.22 *Error rate against SNR in dB for $n_t = 8$, $n_r = 12$ V-BLAST system. Reproduced with permission from Electronics Letters (Figure 2, Golden, G.D., Foschini, C.J., Velenzuale, R.A. and Wolniansky, P.W. (1999) Detection algorithm and initial laboratory results using V-BLAST space-time communication architecture, IEE Electronics Letters, 35(1), 14–16.)*

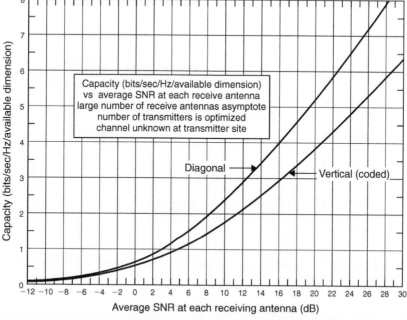

Figure 7.23 *Comparison of capacity in b/s/H/available dimension for D-BLAST and V-BLAST versus average SNR at each receiving antenna. Reproduced with permission from IEEE (Figure 6, Foschini, G.J., Golden, G.D., Valenzuela, R.A. and Wolniansky, P.W. (1999) Simplified processing for high spectral efficiency wireless communication employing multi-element arrays, IEEE Journal on Select Areas in Communications, 17(11), 1841–1852).*

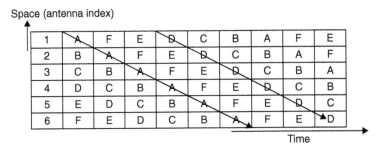

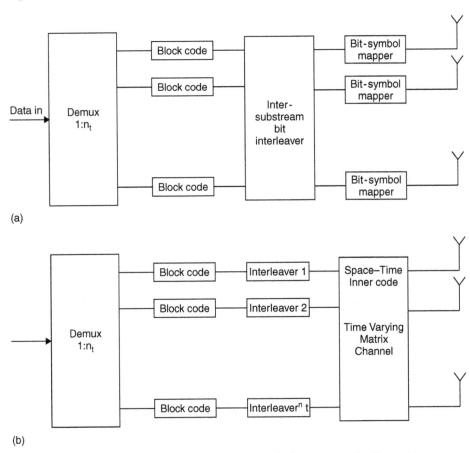

Figure 7.24 *Turbo-BLAST space–time diagonal layers.*

Figure 7.25 *(a) Block diagram of Turbo-BLAST tansmission system; (b) Turbo-BLAST transmitter using serially concatenated codes.*

are bit interleaved using random space–time interleaver and mapped to complex symbols. Each symbol will have bits from more than one substream so that errors occurring in a substream are spread across all substreams. The Turbo-BLAST space–time diagonal layers are shown in Figure 7.24 and the Turbo-BLAST transmitter is shown in Figure 7.25a.

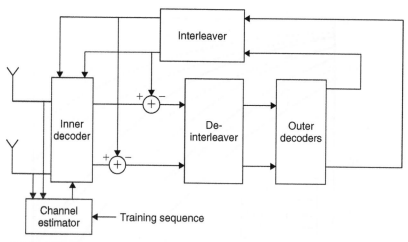

Figure 7.26 *Turbo-BLAST iterative detection.*

The decoding of the bit interleaved and encoded substream symbols has an *exponential computational complexity* in the number of transmit antennas, block size and symbol constellation. An alternative scheme, proposed in Sellathurai and Haykin (2002), is based on decoding serially concatenated Turbo codes that substitute the encoding and interleaving at the transmitter by *serially concatenated code*; the outer code has n_t parallel channel codes and the inner code is the time-varying channel matrix. The inner and the outer codes are connected by n_t parallel interleavers. The Turbo-BLAST transmitter using serially concatenated codes is shown in Figure 7.25b.

The concatenated code is developed to a lower complexity iterative receiver similar to the one proposed for serially concatenated Turbo codes. The Turbo BLAST iterative detection is shown in Figure 7.26.

The BER performance for $n_t = 5, 6, 7$ and 8 and $n_r = 8$ of Turbo-BLAST system using $\frac{1}{2}$ rate convolutional code and constraint length 3 with QPSK modulation is shown in Figure 7.27.

7.5 Interference Cancellation (IC) techniques

The Interference Cancellation techniques are based on the principle that it is possible to remove the multiple access interference from each user's received signal before making data decisions. The IC techniques can be grouped into two categories: *successive IC* where the interference is cancelled serially and in stages starting with the strongest interferer. The *parallel IC* which is achieved by cancelling the interference from all users simultaneously and could be carried out in multi-stages as well. The main stages involved in the IC schemes are the estimation of the received signal amplitudes (energies) of the active users, the regeneration of the appropriate interfering signals and the subtraction of the interfering

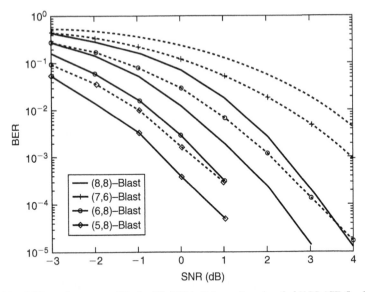

Figure 7.27 *BER performance of Turbo-BLAST (solid trace) and coded V-BLAST (broken trace), using $\frac{1}{2}$ rate convolutional code, versus SNR in dB. Reproduced with permission from IEEE (Figure 6, Sellathurai, M. and Haykin, S. (2002) Turbo-BLAST for wireless communications: Theory and experiments, IEEE Transactions on Signal Processing, 50(10), 2538–2546).*

signal from the received signal. Both IC schemes use the conventional matched filter as a first stage detector.

Let us denote the ratio between the average remaining interference power after subtraction and the total interference power before the subtraction by υ, then clearly $\upsilon = 1$ for the conventional matched filter receiver. It is shown in Jamalipour et al. (2005) that when forward error correction is used, $\upsilon \approx 0.45$ when cancellation is carried out based on the feedback soft value generated before the decoder. Conversely, when IC cancellation is carried out after the decoder, $\upsilon \approx 0.05$.

It must be emphasized that IC techniques deal effectively with intra cell interference and have no control over the interference coming from other cells (inter cell interference) since these interferers cannot be controlled by the serving base station and consequently cannot be cancelled.

7.5.1   Successive Interference Cancellation (SIC)

The successive IC scheme is simpler to implement in hardware than the parallel IC scheme, but more robust in cancelling the interference since it successively subtracts the interference from the received signal. The strongest interferer can be detected in the presence of the weaker interferer, and its removal from the received signal enhances the detection of the weaker signals, so we start with detecting the strongest interferer. This implicitly implies that we first have to rank the received signals according to their strength such that:

$$\sqrt{E_1} > \sqrt{E_2} > \sqrt{E_3}, \ldots \ldots \ldots \ldots \ldots \ldots > \sqrt{E_K} \qquad (7.127)$$

where E_K is the energy of the kth user. For simplicity and without loss of generality, we assume that the signal from user-1 is the largest followed by signal from the user-2 signal, and so on. The implication in (7.127) is that the signal from the kth user is the weakest. Once the ranking of the received signals is achieved, the detection of the user-1 signal using the conventional Matched Filter (MF) receiver can be accomplished. Let user-1 detected symbol be $\hat{b}_1(t)$ which will be used to regenerate user-1 signal $\hat{x}_1(t)$ as:

$$\hat{x}_1(t) = \sqrt{E_1}\hat{b}_1(t)C_1(t) \qquad (7.128)$$

where $C_1(t)$ is signature waveform for user-1. Subtracting (7.128) from the received signal $r(t)$ we get the input to the next stage of the IC as:

$$r(t) - \hat{x}_1(t)$$

Repeating the above procedure on the next strongest user signal (user-2) enables the detection of $\hat{b}_2(t)$ and the generation of the feedback signal $\hat{x}_2(t)$. The input to the next stage IC is:

$$r(t) - \hat{x}_1(t) - \hat{x}_2(t) \qquad (7.129)$$

The procedure can be repeated until the user with the weakest signal (k^{th} user) is detected $\hat{b}_K(t)$ and the input to the final stage is:

$$r(t) - \sum_{i=1}^{K-1} \hat{x}_i(t) \qquad (7.130)$$

The computation complexity of the successive IC algorithm is linear in the number of users. This scheme, though simple, has a number of disadvantages, the most prominent of which is that, since the IC proceeds serially, a delay of the order of K computation stages is required to complete the multi-user detection. Furthermore, an erroneous detection at any stage will be fed back to cause error propagation in the following stages which increases, rather than decreases, the interference level. Another disadvantage is that the first user to be processed sees the interference from K − 1 users whereas each user down stream sees less and less interference so that the service quality of the system is not the same. A high level system view of the successive IC scheme is shown in Figure 7.28.

The BER performance of a CDMA system operating in Rayleigh fading channel versus $\frac{E_b}{N_0}$ using a successive IC cancellation, compared with single-user detector and conventional MF detector BER performances, is shown in Figure 7.29. For $\frac{E_b}{N_0} = 20\,\text{dB}$ under Rayleigh

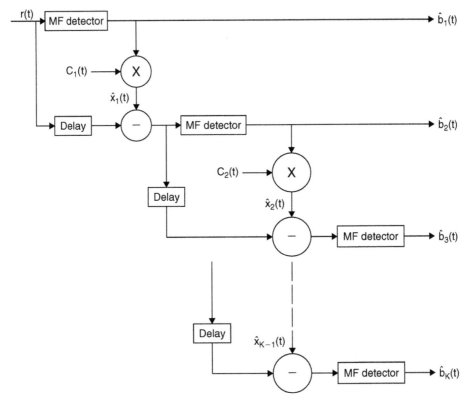

Figure 7.28   *Schematic showing the successive interference cancellation.*

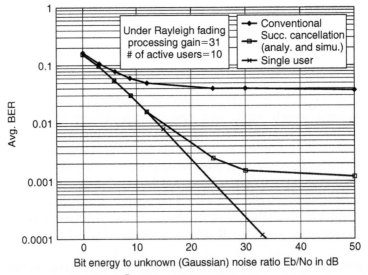

Figure 7.29   *Average BER versus $\frac{E_b}{N_o}$ under Rayleigh fading. Reproduced with permission from IEEE (Figure 7, Patel, P. and Holtzman, J. (1994) Analysis of a simple successive interference cancellation scheme in a DS/CDMA system, IEEE Journal on Select Areas in Communications, 12(5), 796–807).*

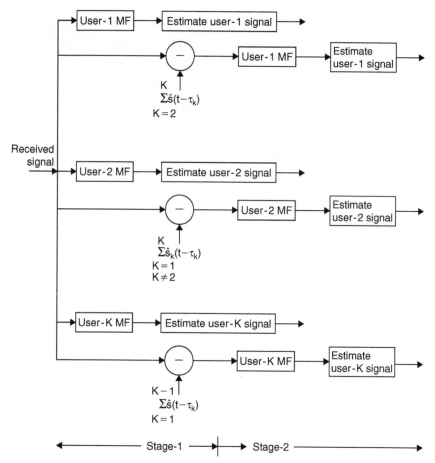

Figure 7.30 *Parallel interference cancellation.*

fading channel, the average BER of about 3×10^{-2} for matched filter receiver, 5×10^{-3} for successive IC receiver, and 2.5×10^{-3} for a single-user receiver.

7.5.2 Parallel Interference Cancellation (PIC)

The parallel IC scheme accomplishes parallel processing of the access interference, and removes the interference from all users simultaneously. Since the IC is performed in parallel, the delay required for interference removal is, at most, of a few bits duration. In order to cancel the interference, an estimate of the interference is required. However, such estimate is poor in the early stages of multistage PIC process. Therefore, it is preferable to use '*partial IC*' and to increase the portion of the IC as the interference estimation improves in the later stages. In the parallel iterative scheme, each stage of the iteration produces a new and better estimate of user bits based upon those obtained in the previous stage which improves the interference estimates. Multistage parallel interference cancellation is shown in Figure 7.30.

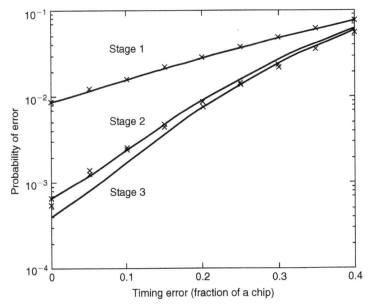

Figure 7.31 *Probability of error of PIC versus timing error; X = simulated. Reproduced with permission from IEEE (Figure 2, Buehrer, R.M., Kaul, A., Striglis, S. and Woerner, B.D. (1996) Analysis of DS-CDMA parallel interference cancellation with phase and timing errors, IEEE Journal on Select Areas in Communications, 14(8), 1522–1535.*

The probability of error performance of parallel interference cancellation scheme versus timing error for $\frac{E_b}{N_0} = 8$ dB with number of users = 10 and spreading gain 31 is shown in Figure 7.31 for timing error varying from zero to $0.4\,T_c$. The curves show that as the timing error increases, the gain from the IC degraded.

Figure 7.32 shows the probability of error performance of the parallel IC scheme versus phase errors for $\frac{E_b}{N_0} = 8$ dB with number of users = 10 and spreading gain 31. As in the timing error case, the gain of the IC is reduced with increasing phase errors.

The complexity (in instruction cycles per bit) for SIC and PIC receivers is compared when spreading gain = 31, samples per chip = 4 and the parallel cancellation stages = 3 in Figure 7.33. The figure shows the complexity of PIC receiver is higher than the complexity of SIC receiver.

7.5.3 Hybrid successive and parallel cancellation

We now summarize the shortcomings of the previous interference cancellation techniques before considering a possible hybrid interference cancellation scheme comprising of SIC and PIC. The SIC has complexity and latency proportional to number of users (K) and, for large K, this latency may be prohibitive for use with real time applications. PIC, on

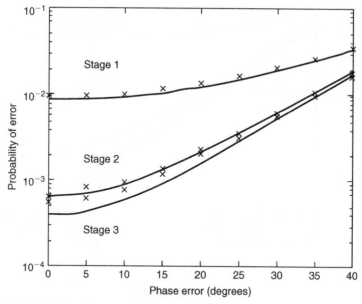

Figure 7.32 *Probability of error of PIC versus phase error; X = simulated. Reproduced with permission from IEEE (Figure 6, Buehrer, R.M., Kaul, A., Striglis, S. and Woerner, B.D. (1996) Analysis of DS-CDMA parallel interference cancellation with phase and timing errors, IEEE Journal on Select Areas in Communications, 14(8), 1522–1535).*

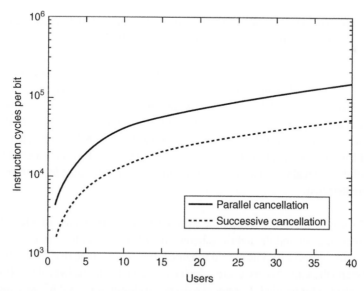

Figure 7.33 *Comparison of complexity of SIC with PIC receivers. Reproduced with permission from IEEE (Figure 12, Buehrer, R.M., Kaul, A., Striglis, S. and Woerner, B.D. (1996) Analysis of DS-CDMA parallel interference cancellation with phase and timing errors , IEEE Journal on Select Areas in Communications, 14(8), 1522–1535).*

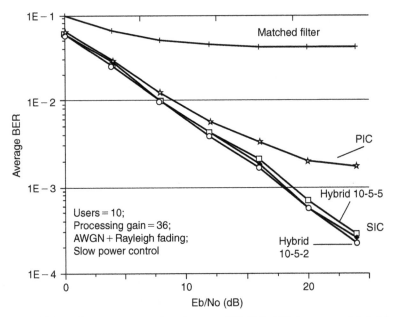

Figure 7.34 *BER performance comparison between SIP, PIC, MF detector and hybrid scheme. Reproduced with permission from IEEE (Figure 2, Koulakiotis, D. and Aghvami, A.H. (1998) Evaluation of a DS/CDMA multiuser receiver employing a hybrid form of interference cancellation in Rayleigh-fading channels, IEEE Communications Letters, 2(3), 61–63).*

the other hand, has lower latency but higher overall complexity (Andrews, 2005) due to detection of K users in parallel and the multistage cancellation.

Therefore, to get the best from each scheme, the hybrid schemes have to cancel the interference partially in parallel and partially in serial (successive). At each stage, a number of users are cancelled in parallel and serial (Koulakiotis and Aghvami, 1998). The received signal is first detected using a bank of matched filters and buffered. The first P strongest users are selected to perform PIC between them and their signals are reconstructed and subtracted from the buffered matched filter outputs. Then the P signals are ranked again and S out of the P signals is selected for serial cancellation. The procedure is repeated until all users are detected.

Figure 7.34 shows a comparison of the average BER performance of SIC, PIC using matched filter detector and the hybrid scheme under Rayleigh fading. In this scheme, the number of active users is 10. The hybrid scheme (10-5-5) select the five most reliable users for PIC first and the same five users are selected for serial IC. In the hybrid IC (10-5-2), out of the five selected for parallel IC, only two users are chosen for serial IC. As expected, the matched filter detector in Rayleigh fading environment gives the worst BER performance and SIC gives better performance than PIC. In fact, the hybrid scheme is not much better than the performance of SIC. However, unlike the SIC the hybrid offers lower latency and complexity.

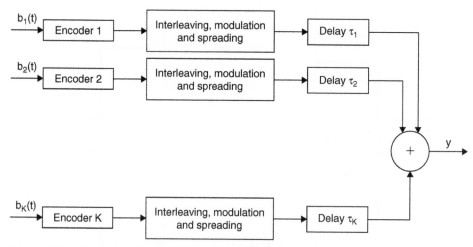

Figure 7.35 *Turbo encoded CDMA multi-user channel.*

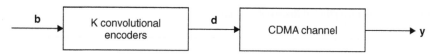

Figure 7.36 *Simplified representation of the encoded CDMA multi-user channel.*

7.5.4   Iterative (Turbo) interference cancellation

The iterative interference cancellation scheme gained an increasing interest (Wang and Poor, 1999; Moher, 1998) as an appropriate technique to mitigate intra cell interference over multipath channels since it offers low computing cost and near single-user performance. The interference is estimated and subtracted from the received signal to produce a new, cleaner signal for the desired user after each iterative process. An improved performance could be obtained by using channel coding and performing turbo decoding inside each detection iteration. A block diagram of the multi-user transmitter on the down link is shown in Figure 7.35.

The transmission system consists of K active users. The kth user's bits $b_k(t)$ are convolutionally encoded and interleaved by the kth user specific interleaver to guarantee independence between different users' codes. The interleaved data are QSK modulated and finally spread using k^{th} user's signature waveform. The encoded CDMA channel can be simplified as shown in Figure 7.36.

We can view the CDMA channel (Alexander, 1999) as a convolutional code with the number of states being exponential in the number of users. Consequently, the encoded channel can be thought of as a concatenation of two convolutional codes; the outer code pertains to the single-user encoders and the inner code is due to the time varying CDMA channel. A block diagram of the receiver structure is shown in Figure 7.37.

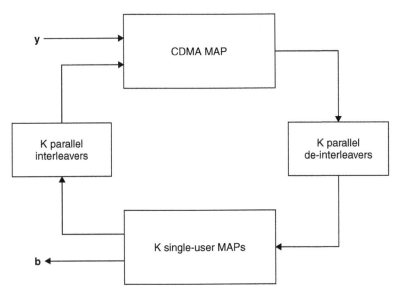

Figure 7.37   *Iterative (Turbo) Receiver structure.*

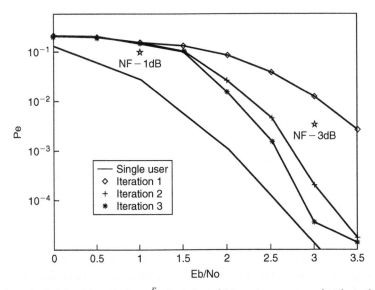

Figure 7.38   *Probability of error versus $\frac{E_b}{N_0}$ for 1, 2, and 3 iterations compared with single-user (no interference) performance. Reproduced with permission from IEEE (Figure 4, Reed, M.C., Schlegel, C.B., Alexander, P.D. and Asenstorfer, J.A. (1998) Iterative multiuser detection for CDMA with FEC: Near-single-user Performance, IEEE Transactions on Communications, 46(12), 1693–1699).*

The receiver operates on the serial Turbo decoding principle and the iterative process continues until further iterations yield minimal improvement.

Figure 7.38 shows the probability of error versus $\frac{E_b}{N_0}$ for 1, 2, and 3 iterations compared with single user (no interference) performance. The figure shows after three iterations, the BER is within 0.3 dB of single-user performance at BER of 10^{-4}. The N-F performance was investigated by setting three users to 1 dB and two users to 3 dB. The average performance of the 1 db users (labelled NF-1 dB) is improved by about 0.5 dB and the strong power users (labelled NF-3 dB) is degraded by about 0.7 dB. Therefore, the *receiver is not N-F resistant*.

7.6 Summary

In Chapter 6, we presented techniques for improving the performance of the conventional matched filter receivers by reducing the interference input to the matched filter detectors. In this chapter, we considered techniques and algorithms which minimize the multiple access interference in group detection. We started by distinguishing between two strategies of optimization in Section 7.2: the individually optimum strategy which minimizes the BER for individual users, and the jointly optimum strategy which optimizes the detection performance of a group of users. Since the individually optimum strategy is relatively more complex as compared with a joint scheme, we opted for the jointly optimum schemes which we considered in detail for synchronous and asynchronous channels. From the analysis, we found that the cost of the computation of the synchronous optimum detectors is exponential to the number of users. We also found that the computation cost of the asynchronous optimum detectors is exponentially dependent on the number of users, and on the data block length transmitted by each user. We concluded that the cost of computation in each of these systems (optimum synchronous and optimum asynchronous detectors) is too excessive for contemporary CDMA wireless systems since both the number of users and data block length is usually too large.

The past decade has witnessed a tremendous increase in research in order to facilitate and develop sub-optimal detection alternatives. We have compared the performance of these sub-optimal detectors using the performance of the optimum detectors as a benchmark.

We classified the sub-optimal detectors into two categories: *linear detectors* and *interference cancellation detectors*. The linear detectors employ linear mapping (transformation) at the output of the conventional detector to reduce the access interference and provide improved BER performance. The linear schemes discussed in Section 7.3 were the decorrelator detection and the Minimum Mean Square Error (MMSE) detection.

The decorrelating receiver greatly reduces the multiple access interference associated with the matched filter detection, but changes (generally enhances) the level of background noise. The significant advantages of the decorrelating receiver is its low complexity (increases linearly with number of users) compared to the optimum receiver (complexity increases exponentially with number of users). However, the decorrelating receiver needs the knowledge of the signature waveforms, timing information and energies of all users in order to operate correctly.

The MMSE receivers reduce the impact of background noise at the expense of a reduction in interference mitigation compared with the decorrelator detectors. From the MMSE receiver's analysis, we conclude the MMSE receiver substitutes the linear transformation R^{-1} in the decorrelator receiver by:

$$[\mathbf{R} + \sigma^2 \mathbf{A}^{-2}]^{-1}$$

Having explored the potential capabilities of linear multi-user detectors, we moved on to search for techniques to combat the interference in Section 7.4 independent of the detection process. These techniques included using multiple antennas at the transmitter and/or receiver to reduce interference and consequently exhibit increased system capacity and network coverage. An important scheme in such techniques is the *'smart' antenna* which intelligently controls the radiation towards the desired user and significantly reduces the interference to other users. Section 7.4 also presented the space diversity algorithms by considering a system with a single transmit antenna but received by two antennas and a system with two transmit antennas but received by a single antenna. This deliberation led to an important space–time coding scheme used in MIMO systems (known as Alamouti's algorithm) considering a system with two antennas at both transmit and receive ends.

The other effective technique presented in Section 7.4 was the beam forming technology that offered a significantly improved solution to reduce interference levels. The principle behind the technology is that each user's signal is transmitted by the base station only in the direction of the target user. Consequently, the overall interference in the system is reduced. Two beam forming schemes were presented: the 'delay and sum' scheme and the null steering scheme. A method to optimize the beam former using a reference signal was also described.

A radical change to the digital transmission in fading channels was published in 1996 describing a system that uses multiple antennas at the transmitter and receiver with the transmission speed many times the conventional system speed in fading channels. The new technique applies a layered space–time coding architecture and is commonly known as the Bell Labs Layered Space–Time (BLAST) architectures.

BLAST is a wireless architecture capable of realizing very high spectral efficiencies over *rich-scattering wireless channel*. Three structures of Blast were described in detail in Section 7.4, the Diagonal BLAST (D-BLAST), the Vertical BLAST (V-BLAST), and the Turbo-BLAST.

In the D-BLAST, originally proposed by Foschini, the symbols of each substream are dispersed across diagonals in the space–time scheme. Consequently, an average SNR (the same for all layers), is achieved for the operating channel and the worst probability of error is no longer dependent on the weakest substream and so the effect of the deep fades is reduced. However, due to the diagonal arrangement of the symbols of individual layers, space–time at the burst edge is wasted reducing the efficiency of the D-BLAST scheme.

The V-BLAST is a simplified version of D-BLAST to reduce the computation cost. The input data is separated in substreams but the substreams are tied up to fixed antennas, thus loosing the transmit diversity found in the D-BLAST. The received 'vertical column' of substreams is re-ordered according to the strength of their SNRs. The detector estimates the substream with the strongest SNR and subtract it from the received vector. After that, the substream with the next highest SNR is estimated and subtracted and so on.

The Turbo-BLAST was also presented in Section 7.4. We showed that there is an iterative interference cancellation scheme and performing turbo decoding inside each detection iteration to mitigate intra cell interference over multipath channels. Finally, the interference cancellation approaches are considered in Section 7.5 where we introduced and compared the performance of various techniques to remove multi-access interference (such as SIC, PIC), and hybrid successive and parallel interference cancellations and Iterative (Turbo) interference cancellation schemes.

Problems

7.1 Consider a 2-user CDMA communications system operating in a Gaussian noise environment and the a priori probability takes the following values within the symbol duration T such that for any time t where $0 \leq t \leq T$:

$$P[(+1,+1)|\{y(t)] = 0.05 \quad P[(+1,-1)|\{y(t)] = 0.25$$
$$P[(-1,+1)|\{y(t)] = 0.30$$

Determine the following under optimal detection:
  i. A posteriori probability $P[(-1,-1)|\{y(t)]$
     The estimated data with
 ii. Individually optimum decisions
iii. Jointly optimum decisions.

7.2 Consider a 3-user CDMA communications system operating in Gaussian noise environment such that the a priori probability takes the following values within the symbol duration T such that for any time t where $0 \leq t \leq T$:

$$P(+1,+1,+1)|y(t)] = 0.05 \quad P(+1,+1,-1)|y(t)] = 0.12$$
$$P(+1,-1,+1)|y(t)] = 0.18 \quad P(+1,-1,-1)|y(t)] = 0.25$$
$$P(-1,+1,+1)|y(t)] = 0.2 \quad P(-1,+1,-1)|y(t)] = 0.08$$
$$P(-1,-1,+1)|y(t)] = 0.02 \quad P(-1,-1,-1)|y(t)] = 0.1$$

Determine the estimated data under optimal detection:
  i. Individually optimum decisions
 ii. Jointly optimum decisions.

7.3 Consider 3 users sharing a CDMA communications channel, each transmitting one symbol with received normalized energy of each user is set to unity. The signature waveforms used by the users are:

$$C_1(t) = [1, \ -1, \quad 1, \ -1, \ 1, \ -1, \ 1, \ -1]$$
$$C_2(t) = [1, \quad 1, \quad 1, \ -1, \ 1, \ -1, \ 1, \quad 1]$$
$$C_3(t) = [1, \ -1, \ -1, \ -1, \ 1, \ -1, \ 1, \quad 1]$$

The samples of received signals y_1, y_2, y_3 are given by:

$$\mathbf{y} = \{0.25, 0.42, 0.75\}^T$$

Tabulate values of $\Omega(\mathbf{b})$ versus $\mathbf{b}$ and hence determine an estimate for the users symbols $\hat{\mathbf{b}}$.

7.4 Consider three users sharing a CDMA channel, each transmitting one symbol with received normalized energy of each user is unity. The signature waveforms used by the users are:

$$C_1(t) = [1, \ -1, \quad 1, \ -1, \ 1, \ -1, \ 1, \ -1]$$
$$C_2(t) = [1, \quad 1, \quad 1, \ -1, \ 1, \ -1, \ 1, \quad 1]$$
$$C_3(t) = [1, \ -1, \ -1, \ -1, \ 1, \ -1, \ 1, \quad 1]$$

Let the transmitted symbols from three users be $\mathbf{b}$ and the detected symbols be $\hat{\mathbf{b}}$ the error vector $\in$ where

$$\mathbf{b} = [-1 \ +1 \ -1]^T$$
$$\hat{\mathbf{b}} = [+1 \ +1 \ +1]^T$$

Calculate the upper and lower limits of the probability of minimum errors.

7.5 Consider two users sharing the CDMA asynchronous communications channel, each user is transmitting three symbols at a time $i = -1, 0, 1$ where $M = 1$. Let the two matched filter detectors hard decision output $\mathbf{y}$ be:

$$\mathbf{y} = 01 \ 10 \ 11$$

Assuming the energies of the two users are the equal ($\mathbf{E}_1 = \mathbf{E}_2$). Using the following simplified expression for the correlation metric $\Omega(\mathbf{b})$:

$$\Omega(\mathbf{b}) = \sum_{j=-1}^{4} \lambda_j(b_{j-1}, b_j)$$

where

$$\lambda_j(b_{j-1}, b_j) = 2b_j y_j - 2\rho_{12} b_j b_{j-1} - 1 \quad \text{if } j \text{ is even}$$
$$= 2b_j y_j - 2\rho_{21} b_j b_{j-1} - 1 \quad \text{if } j \text{ is odd}$$

$$b_{-2} = 0, \ \rho_{12} = 0.235, \ \rho_{21} = -0.235$$

Determine the estimated symbols $\hat{\mathbf{b}}$ using the maximum likelihood Viterbi decision algorithm.

7.6 Consider the decorrelator receiver used in a synchronous CDMA communications system shared by three users, each transmitting three symbols defined by vector **b** where:

$$\mathbf{b} = \begin{matrix} -1 & -1 & 1 \\ -1 & 1 & -1 \\ 1 & 1 & 1 \end{matrix}$$

Assume the received energy from each user is set to unity. The signature sequences are of one byte for each user given by:

$$C_1 = [-1 \quad 1 \quad 1 \quad -1 \quad 1 \quad -1 \quad 1 \quad 1]$$

$$C_2 = [-1 \quad 1 \quad -1 \quad 1 \quad -1 \quad 1 \quad 1 \quad -1]$$

$$C_3 = [\ 1 \quad -1 \quad 1 \quad 1 \quad -1 \quad 1 \quad 1 \quad 1]$$

The noise has zero mean, a variance of unity and vector **n** at the matched filter detectors is given by:

$$\mathbf{n} = \begin{matrix} 0.2 & 0 & 0.5 \\ 0 & 0.5 & 0.1 \\ 0.6 & 0.2 & -0.2 \end{matrix}$$

Analyse the decorrelator to estimate its output $\hat{\mathbf{b}}$.

7.7 Consider the synchronous CDMA communications channel shared by the three users described in 7.6. Assume all users are received with equal power, compute the BER for each user.

7.8 Consider the CDMA communications system described in 7.6 but with the decorrelator receiver is replaced with MMSE receiver so we can compare their performance. The users energy matrix is given by

$$\mathbf{A} = \begin{matrix} 1 & 0 & 0 \\ 0 & 1 & 0 \\ 0 & 0 & 1 \end{matrix}$$

Compute the estimated symbols for the three users at the output of the MMSE detector.

7.9 Consider a complex flat Rayleigh fading SISO channel as a media for transmission system with AWGN of zero mean and unity variance. The sampled complex gain of the channel h(n) at discrete time n = 1,2,3,...,8 are:

$$\begin{matrix} 1.0424 + j0.0911 & 0.0395 + j0.4647 & -0.8618 - j0.8268 \\ -0.0292 - j0.3261 & -0.7988 - j0.1858 & -0.9552 - j0.8588 \\ -0.1848 - j0.9341 & 0.6750 + j0.6593 \end{matrix}$$

Assuming $\frac{\sigma_s^2}{\sigma_n^2} = 2$, calculate the capacity of the channel in b/s/H.

7.10   Consider a complex flat Rayleigh fading MIMO channel as a media for transmission system with $n_t = n_r = 2$. The AWGN has zero mean and unity variance. The sampled complex gain of the channel h(n) at discrete time n are:

$$H1 = \begin{matrix} 0.0080 + j0.8057 & -0.4567 - j0.4843 \\ -0.4567 - j0.4843 & -0.5995 - j0.4216 \end{matrix}$$

$$H2 = \begin{matrix} 0.5704 - j0.9146 & 0.1640 - j0.0516 \\ 0.1640 - j0.0516 & 0.4694 - j0.3078 \end{matrix}$$

$$H3 = \begin{matrix} -0.7007 - j0.2340 & 0.9484 - j0.5972 \\ 0.9484 - j0.5972 & -0.8505 + j1.0868 \end{matrix}$$

$$H4 = \begin{matrix} 0.2050 + j0.3524 & 1.0470 + j1.0538 \\ 1.0470 + j1.0538 & -0.0462 - j0.9539 \end{matrix}$$

Assuming $\frac{\sigma_s^2}{\sigma_n^2} = 2$, calculate the capacity of the channel in b/s/H.

References

Alamouti, S.M. (1998) A simple transmit diversity technique for wireless communications, *IEEE Journal on Select Areas in Communications,* 16(8), 1451–1458.

Alexander, P.D., Reed, M.C., Asenstorfer, J.A. and Schlegel, C.B. (1999) Iterative multiuser interference reduction: Turbo CDMA, *IEEE Transactions on Communications,* 47(7), 1008–1014.

Alexiou, A. and Haardt, M. (2004) Smart Antenna technologies for future wireless systems: Trends and challenges, *IEEE Communications Magazine,* pp. 90–97.

Andrews, J.G. (2005) Interference cancellation for cellular systems: A contemporary overview, *IEEE Wireless Communications,* 12(2), 19–29.

Bartlett, M.S. (1956) *An Introduction to the Stochastic Process,* Cambridge University Press.

Buehrer, R.M., Kaul, A., Striglis, S. and Woerner, B.D. (1996) Analysis of DS-CDMA parallel interference cancellation with phase and timing errors, *IEEE Journal on Select Areas in Communications,* 14(8), 1522–1535.

Foschini, G.J. (1996) Layered space-time architecture for wireless communication in fading environment when using multi-element antennas, *Bell Labs Technical Journal,* 1(2), 41–59.

Foschini, G.J., Golden, G.D., Valenzuela, R.A. and Wolniansky, P.W. (1999) Simplified processing for high spectral efficiency wireless communication employing multi-element arrays, *IEEE Journal on Select Areas in Communications,* 17(11), 1841–1852.

Foschini, G.J., Chizhik, D., Gans, M.J., Papadias, C. and Valenzuela, R.A. (2003) Analysis and performance of some basic space-time architectures, Invited Paper, *IEEE Journal on Select Areas in Communications,* 21(3), 303–320.

Godara, L.C. (1997) Application of antenna arrays to mobile communications. Part II: Beam-forming and direction of arrival considerations, *Proceedings of the IEEE,* 85(8), 1195–1245.

Golden, G.D., Foschini, C.J., Velenzuale, R.A. and Wolniansky, P.W. (1999) Detection algorithm and initial laboratory results using V-BLAST space-time communication architecture, *IEE Electronics Letters,* 35(1), 14–16.

Haykin, S. and Moher, M. (2005) *Modern Wireless Communications,* International Edition, Pearson Prentice Hall.

Iltis, R.A. and Mailaender, L. (1996) Multiuser detection of quasi synchronous CDMA signals using linear decorrelators, *IEEE Transactions on Communications,* 44(11), 1561–1571, Nov. 1996.

Jamalipour, A., Wada, T. and Yamazato, T. (2005) A Tutorial on multiple access techniques for beyond 3G mobile networks, *IEEE Communication Magazine*, pp. 110–117.

Koulakiotis, D. and Aghvami, A.H. (1998) Evaluation of a DS/CDMA multiuser receiver employing a hybrid form of interference cancellation in Rayleigh-fading channels, *IEEE Communications Letters*, 2(3), 61–63.

Lehne, P.H. and Pettersen, M. (1999) An overview of smart antenna technology for mobile communications systems, *IEEE Communications Surveys*, Fourth Quarter, 2(4), 875–883.

Luo, W. and Ephremides, A. (2001) Indecomposable error sequence in multiuser detection, *IEEE Transactions on Information Theory*, 47(1), 284–294.

Ma, W.-K., Wong, K.M. and Ching, P.C. (2001) On computing Verdu's upper bound for a class of Maximum-Likelihood multiuser detection and sequence detection problems, *IEEE Transactions on Information Theory*, 47(7), 3049–3053.

Moher, M. (1998) An Iterative multiuser decoder for near-capacity communications, *IEEE Transactions on Communications*, 46(7), 870–880.

Patel, P. and Holtzman, J. (1994) Analysis of a simple successive interference cancellation scheme in a DS/CDMA system, *IEEE Journal on Select Areas in Communications*, 12(5), 796–807.

Pedersen, K.I., Mogensen, P.E. and Ramiro-Moreno, J. (2003) Application and performance of downlink beamforming techniques in UMTS, *IEEE Communications Magazine*, 41(10), 134–143.

Proakis, J.G. (1995) *Digital Communications*, McGraw-Hill, 3rd edn.

Razavilar, J., Farrokhi, F.R. and Liu, K.J.R. (1999) Software radio architecture with smart antennas: A tutorial on algorithms and complexity, *IEEE Journal on Select Areas in Communications*, 17(4), 662–675.

Reed, M.C., Schlegel, C.B., Alexander, P.D. and Asenstorfer, J.A. (1998) Iterative multiuser detection for CDMA with FEC: Near-single-user Performance, *IEEE Transactions on Communications*, 46(12), 1693–1699.

Schmidt, R.O. (1986) Multiple emitter location and signal parameter estimation, *IEEE Transactions on Antennas and Propagation*, AP–34, 276–280.

Sellathurai, M. and Haykin, S. (2002) Turbo-BLAST for wireless communications: Theory and experiments, *IEEE Transactions on Signal Processing*, 50(10), 2538–2546.

Sfar, S., Murch, R.D. and Latief, K.B. (2003) Layered space-time multi user detection over wireless uplinks system, *IEEE Transactions on Wireless Communications*, 2(4), 653–668.

Thompson, J.S., Grant, P.M. and Mulgrew, B. (1996) Smart antenna arrays for CDMA systems, *IEEE Personal Communications*, 3(5), 16–25.

Verdu, S. (1998) *Multiuser Detection*, Cambridge University Press.

Wang, X. and Poor, V. (1998) Iterative (Turbo) soft interference cancellation and decoding for coded CDMA, *IEEE Transactions on Communications*, 47(7), 1046–1061.

Winters, J.H. (1998) Smart antennas for wireless systems, *IEEE Personal Communications*, 1, 23–27.

Appendix 7.A

$$
\begin{aligned}
\text{cov}\{\mathbf{b} - \mathbf{Wy}\} &= E[(\mathbf{b} - \mathbf{Wy})(\mathbf{b} - \mathbf{Wy})^{\mathrm{T}}] \\
&= E[\mathbf{bb}^{\mathrm{T}}] - E[\mathbf{by}^{\mathrm{T}}]\mathbf{W} - \mathbf{W}E[\mathbf{yb}^{\mathrm{T}}] + \mathbf{W}E[\mathbf{yy}^{\mathrm{T}}]\mathbf{W}^{\mathrm{T}} \quad (\text{A.1})
\end{aligned}
$$

$$
E[\mathbf{bb}^{\mathrm{T}}] = I \quad (\text{A.2})
$$

$$
E[\mathbf{by}^{\mathrm{T}}] = E[\mathbf{b}[\mathbf{RAb} + \mathbf{n}]^{\mathrm{T}}] = E[\mathbf{bb}^{\mathrm{T}}\mathbf{A}^{\mathrm{T}}\mathbf{R}^{\mathrm{T}}] + E[\mathbf{bn}^{\mathrm{T}}]
$$

The noise vector $\mathbf{n}$ has zero mean so that

$$E[\mathbf{bn^T}] = 0$$

Also $\mathbf{R}$ is symmetric so $\mathbf{R} = \mathbf{R^T}$, and $\mathbf{A}$ is a diagonal matrix so $\mathbf{A} = \mathbf{A^T}$. Therefore,

$$E[\mathbf{by^T}] = \mathbf{AR} \tag{A.3}$$

$$E[\mathbf{yb^T}] = E[[\mathbf{RAb} + \mathbf{n}]\mathbf{b^T}] = \mathbf{RA} \tag{A.4}$$

$$E[\mathbf{yy^T}] = E[[\mathbf{RAb} + \mathbf{n}][\mathbf{RAb} + \mathbf{n}]^T] = E[\mathbf{RAbb^T A^T R^T}] + E[\mathbf{nn^T}] + [\mathbf{RAbn^T}]$$
$$+ E[\mathbf{nb^T A^T R^T}]$$

Now the last two terms $E[\mathbf{Ran^T}] + E[\mathbf{nb^T A^T R^T}]$ are multiplication by noise which has an average of zero so these two terms are zero.

The noise covariance matrix $E[\mathbf{nn^T}]$ equal to $\sigma^2\sigma^2\mathbf{R}$ (see Section 6.4.2.2).

$$E[\mathbf{yy^T}] = E[\mathbf{RAbb^T A^T R^T}] + \sigma^2\mathbf{R} = \mathbf{RA^2R} + \sigma^2\mathbf{R} \tag{A.5}$$

Substituting [A.2, A.3, A.4, A.5] in [A.1], we get

$$cov\{\mathbf{b} - \mathbf{Wy}\} = \mathbf{I} - \mathbf{ARW^T} - \mathbf{WRA} + \mathbf{W}[\mathbf{RA^2R} + \sigma^2\mathbf{R}]\mathbf{W^T}$$

Let

$$\overline{\mathbf{W}} = \mathbf{A^{-1}}[\mathbf{R} + \sigma^2\mathbf{A^{-2}}]^{-1} \tag{A.6}$$

$$\mathbf{Y} = \mathbf{I} - \mathbf{ARW^T} - \mathbf{WRA} + \mathbf{W}[\mathbf{RA^2R} + \sigma^2\mathbf{R}]\mathbf{W^T} \tag{A.7}$$

Where $\mathbf{R}$ is symmetric and $\mathbf{A}$ is diagonal, we want to show that

$$\mathbf{Y} = [\mathbf{I} + \sigma^{-2}\mathbf{ARA}]^{-1} + (\mathbf{W} - \overline{\mathbf{W}})(\mathbf{RA^2R} + \sigma^2\mathbf{R})(\mathbf{W} - \overline{\mathbf{W}})^T \tag{A.8}$$

Let

$$\mathbf{X} = \mathbf{RA^2R} + \sigma^2\mathbf{R} \tag{A.9}$$

Then from A.6

$$\begin{aligned}
\overline{\mathbf{W}}^{-1} &= [\mathbf{R} + \sigma^2\mathbf{A^{-2}}]\mathbf{A} \text{ from } [\mathbf{PQ}]^{-1} = \mathbf{Q^{-1}P^{-1}} \\
&= \mathbf{RA} + \sigma^2\mathbf{A^{-1}} \\
&= [\mathbf{RA^2} + \sigma^2\mathbf{I}]\mathbf{A^{-1}} \\
&= [\mathbf{RA^2R} + \sigma^2\mathbf{R}]\mathbf{R^{-1}A^{-1}} \\
&= \mathbf{XR^{-1}A^{-1}} \tag{A.10}
\end{aligned}$$

Therefore,

$$\overline{\mathbf{W}} = \mathbf{A}\mathbf{R}\mathbf{X}^{-1} \tag{A.11}$$

Using $[\mathbf{PQR}]^{\mathrm{T}} = \mathbf{R}^{\mathrm{T}}\mathbf{Q}^{\mathrm{T}}\mathbf{P}^{\mathrm{T}}$ and that $\mathbf{R}, \mathbf{A}$, and hence $\mathbf{X}$ are all symmetric, then

$$\overline{\mathbf{W}}^{\mathrm{T}} = \mathbf{X}^{-1}\mathbf{R}\mathbf{A} \tag{A.12}$$

Starting with A.8:

$$
\begin{aligned}
\mathbf{Y} &= [\mathbf{I} + \sigma^{-2}\mathbf{ARA}]^{-1} + (\mathbf{W} - \overline{\mathbf{W}})\mathbf{X}(\mathbf{W} - \overline{\mathbf{W}})^{\mathrm{T}} \\
&= [\mathbf{I} + \sigma^{-2}\mathbf{ARA}]^{-1} + \mathbf{WXW}^{\mathrm{T}} - \overline{\mathbf{W}}\mathbf{XW}^{\mathrm{T}} - \mathbf{WX}\overline{\mathbf{W}}^{\mathrm{T}} + \overline{\mathbf{W}}\mathbf{X}\overline{\mathbf{W}}^{\mathrm{T}} \\
&= [\mathbf{I} + \sigma^{-2}\mathbf{ARA}]^{-1} + \mathbf{WXW}^{\mathrm{T}} - \mathbf{ARX}^{-1}\mathbf{XW}^{\mathrm{T}} - \mathbf{WXX}^{-1}\mathbf{RA} + \mathbf{ARX}^{-1}\mathbf{XX}^{-1}\mathbf{RA} \\
&= [\mathbf{I} + \sigma^{-2}\mathbf{ARA}]^{-1} + \mathbf{WXW}^{\mathrm{T}} - \mathbf{ARW}^{\mathrm{T}} - \mathbf{WRA} + \mathbf{ARX}^{-1}\mathbf{RA} \tag{A.13}
\end{aligned}
$$

To get (A.7) we just need to show that the first and last terms give the identity matrix. These two terms are:

$$
\begin{aligned}
[\mathbf{I} + \sigma^{-2}\mathbf{ARA}]^{-1} + \mathbf{ARX}^{-1}\mathbf{RA} &= [\mathbf{I} + \sigma^{-2}\mathbf{ARA}]^{-1} + \mathbf{AR}[\mathbf{RA}^2\mathbf{R} + \sigma^2\mathbf{R}]^{-1}\mathbf{RA} \\
&= [\mathbf{I} + \sigma^{-2}\mathbf{ARA}]^{-1} + [(\mathbf{RA}^2\mathbf{R} + \sigma^{-2}\mathbf{R})\mathbf{R}^{-1}\mathbf{A}^{-1}]^{-1}\mathbf{RA} \text{ from } \mathbf{PQ}^{-1} = [\mathbf{QP}^{-1}]^{-1} \\
&= [\mathbf{I} + \sigma^{-2}\mathbf{ARA}]^{-1} + [\mathbf{RA} + \sigma^{-2}\mathbf{A}^{-1}]^{-1}\mathbf{RA} \\
&= [\mathbf{I} + \sigma^{-2}\mathbf{ARA}]^{-1} + [\mathbf{A}^{-1}(\mathbf{ARA} + \sigma^{-2}\mathbf{I}]^{-1}\mathbf{RA} \\
&= [\mathbf{I} + \sigma^{-2}\mathbf{ARA}]^{-1} + [\mathbf{ARA} + \sigma^{-2}\mathbf{I}]^{-1}\mathbf{ARA} \text{ from } [\mathbf{P}^{-1}\mathbf{Q}]^{-1} = \mathbf{Q}^{-1}\mathbf{P} \\
&= [\mathbf{I} + \sigma^{-2}\mathbf{ARA}]^{-1} + [\sigma^{-2}\mathbf{ARA} + \mathbf{I}]^{-1}\sigma^{-2}\mathbf{ARA} \\
&= [\mathbf{I} + \sigma^{-2}\mathbf{ARA}]^{-1}[\mathbf{I} + [\sigma^{-2}\mathbf{ARA}] \\
&= \mathbf{I}
\end{aligned}
$$

8

CDMA Wireless Communication Standards

8.1 Introduction

The development of the cell phone systems has witnessed a tremendous improvement in technology since the first cell phones were introduced at the beginning of the 1980s. The first generation (1G) of mobile phone systems were based on analogue technology and provided voice communications only. The first commercial analogue system to be launched was the Nordic Mobile Telephone (NMT) and shortly after, the Advanced Mobile Phone Service (AMPS) was commercially launched in the US. The Total Access Communication System (TACS) developed by Motorola was introduced in the UK and many other countries. These were the main mobile systems of the 1G that were developed around the globe but various variants were developed in other individual countries to suit their needs.

The 1G mobile phone systems were very similar in concept; the voice information was carried on a frequency modulated carrier and a control channel enables the voice signal to be routed to an available channel. The channel spacing used in these systems was different: NMT used a channel spacing of 12.5 kHz, AMPS a 30 kHz spacing, and TACS used a 25 kHz spacing.

The penetration rate of mobile phone subscription was increasing rapidly around the world. However, since a limited spectrum was available for the mobile phone provision, the analogue technology could not meet the demand because of its inherent shortcomings. There was a great deal of research into the application of digital technology into mobile phone systems. The research has led to the development of the second generation (2G) wireless mobile phone system that has been deployed in the early 1990s.

In a meeting in 1982, the European Conference of Postal and Telecommunications Administrations (CEPT), as a coordinating body for European state owned telecommunications organizations, decided to work towards a digital system to replace the multi-standards systems of 1G operation within the Europe at that time. To this end, CEPT established the European Telecommunications Standards Institute (ETSI) in 1988 to lay down the specifications and recommendations for the new pan-European mobile system. This was known as the Global System for Mobile (cellular) communications (GSM) devised by the Special Mobile Group of ETSI. The GSM system operates in Europe within the 900 MHz band with channels spaced 200 kHz. It employs Time Division Multiple Access (TDMA) to carry data from up to eight users on each of the available time slots. Other bands in the 1800 and 1900 MHz bands are also used. The basic service offered by GSM system is essentially voice with low data rate communications.

Mobile system development in North America took a slightly different direction towards the evolution of mobile communications. Unlike the Europeans who opted for a new 2G mobile system, the North Americans adopted the Interim Standard 54 (IS-54) and later updated to Interim Standard 36 (IS-36), for Analogue AMPS and Digital AMPS (D-AMPS) systems to work side by side with the 2G mobile phone systems. The D-AMPS employ TDMA technology with 30 kHz channel spacing to be compatible with existing AMPS. Again, as in GSM, D-AMPS provide voice services plus low rate data communications.

An important development took place in the mid 1990s in North America which allowed a major leap in mobile technology. A US firm called Qualcomm pioneered a totally new concept in mobile technology based on direct sequence spread-spectrum techniques, previously used for military communications. The multiple access used in this technology is called Code Division Multiple Access (CDMA). The new digital system was defined under Interim Standard 95 (IS-95A). Each channel has a bandwidth of 1.25 MHz and many users can communicate using the same bandwidth. The current mobile technologies, the wideband CDMA and the CDMA2000 standards are all natural migrants from earlier IS-95 system technologies.

This chapter starts with an introduction to the developments that led into the various generations of wireless mobile communications in Section 8.1. The development of the CDMA systems based on IS-95 is presented in Section 8.2. The channels used in a forward link of the cdmaone are presented in Section 8.3. Section 8.4 deals with the channels operating in the reverse link. The mobility issues in cdmaone systems are discussed in Section 8.5. The evolution of IS-95A standards to IS-95B is presented in Section 8.6. The cdma2000 standard is introduced in Section 8.7. The wideband CDMA systems are introduced in Section 8.8. The physical channels spreading and frame structures used in wideband CDMA are presented in Section 8.9. The physical channels used in the forward link in the wideband systems are discussed in Section 8.10. Section 8.11 presents the rate matching in wideband CDMA systems. Various issues such as carrier modulation, service multiplexing, power control, and mobility issues in the wideband CDMNA systems are given in Sections 8.12–8.18, while Section 8.19 presents the *high speed downlink packet access* (HSDPA) and Section 8.20 deals with the *High Speed Uplink Packet Access* (HSUPA). After a summary in Section 8.21, the Appendices include a list of Standards mentioned throughout the chapter.

8.2 IS-95A standard

IS-95A is the interim standard used in North America and defines a digital cellular system based on CDMA direct sequence spread spectrum which interoperates with the analogue cellular system AMPS. The IS-95A describes an air interface and a set of protocols used between mobile terminals and the network, and can be illustrated by a three-layer stack. Layer 1 is related to the physical layer of the OSI model, layer 2 corresponds to the two sublayers of the data link layer: the Media Access Control (MAC) and the Link Access Control (LAC). In the physical layer, the IS-95A describes the transmission from the network to the mobile terminal on the forward link (down link) and the transmission from the mobile terminal to the network on the reverse link (uplink). There are two sets of channels which can be used over the forward link; broadcast channels transmitted information for the mobile units and include the pilot channel, the sync channel and the paging channels. The other forward channels are the Traffic channels carrying data to the mobile terminals. Layer 3 corresponds to the network layer.

The mobile systems built according to IS-95A are called *cdmaone*, which include mobile station and base station systems operating on a dual-mode (analogue and digital) cellular system. The analogue technique refers to the AMPS FM system and the digital technique refers to the cdmaone system. Consequently, IS-95A is made up of two parts; one deals with the interface to the AMPS system, and the other deals with the cdmaone system. Our attention in this chapter focuses on the cdmaone system.

When a cellular phone user makes a telephone call, the cellular phone communicates with the serving base station (base station with the strongest signals). This base station is connected to a base station controller which forwards the call to the mobile switching centre. Calls destined to subscribers of the serving base station are relayed by the mobile switching centre to a destination base station. However, calls destined to the fixed telephone networks are relayed by the mobile switching centre to the PSTN. A flow chart for wired and wireless communications is shown in Figure 8.1.

8.3 IS-95A Forward link channels (IS-95)

The forward link comprises of 64 logic channels represented by 64 Walsh codes $(W_0, W_1, \ldots, W_{63})$. Each code is made of 64 chips which are transmitted at the chip rate of 1228800 chip/s. These codes are used to modulate (spread) the digital information signals that are transmitted on the forward channel. Since Walsh codes are perfectly orthogonal to each other, they provide channelization scheme on the forward channel so that all these signals (broadcast and traffic) are transmitted over the same bandwidth of the forward channel, hopefully with little or no mutual interference between them when used in a synchronous IS-95 CDMA system. The overall structure of the forward channel is shown in Figure 8.3.

The forward link consists of the following logic channels: one pilot channel using Walsh code W_0 for spreading all 0's signals; one synch channel using Walsh code W_{32} for spreading the synch data transmitted at data rate 1200 bps; up to seven paging channels

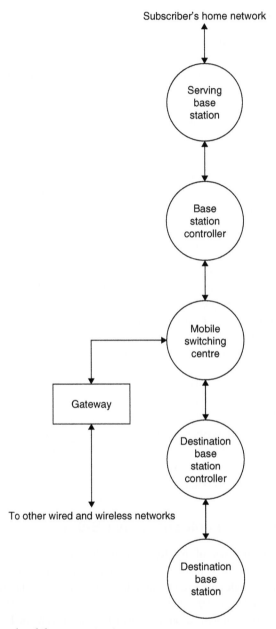

Figure 8.1   *Fixed and mobile communications.*

using Walsh codes W_1 to W_7 for spreading the 9600 bps and 4800 bps paging data; and between 55 up to 63 Traffic channels depending on how many paging channels are in use at that particular instant. There are two sets of rates for the traffic data: rate set 1 provides rates of 9600, 4800, 2400, and 1200 bps; and rate set 2 supports rates 14,400, 7200, 3600, and 1800 bps. The 64-ary Walsh code functions are shown in Appendix 8.A.

The base station, the mobile switching centre, the base station controller and the fixed line PSTN networks are all connected and are also connected to other wireless networks by fibre optical cable or microwave links. The base station controller is responsible for forwarding messages that are exchanged between the serving base station and the mobile switching centre. Each base station controller is connected to a number of base stations. In practise, populated areas are served by both the fixed PSTN networks and cellular phone networks. In general, a number of wire subscribers are connected to a wire line local switch that is connected to a local access network node. These nodes are connected together and to the serving base station as shown in Figure 8.2.

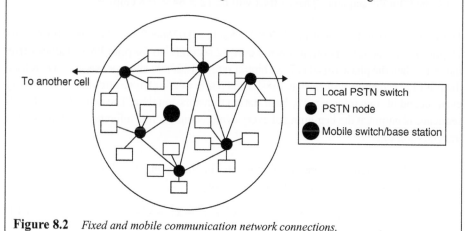

Figure 8.2 *Fixed and mobile communication network connections.*

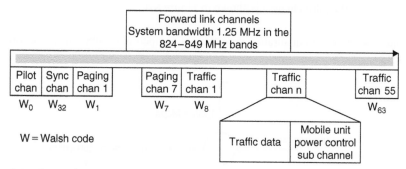

Figure 8.3 *Forward IS-95 CDMA link channels.*

8.3.1 Pilot channel

The pilot signal is transmitted over the pilot logic channel at all time by the base station on each forward link and carries all 0's signal modulo -2 added to W_0 (an all zero Walsh code) to identify the pilot logic channel. The output of this process is spread by a pair of short pilot PN sequences (each of length 32,768 chips) for quadrature spreading, a process which is carried out on all logic channels. The same pilot PN sequences are used in all base stations but each base station (cell) in the network is assigned with a signature generated by phase offset in steps of 64 chips of the pair of the pilot PN sequences. Consequently, the pilot PN sequences are used to isolate different base stations that are transmitting on the same frequency spectrum.

The pilot PN sequences used in forward and reverse channels are generated by binary m sequences of length $2^{15} - 1$ chips ($\cong 2^{15} = 32{,}768$ chips). The sequence length is also called the period of the pilot PN sequences. Since each phase shift $= 64$ chips, there are $32{,}768/64 = 512$ offset indices, from 0 up to 511, which can be used to identify base station (cells) in any CDMA wireless network. For example, if the pilot PN sequence offset index is 12, the pilot PN sequence phase offset will be $12 \times 64 = 768$ chips.

The pilot PN period is equal to $32{,}768/1228800 = 0.0266667$ sec. The time for exactly 75 pilot PN sequence repetitions occurs every 2 seconds, so for pilot PN sequence offset index 12, then the phase offset is 768 chips at rate of 1228800 chips per second. Therefore, the pilot PN sequence will start at $768/1.2288\,\mu\text{sec} = 625.00\,\mu\text{sec}$ after the start of every even second of time referenced to base station transmission time. The zero offset pilot PN sequence is output at the beginning of every even second in time referenced to base station transmission time.

The pilot PN sequence generator polynomials for the in-phase and quadrature PN pilot sequences are given by (8.1) and (8.2), respectively:

$$P_I(x) = x^{15} + x^{13} + x^9 + x^8 + x^7 + x^5 + 1 \tag{8.1}$$
$$P_Q(x) = x^{15} + x^{12} + x^{11} + x^{10} + x^6 + x^5 + x^4 + x^3 + 1 \tag{8.2}$$

The structure of the pilot channel is shown in Figure 8.4.

The binary output of the quadrature spreading is mapped into QPSK phase according to Table 8.1.

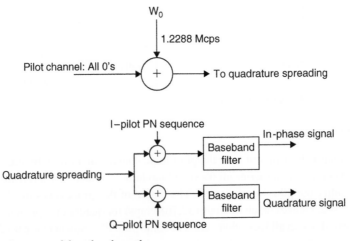

Figure 8.4 *Structure of the pilot channel.*

The spread I and Q data are applied to the input of I and Q Finite Impulse Response (FIR) baseband filters to shape the pulse waveforms to constrain the bandwidth while minimizing the intersymbol interference (ISI). These filters are used at both transmitter and receiver sides. The frequency response of each of the filters is contained within $\pm\delta_1$ in the passband $0 \leq f \leq 590$ kHz and is less than or equal to δ_2 in the stopband $f \geq 740$ kHz as shown in Figure 8.5.

The impulse response of the baseband filter h(k) for k = 0 to 47 is given in Table 8.2 and plotted in Figure 8.6.

The pilot channel operates without power control and provides a reference signal for all mobile stations for acquisition and timing and phase reference for coherent demodulation. The pilot signal is about 4–6 dB stronger than all other channels and once the mobile station has acquired the pilot channel, it can lock onto other channels and distinguish different base stations for handoff (see Section 8.5.2).

Table 8.1 *Quadrature spreading data mapped to QPSK carrier phase*

In-phase data	Quadrature data	QPSK carrier phase
0	0	$\pi/4$
1	0	$3\pi/4$
1	1	$-3\pi/4$
0	1	$-\pi/4$

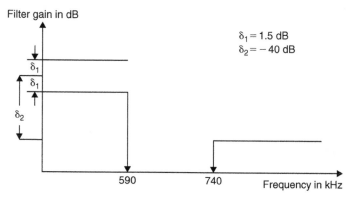

Figure 8.5 *Frequency response of the baseband filters of cdmaone systems.*

8.3.2 Sync channel

Every base station is synchronized with a GPS receiver, and transmission on the forward channels is tightly controlled in time. The sync channel is used by the mobile terminal to acquire initial time synchronization. Each base station continually transmits a single sync

Table 8.2   *Coefficients h(k) of the impulse response of the baseband filters*

k		h(k)
0,	47	−0.025288315
1,	46	−0.034167931
2,	45	−0.035752323
3,	44	−0.016733702
4,	43	0.021602514
5,	42	0.064938487
6,	41	0.091002137
7,	40	0.081894974
8,	39	0.037071157
9,	38	−0.021998074
10,	37	−0.060716277
11,	36	−0.051178658
12,	35	0.007874526
13,	34	0.084368728
14,	33	0.126869306
15,	32	0.094528345
16,	31	−0.012839661
17,	30	−0.143477028
18,	29	−0.211829088
19,	28	−0.140513128
20,	27	0.094601918
21,	26	0.441387140
22,	25	0.785875640
23,	24	1.0

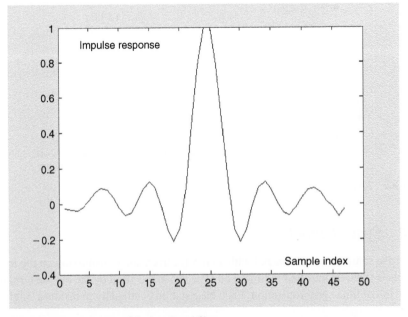

Figure 8.6   *Impulse response of the baseband filters.*

channel which contains information about the network including the system ID, network ID, paging data rates (9600 or 4800 bps) and pilot PN sequence offset used by the base station. Each base station transmits a sync channel data spread with Walsh code W_{32} and is divided into a number of super frames, each of 80 ms (96 bits) duration.

Each super frame is made up of three frames of 32 bits (26.67 ms) each. Each frame comprises of the frame body (31 bits) and a single bit at the start of the message (SoM). The start of the sync channel message is preceded with SoM bit '1' and all other SoM bits are set to '0' as set by the base station.

The base station transmits the sync channel message in consecutive frames with sufficient padding zeros to extend the message to the beginning of the next super frame. The sync channel message consists of 'message length' field of 8 bits (i.e. one octet). The maximum sync channel is equal 256 octets, i.e. the maximum field length = 255 octets = 2040 bits. The message body could be between 2 up to 2002 bits followed by *Cyclic Redundancy Code* (CRC) field of 30 bits. The sync channel message length in bits is an integer multiple of number of octets.

The sync channel message consists of 13 fixed length message formats given in Table 8.3. The sync channel structure is shown in Figure 8.7.

Suppose the number of Super Frames = 23, then number of zeros padded at the end of the sync channel message = $23 \times 93 - 2040 = 99$ zero bits.

Table 8.3 *Sync channel message formats*

Field	Length in bits
Message type	8
Common air interface revision level	8
Minimum common air interface revision level	8 (minimum revision level it supports)
Cellular System identification	15 (identification number max. 32,767)
Network identification	16 (max. identification number is 65,535)
Pilot PN sequence offset index	9 (offset in units of 64 PN chips)
Long code state	42 (long code state at the 'system time')
System time	36 (in units of 80 ms)
Number of leap seconds since the start of the system time	8
Offset of local time from system time	6 (current local time = system time + number of leap second + offset of local time. Set this field to two's complement offset of local time from system time.
Daylight savings time indicator	1 (if daylight saving is in effect set to '1' otherwise set to '0')
Paging channel data rate	3 (000 for 9600 bps) (001 for 4800 bps) (010 for 2400 bps)
Reserved bits	2 (00)

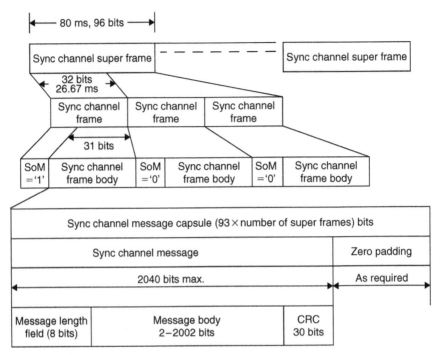

Figure 8.7 *Sync channel frame structure.*

The 30-bit CRC is computed from both the 'message length' field and the 'message body' field using the following generator polynomial:

$$g(x) = x^{30} + x^{29} + x^{21} + x^{20} + x^{15} + x^{13} + x^{12} + x^{11} + x^{8} + x^{7} + x^{6} + x^{2} + x + 1 \quad (8.3)$$

Computation of the CRC is carried out using multiple elements of shift register that are initialized to logical '1'. The generator used for calculating the CRC bits is shown in Figure 8.8. The register elements are all initialized to logical 1 and with the switches set in the up position, the generator is clocked k times where k is defined as the length (8 bits) plus message body, then the switches are set down and the generator is clocked an additional 30 times to output the CRC bits that are transmitted in the order in which they are computed.

The sync channel structure is shown in Figure 8.9. The sync channel message is 32 bits per frame transmitted within 80/3 ms (26.67 ms) yielding a sync channel data rate of 1200 bps. The sync channel data is convolutionally encoded by rate $\frac{1}{2}$ with constraint length (K) of 9 shown in Figure 8.10. The encoded symbols are repeated one time (each symbol occurs 2 consecutive times) so that the number of symbols per frame that is input to the block interleaver is 128.

Channel interleavers are important for communications over fading channels. Block interleavers are commonly used because of their simplicity where data is divided into blocks or

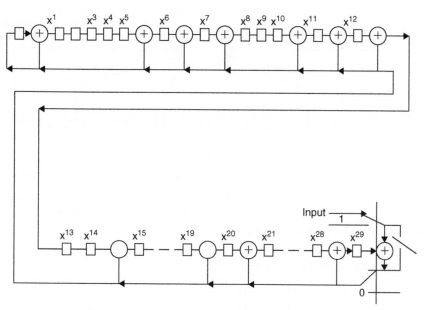

Figure 8.8 *Sync channel CRC generator.*

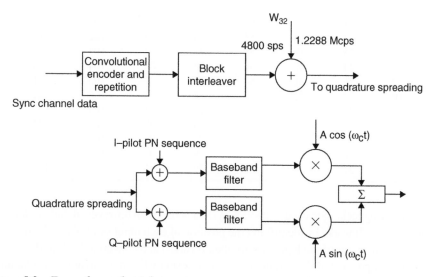

Figure 8.9 *Forward sync channel structure.*

frames. The main objective of the interleaver is to disperse bursts of error in time. The sync channel data consists of eight columns and 16 rows and is written into the block interleaver column by column from left to right. That is the first input symbol to the interleaver is indexed 0 followed by symbol 1 then symbol 2 followed by symbol 3 and so on as shown in Table 8.4.

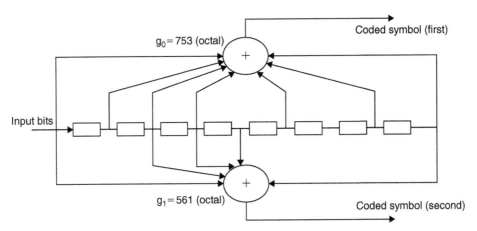

Figure 8.10   *Rate $\frac{1}{2}$ convolutional encoder $k = 9$.*

Table 8.4   *Sync channel interleaver input array*

0	16	32	48	64	80	96	112
1	17	33	49	65	81	97	113
2	18	34	50	66	82	98	114
3	19	35	51	67	83	99	115
4	20	36	52	68	84	100	116
5	21	37	53	69	85	101	117
6	22	38	54	70	86	102	118
7	23	39	55	71	87	103	119
8	24	40	56	72	88	104	120
9	25	41	57	73	89	105	121
10	26	42	58	74	90	106	122
11	27	43	59	75	91	107	123
12	28	44	60	76	92	108	124
13	29	45	61	77	93	109	125
14	30	46	62	78	94	110	126
15	31	47	63	79	95	111	127

The interleaving technique used in IS-95A system is based on bit-reversal that randomizes the burst errors. It works by identifying the position of each input symbol in a 7-bit binary number and the order of the logic bits are then reversed and converted back to decimal to define the new position of the output symbol. Let's explain the interleaving by an example: Consider the input symbols at position 0, 1, 9, 43, and 76. We will compute the new positions of these symbols at the output of the block interleaver as follows:

Decimal	0	1	9	43	76
Binary	0000000	0000001	0001001	0101011	1001100
Bit-reversed	0000000	1000000	1001000	1101010	0011001
Decimal	0	64	72	106	25

Table 8.5 *Forward sync channel interleaver output array.*

0	4	2	6	1	5	3	7
64	68	66	70	65	69	67	71
32	36	34	38	33	37	35	39
96	100	98	102	97	101	99	103
16	20	18	22	17	21	19	23
80	84	82	86	81	85	83	87
48	52	50	54	49	53	51	55
112	116	114	118	113	117	115	119
8	12	10	14	9	13	11	15
72	76	74	78	73	77	75	79
40	44	42	46	41	45	43	47
104	108	106	110	105	109	107	111
24	28	26	30	25	29	27	31
88	92	90	94	89	93	91	95
56	60	58	62	57	61	59	63
120	124	122	126	121	125	123	127

Table 8.6 *Sync channel parameters*

Data rate	1200	bps
PN chip rate	1.2288	Mcps
Code rate	1/2	Bits/code symbol
Code repetition	2	Mod symbol/code symbol
Modulation symbol rate	4800	Symbol per sec
PN chips/Mod symbol	256	PN chips/Mod symbol
PN chips/bit	1024	PN chips/bit

Applying this interleaving scheme to the input symbols in Table 8.4, we get the output symbols (Read column by column) are shown in Table 8.5. The sync channel parameters are given in Table 8.6.

The modulation symbol rate $= 1200 \times 2$ (at rate $\frac{1}{2}$ encoder output) $\times 2$ (repetitions) $= 4800$ symbols per sec

Number of PN chips per symbol $= 1228800/4800 = 256$

Number of chips per bit $= 1228800/1200 = 1024$.

8.3.3 Paging channel

The IS-95 base station transmits none or as many as seven paging channels at one time using Walsh codes W_1–W_8. The paging channel is used by the base station to page an idle mobile terminal in the case of an incoming call, or to carry the control messages for call set-up. When a mobile terminal is idle, it is mostly listening to a paging channel.

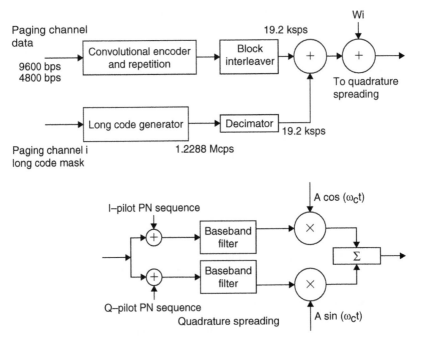

Figure 8.11 *Paging channel structure.*

The base station transmits paging information at a fixed data rate of 9600 or 4800; the same data rate is used on all paging channels in a given system. The paging channel frame is 20 ms in length and is time aligned to the IS-95 system 2 seconds roll-over using the GPS system.

The paging channel uses the same pilot PN sequences, chip rate and time offset as the pilot channel for a given base station. The paging channel data are convolutionally encoded using rate $\frac{1}{2}$ encoder shown in Figure 8.10. The encoded symbols are repeated prior to block interleaving if the used data rate is lower than 9600 bps. The encoded symbols are repeated such that each symbol occurs two consecutive times when symbol rate is 4800 bps. The structure of a paging channel is shown in Figure 8.11.

Signalling on the paging channel is carried out in cycles of 128 second. Each cycle is divided into 640 slots (from 0 to 639). Each slot has a duration of 200 ms and is composed of ten paging frames, each of 20 ms. Each frame is divided into 10 ms long paging channel half frames. The first bit in each half frame is the SoM (Start of Message) bit. A paging channel message capsule is composed of a paging message and padding. The paging message is made of the 'message length' field, a message body and a CRC field. The Paging channel signalling structure is shown in Figure 8.12.

The paging message capsule may be transmitted by base station either synchronized or unsynchronized. A synchronized message starts at the second bit of the paging half frame while the unsynchronized message starts immediately after the previous message. The

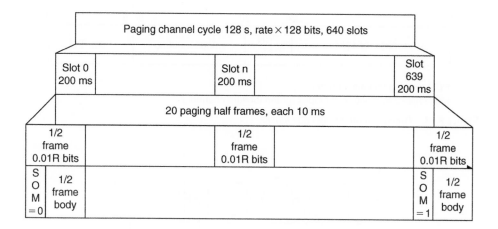

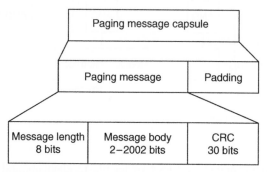

Figure 8.12　*Paging channel signalling structure.*

CRC bits are generated by shift register generator shown in Figure 8.8 and connected using polynomial given in (8.3).

The paging channel data at 9600 bps or 4800 bps are encoded and repeated (only for rate 4800 bps) to generate 19.2 k symbol per sec at the input of the block interleaver. As in the synch channel, the interleaver uses the bit-reversal technique to change the position of each interleaved output symbol in a method similar to the scheme described for the sync channel. The interleaver operates on each paging frame so that the number of symbols at the input of the block interleaver during each operation is:

$$19.2\,\text{ks/s} \times 20\,\text{ms} = 384\,\text{symbols}$$

Page channel interleaving at rate 9600 b/s and 4800 b/s can be found in Lee and Miller (1998). The paging data at the output of the block interleaver is scrambled using long PN code mask. The long PN code generator polynomial is:

$$P(x) = x^{42} + x^{35} + x^{33} + x^{31} + x^{27} + x^{26} + x^{25} + x^{22} + x^{21} + x^{19} + x^{18}$$
$$+ x^{17} + x^{16} + x^{10} + x^{7} + x^{6} + x^{5} + x^{3} + x^{2} + x + 1 \tag{8.4}$$

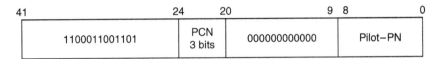

PCN—Paging channel number
Pilot—PN—Sequence offset index for the forward channel

Figure 8.13   *Paging channel long code mask.*

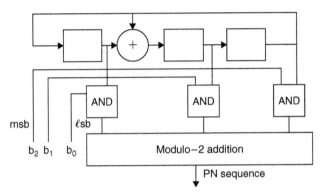

Figure 8.14   *Mask code generator.*

The long PN code is generated by LFSR of length 42 elements and has a period of $2^{42} = 4,398046511 \times 10^3$ chips which repeats itself every 41.43 days. The long PN code is operated on by a 42-bit mask to generate different phases of the long code that are used to page mobile terminals. The paging channel long code mask is shown in Figure 8.13.

Let us demonstrate by an example how the masking generates the phase shifted PN sequence and for clarity we will consider the following simple code generator polynomial:

$$P(x) = x^3 + x + 1 \tag{8.5}$$

which has a period of 7. Therefore, the PN sequence with zero shift will consist of 7 bits. A code mask of three bits from 000 to 111 should generate all eight possible phases of the PN sequence. The mask code generator is shown in Figure 8.14.

Let the initial state of the generator be 001. It can be shown that the generator without masking, and after seven clock cycles from its initial state, outputs the sequence 1001011. If there are any further clock cycles, the generator repeats this sequence.

Now we operate the generator with mask bits and initial state as shown below. It can be shown that the PN sequences are given in Table 8.7.

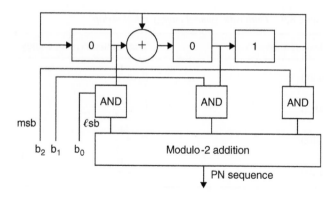

Table 8.7 *Masked PN sequence*

Mask	PN sequence
001	0101110 01...
100	1011100 10...
110	1100101 11...
111	1001011 10...
011	0010111 00...
101	111010 11...
010	0111001 01...

Table 8.8 *Paging channel modulation parameters*

Data rate	9600	4800	bps
PN chip rate	1.2288	1.2288	Mcps
Code rate	1/2	1/2	Bits per symbol
Code repetition	1	2	Mod sym/code sym
Mod sym rate	19200	19200	sps
PN chips/Mod sym	64	64	Chips/Mod sym
PN chips/bit	128	256	Chip/bit

It can be seen from Table 8.8, that the PN sequences are a phase shifted version of the original sequence. This masking method described in the example can be applied to the I-channel and Q-channel of the pilot PN code and the long code to get the selected offset using 15-bit mask and 42-bit mask, respectively. The paging channel parameters are shown in Table 8.8.

8.3.4 Traffic channel frame structure

The traffic channel is used for transmitting digital voice and data services on a frame by frame basis, each frame with a duration of 20 ms. The digital information is transmitted on two sets of rates; rate set 1 provides a maximum rate of 9600 b/s and rate set 2 supports a maximum rate of 14,400 b/s. Lower data rates are transmitted using lower transmit power.

Table 8.9 *Data rate–symbol energy relationship*

Data rate bps	Transmitted symbol energy
9600	$E_s = \dfrac{E_b}{2}$
4800	$E_s = \dfrac{E_b}{4}$
2400	$E_s = \dfrac{E_b}{8}$
1200	$E_s = \dfrac{E_b}{16}$

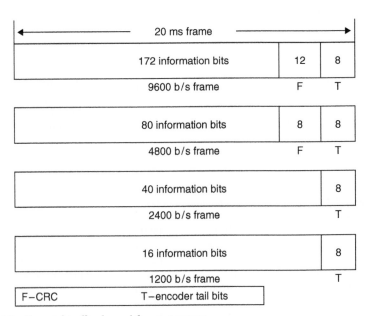

Figure 8.15 *Forward traffic channel frame structure.*

Let the energy per symbol and energy per bit be denoted as E_s and E_b, respectively. All symbols transmitted from the same frame are interleaved and transmitted at the same power. The transmitted symbol energy versus data rate is shown in Table 8.9.

There are four traffic channel frames to carry the four data rates as shown in Figure 8.15.

The traffic channel frames for data rate 9600 b/s and 4800 b/s include CRC which is used at the receiver to determine the transmission rate and whether the frames are in error. The 9600 b/s transmission rate frame includes 12-bit CRC generated by the following polynomial:

$$g(x) = x^{12} + x^{11} + x^{10} + x^9 + x^8 + x^4 + x + 1 \qquad (8.6)$$

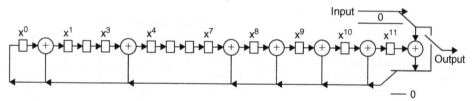

Figure 8.16 *Traffic channel CRC at 9600 b/s.*

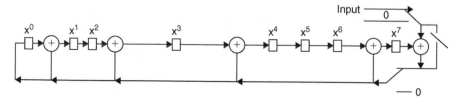

Figure 8.17 *Forward Traffic channel CRC at 4800 b/s.*

The CRC generator for 9600 b/s rate is shown in Figure 8.16.

It is clear that the CRC shown in Figure 8.16 is a linear cyclic code with $(n,k) = (172 + 12, 172) = (184, 172)$. The 4800 b/s transmission rate includes an 8-bit CRC generated using the following polynomial:

$$g(x) = x^8 + x^7 + x^4 + x + 1 \tag{8.7}$$

The CRC generator for 4800 b/s rate is shown in Figure 8.17.

The CRCs bits are computed on all bits (except the F and T bits) within the frame. The following procedure is used with the CRC generators to produce the CRC bits: with the switches in the up position, and all the generator elements set to logic one, the generator is clocked 172 times for the 9600 b/s generator and 80 times for the 4800 b/s generator. The switches are then set in the down position, and the generator is clocked an additional 12 times for the 9600 b/s generator and 8 times for the 4800 b/s generator to output the CRC bits which are transmitted in the order calculated.

8.3.5 Traffic channel signal processing

There are 55 traffic logical channels using Walsh codes W_8 to W_{63} and these can increase up to 63 channels minus the number of sync and paging channels operating on the same forward channel. The traffic channel structure is shown in Figure 8.18.

The traffic frame is encoded using a rate $\frac{1}{2}$ convolutional encoder described previously. Each encoded symbol is repeated twice, four times and eight times for data rates 4800 b/s, 2400 b/s and 1200 b/s, respectively. The Traffic channel parameters are given in Table 8.10.

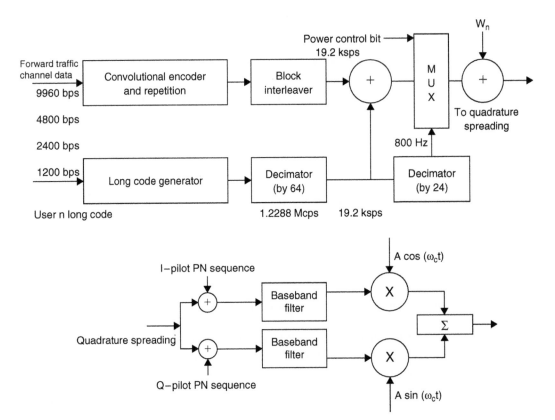

Figure 8.18   *Traffic channel structure.*

Table 8.10   *Traffic channel parameters*

Data rate	9600	4800	2400	1200	bps
PN chip rate	1.2288	1.2288	1.2288	1.2288	Mcps
Code rate	1/2	1/2	1/2	1/2	Bits/code symbol
Code repetition	1	2	4	8	Mod sym/code sym
Symbol rate	19200	19200	199200	19200	sps
PN chips/Mod sym	64	64	64	64	PN chip/Mod sym
PN chips/bit	128	256	512	1024	

The encoded and repeated symbols are input into the block interleaver column by column; each column consists of 24 symbols and each interleaving operates on 16 columns. The interleaving uses bit-reversal method to map the position of the symbols as explained earlier. The interleaver input–output of the traffic channels for data rate 2400 b/s and 1200 b/s, as well as 9600 b/s and 4800 b/s can be found in Lee and Miller (1998).

The traffic channel long code mask shown in Figure 8.19 of 42 bits made up of two types: public long code mask used for identifying the mobile terminal and subscriber private long code mask to identify the subscriber.

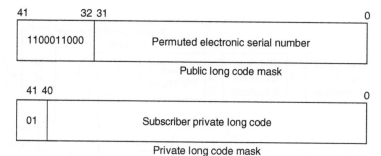

Figure 8.19 *Traffic channel long code mask.*

The traffic channel data at the output of the block interleaver is scrambled by a modulo-2 addition with the masked long code decimated by 64. A further decimation by 24 of the already decimated masked long code to multiplex with the power control bit and scrambled data.

We now discuss techniques used in the cellular wireless network for controlling the transmit power to limit the multiple access interference and increase the system capacity. In IS-95A systems, a combination of open-loop and closed-loop power control is used. In the open-loop, the mobile terminal measures the power received from the base station pilot signal and adjust its transmitter power accordingly. A decrease in the average received power is an indication that the mobile terminal needs to increase its transmitted power in order to keep the same signal quality assuming the reverse and forward channels have reciprocal characteristics. Consequently, the open-loop provides a quick power control to variations in the received signal. However, for fine control of the terminal power, the base station measures the terminal received power on the reverse link and feeds back the measurement to the mobile terminal in a closed-loop scheme. The base station transmits a power control command of a single bit at the rate of 800 Hz to continuously adjust (increasing/decreasing) the mobile terminal power in steps of 1 dB.

The base station transmits a *Null* traffic frame when no service available. The Null frame contains 16 ones followed by 8 zeros for encoder tail and is transmitted at 1200 b/s. The mobile terminal keeps its connection by maintaining synchronization with the base station.

So far, the forward traffic channel processing is presented for traffic data transmission at rate set 1. We now modify our presentation for the data transmission rate set 2 as shown in Figure 8.20.

Note that we have introduced a processing block at the input of the interleaver to adjust (puncture) the data rate to 19,200 sps when data is transmitted at data rate 2. The block threw away two symbols for every six symbols input and, therefore, the symbol rate to the interleaver is kept fixed at 19,200 sps as for rate set 1.

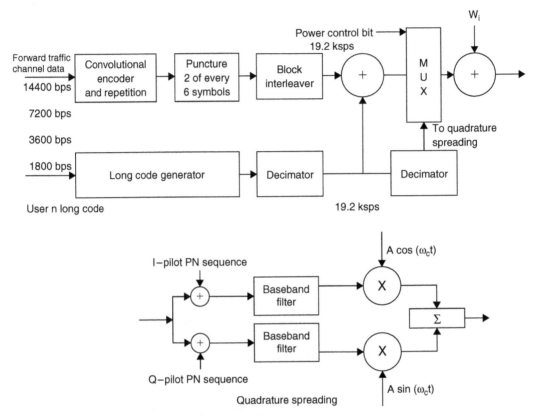

Figure 8.20 *Traffic channel for data rate set 2.*

8.3.6 Traffic channel signalling

Signalling bits are usually multiplexed with information bits or use the whole of the traffic frame transmitted by the base station on the forward traffic channels on the 9600 b/s frame. Considering the 9600 b/s traffic frame shown in Figure 8.15, the mixing of the 172 bits between information and signalling have the options shown in Figure 8.21.

For signalling bits only on the forward traffic frame above, option (e) frame in Figure 8.21 is used. The signalling traffic frame body of 168 bits is combined with other frames (if needed) for signalling message transmission. Let the number of traffic frames needed for signalling message transmission be N_r, then the traffic signalling message capsule will look like the format shown in Figure 8.22.

The other option containing a mixture of information and signalling has a signalling message capsule similar to that shown in Figure 8.22. There are 15 types of signalling messages on the forward Traffic channel which are outlined in the CDMA standard IS-95A. These signalling messages include authentication, handoff, and mobile terminal registration. Readers who need more information about these messages should consult the IS-95A document.

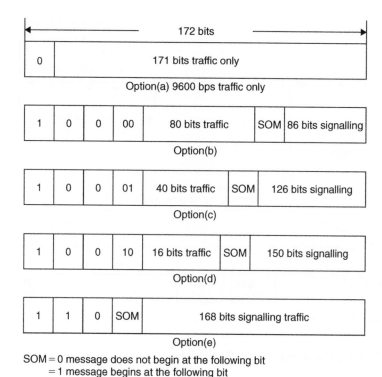

Figure 8.21 *Mixing the information bits with the signalling bits in the 9600 b/s traffic frame.*

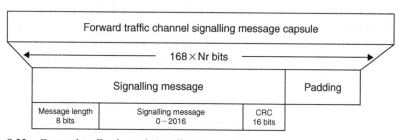

Figure 8.22 *Forward traffic channel signalling option (e).*

The 16-bit CRC is computed for each signalling message from the 'signalling message' and the 'message length' field and using the following generator polynomial:

$$g(x) = x^{16} + x^{12} + x^5 + 1 \tag{8.8}$$

The signalling CRC generator is shown in Figure 8.23.

To compute the CRC bits, the register elements are initialized with a logical 1. With the switches in the up position, the generator clocked k times where $k = 8 +$ signalling message length in bits. With the switches set to the down position, the generator clocked 16 times to output the CRC bits.

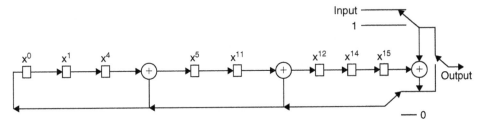

Figure 8.23   *Traffic channel signalling CRC generator.*

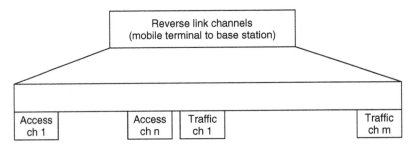

Figure 8.24   *Reverse link logical channels.*

8.4   IS-95A Reverse link channels

The reverse link is composed of access channels and reverse traffic channels sharing the same CDMA spectrum, which is 45 MHz lower than the frequency allocated to the forward link channels. The frequency band allocated to the forward link channels is 869–894 MHz and 824–849 MHz for the reverse link channels and carrier spacing 1.25 MHz. Each reverse traffic channel/access channel is identified by a distinct user long code. The logical channels over the reverse CDMA link are shown in Figure 8.24 which consists of n access channels and m traffic channels.

The reverse traffic channel is shown in Figure 8.25 for rate set 1 with peak data rate of 9600 b/s. Comparing the reverse traffic channel with the forward traffic channel shown in Figure 8.18, we can see many similarities between the two channels. But the differences are that the reverse traffic channel, unlike the forward, does not carry power control information, and the channelization is carried out by 64-ary orthogonal modulation by using a set of 64 Walsh functions.

The modulation by the 64-ary Walsh functions is achieved as follows: the interleaver output is partitioned into 6-symbol blocks used to select one of 64 orthogonal Walsh functions. If the 6-symbol block is converted into a decimal number k such that $0 \le k \le 63$ then the Walsh function with index k is selected for transmission. Since six interleaved symbols are modulated as one Walsh sequence of 64 chips, there are 4800 Walsh sequences per second, which imply a chip rate of $4800 \times 64 = 307.2$ kc/s. The Walsh index i is computed as follows:

$$i = c_0 + 2c_1 + 4c_2 + 8c_3 + 16c_4 + 32c_5 \tag{8.9}$$

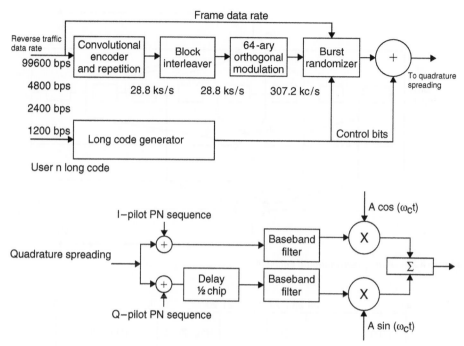

Figure 8.25 *Reverse traffic channel for data rate set 1.*

where i is a row in the 64×64 Hadamard matrix (Appendix A) and $\{c_k\}$ are symbols $\{0, 1\}$ at the interleaver output. For example, if the symbols of block i were [011 101], then we send the Walsh function with index 53. Furthermore, while the carrier modulation on the forward link channels are QPSK, the carrier modulation on the reverse link traffic channel is Offset QPSK (O-QPSK). The data is transmitted in 20 ms frames. The reverse link traffic channels for data rate set 2 is similar to the forward traffic channel in Figure 8.20.

The Access channel has a fixed data rate of 4800 b/s and is repeated once (each symbol occurs two consecutive times) and both repeated encode symbols are transmitted on the Access channel.

8.4.1 Traffic channel coding

The input data is convolutionally encoded using rate $\frac{1}{3}$ encoder constraint length (K) 9 shown in Figure 8.26. The generator polynomials used for the convolutional code rate $\frac{1}{3}$ are:

$$g_0 = 557 \,(\text{Octal})$$
$$g_1 = 663 \,(\text{Octal})$$
$$g_2 = 711 \,(\text{Octal})$$

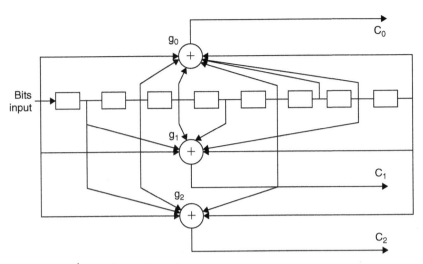

Figure 8.26 *Rate $\frac{1}{3}$ convolutional encoder.*

41	29 28	24 23	21 20	9 8	0
1100011001101	ACN	PNC	Reg_Zone	Pilot_PN	

ACN–Access channel number
PCN–Paging channel number
Reg_Zone–Registration zone for forward channel
Pilot_PN–PN offset of the pilot channel

Figure 8.27 *Access channel mask.*

8.4.2 Reverse link long code masking

The reverse traffic and access channels are addressed by masking the long code. When the mobile terminal is transmitting on the access channel, the mask is shown in Figure 8.27.

As in the traffic channel long code mask (see Figure 8.19), the mobile terminal can use one of two long code masks for transmitting on the reverse link Traffic channel: a public long code mask unique to the mobile station ESN or a private long code mask unique for each subscriber. These masks are shown in Figure 8.28.

8.4.3 Reverse link interleaving

The block interleaver is made up of an array of 32 rows and 18 columns containing 576 encoded symbols and spans 20 ms. The data is written into the interleaver by columns and read out by rows. The interleaver columns are written as in Appendix B for data rate 9600 b/s, 4800 b/s, 2400 b/s and 1200 b/s, respectively.

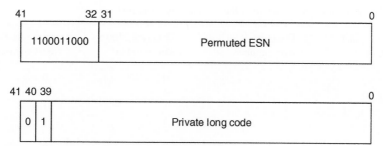

Figure 8.28 *Long code mask for reverse link traffic channel.*

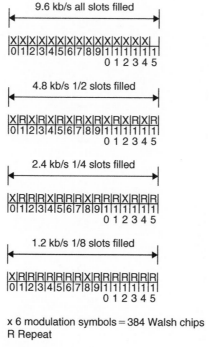

Figure 8.29 *Repetition in traffic frame.*

Unlike the forward traffic channel, the reverse traffic channel gates the amplifier to eliminate the repeated symbols. The 20-ms frame is divided into 16 equal time slots called power control groups. Each slot has a duration of $20\,\text{ms}/16 = 1.25\,\text{ms}$. The symbol rate at the interleaver output is 28.8 ks/s so the interleaver size $= 20\,\text{ms} \times 28.8\,\text{ks/s} = 576$ symbols and number of symbols per slot $= 576/16 = 36$ symbols, which represents six modulation symbols as shown in Figure 8.29.

At speed 9600 b/s, all time slots are filled with modulation symbols (six symbols each) and none are gated. At speed 4800 b/s, half of the slots would be gated, i.e. slot 0 or 1, slot 2 or 3, 4 or 5, and so on. Similarly, for speed 2400 b/s, a quarter of the slots filled and speed 1200 b/s $\frac{1}{8}$ are filled.

Table 8.11 *Interleaver rows output of the reverse traffic channel*

Data rate 9600 b/s Row number	Data rate 4800 b/s Row number	Data rate 2400 b/s Row number	Data rate 1200 b/s Row number
0	0	0	0
1	2	4	8
2	1	1	1
3	3	5	9
4	4	2	2
5	6	6	10
6	5	3	3
7	7	7	11
8	8	8	4
9	10	12	12
10	9	9	5
11	11	13	13
12	12	10	6
13	14	14	14
14	13	11	7
15	15	15	15
16	16	16	16
17	18	20	24
18	17	17	17
19	19	21	25
20	20	18	18
21	22	22	26
22	21	19	19
23	23	23	27
24	24	24	20
25	26	28	28
26	25	25	21
27	27	29	29
28	28	26	22
29	30	30	30
30	20	27	23
31	31	31	31

The order of the reverse traffic channel rows at output is shown in Table 8.11.

Clearly, the slots of data are grouped in twos following a group of their repeats. This regrouping of the repeat symbols is to enable deletion of the repeat symbols.

The access channel interleaver uses bit-reversed readout of the row of addresses as in the forward Traffic channel. For example:

Row 0 address is 00 00 0. The bit-reversed is 00 00 0 = 0 out Row = 0 + 1 = 1
Row 1 address is 00 00 1. The bit-reversed is 1 00 00 = 16 and out Row is 16 + 1 = 17

Row 2 address is 00 01 0. The bit-reversed is 0 10 00 = 8 and out Row is 8 + 1 = 9
Row 3 address is 00 01 1. The bit-reversed is 11 00 0 = 24 and output Row is 24 + 1 = 25
. . . . .
. . . . .
Row 31 address is 11 11 1. The bit-reversed is 11 111 = 31 and output Row is 31 + 1 = 32

Following the same method, the Access channel output rows are in the following order: 1, 17, 9, 25, 5, 21, 13, 29, 3, 19, 11, 27, 7, 23, 4, 20, 12, 28, 8, 24, 16, and 32.

8.4.4 Link power control

We have indicated in the previous section that the reverse Traffic channel 9600 b/s frame spans 20 ms comprising 192 information bits (i.e. 576 encoded symbols using $\frac{1}{3}$ rate encoder), and the encoded symbols are partitioned into 16 (0 to 15) power control groups. The duration of each power control group is 1.25 ms and contains 12 information bits, which is 36 encoded symbols or 6 Walsh words. The 9600 b/s reverse Traffic channel frame is shown in Figure 8.30 where time slot 0 contains 36 symbols consisting of 18 symbols from first row and 18 symbols from second row of the interleaver block. Similarly, slot 1 contains row 3 and row 4 and so on.

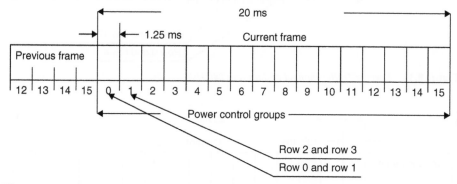

Figure 8.30 *Reverse Traffic channel at data rate 9600 b/s.*

The 4800 b/s reverse Traffic channel frame is shown in Figure 8.31 where slot 0 contains 18 symbols from row 0 and 18 symbols from row 2, slot 1 contains the repeated symbols in row 1 and row 3 which have been gated and so on for the rest of the slots in the frame.

The 2400 b/s Reverse Traffic channel frame is shown in Figure 8.32 showing slot 2 containing row 4 and row 8 and slot 5 containing row 8 and row 12 and so on.

The 1200 b/s Reverse Traffic channel frame is shown in Figure 8.33.

The data burst randomizer removes the repetitive encoded symbols by generating a binary masking pattern of the 14 bits taken from the long code used for spreading previous frame.

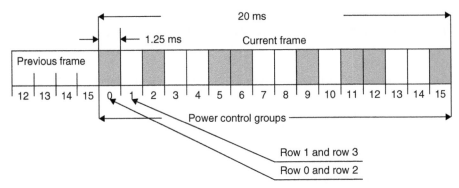

Figure 8.31 *Reverse Traffic channel at data rate 4800 b/s.*

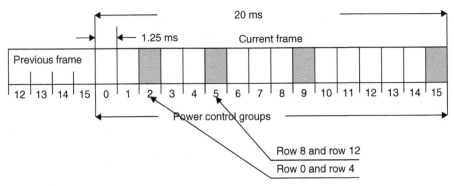

Figure 8.32 *Reverse Traffic channel at data rate 2400 b/s.*

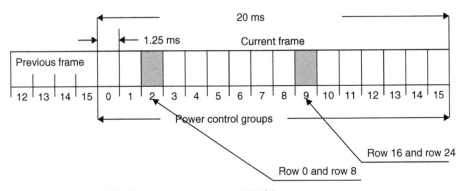

Figure 8.33 *Reverse Traffic channel at data rate 1200 b/s.*

8.4.5 Traffic channel modulation

The quadrature spread signal is modulated on an Offset–QPSK (OQPSK) carrier to be transmitted on the reverse traffic channel. The OQPSK modulation was presented in Section 2.4, Chapter 2. The OQPSK signal constellation is shown Figure 8.34.

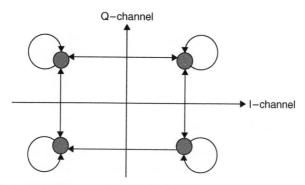

Figure 8.34 *Reverse Traffic channel carrier modulation.*

Table 8.12 *Reverse traffic data–carrier phase mapping*

I	Q	phase
0	0	$\pi/4$
1	0	$3\pi/4$
1	1	$-3\pi/4$
0	1	$-\pi/4$

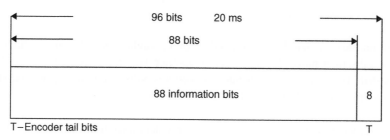

Figure 8.35 *Access channel frame structure.*

The I and Q binary data is mapped to the carrier phase for {I, Q} according to Table 8.12. For example, the symbol of (0, 0) mapped to carrier phase is $\frac{\pi}{4}$. Similarly, the other symbol combinations are mapped to the corresponding carrier phase.

8.4.6 Link frame structures

The access channel frame structure is shown in Figure 8.35. The reverse traffic channel frame structure is similar to that shown in Figure 8.19 for the forward traffic channel frame.

8.4.7 Traffic channel preamble

The base station is aided, to acquire the asynchronous reverse traffic channel, by the mobile terminal transmitting the channel preamble which consists of 192 zeros transmitted at the 9600 b/s rate. The preamble does not include the CRC bits.

8.4.8   Signalling on the reverse traffic channel

Signalling bits in the reverse traffic channel are mixed with the information bits in the 9600 b/s traffic frame in a manner similar to that used in the forward traffic channel shown in Figure 8.21.

8.5   IS-95A Mobility issues

Mobility in wireless communication systems is a unique characteristic which enables roaming subscribers to receive the same services from foreign systems while on the move, a characteristic not available to subscribers of the wired public switched telephone networks. Mobility is concerned with issues allowing the serving network to be always in contact with the subscribers and tracking the up-to-date locations of the subscribers to relay their incoming calls. Mobility is also concerned with subscriber billing for the services provided by the foreign systems.

Mobility includes two essential tasks: location registration and handoff procedures. Registration is the process by which the mobile terminal notifies the home network of its location in order to enable it to efficiently page the terminal for any incoming calls. Registration also allows the home network to determine which paging channel the mobile terminal is monitoring and the capabilities of the mobile terminal.

Handoff, on the other hand, is the procedure of automatic channel change which occurs when the subscriber moves from one cell to another. However, since cells in the IS-95A systems share the same channel, handoff procedures only require changing base stations as the subscriber moves from an old to new location area.

Roaming agreement contracts between service providers and billing mechanisms are essential in providing services such as the ability to make/receive calls, i.e. the inter-networking allowing the call to continue uninterrupted when the mobile subscriber crosses the boundary between two networks; provision for Short Message Service (SMS); provision for calling number/name presentation, international dialling, etc.

8.5.1   IS-95A Registration

The IS-95A supports a number of different forms of registration. Certain forms of registration are called autonomous registrations, since the mobile terminal initiates registration in response to certain events (i.e. without being directed by the network) like *power-up registration* (a mobile terminal registers when it powers on in order to access a new serving base station); *power-down registration* (mobile terminal registers when it powers off); *parameter-change registration* (a mobile terminal registers when one or more of its stored parameters has changed); *distance-based registration* (a mobile terminal registers when the distance between the current serving base station and the base station in which the terminal last registered exceeds a certain threshold); *timer-based registration* (a mobile terminal registers when a timer expires) and *ordered registration* (a mobile terminal registers when the base station requests it). *Implicit registration* carried out implicitly by the network when mobile terminal successfully uses the access channel. The traffic channel registration is carried out by the base station when the mobile terminal is assigned a traffic channel.

8.5.2 Handoff procedures

The IS-95A base station supports two types of handoff procedures: *soft handoff* in which a new base station commences communicating with the mobile terminal concurrently and without interrupting the ongoing communications from the old base station, thus providing diversity on both the forward and reverse traffic channels. The *hard handoff* takes place when the mobile terminal moves to a different network or a different frame offset.

Both soft handoff and hard handoff are typically initiated by the mobile terminal when it searches for pilots and measures their strengths. When the mobile terminal detects a pilot of sufficient strength is not associated with the forward channels assigned to it, it sends a *pilot strength measurement message* to the base station. The base station can then assign a forward traffic channel associated with that pilot to the mobile terminal and direct the mobile terminal to perform a handoff.

During the terminal pilot search, the base station sends a number of related messages: a *system parameters message* containing handoff parameters sent on the paging channel; an *in-traffic system parameters message* when the base station revises handoff parameters and the mobile terminal receives the forward traffic channel; and a *handoff direction message* when the base station modifies the handoff direction parameters. The base station may also send a *neighbour list message* on the paging channel and can update the neighbour list for a mobile terminal by sending a *neighbour list update message*.

The IS-95A base station also supports *CDMA to analogue handoff* for a dual-mode terminal but this mode of operation is not the main focus of this chapter. A typical handoff process is shown in Figure 8.36.

Considering Figure 8.36, the messages exchanged between the mobile terminal and the base station are: at time (1) the mobile terminal sends a *pilot strength measurement message* and transfer pilot to the candidate set when the pilot strength exceeds threshold Th1; at time (2) the base station sends a *handoff direction message*; and at time (3) the mobile terminal transfers the pilot to the *active set* and sends a *handoff completion message*. When the pilot drops below threshold Th2 the mobile terminal starts the *handoff drop timer* at time (4); at time (5) the handoff drop timer expires and the mobile sends a *pilot strength measurement message*; at time (6) the base station sends a *handoff direction message*; and finally, the mobile terminal moves a pilot from the active set to the neighbour set and the handoff procedure is repeated.

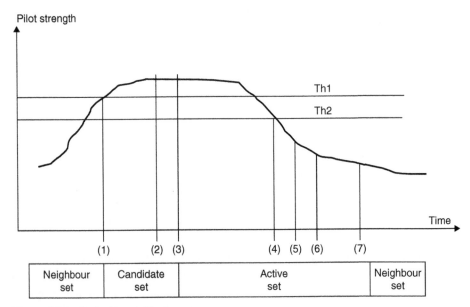

Figure 8.36   *Typical handoff process.*

8.6   Evolution of IS-95A standards to IS-95B (Kumar and Nanda, 1999)

The rapid growth of the Internet and the increasing demand for higher data rates in the wireless multimedia services has accelerated the development of IS-95A, coupled with the enthusiastic interest in protecting the substantial investment in the already deployed IS-95A networks have developed the work on IS-95A revision B (IS-95B). Consequently, the IS-95A enhancements have developed an air interface which is backward compatible to the current IS-95A air interface, i.e. no change to the physical layer of IS-95A.

Conventionally, IS-95A is categorized as the standard for the second generation networks (2G) paired with the GSM. On the other hand, the IS-95B standard offers a 2.5G technology. It was first deployed in September 1999 in Korea and followed by operators in Japan.

The IS-95B systems provide a higher data rate service through code aggregation (accumulation). An active high speed data mobile terminal always has a Fundamental Code cHannel (*FCH*) operating on 9.6 kb/s. When high data rates are needed, the base station assigned to the mobile terminal up to seven channels, known as Supplemental Code cHannels (*SCH*) on the forward or reverse links, making the maximum code channels available for the mobile terminal for the duration of a data burst being eight code channels. A burst is defined as a short duration over which a high data rate is transmitted. Each SCH supports 9.6 kb/s, making the maximum data rate equal $9.6\,\text{kb/s} \times 8 = 76.8\,\text{kb/s}$ at rate set 1 and $14.4\,\text{kb/s} \times 8 = 115.2\,\text{kb/s}$ at rate set 2. The SCHs are generated using a fundamental code mask where each supplemental code corresponds to a different phase shift of the fundamental code.

The FCH has soft handoff support and is used for signalling, power control, and user data. FCH is assigned for the call duration while the SCHs are assigned for the burst duration. Data rates are controlled and assigned by the base station. To help the base station in the data rate decision, the mobile provides the base station with information about the channel quality by sending pilot strength measurements on the reverse link. Power control for the SCHs are derived from the FCH, i.e. there is no power control loop for the SCHs. The IS-95B is capable of supporting asymmetric data service meaning different data rates on the forward and reverse links.

8.6.1 Burst-mode high rate data

The main drawback in supporting high rate data using CDMA systems is the bandwidth spreading given the limited wireless spectrum. IS-95B defined a new burst mode where the base station dynamically controlled the allocation of data burst to mobiles, given the instant load and access interference in the cell and adjacent cells, in a technique known as *Load and Interference based Demand Assignment* (LIDA) (I and Nanda, 1996). LIDA is a solution which allows several high rate data mobile terminals to share the available bandwidth without degrading the quality of voice services.

LIDA uses the FCH for CDMA system acquisition and tracking, thereby reducing the acquisition delay and the signalling overhead. Furthermore, different data rates and Quality of Services (QoSs) are achieved through variable spreading gain, code aggregation and management of interference. Clearly, a specific data rate assigned to the mobile terminal at a given instant is dynamically adjusted, since it is dependent on the path loss, the fading radio environment, and user mobility. Thus, it is advantageous to allocate a short duration (i.e. burst) for high rate data transmission on either forward or reverse links when the radio channel states are favourable.

On the call origination, a terminal is assigned an FCH and the parameters of the high data service are negotiated with the serving base station. If the terminal has no immediate data to transmit, it goes into a dormant state while still able to communicate at a low rate (i.e. on $\frac{1}{8}$ of the basic rate 9.6 kb/s or 14.4 kb/s = 1.2 or 1.8 kb/s) to maintain the synchronization and power control. For transmission of high rate data on the reverse link, the terminal sends a request to the base station indicating its data backlog and the maximum data rate requested. As mentioned above, to help the base station decide on the requested rates, the mobile terminal includes its measurement of the pilot strength in the neighbouring cells. The base station converts the pilot strength into interference levels seen in the terminal's cell and its neighbouring cells due to the high rate data transmission by the terminal.

For burst transmission on the forward link, the base station requests the mobile to report the pilot strength measurements prior to burst allocation in order to determine the interference seen by the terminal. Following receipt of the burst request by the Mobile Switching Centre (MSC), the network uses burst admission control algorithms to determine the start and duration of the burst and the number of codes (data rate) assigned by using a procedure known as *Burst Admission based Load and Interference* (BALI). (Ejzak et al., 1997)

8.6.2   MAC sub-layer protocol services

The link layer is subdivided into the Link Access Control (LAC) and Medium Access Control (MAC) sublayers. The LAC sublayer provides services such as signalling, voice, packet-switched data and circuit-switched data applications to the upper OSI layers (3–7). IS-95B defines an option called *high-speed packet data service* which the mobile terminal uses to request a high data speed packet mode service from the base station. In the request, the terminal specifies its high-speed data capability, i.e. number of SCHs on the forward and reverse links. The base station responds with the number of SCHs it provides to the mobile terminal. The MAC protocol service has two states: an *Active state* in which a Traffic channel is assigned to the terminal, and a point to point connection, is established between the mobile terminal and the network. User data is transmitted only when MAC is in the active state. The terminal usually remains in the active state for a short duration after completion of the burst transmission in order to minimize the access time for the next burst. An inactivity timer is defined and, if it is expired, the mobile terminal is transferred to the *Dormant state* in which no traffic channel is assigned to the terminal. But the terminal registration for high-speed data service is maintained together with the point to point protocol connection which remains intact. In the dormant state, very few resources are allocated to the terminal to enable it to maintain signalling at a low rate channel for synchronization and power control, as mentioned previously. The MAC transition states in IS-95B are shown in Figure 8.37.

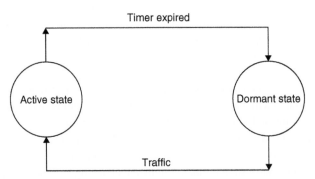

Figure 8.37   *MAC transition states in IS-95B.*

8.6.3   IS-95B system performance trade-offs

High-speed data users have diverse affects on the capacity and coverage on both forward and reverse links. In a cell with a mixture of voice and high-speed data users, the system capacity over a given cell is defined by the average throughput, i.e. average number of users supported by the cell multiplied by the data rate used. On the forward link, a large capacity cannot be achieved under total cell coverage, since high-speed data contributes high levels of access interference to the serving cell and its neighbouring cells, especially when such mobile terminals are located on or near the cell boundary. Consèquently, a higher capacity can only be achieved by relaxing the cell coverage requirements.

On the other hand, the system capacity over the reverse link is smaller when the system supports only high-speed data terminals than for one with only voice users, since the number of high-speed data users that can be supported is smaller than the corresponding voice users. The burst access schemes applied in IS-95B use the system current load, an estimate of the resource usage of the terminal, together with pilot strength measurements received from the terminal so that the network decides on the permitted data rate for terminal data burst. Such schemes provide a dynamic mechanism to trade off bandwidth between voice users and high-speed data users.

When high-speed data terminals are accessing the reverse link, the same coverage as for voice users can be achieved by increasing the transmit power linearly with data rates. For example, if M codes are assigned per user, the average transmit power of the user has to be increased by M times. Consequently, the peak to average ratio of the transmit power also increases by the same ratio and the amplifier power has to peak slightly more than M times to achieve the expected quality of service. However, if the same power amplifier is used for voice and high-speed data users, the coverage area for the high-speed data users must be reduced compared to that of voice users. Alternatively, the assigned data speed is reduced to a level suitable to the mobile terminal peak power.

The data burst length, and number of codes assigned to a packet data user, are functions of the user mobility. The IS-95B assigns more codes to the fixed or low mobility users than high mobility users. The allocation of resources to a mobile terminal is triggered by the hands off activity and so a running average of the number of hands off events can be used to optimize the allocation of resources.

8.7 cdma2000 standards (IS-2000; Kinsely et al., 1998)

8.7.1 Introduction

The International Telecommunication Union (ITU) produced its first document in 1999 to define future global requirements for the next generation for wireless telecommunications (known as the third generation (3G)) in a document called the International Mobile Telecommunication specifications (IMT2000). The release in 1999 was then developed, as we will explain in this chapter. Furthermore, the ITU allocated the spectrum 1885–2025 MHz for the reverse link channels, including the satellite component band 1980–2010 MHz in the new system. The band allocated for the forward link channels in the new system was 2110–2200 MHz, including the satellite component band 2170–2200 MHz.

The most important 3G technology compatible with IMT2000 standards was the wideband direct sequence spread spectrum CDMA, which was developed by an International Industrial Consortium and known as 3GPP (Third Generation Partner Project), details of the 3GPP specifications mentioned in this chapter are located at the end of the chapter). The 3G wideband CDMA will be presented in Section 8.8. The other important system developed by another International Industrial Consortium known as the 3GPP2 was the cdma2000. We will focus in this section on the cdma2000 as an evolution of IS-95 to 3G systems.

In the US, most of the spectrum allocated by the ITU for use by the IMT2000 was already allocated to the 2G systems at the time. The total 3G spectrum allocation was 230 MHz. The World Radio Communication Conference 2000 (WRC2000), considering the tremendous future growth in mobile communications, added the following new bands: 806–960 MHz (154 MHz), 1710–1885 MHz (175 MHz) and 2500–2690 MHz (190 MHz), a total of 519 MHz. These band allocations are very optimistic and represent the enormous expected growth in future cellular wireless communications.

3GPP2 produced the specifications for the 3G systems cdma2000. These specifications define the radio transmission technology designed to meet the requirements of the IMT2000 and are backward compatible with IS-95 systems. Both systems share a common 1.25 MHz channel as a migration path from IS-95 to the 3G wideband systems. Consequently, there is a graceful transition from the 2G networks IS-96A or the 2.5G networks IS-95B to the high-speed data 3G systems (cdma2000) enabling service providers to gradually build up their networks in a seamless way in selected areas to provide additional services where and when needed.

Additionally, some components of the IS-95 systems can be re-used in cdma2000, e.g. IS-95 hand sets can operate in a dual mode cdma2000 environment and the cdma2000 system can be overlaid in the same bandwidth with IS-95 system. The forward link in the cdma2000 has multiple CDMA carriers, each with a band of 1.25 MHz, while the reverse link has a single wideband carrier as shown in Figure 8.38.

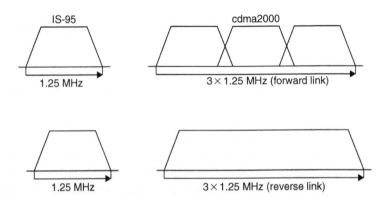

Figure 8.38   *Bandwidths of the IS-95 and cdma2000 systems.*

However, there seems to be little commercial interest in the multicarrier system, so far and service providers' interests are aimed at the development of the single-carrier (1x) system. The most prominent of these is the Qualcomm single-carrier system for high data rates known as the *single carrier Evolution Data Only (1xEV-DO)*, which is discussed in more detail in Section 8.7.15.

8.7.1.1 *The physical channels*

The physical channels used in cdma2000 can be classified into a number of categories. The first category is called *dedicated channels*. Examples of such channels would be: data transportation and control signalling on both forward and reverse links. The second category is called *common channels* and are used to broadcast the pilot, paging, synch, and control information on the forward link or used by a user to send access request on the reverse link. The physical common channels are shown in Figure 8.39 where channels on the forward link are prefixed with F and with R for the reverse link. There are five forward common channels: the *Pilot* (F-PICH), the *Common Auxiliary Channel* (F-CAPICH), the *Paging Channel* (F-PCH), the *Common Control Channel* (F-CCCH), and *the Sync Channel* (F-SYNC). There are also three reverse common channels: the *Access Channel* (R-ACH), the *Enhanced Access Channel* (R-EACH), and the *Common Control Channel* (R-CCCH).

The pilot channels F-PICH and F-CAPICH provide the cdma2000 system with the capability for soft handoff and coherent detection on the forward link. Paging the mobile terminals can be enabled using the F-PCH channel. Synchronization of the mobile terminals is enabled by applying information provided by the cdma2000 system using the F-SYNC channel. Signalling control information on the forward link is carried out via channel F-CCCH. The R-ACH is used by the mobile terminals to communicate with the base station for access permission while the R-CCCH channel is utilized by the terminal for signalling control information to the base station.

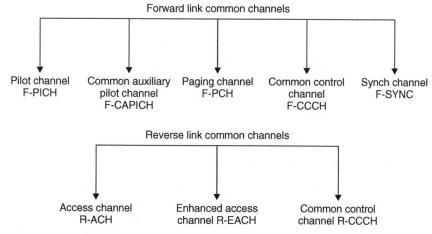

Figure 8.39 *cdma2000 common channels.*

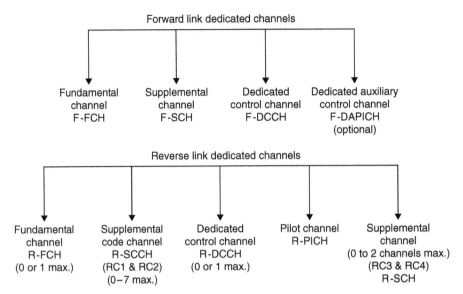

Figure 8.40 *cdma2000 dedicated channels.*

The physical dedicated channels are shown in Figure 8.40. The forward dedicated channels include either a *Fundamental Channel* (F-FCH) or a *Dedicated Control Channel* (F-DCCH). The F-FCH is dedicated for data and voice traffic and some signalling multiplexed with the traffic at any rate up to 14.4 kb/s. The F-FCH channel corresponds in functionality to the *Traffic Channel* (TCH) in IS-95. The F-SCH operates in conjunction with the F-FCH or the F-DCCH to provide higher data rate services. The packet data is transmitted in bursts, i.e. short durations of high rate traffic separated by large durations of no data traffic. A single-carrier cdma2000 carries voice and high rate data simultaneously with a negligible impact on the quality of both services. The F-DCCH is used for transmission of data, control information, and power control information from the base station to the mobile terminal.

Physical dedicated channels in the reverse link include the *Reverse Fundamental Channel* (R-FCH), the *Reverse Pilot Channel* (R-PICH), and the *Reverse Dedicated Control Channel* (R-DCCH) and up to two *Reverse Supplemental Code Channels* (R-SCCH). While the standard supports a maximum data rate over 1 Mb/s, the realistic peak rate achieved in practical system is not more than 307 kb/s.

8.7.2 The spreading chip rates

A range of spreading chip rates is used in cdma2000 and these rates are expressed as integer multiple of the IS-95 basic chip rate, i.e. $N \times 1.21288$ Mc/s where $N = 1, 3, 6, 9$ and 12. Each of these spreading rates is denoted by a particular value of N, i.e. Spreading Rate 1 (SR1) for $N = 1$, Spreading Rate 3 (SR3) for $N = 3$, etc. The cdma2000 system

Table 8.13 *Parameters of the reverse traffic channel*

Radio Configuration	Spreading rate	Data rate kb/s	Coding rate	Modulation
1	SR1	1.2, 2.4, 4.8, 9.6	1/3	64-ary orthogonal modulation
2	SR1	1.8, 3.6, 7.2, 14.4	1/2	64-ary orthogonal modulation
3	SR1	1.2, 1.35, 1.5, 2.4, 2.7, 4.8, 9.6, 19.2, 38.4, 76.8, 153.6	1/4	BPSK modulation with a pilot
		307.2	1/2	
4	SR1	1.8, 3.6, 7.2, 14.4, 28.8, 57.6, 115.2, 230.4	1/4	BPSK modulation with a pilot
5	SR3	1.2, 1.35, 1.5, 2.4, 2.7, 4.8, 9.6, 19.2, 38.4, 76.8, 153.6	1/4	BPSK modulation with a pilot
		307.2, 614.4	1/3	
6	SR3	1.8, 3.6, 7.2, 14.4, 28.8, 57.6, 115.2, 230.4, 460.8	1/4	BPSK modulation with a pilot
		1036.8	1/2	

employs the multicarrier technology for transmission on the forward link and each of the multi-stream data is spread on a different carrier with spread rate 1. The reverse link uses a single carrier and spreads the data with a chip rate of $N \times 1.2288$ Mc/s.

8.7.3 The reverse link radio configurations

There are six *Radio Configurations* (RCs) for the reverse traffic channels, four are associated with Spreading Rate 1 (SR1) and two associated with Spreading Rate 3 (SR3). Each mobile terminal supports configurations 1, 3 and 5 while configurations 2, 4 and 6 are optional. Mobile terminals usually support radio configurations in pairs, i.e. 1 & 2, 3 & 4, and 5 & 6. Table 8.13 shows data rates, channel coding rate and type of modulation for the traffic reverse channels.

8.7.4 The long code generator

Considering the cdma2000 spreading rate RS1, the long code access channel mask is similar to that used in IS-95 and is shown in Figure 8.27. The long code mask for the reverse fundamental and the reverse supplemental channels are similar to those shown in Figure 8.28 for IS-95, except that the code channel number index bits from 37 to 39 are '000' for a reverse fundamental channel, i.e. compatible with IS-95; and '001–111' for reverse supplemental code channels. The private long code masks of both IS-95 and cdma2000 RS1 are the same.

The in-phase long code for cdma2000 SR3 consists of three multiplexed sequences, each at the chip rate of 1.2288 Mc/s. The first sequence represents the in-phase long code at SR1. The second sequence is the mod-2 addition of the first sequence with a $\frac{1}{1.2288}\mu$s delayed version of the first sequence. The third sequence is the mod-2 addition of the first sequence and a $\frac{2}{1.2288}\mu$s delayed version of the first sequence. The chip rate of the in-phase long code for SR3 is $3 \times 1.2288\,\text{Mc/s} = 3.6864\,\text{Mc/s}$. The quadrature long code for SR3 is similar to the in-phase long code which is delayed by one chip duration. The in-phase long code generator is shown in Figure 8.41.

The long code mask for spreading rate SR3 depends on the channel type on which the mobile terminal is transmitting, as shown in Figure 8.42.

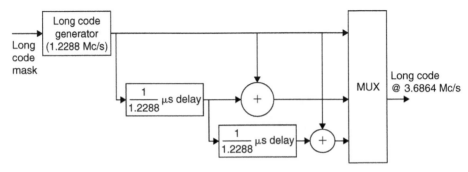

Figure 8.41 *In-phase long code for SR3.*

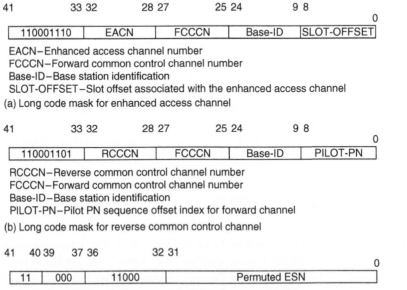

EACN–Enhanced access channel number
FCCCN–Forward common control channel number
Base-ID–Base station identification
SLOT-OFFSET–Slot offset associated with the enhanced access channel

(a) Long code mask for enhanced access channel

RCCCN–Reverse common control channel number
FCCCN–Forward common control channel number
Base-ID–Base station identification
PILOT-PN–Pilot PN sequence offset index for forward channel

(b) Long code mask for reverse common control channel

(c) Public long code mask for reverse fundamental channel, reverse supplemental channels, and reverse dedicated control channel

Figure 8.42 *Long code mask for RS3.*

Table 8.14 *Coefficients h(k) of the impulse response of the baseband filter for RS3*

k	h(k)	k	h(k)
0, 107	0.005907324	27, 80	0.036864993
1, 106	0.021114345	28, 79	0.032225981
2, 105	0.017930022	29, 78	0.007370446
3, 104	0.019703955	30, 77	−0.025081919
4, 103	0.011747086	31, 76	−0.046339352
5, 102	0.001239201	32, 75	−0.042011421
6, 101	−0.008925787	33, 74	−0.011379513
7, 100	−0.013339137	34, 73	0.030401507
8, 99	−0.009868192	35, 72	0.059332552
9, 98	−0.000190463	36, 71	0.055879297
10, 97	0.010347710	37, 70	0.017393708
11, 96	0.015531711	38, 69	−0.037885556
12, 95	0.011756251	39, 68	−0.078639005
13, 94	0.000409244	40, 67	−0.077310571
14, 93	−0.012439542	41, 66	−0.027229017
15, 92	−0.019169850	42, 65	0.049780118
16, 91	−0.015006530	43, 64	0.111330557
17, 90	−0.001245650	44, 63	0.115580285
18, 89	0.014862732	45, 62	0.046037444
19, 88	0.023810108	46, 61	−0.073329573
20, 87	0.019342903	47, 60	−0.182125302
21, 86	0.002612151	48, 59	−0.207349170
22, 85	−0.017662720	49, 58	−0.097600349
23, 84	−0.029588008	50, 57	0.148424686
24, 83	−0.024933958	51, 56	0.473501031
25, 82	−0.004575322	52, 55	0.779445702
26, 81	0.020992966	53, 54	0.964512513

We can see that the long code mask for the Traffic channel in IS-95 is similar to the long code mask for cdma2000 by comparing Figure 8.42 with Figure 8.28.

8.7.5 Baseband filtering

The RS3 baseband filter would cover a wide spectrum compared with RS1 and thus it has a different impulse response. The coefficients of the baseband filters used in cdma2000 with spreading rate RS1 are given in Table 8.2. The coefficients of the impulse response of the baseband filters used for spreading rate RS3 are given in Table 8.14.

8.7.6 Reverse link frames

The reverse link in cdma2000 introduces two important techniques which are not available to the IS-95, namely the reverse link coherent detection using the reverse link

R/E	Data bits	F	T

R/E–Reserved/erasure indicator bit
F–Frame quality indicator (CRC)
T–Encoder tail bits

Figure 8.43 *Reverse fundamental channel frame structure.*

R	Data bits	F	R/T

R–Reserved bit
F–Frame quality indicator (CRC)
R/T–Reserved/encoder tail bits

Figure 8.44 *Reverse supplemental channel frame structure.*

pilot channel and the fast forward link power control using the reverse link. The main backwards-compatible reverse link common channel in cdma2000 is identical to IS-95 is the R-ACH.

8.7.6.1 The fundamental channel

The reverse link contains one fundamental channel if the terminal has data transmission and signalling information to the base station during a call. The duration of the fundamental channel's frame is 20 ms and the data rates vary on a frame by frame basis but the modulation symbol rate is kept fixed by repetition. The fundamental channel frame structure is shown in Figure 8.43. The encoder tail bits are 8 bits and the quality indicator bits vary between 6 and 16 bits according to the radio configuration. The number of data bits per frame depends on the radio configuration as well and the transmission data rate. The reserved/erasure indicator (R/E) bit is either 0 or 1. The generator polynomials for the quality indicator (also called the Cyclic Redundancy Code [CRC]) are:

$$g(x) = x^{16} + x^{15} + x^{14} + x^{11} + x^6 + x^5 + x^2 + x + 1$$
$$g(x) = x^{12} + x^{11} + x^{10} + x^9 + x^8 + x^4 + x + 1$$
$$g(x) = x^{10} + x^9 + x^8 + x^7 + x^6 + x^4 + x^3 + 1$$
$$g(x) = x^8 + x^7 + x^4 + x^3 + x + 1$$
$$g(x) = x^6 + x^2 + x + 1 \qquad\qquad\qquad \text{RC2}$$
$$g(x) = x^6 + x^5 + x^2 + x + 1 \qquad\qquad \text{RC3} - 6$$

8.7.7 Supplemental code channel (RC3–RC6)

The reverse supplemental code channel is used in radio configuration 3 through 6 only in association with the reverse fundamental channel. The mobile terminal transmits data to the base station on up to two supplemental channels during a call. The data rate used depends on the radio configuration of the system. The reverse supplemental channel frames are 20, 40 or 80 ms in duration. The frame structure is shown in Figure 8.44. The reserved bit is either 0 or 1 and the number of the encoder tail bits is 8 bits. The number of data bits

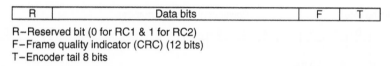

R—Reserved bit (0 for RC1 & 1 for RC2)
F—Frame quality indicator (CRC) (12 bits)
T—Encoder tail 8 bits

Figure 8.45 *Reverse supplemental channel frame structure.*

per frame depends on the type of radio configuration used in the system. The generator polynomials for the frame quality indicator are:

$$
\begin{aligned}
g(x) &= x^{16} + x^{15} + x^{14} + x^{11} + x^6 + x^5 + x^2 + x + 1 \\
g(x) &= x^{12} + x^{11} + x^{10} + x^9 + x^8 + x^4 + x + 1 \\
g(x) &= x^{10} + x^9 + x^8 + x^7 + x^6 + x^4 + x^3 + 1 \\
g(x) &= x^8 + x^7 + x^4 + x^3 + x + 1 \\
g(x) &= x^6 + x^2 + x + 1 \\
g(x) &= x^6 + x^5 + x^2 + x + 1
\end{aligned}
\tag{8.10}
$$

8.7.8   Supplemental code channel (RC1–RC2)

The reverse supplemental code channels (up to 7 maximum) are used with Radio Configuration 1 (RC1) at a transmission data rate of 9.6 kb/s and Radio Configuration 2 (RC2) at a transmission data rate 14.4 kb/s in association with the fundamental channel to provide higher data rate services (Figure 8.45). The number of bits per frame is 192 bits for RC1 of which 172 bits are data. The number of bits per frame is 288 bits for RC2 of which 267 bits are data. The generator polynomial for CRC is:

$$
g(x) = x^{12} + x^{11} + x^{10} + x^9 + x^8 + x^4 + x + 1
\tag{8.11}
$$

8.7.8.1   Pilot channel gating

The reverse pilot is an unmodulated signal that is spread with W_0^{32} and carries the reverse power control information used to control the power of the base station in the forward link with radio configuration 3 through 6. The reverse pilot also assists the base station in detecting the mobile transmission by providing a phase reference for coherent demodulation. Gating patterns for the power control groups on the reverse pilot channel with gating rates 1, $\frac{1}{2}$, and $\frac{1}{4}$ are shown in Figure 8.46. The reverse pilot gating frame is 20 ms duration, each group is 1.25 ms. There are 16 power control groups (0–15).

When gating rate 1 is used, every Power Control (PC) group is transmitted, while in gating rate $\frac{1}{2}$, only odd numbered power control groups are transmitted. When $\frac{1}{4}$ rate gating is used, only power control groups 3, 7, 11 and 15 are transmitted.

8.7.8.2   Dedicated control channel

The reverse dedicated control channel is used by the mobile terminal to transmit data and signalling information to the base station during a call. The reverse traffic channel contains

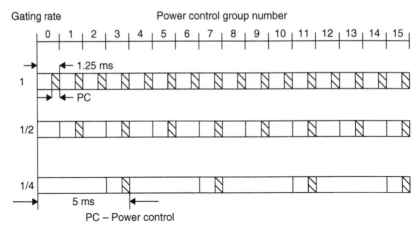

Figure 8.46 *Reverse pilot gating.*

R	Data bits	F	T

R–Reserved bit
F–Frame quality indicator (RC)
T–Encoder tail bits

Figure 8.47 *Reverse dedicated control channel frame structure.*

Data bits	F	T

F–Frame quality indicator (CRC)
T–Encoder tail bits

Figure 8.48 *Reverse common control channel frame structure.*

one reverse dedicated control channel. Data bits are transmitted on the reverse dedicated control channel on a fixed rate of 9.6 kb/s or 14.4 kb/s. The frame duration is 20 ms of which 5 ms is reserved for data at the rate of 9600 b/s. The number of encoder tail bits is 8 and one reserved bit (0 or 1) (Figure 8.47).

8.7.8.3 Common control channel

The common control channel is used for the transmission of data and signalling information from the terminal to the base station when reverse traffic channels are not in use. The channel transmission is coded, interleaved and modulated as spread spectrum signal. The mobile terminal transmits during intervals specified by the base station. Data is transmitted at 9.6, 19.2, and 38.4 kb/s in frames of duration 20, 10 or 5 ms in respect of the speed of transmission. The frame structure is shown in Figure 8.48. Transmission starts with preamble followed by reverse common control channel transmission.

The number of the encoder tail bits is 8 and the number of the frame quality indicator (CRC) is either 12 or 16 bits depending on the transmission speed. The generator polynomials for the frame quality indicator are:

$$g(x) = x^{16} + x^{15} + x^{14} + x^{11} + x^6 + x^5 + x^2 + x + 1$$
$$g(x) = x^{12} + x^{11} + x^{10} + x^9 + x^8 + x^4 + x + 1 \qquad (8.12)$$

8.7.8.4 Enhanced access channel

The enhanced access channel is used by the terminal to initiate communication with the base station or to reply to a directed message. The transmission starts with a channel preamble followed by either an access header or access data, depending on whether the enhanced access channel is in the basic mode or in the enhanced access mode. The frame structure for the enhanced access channel is shown in Figure 8.49.

The number of bits in the frame quality indicator is 8, 12 or 16 depending on the data transmission speed, and the number of bits for the encoder tail is 8. The generator polynomials are:

$$g(x) = x^{16} + x^{15} + x^{14} + x^{11} + x^6 + x^5 + x^2 + x + 1$$
$$g(x) = x^{12} + x^{11} + x^{10} + x^9 + x^8 + x^4 + x + 1$$
$$g(x) = x^8 + x^7 + x^4 + x^3 + x + 1 \qquad (8.13)$$

8.7.8.5 Access channel

The access channel is used by the mobile terminal to initiate communication with the base station or to respond to a paging message. The access channel transmission is a coded, interleaved and modulated spread-spectrum signal. An access channel consists of an access preamble followed by a series of access frames.

The access channel transmits at a fixed data rate of 4.8 kb/s (same data rate as in IS-95) in frames of 20 ms duration each. The preamble consists of frames of 96 zeros transmitted at 4.8 kb/s. The access channel frame structure is shown in Figure 8.50.

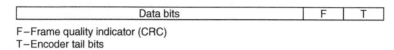

F−Frame quality indicator (CRC)
T−Encoder tail bits

Figure 8.49 *Enhanced access channel frame structure.*

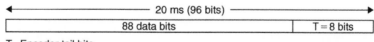

T−Encoder tail bits

Figure 8.50 *Access channel frame structure.*

8.7.9   Complex spreading and modulation system

A complex spreading, shown in Figure 8.51, is used in cdma2000 to reduce the peak to average signal power ratio and to improve the power efficiency. When fed into a power amplifier, a signal with a large peak to average power ratio impairs the amplifier output with distortion. Consequently, complex spreading is essential to the cdma2000 system spatially when the multicarrier technique is used.

The block diagram of the reverse link channels modulation is depicted in Figure 8.52. The logical channels A–E in Figure 8.52 are given in Table 8.15. These reverse channels are considered in Sections 8.7.7–8.7.8.

Walsh code W_n^N represents a Walsh function of length N constructed from the nth row of an N × N Hadamard matrix with zeroth row being Walsh function 0, the first row being Walsh function 1, etc. The Walsh functions used by the mobile in the reverse link channels are given in Table 8.15.

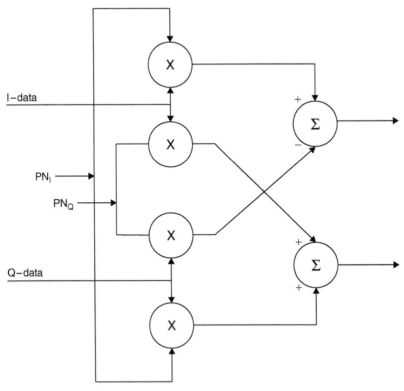

Figure 8.51   *Complex spreading.*

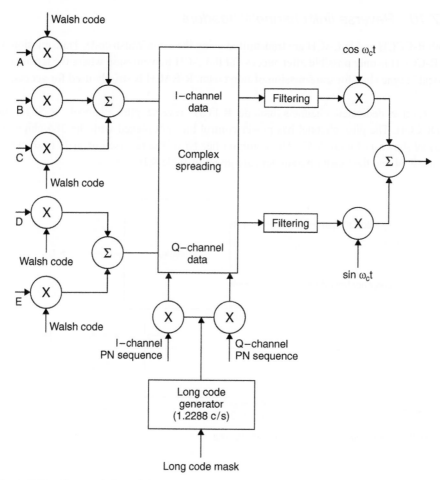

Figure 8.52 *Reverse link modulation systems.*

Table 8.15 *Spreading codes used for reverse link channels*

Reverse link channel	Walsh code
Pilot channel (B)	W_0^{32}
Enhanced access channel (E)	W_2^8
Common control channel (E)	W_2^8
Dedicated control channel (C)	W_8^{16}
Fundamental channel (D)	W_4^{16}
Supplemental channel 1 (E)	W_1^2 or W_2^4
Supplemental channel 2 (A)	W_2^4 or W_6^8

8.7.10 Reverse link channels' headers

Both R-CCCH and R-EACH are transmitted using the same Walsh code. Transmission on an R-CCCH is only possible after successful R-EACH transmission where the terminal is allocated time slots for transmission of burst data. R-EACH is strictly used for access.

The reverse dedicated channels include R-FCH, reverse pilot channel, R-SCH and R-DCCCH. The pilot channel has power control bit multiplexed with the 384 chips (all '0's) as shown in Figure 8.53. The common header of the reverse channels is shown in Figure 8.54 and the header parameters are given in Table 8.16.

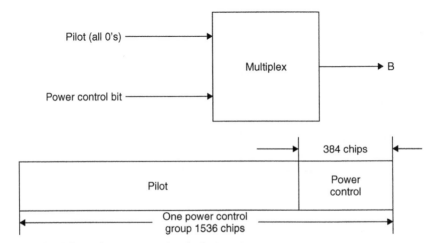

Figure 8.53 *Pilot and power control multiplexing.*

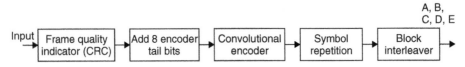

Figure 8.54 *Reverse link channel header.*

8.7.11 Error correction coding in the reverse channels

A $\frac{1}{4}$ rate convolutional coding is employed in cdma2000 in addition to the $\frac{1}{3}$ rate and $\frac{1}{2}$ rate convolutional codes used in IS-95 systems. The error correction codes used in the enhanced access, common control, dedicated control and fundamental channels for spreading rates SR1 and SR3 are rate $\frac{1}{4}$, $\frac{1}{2}$ or $\frac{1}{3}$ convolutional codes with constraint length 9. The reverse supplemental channel uses either convolutional code or turbo code for SR1 and SR3. Turbo coding is presented in Chapter 2 Section 2.8. Rates $\frac{1}{2}$ and $\frac{1}{3}$ convolutional codes are shown in Figures 8.10 and 8.26 and rate $\frac{1}{4}$ convolutional coding is shown in Figure 8.55.

Table 8.16 *Reverse link header parameters*

Channel	CRC	Convolutional/ Turbo encoder	Symbol repetition	Block interleaver
Enhanced access (E)	8	1/4 rate $K = 9$	$4\times$	768
Common control (E)	16 or 12	1/4 rate , $K = 9$	$1\times, 2\times, 4\times$	768, 1536, 3072,1536
Dedicated control RC3 (C)	16 or 12	1/4 rate, $K = 9$	$2\times$	384, 1536
Fundamental channel and Supplemental code RC1, RC2 (D)	8 or 12	1/3 rate, $K = 9$	1, 2, 4, or 8	576
Fundamental channel and Supplemental code RC3 (D)	6, 8, 12 or 16	(Convolutional or turbo)	$1\times, 2\times, 4\times,$ $8\times, 16\times$	384, 6144, 12288, 1536, or 3,072
Fundamental channel and Supplemental code RC4 (D)	6, 8, 10, 12, or 16	1/4 rate (Convolutional or turbo)	$1\times, 2\times, 4\times,$ $8\times,$ or $16\times$	384, 6144, 1536, 12288, or 3072

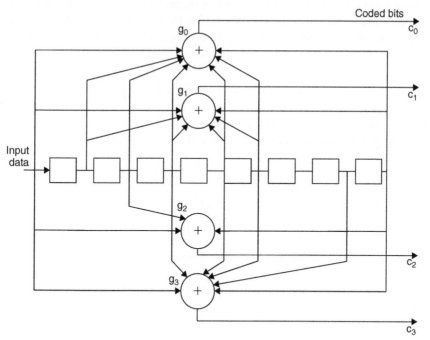

Figure 8.55 *Rate $\frac{1}{4}$ convolutional code constraint length 9.*

8.7.12 Reverse channels block Interleaving

The access channel, the enhanced access channel, the common control channel and the Traffic channel are all interleaved before modulation and transmission. The block inter-leavers for the reverse Traffic channel for RC1 and RC2 are similar to those used for

Table 8.17 *Interleaver parameters for reverse link channels*

Interleaver size	m	J
384	6	6
768	6	12
1536	6	24
3072	6	48
6144	7	48
12288	7	96
576	5	18
2304	6	36
4608	7	36
9216	7	72
18432	8	72
36864	8	144

the reverse Traffic channel in IS-95 and shown in Table 8.12. The access channel, the enhanced access channel, the common control channel and the Traffic channel for RC3–RC6 are interleaved such that the symbols input to the interleaver are written sequentially (one after the other) at addresses 0 to the block size (N-1). The interleaver symbols are read out with the ith symbol being read from address A_i where:

$$A_i = 2^m(i \bmod J) + BRO_m\left(\left\lfloor \frac{i}{J} \right\rfloor\right) \qquad (8.14)$$

where $i = 0$ to $N - 1$, $\lfloor x \rfloor$ equal the largest integer $\leq x$ and $BRO_m(y)$ is the bit-reversed m-bit value of y. For example consider $m = 6$ and $J = 6$, then for $i = 23$:

$$A_{23} = 2^6(23 \bmod 6) + BRO_6\left(\left\lfloor \frac{23}{6} \right\rfloor\right) = 64 \times 5 + BRO_6(3)$$

$$BRO_6(3) = BR(000\,011) = 110\,000 = 48$$

$$A_{23} = 64 \times 5 + BRO_6(3) = 320 + 48 = 368$$

Therefore, symbol 23 is read from address 368 input symbols.

The interleaver parameters are shown in Table 8.17 where the interleaver size is given by $2^m * J$.

8.7.13 Forward channels' headers

The cdma2000 forward channels have to provide a higher data rate to satisfy the requirement of the IMT2000 for multimedia services and at the same time the channels have to be backwards compatible to operate seamlessly with IS-95 systems. In particular, forward channels pilot, paging and synch channels of cdma2000 and IS-95 are identical but the dedicated channels for RC3–6 have different structures.

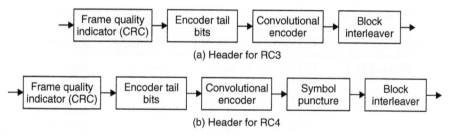

(a) Header for RC3

(b) Header for RC4

Figure 8.56 *Headers for F-DCCH for RC3 and RC4.*

Table 8.18 *Parameters for F-DCCH headers*

RC type	Bits /frame	CRC bits	Puncture (deletion)	Interleaver size
RC3	24/5 ms	16	None	192
	172/20 ms	12	None	768
RC4	24/5 ms	16	None	96
	172/20 ms	12	None	384
RC5	24/5 ms	16	None	192
	267/20 ms	12	4 or 12	768

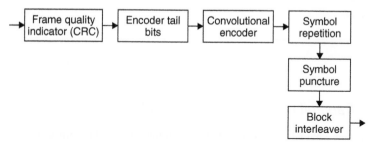

Figure 8.57 *Headers for F-FCH and F-SCH for RC3 and RC4.*

The headers of F-DCCH for RC3 and RC4 contain eight added encoder tail bits, the data rate is 9.6 kb/s for RC3 and RC4 and 14.4 kb/s for RC5. The forward error correction coding is $\frac{1}{4}$ rate convolutional encoder for RC3 and $\frac{1}{2}$ rate convolutional encoder for RC5. The headers are shown in Figure 8.56 and their parameters are given in Table 8.18.

The headers for F-FCH and F-SCH for RC3 and RC4 contain eight added encoder tail bits but the forward error correction coding is either convolutional or a turbo encoder with a rate $\frac{1}{4}$ for RC3 and rate $\frac{1}{2}$ for RC4. The headers are shown in Figure 8.57 and their parameters are shown in Table 8.19. The interleaver size Table 8.19(b) corresponds to header parameters in Table 8.19(a).

Table 8.19   *Parameter values for F-FCH and F-SCH headers for RC3 and RC4*

(a) RC type	Bits/frame	CRC bits	Repetition	Puncture
RC3 and RC4	24/5 ms	16	1×	None
	16/20 ms	6	8×	1 of 5
	40/20 ms	6	4×	1 of 9
	80/20 ms	8	2×	None
	172/20 ms	12	1×	None
	360/20 ms	16	1×	None
	744/20 ms	16	1×	None
	1512/20 ms	16	1×	None
	3048/20 ms	16	1×	None
RC4	6120/20 ms	16	1×	None

(b)	Interleaver size	
RC3		**RC4**
192		96
768		384
768		384
768		384
768		384
1536		768
3072		1536
6144		3072
12288		6144
–		12288

8.7.14   Transmit diversity over forward link

One of the new characteristics introduced in cdma2000 is the ability of the system to transmit over two antennas in the forward channels. This feature is known as *transmit diversity* which was not available in IS-95 systems. All forward channels except F-pilot channel and F-SYN can be transmitted using the transmit diversity. The various space diversity schemes are presented in Section 7.4.2, Chapter 7.

The transmit diversity technique in cdma2000 increases the forward link capacity and can be performed using either of two methods: *Orthogonal Transmit Diversity* (OTD) and *Space–Time Spreading* (STS). In multicarrier systems, different carriers can be mapped into different antennas.

A single stream of data (input x) can be split (demultiplexed) into four parallel streams. Alternatively, the single stream can be split into two data streams first (input x_I, input x_Q) and then into four parallel streams, i.e. each stream is split into two parallel streams as shown in Figure 8.58.

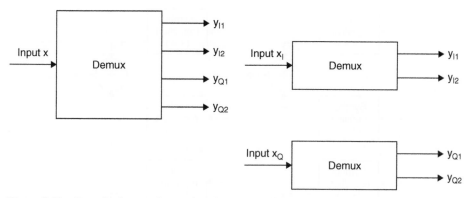

Figure 8.58 *Demultiplexing of input data for transmit diversity.*

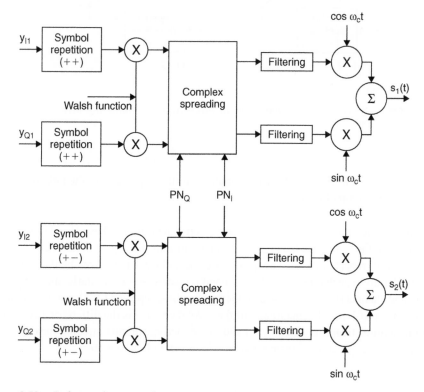

Figure 8.59 *Orthogonal transmit diversity.*

The orthogonal transmit diversity is shown in Figure 8.59. The input data is split into y_{I1}, y_{Q1}, y_{I2} and y_{Q2}. The signals y_{I1} and y_{Q1} are combined to form the in-phase and quadrature arms of the first channel. Similarly, the second channel is formed by combining y_{I2} and y_{Q2}. The data in the first channel is repeated $(++)$, i.e. the symbol is repeated without

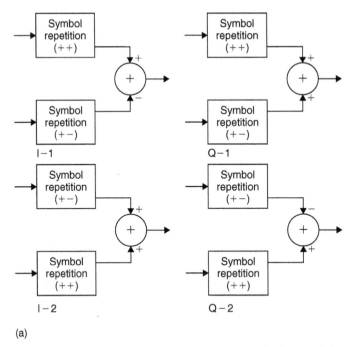

(a)

Figure 8.60a *Space–time spreading transmit diversity. (a) Multiplexing of the in-phase and quadrature streams.*

polarity change. The data on the second channel is also repeated but the polarity symbol of the repeated symbol is changed, i.e. $(+-)$.

The space–time spreading transmit diversity constructs the in-phase $(I-1)$ of the first channel from the difference between the in-phase first stream repeated $(++)$ and in-phase second stream repeated $(+-)$. The quadrature channel $(Q-1)$ is given by the sum of the first quadrature repeated $(++)$ and the second repeated $(+-)$. Similarly, the in-phase of the second channel $(I-2)$ is the sum of the first in-phase repeated $(+-)$ and the second in-phase repeated $(++)$. The quadrature of the second channel is the difference between the first quadrature stream repeated $(+-)$ and the second quadrature stream repeated $(++)$. The space–time spreading transmit diversity is shown in Figure 8.60.

8.7.14.1 *MC mode cdma2000 modulation and spreading the forward channels*

The input symbols to the *MultiCarrier* (MC) system are sequentially distributed from top to the bottom output. A single input stream or two different input streams are demultiplexed into six parallel output streams as shown in Figure 8.61 for use in a 3-carrier cdma2000 forward channels.

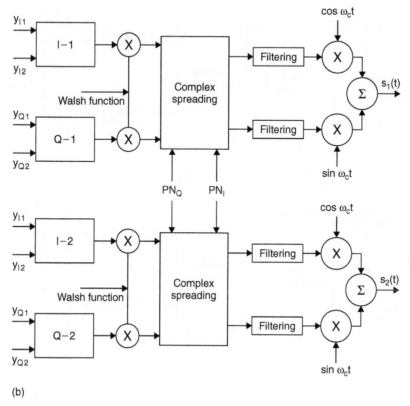

(b)

Figure 8.60b *Space–time spreading transmit diversity. (b) Constructing the in-phase and quadrature channels.*

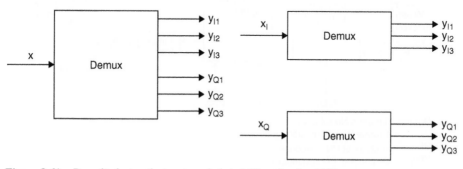

Figure 8.61 *Demultiplexing the input symbols in MC mode cdma2000 system.*

The multicarrier forward channels modulation and spreading for three carrier system is shown in Figure 8.62. The Carrier 1 channel is constructed by using the y_{I1} and y_{Q1}, and rotated by 90° when enabled. Carrier 2 and Carrier 3 channels are constructed in a similar method.

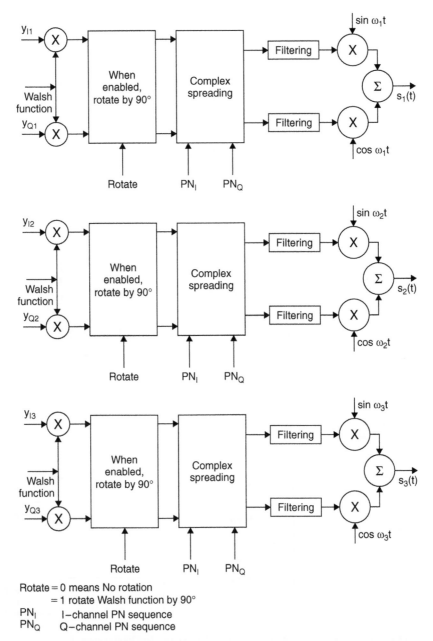

Rotate = 0 means No rotation
       = 1 rotate Walsh function by 90°
PN$_I$ I−channel PN sequence
PN$_Q$ Q−channel PN sequence

Figure 8.62   *MC mode cdma2000 modulation and spreading for 3-carrier system.*

8.7.15   cdma2000 Revisions

At the time of writing, the 3GPP2 has, so far, launched five revisions of cdma2000 in an effort to satisfy users demands for high rate data on both the reverse and forward links to facilitate multimedia, broadcast and multicast services. A brief introduction of these

Table 8.20 *Single-user data rates supported by cdma2000*

Channel bandwidth MH/s	Data rate kb/s (Forward link)	Data rate kb/s (Reverse link)
1 × 1.25	**9.6–307**	**9.6–307**
3 × 1.25	9.6–1,037	9.6–1,037
6 × 1.25	**9.6–2,074**	**9.6–2,074**
9 × 1.25	9.6–2,458	9.6–2,074
12 × 1.25	9.6–2,458	9.6–2,074

revisions is presented in the following sections. Readers are encouraged to consult the specific standard for further information.

8.7.15.1 cdma2000 Rev. 0 (Bi, 2005)

This revision is also known as the 1xEV-DO (*one-carrier Evolution–Data Only*) standard previously introduced in Section 8.7.1. The marketing negativity associated with the expression '*Data Only (DO)*' is of concern to service providers so the '*DO*' was changed to '*Data Optimized*', highlighting the fact that the technology was optimized for data transmission.

Rev. 0 offers multicast services which enable transmission of the same information to an unlimited number of users. Rev. 0 was driven by a demand for high rate packet data on the forward link to provide users with multimedia services such as real time TV broadcast and movies, etc. The peak single user data rates supported by cdma2000 are given in Table 8.20. A data rate of up to 2 Mb/s on the forward/reverse link requires the use of six carriers in a multicarrier cdma2000 system. Recall from Section 8.7.1 that at the launch of the cdma2000 standard little commercial interest has been shown in the multicarrier system and the commercial interest was invested in the single-carrier (1x) system. The peak data rate on the forward/reverse link channels provided by a single-carrier (1x cdma2000) is 307 kb/s which are not fast enough to provide high QoS multimedia applications.

In addition to the demand for the provision of multimedia services on the forward link channels, service providers wish to increase their revenue through enhancing network capacity, network coverage and provide for applications with different throughput, QoS and latency requirements.

To achieve these improvements, Rev. 0 uses a bandwidth of 1.25 MHz and applies direct spread spectrum waveform at chip rate of 1.2288 Mc/s. Furthermore, the Rev. 0 standard uses a number of efficient techniques including turbo coding, data rate adaptation, reduced latency, fast packet scheduling, and receive diversity. These same advanced techniques are used in most of the revisions that have been launched subsequently.

Rev. 0 divides the time into many time slots such that each second of transmission time is divided into 600 time slots making the duration of each time slot 1.6666 ms. The frame used in Rev. 0 is 26.66 ms containing 16 time slots compared to the single-carrier cdma2000 frame size of 20 ms. Each subscriber uses one or more time slots to send the payload.

Table 8.21   *Rev. 0 channel parameters*

Data rate index	Data rate kb/s	Bits/packet	Number of slots	Code rate	Modulation
0	0	0	0	–	–
1	38.4	1024	16	1/5	QPSK
2	76.8	1024	8	1/5	QPSK
3	153.6	1024	4	1/5	QPSK
4	307.2	1024	2	1/5	QPSK
5	**307.2**	**2048**	**4**	**1/5**	**QPSK**
6	614.4	1024	1	1/3	QPSK
7	**614.4**	**2048**	**2**	**1/3**	**QPSK**
8	921.6	3072	2	1/3	QPSK
9	1228.8	2048	1	1/3	8-PSK
10	**1228.8**	**4096**	**2**	**1/3**	**8-PSK**
11	1843.2	3072	1	1/3	16-QAM
12	2457.6	4096	1	1/3	16-QAM

However, each slot will be used by only one user. Multiple users are accommodated using a TDMA scheme to allow each user to use the maximum power available to the serving base station on the forward link. The Rev. 0 standard was approved in 2002 and the first commercial deployment took place in Korea in 2002. Currently, there are more than 10 million subscribers to the service worldwide.

Rev. 0 supports thirteen data rates, as given in Table 8.21. For each data rate, the standard provides the number of bits per packet, the number of slots needed, the code rate and the type of carrier modulation used. The rows in Table 8.21 shown in bold imply frequently used data rates in Rev. 0 enabled systems.

Rev. 0 applies low rate ($\frac{1}{3}$ or $\frac{1}{5}$) turbo codes in the forward traffic channel and rate $\frac{1}{4}$ or $\frac{1}{2}$ turbo encoder in the reverse Traffic channel. Unlike cdma2000 (which uses the BPSK carrier modulation) Rev. 0 uses an adaptive QPSK, 8-PSK or 16-QAM carrier modulation and a *Hybrid Automatic Repeat Request* (HARQ) for high-speed transmission on the forward Traffic channels. Similar HARQ technique is used in WCDMA systems and presented in detail in Section 8.18. Rev. 0 uses forward multi-user diversity together with antenna diversity at the receiver. Furthermore, it supports per flow QoS on forward link and per terminal QoS on the reverse link with latency of about 150 ms.

The structure of the reverse link traffic channel is shown in Figure 8.63. The pilot channel is used for coherent demodulation and synchronization at the base station. The Reverse Rate Indicator (RRI) channel is used to inform the base station about the data rate used for transmission on the reverse link. The *Acknowledgement* (ACK) channel is used to support the early completion of the packet transmission on the forward link, a technique called *incremental redundancy* and used in WCDMA systems as well. The base station transmits a packet to a mobile terminal at the data rate specified in the *Data Rate Control* (DRC) request message from the terminal.

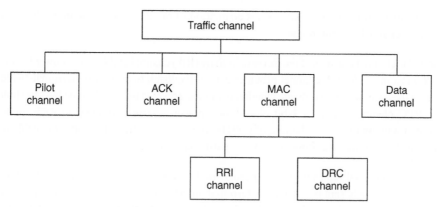

Figure 8.63   *Reverse Traffic channel structure in Rev. 0.*

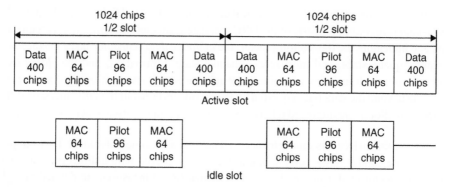

Figure 8.64   *Forward slot structure for Rev. 0.*

In the early completion scheme, the total packet energy is divided into several portions and a packet is incrementally transmitted using separate slots. The repetition of the packet bits is accompanied with incremental redundancy to improve the chance of correctly decoding the packet and terminating the transmission. To allow time for the terminal to process each subpacket and send ACK information to the base station, subpackets transmission is separated by three time slots.

The base station schedules the payload from the users, starting with the users who have better quality channel conditions then followed by the other users scheduled when their radio conditions are improved. This scheduling scheme provides multi-user diversity gain. The structure of the forward link time slot is shown in Figure 8.64.

The pilot channel is transmitted at full power for 96 chips every $\frac{1}{2}$ slot to provide reference for coherent detection of traffic as mentioned before; and to provide prediction of the received *Signal to Interference plus Noise Ratio* (SINR) at the terminal to aid the determination of the maximum data rate which can be supported on the forward link.

The channel estimation enhances a fast adaptation of modulation and coding schemes to different channel environments.

The MAC channel consists of the *Reverse Activity* (RA) channel and *Reverse Power Control* (RPC) channel. The RA channel provides a 1-bit feedback to all mobile terminals which receive the base station to indicate whether or not their reverse link load exceeds a threshold. The RPC channel carries a 1-bit closed-loop power control command to each mobile terminal with an update at the rate of 600 Hz. It also carries the addressing to enable the identification of the data transmission to a particular terminal.

The Traffic channel data rate to a particular mobile terminal is determined by the DRC message previously sent by the mobile terminal on the reverse link. The message also indicates the type of carrier modulation, the channel coding rate, the preample length, the data rate, in addition to the maximum number of slots required to achieve the desired error rate.

Rev. 0 has three mechanisms to control the data rate on the forward link Traffic channel. The *open-loop rate control* consists of the DRC message from mobiles requesting certain data rate and indicating the transmitting sector; the *adaptive data scheduler* takes into account fairness, queue sizes, and most recent channel state information; and the *closed-loop rate control* which includes a fast feedback acknowledgement to effectively increase the data rate beyond the data rate requested by the mobile terminal in the DRC message if the channel conditions are improved relative to the channel estimate used to generate the DRC.

The reverse link in Rev. 0 is similar to that in cdma2000 with some differences. For example, the data rate control in the reverse link in Rev. 0 is achieved with a direct measurement of the cell/sector load using the Rise-over-Thermal (RoT) measurement defined by $\frac{S+I}{N}$. Where $S + I$ is the total received power of the desired signal (S), the total interference (I), and N is the background noise measured during a silent interval when all mobile terminals in the system do not transmit.

Each mobile terminal receives an RA bit from the base station on the forward MAC channel indicating whether or not the RoT exceeds a predetermined threshold. This information helps the mobile terminal to either increase or decrease its data rate relative to the maximum data rate at which it may transmit.

Two types of control channels are supported on the forward link of Rev. 0: *Synchronous Control* (SC) and *Asynchronous Control* (AC). The SC is transmitted once every 256 slots. The AC is used when transmitting delay-sensitive signalling to the mobile terminal during active connection or during the establishment of an active connection. The AC is used to send the ACK to the terminal whenever it is needed but it must not overlap with the SC. The SC is used by the base station to page the terminals. Both SC and AC are transmitted at 38.4 kb/s or 76.8 kb/s.

A field measurement on the forward traffic channel of Rev. 0 was carried out by Bell Labs in 2004 (Bi, 2005; 3GPP2 C.S0024, 2002) who reported a peak data rate of about 850 kb/s in a mobile environment being achieved with a data speed 20% higher when the mobile speed was less than 10 miles per hour. Furthermore, to achieve data speed on the cell's border at least 200 kb/s, the cell radius was about 3 km for large cities and 6.5 km for suburban environments.

8.7.15.2 cdma2000 Rev. A (Bhushan, 2006; Yavus, 2006; 3GPP2 C.S0024-A, 2004)

The high-speed data transmission on the forward traffic channels enabled by Rev. 0 provided users with fast internet access which encouraged service providers to consider applications such as *Voice over IP* (VoIP), video telephony and on-line gaming. All of these applications required high-speed data transmission not only on the forward link but on the reverse channels as well. To meet these demands, the 3GPP2 approved Rev. A, also known as *1xEV–DO Rev. A* standard in 2004 to support the required enhancements on the reverse link of Rev. 0. Although Rev. 0 does not include voice provision, Rev. A is fast enough to support VoIP services with quality better than that offered by single-carrier cdma2000 systems.

The synchronous and the asynchronous control in Rev. A are transmitted as packets of lengths 128, 256 or 512 bits and with minimum inter-transmit interval of four slots. Consequently, page transmission and access ACK take a maximum of just four slots each. Furthermore, Rev. A reduces the connection set up using high data rate on the access channel (up to 38.4 kb/s), while in Rev. 0 the data rate is restricted to 9.6 kb/s. The amount of data the access terminal is allowed to transmit on the access channel is also controlled by the network.

Rev. A uses a shorter preamble of four slots compared to 16 slots in Rev. 0. In addition, Rev. A achieves a significant reduction in reverse link access latency and permits access terminals to adjust their transmit power level as a function of the forward pilot strength of the serving cell/sector. The network has control over the access channel performance, data rate and the maximum payload per terminal.

The peak data rate on the forward link in Rev. A is 3.1 Mb/s compared with 2.4 Mb/s Rev. 0. Data rates of 1.8 Mb/s on the reverse link (compared to 153.6 kb/s in Rev. 0) and 3.1 Mb/s on the forward link enable services such as video messaging, mobile video gaming, streaming video and Instant Multimedia Messaging (IMM) and the integration of voice and data services such as video telephony.

Physical layer forward link in Rev. A: Rev. A supports shorter packets on the forward link, i.e. 128, 256, 512 bits per packet, in addition to packets introduced in Rev. 0. The format of each packet is defined by the packet size in number of slots, nominal packet duration, and the preamble length in chips. For example, a packet (256, 16, 1024) indicates that the packet has 256 bits, its duration is 16 slots and the preamble is 1024 chips. The support of relatively smaller packet size on the forward link in Rev. A is highly efficient, particularly for terminals in poor channel conditions since in such circumstances only a small number of bits can be sent and the rest of the packet is padded.

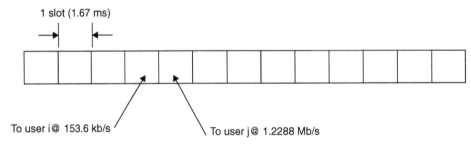

1 slot (1.67 ms)

To user i @ 153.6 kb/s   To user j @ 1.2288 Mb/s

Figure 8.65   *Rev. A forward link time multiplexed data to terminals.*

Transmissions on the forward link are time multiplexed with slot durations of 1.67 ms and all transmissions are at the same power; but the data rate can change from slot to slot from 38.4 kb/s to 2.45 Mb/s, as shown in Figure 8.65. Transmission in each slot is directed to only one terminal.

The forward MAC layer in Rev. A uses single-user packets in addition to a single packet which contains data to multiple terminals (maximum of eight terminals per packet). Unlike Rev. 0, which uses a one-to-one mapping between requested DRC and used data rates/packet sizes, Rev. A uses a more flexible mapping which allows for improved data packing efficiency. Each DRC index in Rev. A is associated with a set of transmission formats for a single-user packet and *Multi-User Packet* (MUP).

The soft handoff in Rev. 0 causes a packet transmission delay due to a base station change. Such delay degrades the performance applications characterized as delay sensitive. Rev. A eliminates this delay using the *Data Source Control* (DSC) channel. A mobile terminal using the DRC channel provides the network with an exact instant of time when the change in forward link from the old to the new base station would take place.

A mobile terminal supporting MUPs may send a null-rate DRC (DRC index 0) message to the base station which will be mapped to the transmission format (1024, 16, 1024) resulting in an effective data rate exceeding 38.4 kb/s. The message may equally be mapped to transmission formats (512, 16, 1024), (256, 16, 1024).

Physical layer reverse link in Rev. A: The reverse traffic channels in Rev. A support data rates up to 1.8 Mb/s compared with peak rate in Rev. 0 of 153.6 Mb/s. The reverse link physical layer channels are shown in Figure 8.66 and the physical layer subframe structure is shown in Figure 8.67.

The reverse link supports a payload from 256 to 12,288 bits and the network sets two latency targets for the terminal for each payload. A latency target is defined by one of two possible transmission modes: *low latency mode* or *high capacity mode*. In general, the packet latency is within the range 6.666–66.66 ms and it is typically 50 ms compared to 150 ms in Rev. 0. The modulation supported by tRev. A are the BPSK with quartenary/binary Walsh cover, QPSK, 8-QPSK with quartenary/binary Walsh cover modulated symbols.

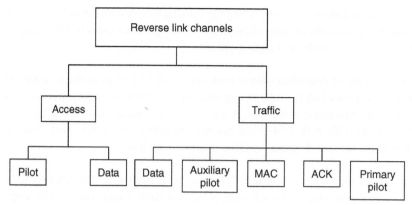

Figure 8.66 *Rev. A reverse link channels.*

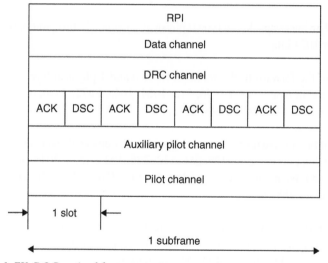

Figure 8.67 *1xEV–DO Rev. A subframe structure.*

The pilot, the Reverse Rate Indicator (RPI) channel, ACK/DSC, DRC, Data, and auxiliary pilot channels are orthogonally spread by Walsh functions of length 2, 4, 8, 16 or 32.

Novatel Wireless Inc. announced that it had developed Rev. A modems and expected to make the modem commercially available by the end of 2006. KDDI announced that it would deploy Rev. A by the end of 2006.

8.7.15.3 cdma2000 Rev. B (3GPP2 C.S0024-B, 2006)

Rev. B is a scalable aggregation of multiple carriers of Rev. A, which is also known as $N \times EV–DO$ or *EV–DO Multicarrier*. The text for the Rev. B standard was published in May 2006. A number of mobile wireless companies, such as Qualcom and Motorola, have shown an interest in the commercialization of Rev. B devices and systems. Rev. B can be implemented incrementally by upgrading the existing Rev. A with the addition of a second carrier or by using a software upgrade so that a service provider can select the geographical areas and hot spots which need high-speed data.

Rev. B is capable of delivering data at peak speeds of 73.5 Mb/s in the forward link and 27 Mb/s in the reverse link through the aggregation of 15 carriers, each carrier bandwidth of 1.25 MHz. The first phase of the Rev. B system may include aggregation of three carriers in 5 MHz bandwidth with a 64-QAM modulation scheme to deliver peak data rates of 9.3 Mb/s in the forward link and 5.4 Mb/s in the reverse link.

Rev. B defines four types of physical layer protocols to provide for different data rates and access and guarantee backwards compatibility with previous systems. The length of a control packet in types 0 and 1 physical layer is 1024 bits and in types 2 and 3 is 128, 256, 512, or 1024 bits.

The length of an access packet in type 0 physical layer is 256 bits, and in types 1, 2 and 3 is 256, 512 or 1024 bits.

The lengths of the forward traffic frame in types 0 and 1 physical layers are 1024, 2048, 3072, 4096 bits; in type 2 is 128, 256, 512, 1024, 2048, 3072, 4096, or 5120 bits; and in type 3 are 128, 256, 512, 1024, 2048, 3072, 4096, 5120, 6144, 7158, or 8192 bits.

The lengths of the reverse traffic frame in types 0 and 1 physical layers are 256, 512, 1024, 2048, or 4096 bits; in type 2 are 128, 256, 512, 768, 1024, 1536, 2048, 3072, 4096, 6144, 8192, or 12,288 bits; and in type 3 are 128, 256, 512, 768, 1024, 1536, 2048, 3072, 4096, 6144, 8192, or 12,288 bits.

The forward traffic rate in types 0 and 1 physical layer varies up to 2.4576 kb/s, in types 2 and 3 the physical layer varies up to 3072 kb/s. The type 3 physical layer offers optionally a data rate of 4915.2 kb/s using 64-QAM modulation.

The reverse traffic rate in types 0 and 1 physical layers varies up to 153.6 kb/s, in type 2 it is 1228.8 kb/s and in type 3 physical layer is up to 1843.2 kb/s.

The packet application in Rev. B provides an octet stream to carry packets between the mobile terminal and the network. The functionality of the packet application is supported by three protocols: the *Radio Link Protocol* (RLP) which provides retransmission and duplicate detection to reduce error rate; the *packet location update protocol* to support user mobility; and the *flow control protocol* to provide data flow control. The encapsulation of the packet application consists of the octet stream from the upper layer, an RLP packet, and a Stream layer payload as shown in Figure 8.68.

The stream layer provides multiplexing of the streams for one mobile terminal. Stream 0 is always assigned to the signalling Application. The other streams can be assigned to

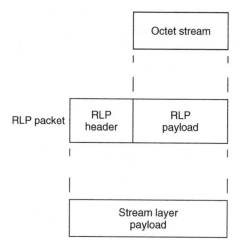

Figure 8.68 *Packet application encapsulation.*

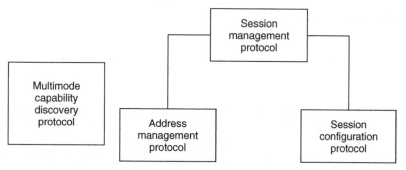

Figure 8.69 *Session layer protocols.*

applications with different *Quality of Service* (QoS) requirements, or applications of the same QoS. The stream layer also provides configuration messages that map applications to streams.

The session layer contains protocols which are used to negotiate a session between the mobile terminal and the network. A session includes information such as: *Unicast Address* (UATI) assigned to the terminal, the set of protocols used by the terminal and the network to communicate over the air-link, configuration settings for these protocols (e.g. authentication keys, parameters for connection layer and MAC layer protocols, etc.), and an estimate of the current terminal location. The session layer protocols are shown in Figure 8.69.

The session management protocol controls the activation of other session layer protocols. In addition, it ensures that the session is still valid and manages the closing of the session. The address management protocol specifies the procedures for the initial UATI assignment and maintains the terminal addresses. The session configuration protocol supports the negotiation for the parameters and the protocols used during the session. The multimode

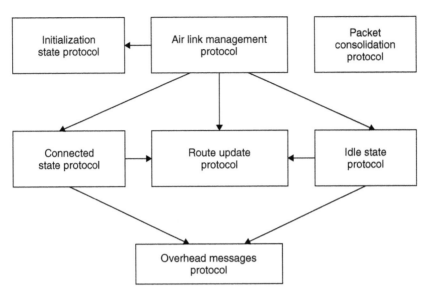

Figure 8.70   *Connection layer protocols.*

capability discovery protocol allows the network to discover the multimode capabilities of the terminal.

The connection layer prioritizes the traffic for transmission and controls the state of the link. The mobile terminal and the network maintain a connection that can be either closed or open. A terminal with a closed connection will not be assigned any dedicated link resources. Consequently, any communication between the terminal and the network would be conducted over either the access channel or the control channel. On the other hand, with the open connection (when the connection is open), the terminal is assigned channels for traffic transmission on the forward and reverse links as well as being able to communicate with the network using the control channel. The connection layer performs these functions using the protocols shown in Figure 8.70. The air link management protocol maintains the overall connection between the terminal and the network. The initialization state protocol performs the actions associated with acquiring access to the network. The idle state protocol performs the actions associated with keeping track of the terminal's approximate location in support of efficient paging and carrying out the procedures leading to the opening of a connection.

The MAC layer contains the rules governing the operation of the control channel, the access channel, the forward Traffic channel, and the reverse Traffic channel. It contains the rules concerning network transmission, packet scheduling and terminal acquisition of the control channel. When the forward Traffic channel is in the variable rate state, the network transmits at the rate defined by the *Data Rate Control* (DRC) channel transmitted from the terminal. The terminal uses either a DRC index 0 or the DRC index associated with a sector in its active set. The DRC values with the data rate are given in Table 8.22.

Table 8.22 *DRC values for data rates and packet lengths*

DRC index	Data rate in kb/s	Packet length in slots
0×0	Null rate	–
0×1	38.4	16
0×2	76.8	8
0×3	153.6	4
0×4	307.2	2
0×5	307.2	4
0×6	614.4	1
0×7	614.4	2
0×8	921.6	2
0×9	1228.8	1
$0 \times a$	1228.8	2
$0 \times b$	1843.2	1
$0 \times c$	2457.6	1
$0 \times d$	Invalid	–
$0 \times e$	Invalid	–
$0 \times f$	Invalid	–

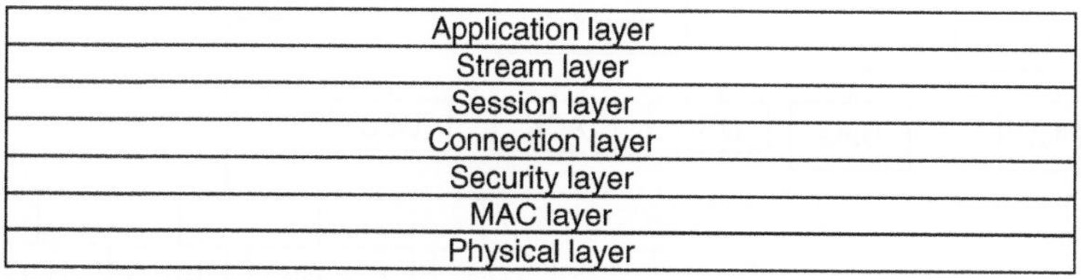

Application layer
Stream layer
Session layer
Connection layer
Security layer
MAC layer
Physical layer

Figure 8.71 *Architecture of the Rev. B air interface layering.*

The architecture of the air interface layering is shown in Figure 8.71.

The physical layer forward channel and reverse channel structures are shown in Figures 8.72 and 8.73.

For type 0 and type 1 physical layers, the first three packets are shown in Figure 8.74 where PAD is set to all '0's and should be ignored by the receiver. The reverse traffic packet is shown in Figure 8.75.

The forward slot structure is shown in Figure 8.76.

The reverse link consists of the access, reverse traffic channels. The access channel consists of a pilot, and data channels. The reverse traffic channel consists of a pilot, *Reverse Rate Indicator* (RRI), *Data Control Channel* (DRC), an *Acknowledgement Channel* (ACK), and data channel. The RRI channel indicates the rate of the data channel transmitted on the reverse link. The DRC is used by the terminal to indicate to the network the preferred data rate on the forward Traffic channel and the chosen serving sector. The ACK channel

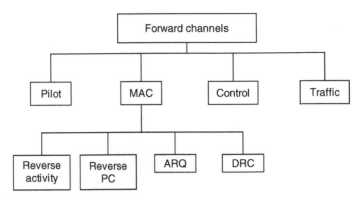

Figure 8.72 *Forward channel structure.*

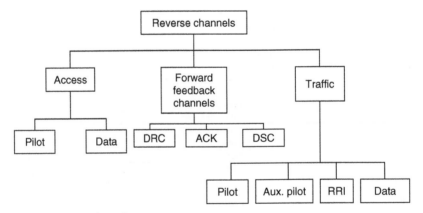

Figure 8.73 *Reverse channel structure.*

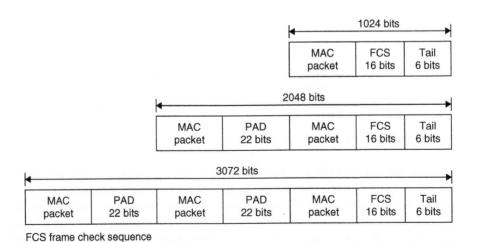

FCS frame check sequence

Figure 8.74 *Forward packet frame.*

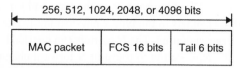

Figure 8.75 *Reverse packet structure.*

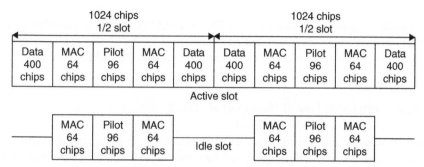

Figure 8.76 *Forward slot structure.*

is used by the terminal to inform the network whether or not the forward packet has been received successfully. The reverse frame is 26.66 ms in duration and consists of 16 slots each of duration 1.66 ms.

The Types 0 and 1 physical layer reverse channels time multiplexed the pilot and the RRI symbols and the multiplexed symbols contained in the Walsh channel W_0^{16} are added to the ACK symbols in the Walsh channel W_4^8 to form the I–channel. The Q–channel consists of the addition of the spread DRC symbols in the Walsh channel W_8^{16} and data symbols in the Walsh channel W_2^4. The I and Q symbols are quadrature spreads and applied to a quadrature modulator for transmission. Types 0 and 1 physical layer forward channels consist of the forward traffic symbols which are scrambled and interleaved and time demultiplexed into 16 logical channels to create a 16-ary Walsh cover. The resulting Walsh sequences are time multiplexed with the pilot symbols and the RPC/RA symbols to form the I and Q–channels which are quadrature multiplexed and applied to a quadrature modulator.

The type 2 physical layer reverse traffic channels consist of the pilot using Walsh channel W_0^{16}, the auxiliary pilot using Walsh channel W_{28}^{32}, the RRI using Walsh channel W_4^{16}, the ACK using Walsh channel W_{12}^{32}, the DRC using Walsh channel W_8^{16}, and the scrambled data. The pilot, the auxiliary pilot, the RRI, the ACK, the DSC symbols together with the data symbols on the data I–channel are added to form the I–channel of the reverse channel. The DRC and the data symbol on the Q–channel are added to form the reverse channel. The reverse channel I and Q–channels are quadrature spread and applied to a quarature modulator. The forward channel in the type 2 physical layer is similar to the forward channel in the type 1 physical layer.

Both the reverse and forward Traffic channels in the type 3 physical layer are similar to the corresponding Traffic channels in the type 2 physical layer. The type 3 physical layer also defines the multicarrier operation. The N carriers can be used by a user to transmit N reverse channels or by the base station to transmit N forward channels. In both systems, the channels are combined to be split into I and Q–channels which are operated on by quadrature spreading by the long code PN sequences and modulated on carrier i where $i = 1, 2, \ldots, N$.

The modulation systems used are QPSK, 8-PSK or 16-QAM carriers for transmission. The 64-QAM modulation is optional to deliver a data rate of 4915.2 kb/s on the forward Traffic channel. Information sent on the access channel is modulated on BPSK carrier.

8.7.15.4   cdma2000 Rev. C (Soong et al., 2003; 3GPP2 C.S0002-C, 2004)

cdma2000 Rev. C, also known as *1×EV–DO Rev. C*, is an optimized standard which provides integrated services of high speed data and voice. It defines a new transmission channel on the forward link called the *Forward Packet Data Channel* (F-PDCH) which can be time-shared among users. Rev. C uses link adaptation by adjusting the modulation and channel coding to best suit the channel conditions as in previous revisions. Furthermore, it introduces a new hybrid known as HARQ type $\prod$.

In Rev. A, power and code resources are allocated only on a dedicated basis for data traffic. This means that dedicated channels are power controlled to compensate for fading and a set of channelization codes are allocated for a relatively long period of time (seconds or even minutes). Such resource allocation (commonly used in circuit switching) is inefficient for packet switching that is characterized by bursty traffic. Furthermore, the unpredictable conditions of the wireless environment dictate the needs for power level adjustment suggesting a poor resource allocation in Rev. A.

Rev. C introduces a more efficient resource allocation scheme which avoids the waste of resources by allocating resources to a mobile terminal over a relatively short period of time (order of ms). In addition, resources are allocated to a terminal experiencing fading when its channel quality is improved and the leftover resources are used for the shared packet data channel so that complete utilization of resources can be realized. Clearly the use of a shared packet channel in place of multiple dedicated Traffic channels permits the full use of the available resources as shown in Figure 8.77.

The shared packet data channel is dedicated to a single mobile only for a short time (1.25, 2.5, or 5 ms) as mentioned before. Rev. C also defines the *Forward Packet Data Control Channel* (F-PDCCH) to operate with the F-PDCH to identify the scheduled mobile terminal for transmission, to specify the Walsh codes used by the F-PDCH and to indicate parameters necessary to demodulate the data carried by the F-PDCH. The mobile terminals always listen to F-PDCCH. The message format used by the F-PDCCH is shown in Figure 8.78.

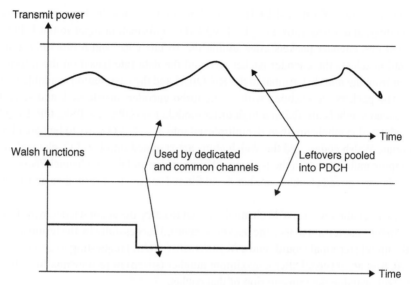

Figure 8.77 *Leftover power and Walsh functions into F-PDCH.*

MAC ID (8 bits)	EP size (3 bits)	ACID (2 bits)	SPID (2 bits)	AISN (1 bit)	LWCI (5 bits)

MAC ID: Mobile MAC identifier
EP size: Encoder packet size
ACID: ARQ channel identification
SPID: Subpacket identification
AISN: ARQ identifier sequence number
LWCI: Last Walsh code index

Figure 8.78 *F-PDCCH message format.*

The information carried by the message is used by the mobile to decode the F-PDCH. The information is about the encoder packet size and subpacket identification.

Rev. C defines a new channel, the *Reverse Channel Quality Indicator Channel* (R-CQICH) to support the fast link adaptation of the F-PDCH. The forward channel quality information feedback from the mobile terminal is used by the base station to select the optimal data rate for the F-PDCH once every 1.25 ms. The mobile terminal transmits the information of the forward link quality using R-CQICH.

The information sent by the mobile terminal consists of the pilot Signal to Interference plus Noise power Ratio (SINR) measured by the mobile terminal from the best base station in the base stations set retained by the terminal and the identity of the best base station received by the mobile. The base station uses the information in R-CQICH to determine the data rate of the F-PDCH and to send the data rate to the mobile terminal on the forward link using the PDCCH.

A number of features for the F-PDCH are used in Rev. C, such as higher order modulation, turbo coding, and a new form of hybrid AQR. The information input to the F-PDCH is formatted into encoder packets after adding the 16 bits CRC and 6 bits turbo tail. The base station selects the encoder packet size and the data rate based on the information received from the mobile terminal in the R-CQICH and the base station available transmit power. The packets are encoded with $\frac{1}{5}$ rate turbo encoder, interleaved, and scrambled. The encoded symbols are fed to a high order modulation (QPSK, 8-PSK, OR 16-QAM). The modulated symbols are then demultiplexed into $1 \ldots N$ I/Q pairs to be spread by the appropriate Walsh codes, and the Walsh chips are summed into I and Q–channels which are complex multiplied with the I–channel PN sequence and the Q–channel PN sequence. The quadrature spread channels are baseband filtered and are transmitted on an RF carrier.

A new channel, the reverse ACK channel, is used to carry the acknowledgement from the mobile to the base terminal once the packet is received successfully by the terminal. Otherwise the mobile terminal would send a *Negative ACK* (NAK) requesting a re-transmission. If the ACK is not received after a maximum number (seven) of re-transmissions, the base station will abandon the transmission of that packet.

The re-transmitted encoder packet size must be kept the same as in the initial transmission. However, depending on the most recent state of the forward link conditions as reported in the R-CQICH, a different data rate (slot length) can be selected for the re-transmission on the forward link. Additional redundant symbols can be added to improve the error correction until a successful decoding occurs. These techniques, together with the provision for asynchronous re-transmission, correspond to a new scheme defined in Rev. C called the *Asynchronous Adaptive Incremental Redundancy* (AAIR).

For synchronous ACK, the mobile terminal sends the ACK or NAK as BPSK modulated signal after a fixed delay. Clearly, the synchronous ACK is a simple stop-and-wait protocol to provide for the hybrid ARQ and is inefficient for a small number of users. The mobile terminal can use multiple ARQ channels to remove this inefficiency. The base station indicates on the ARQ ID field in the F-PDCCH message (as shown in Figure 8.79) which ARQ channels are used.

8.7.15.5   *cdma2000 Rev. D (Derryberry et al.; 3GPP2 C.S0002-D, 2004)*
Rev. C has brought improvements in both the data rates and quality of service on the forward link. Rev. D aims at improving the transmission in the reverse link. Rev. D uses common technologies used in the development of the previous revisions, such as channel-based scheduling, link adaptation, and HARQ. These enhancements enable the delivery of data rates of up to 1.8456 Mb/s. Rev. D reverse channels are shown in Figure 8.79.

Rev. D is supported by a number of new reverse channels. The new Traffic channel referred to as the *Reverse Packet Data Channel* (R-PDCH) supports fixed packet duration of 10 ms. The reverse packet structure is shown in Figure 8.80. A packet with 192 bits uses a

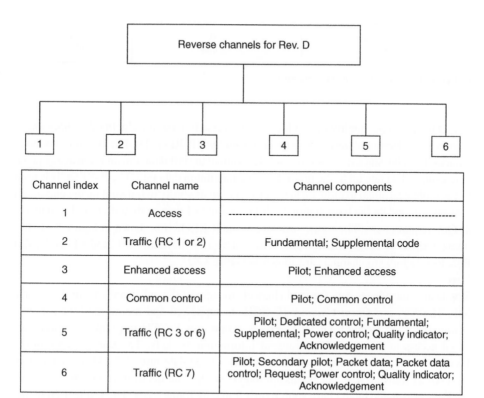

Channel index	Channel name	Channel components
1	Access	---
2	Traffic (RC 1 or 2)	Fundamental; Supplemental code
3	Enhanced access	Pilot; Enhanced access
4	Common control	Pilot; Common control
5	Traffic (RC 3 or 6)	Pilot; Dedicated control; Fundamental; Supplemental; Power control; Quality indicator; Acknowledgement
6	Traffic (RC 7)	Pilot; Secondary pilot; Packet data; Packet data control; Request; Power control; Quality indicator; Acknowledgement

Figure 8.79 *Rev. D reverse channels.*

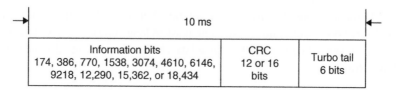

Figure 8.80 *Packet structure of the reverse packet data channel.*

12-bit CRC and the rest of the packet sizes use 16-bit CRC. Each packet is re-transmitted up to a maximum of three times. The packets are turbo encoded with a code rate of $\frac{1}{5}$ and block interleaved. The modulation schemes used are BPSK, QPSK, and 8-PSK.

Rev. D introduces a reverse channel for carrying important information to help the base station decode the signal received from the mobile terminal. This channel is referred to as the *Reverse Packet Data Control Channel* (R-PDCCH) used by the mobile terminal for transmitting control information for the associated R-PDCH to help the base station decode the signal received from the mobile terminal. The Control information consists of six information bits and a mobile status Indicator bit.

Information 12 bits	CRC 12 bits	Encoder tail 8 bits

Figure 8.81 *R-REQCH Frame Structure.*

Rev. D improves the quality of service on the reverse link by introducing the *Autonomous Transmission Mode* and the *Reverse Request Channel* (R-REQCH). The autonomous mode reduces the delay since it allows the mobile terminal to start data transmission at any given time using a data rate up to the pre-authorized maximum data rate. The R-REQCH is used by the mobile terminal to indicate to the base station its ability to transmit on the maximum data rate, to update the base station with the amount of data in its buffer, and to indicate its QoS requirements. The frame structure of the R-REQCH is shown in Figure 8.81. The frame consists of 12 information bits, followed by a 12-bit frame CRC and 8 Encoder Tail Bits. Rev. D forward channels are shown in Figure 8.82.

Rev. D also introduces new forward channels to support the enhancements in the reverse link. The *Forward Grant Channel* (F-GCH) is supported and monitored by all mobile terminals served by the F-GCH. The base station uses F-GCH to grant the mobile terminal access to the network to transmit information packets. The transmission on the F-GCH uses a frame duration of 10 ms and at a fixed data rate of 3.2 kb/s. In addition, the F-GCH transmission can be discontinuous so that the decision to enable or disable the F-GCH is made by the base station on a frame by frame basis. The frame used by the F-GCH is shown in Figure 8.83 which consists of 14 bits data, 10 bits for the CRC, and 8 of '0' bits for the encoder tail.

The F-GCH frame is encoded using $\frac{1}{4}$ convolutional encoder (K $=9$), block interleaved with block size 96 symbols, and spread using masked long code at 1.2288 Mc/s. The long code mask is shown in Figure 8.84. The spread data is then mapped into ± 1 and modulated on QPSK carrier.

The *Forward Indicator Control Channel* (F-ICCH) is used to control the data rate per user (dedicated channel) or per sector data rate and power control (common channel). The reverse HARQ control is supported by the Forward Acknowledgement Channel (F-ACKCH) to provide synchronous acknowledgement to the received R-PDCH transmissions. The F-ACKCH frame is 10 ms and transmitted at a data rate of 19.2 kb/s and it can address 192 different users within a given sector.

Rev. D also enhances the forward F-FCH and the F-SCH to support the *Broadcast and Multicast Service* (BCMCS) for multimedia transmission from a single source to multiple users such as a broadcast video service.

Rev. D supports three types of BCMCSs. In type 1 BCMCS, the F-SCH is shared among idle mobile terminals and a new BCMCS message is periodically sent to the mobile terminals

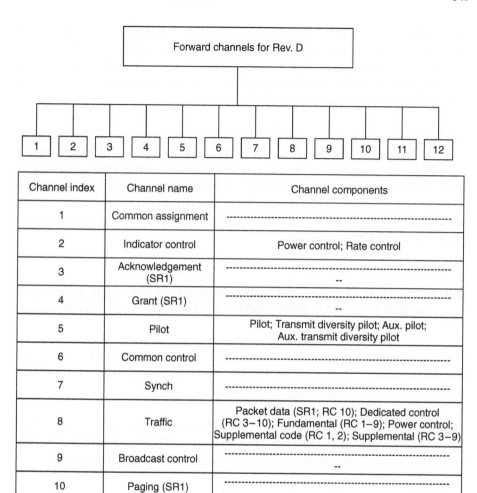

Figure 8.82 *Forward channels for Rev. D.*

14 bits data	10 bits CRC	8 bits encoder tail

Figure 8.83 *Frame structure for the F-GCH.*

41	29	28	24	23	16	15	9	8	0
1100011001101		10100		Walsh code index		0000000		Pilot PN sequence offset	

Figure 8.84 *F-GCH long code mask.*

with the BCMCS parameters and the handoff procedure is autonomous. In type 2 BCMCS, the F-FCH is shared among active mobile terminals and carries common BCMCS contents such as voice streaming to multiple terminals when the terminals are in a traffic state. In type 3 BCMCS, the F-SCH is shared among active mobile terminals in a traffic state using F-SCH signalling procedures.

Rev. D introduces procedures to reduce call set-up delay of as much as 12 seconds which is unacceptable for some applications. These procedures include a reduction in the page delay by allowing the mobile terminal to enter the reduced slotted mode or non-slotted mode which reduces the time between page slots. The new procedures enable the base station to perform channel assignment with or immediately after the page message. The procedures also simplify the Traffic channel set-up by replacing the service negotiation by a new fast service resume procedure.

8.8   Universal Mobile Telecommunications Services (UMTS) (3GPP TS25.215, 2002; 3GPP TS25.301, 2002; 3GPP TS25.215, 2002)

The third generation (3G) wireless networks succeeded the Global System for Mobile telecommunication (GSM) networks to move the wireless technologies from circuit-switched to wideband packet-based transmission.

The 3G standards in Europe are known as UMTS and are based on wideband CDMA defined by the Third Generation Partnership Project (3GPP) which was formed in 1998. In the US, the 3G systems are known as the cdma2000, as discussed in Section 8.7.

The 3GPP is structured in five groups to detail the technical specifications in five main standardization areas: *Radio Access Network* (RAN) specifications, *Core Network* (CN) architectures, *Terminal Equipments* (TE), *services* and *System Aspects* (SA).

The wideband CDMA operates with a bandwidth of 5 MHz to enable multimedia services to be delivered at bit rates up to 384 kb/s for wide area coverage and as high as 2 Mb/s for indoor or fixed applications. In the FDD mode networks use different frequency bands, known as paired bands, which allow for relatively large distances between mobile terminal and base station to provide acceptable coverage requirements. The TDD mode networks operate in an unpaired band as shown in Figure 8.85.

The 3G systems are either designed for *Frequency Division Duplexer* (FDD) or for *Time Division Duplexer* (TDD) operations. The band 1,920–1,980 MHz is allocated to the FDD reverse link (uplink) channels and paired with the band 2,110–2,170 for the FDD forward link (down link) channels as shown in Figure 8.85a. On the other hand, TDD mode networks operate over smaller distances (several hundreds metres) but they allow for asymmetric

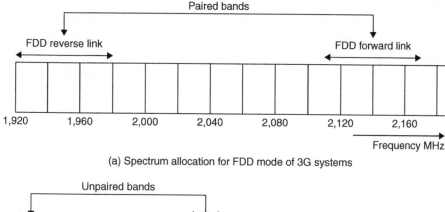

(a) Spectrum allocation for FDD mode of 3G systems

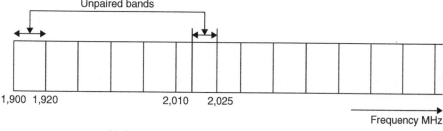

(b) Spectrum allocation for TDD mode of 3G systems

Figure 8.85 *Frequency bands for FDD and TDD modes of 3G systems. (a) Spectrum allocation for FDD mode of 3G systems; (b) Spectrum allocation for TDD mode of 3G systems.*

traffic at high transmission speeds suitable for video on demand applications or down load from the Internet. The bands allocated for TDD UMTS is 1,900–1,920 MHz and a smaller band between 2,020–2,025 MHz. These two bands are not paired, as shown in Figure 8.85b.

The principle of the FDD and TDD modes for 3G operations is defined in Figure 8.86 where the signals on the reverse and forward links are transmitted using different carriers f_r and f_f, respectively, and separated by a frequency band, while the same signals in the TDD mode are transmitted on the same carrier but in different time slots.

8.8.1 Basic system architecture

The UMTS system consists of a number of network elements, each with a defined functionality. These elements can be grouped into sub-networks, such as the *UMTS Terrestrial Radio Access Networks* (UTRAN) that handle all radio functionality, the *Core Network* (CN) which carries out the switching, call routing and connection to the external networks, and the *User Equipment* (UE) which interfaces with the user. Furthermore, there is a logical split into functionalities supporting the transfer of access information and non-access functionalities. The former includes protocols between the mobile terminal and the access network and between the access network and the core network. The latter

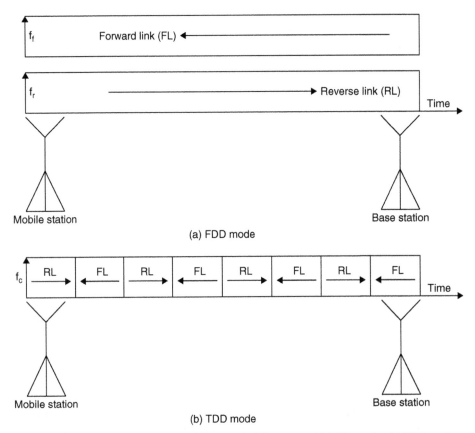

Figure 8.86 *Transmission in FDD and TDD mode 3G systems. (a) FDD mode; (b) TDD mode.*

is independent of the specific access network. Clearly this modular functionality of the UMTS network allows the use of several sub-network elements of the same functionality, but the minimum operational requirement is the use of at least one sub-network element of each type. Furthermore, the modular nature of the UMTS network means that these sub-networks operate on their own in addition to operating with other sub-networks. A system level architecture of the UMTS network is shown in Figure 8.87.

The user sub-network consists of two elements: the *UMTS subscriber identity module* (USIM), which contains a smartcard that holds the subscriber identity and stores the authentication and encryption keys where the subscriber is authenticated when requesting an access to the network. The Cu interface provides the necessary electrical interface between the smartcard (USIM) and the subscriber's *Mobile Equipment* (ME). The open interface Uu provides access to the subscriber equipment (UE) to the fixed elements of the UMTS system particularly when the UE and Node B are produced by different manufacturers. The radio access sub-network UTRAN will be discussed in the next section.

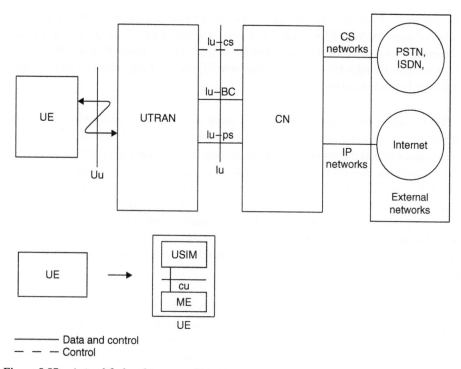

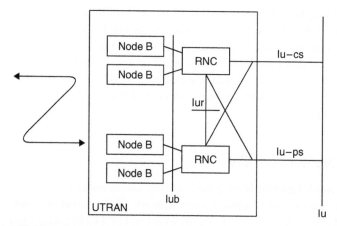

Figure 8.87 *A simplified architecture of UMTS system.*

Figure 8.88 *UTRAN architecture.*

8.8.2 Access network UTRAN

The UTRAN architecture is shown in Figure 8.88. The main characteristics of UTRAN are its support to soft handoff (a mobile station is connected to the UMTS network via more than one active cell), and its support to the WCDMA radio resource management algorithms. UTRAN also supports a unique single air interface which handles both packet-switched from external IP-networks and circuit-switched data from the GSM network. The main transport protocol proposed in the initial UMTS release (Release '99) was the *Asynchronous Transfer Mode* (ATM), a high-speed (up to 155 Mbps), but IP-based transport was added in later releases.

UTRAN consists of two different elements: Node B which is a term used by the 3GPP to define the base station transceiver system and provide the physical radio link between the user equipment and the network. Node B receives the subscriber data at the Uu interface and presents the subscriber data at the Iub interface. The other element is the *Radio Network Controller* (RNC) which controls the radio resources to Node B (base station) connected to it. Each RNC is connected to two or more Node Bs. The Iub interface connects a Node B and an RNC, which is necessary for an open radio interface, since Node Bs and RNCs may be produced by different manufacturers. Furthermore, the open Iur interface between two RNCs permits soft handoff between two RNCs from different manufacturers.

The open Iu interface connecting UTRAN to CN provides switching, routing and service control. The Iu consists of a number of connections: one to connect UTRAN to circuit-switched networks such as PSTN and ISDN and the other to connect UTRAN to packet-switched networks such as the Internet. The third connection is the *Iu Broadcast* (Iu BC) used to connect UTRAN to broadcast service.

It is worth noting that the general protocol model for the Iu, Iub, and Iur interfaces is structured into horizontal layers and vertical planes which are independent of each other. The protocol consists of two horizontal layers: the radio network layer, where UTRAN is visible, and the transport network layer which uses a standard transport technology, such as an IP-based scheme. The vertical planes consist of three main planes: the user plane, the control plane, and the transport network control plane. The interfaces protocol mode is depicted in Figure 8.89.

8.8.3   Core network CN

The core network functionalities cover mobility management issues; core network signalling; and interworking between core network and external networks. The initial launching of CN (Release '99) was mainly based on GSM CN as shown in Figure 8.90. It provides connection to *Circuit-Switched* (CS) and *Packet-Switched* (PS) IP-based external networks and hence it covers the need for different traffic such as real time circuit-switched data and non-real time packet data. The CS domain contains the *Mobile Switching Centre* (MSC) with the *visiting Location Register* (VLR), and the *Gateway MSC* (GMSC). The PS domain contains the serving *General Packet Radio Service* (GPRS) *Support Node* (SGSN) which, for the packet data, covers similar functions as the MSC together with VLR functionality. The *Gate GPRS Support Node* (GGSN) connects the PS CN to the IP-based

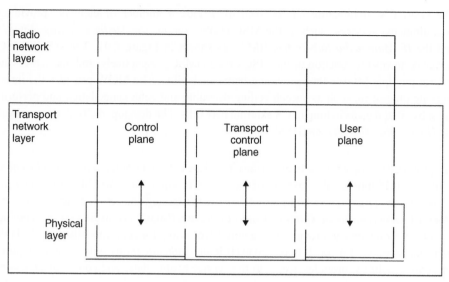

Figure 8.89 *UMTS interfaces protocol mode.*

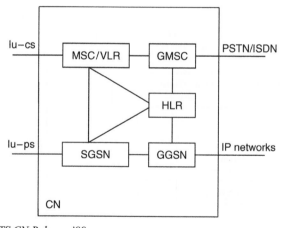

Figure 8.90 *UMTS CN Release '99.*

packet-switched external networks. In addition, the CN contains a *Home Location Register* (HLR), which is a database in the subscriber home network containing information on the permitted services, forbidden services, and the supplementary call services such as call forward, call forwarding number.

The MSC/VLR is a switch/database to switch the CS transactions, and to hold a copy of the service profile for the visiting subscriber as well as the precise location of the UE within the service system. The function of the GMSC is to provide for the incoming and outgoing connection to the CS networks.

The GSM CN architecture has evolved to include a number of elements providing IP multimedia services, such as the MSC/GMSC server, the *Media Gateway* (MGW) and the *IP Multimedia Sub-system* (IMS), as shown in Figure 8.91. The MSC/GMSC server performs the function of the MSC or the GMSC, respectively, and one server can control multiple MGWs thus providing better scalability. The MGW provides the inter-working processing such as speech coding/decoding and echo cancellation and perform the subscriber data switching to CS external networks. The developed CN also consists of the *Home Subscriber Server* (HSS).

The proposed protocol for use between the UE and the IMS is the *Session Initiation Protocol* (SIP). The IMS functionality consists of the *Media Resource Function* (MRF) and the *Call Session Control Function* (CSCF), which handles the session and acts as a firewall towards other networks. The *Media Gateway Control Function* (MGCF) controls a service coming from the CS domain and provides the required processing for echo cancellation. The HSS performs functions previously allocated to HLR and *Authentication Centre* (AuC), supports subscriber mobility, provides access to the service profile data, and communicates with SIP for session establishment and service authorization support.

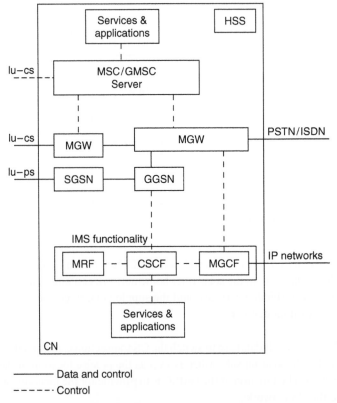

Figure 8.91   *UMTS CN for IP multimedia services.*

8.8.4 Physical layer channels

UTRAN implementation is defined in three parts: radio aspects corresponding to the physical layer, interface aspects defined in link and network layers, and network aspects interworking with the *Core Network* (CN). In this section, we present the radio aspects of the physical layer such as modulation, spreading, and physical channels.

Three types of channels are defined in UTRAN systems: the *logical channels* describe the types of information to be transferred between MAC and the *Radio Link Control* (RLC); the *transport channels* describe how the logical channels are to be transferred between MAC and the physical layer; and the *physical channels* define the transmission media through which the information is actually transferred. The channels concepts are shown in Figure 8.92.

While the physical channels in GSM are organized by the *Base Station Controller* (BSC), the physical channels in UTRAN are defined by the radio interface Uu between UE and the access network UTRAN. Consequently, the RNC is hardly aware of the physical channel's existence. Nevertheless, the RNC sees the transport channels. Furthermore, Node B maps the information flow in the transport channels onto the appropriate physical channels. These channels are shown in Figure 8.93.

Considering the common definition of radio channels, one could argue that logical channels are not real channels in the usual meaning of the word. Rather they are logic functions that both the network and the UE need to perform at different times. The network, for example, has to inform the UE of the radio environment such as cell codes and allowed power levels. It also has to page the UE to relay certain communications. The functions of the channels are discussed in this section.

Two types of physical channels are defined to correspond to two UTRAN operations. In the Frequency Division Duplex (FDD) operation, the channels are characterized by a spreading code and the frequency channel. The TDD mode has, in addition, a time slot characterization.

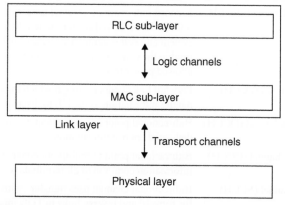

Figure 8.92 *Channel concepts in UMTS.*

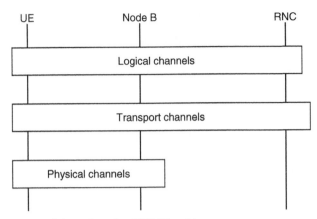

Figure 8.93 *Location of channels within UTRAN architecture.*

UTRAN physical channels can, in general, be grouped into two categories: the *dedicated channels* and the *common channels*. This grouping of UMTS channels is similar to the channel grouping of cdma2000 discussed in Section 8.7. A dedicated channel is used by only one subscriber at a time (i.e. a point to point channel). The common channel is common to all subscribers (i.e. point to multipoint channel).

8.8.4.1 Logical channels

Logic Traffic channels are of two types: *Dedicated Traffic Channel* (DTCH) is a bidirectional point to point channel used to transfer the user's data; the *Common Traffic Channel* (CTCH) is a point to multipoint channel used to transmit information to all terminals or to a specific group of terminals in the cell. The logic control channels are summarized in Table 8.23.

Table 8.23 *Logic control channels*

Logic Control Channels	Applications
Broadcast Control Channel (BCCH)	Forward link common channel used by the network to inform UE of the radio environment by broadcasting system and cell specific information such as codes used in the cell and in the neighbouring cells, allowed power level, etc.
Paging Control Channel (PCCH)	Forward link channel through which the network broadcasts paging information to the UE to discover its exact location.
Dedicated Control Channel (DCCH)	Bidirectional point to point channel to transfer dedicated control information
Common Control Channel (CCCH)	Bidirectional point to multipoint channel to transfer control information common to all terminals residing in the cell
Shared Control Channel (SCCH)	Bidirectional channel uses transfer control information for forward and reverse links in TDD mode only

8.8.4.2 Transport channels

The only dedicated transport channel is the *dedicated channel* (DCH) which is Mandatory, and is used in the forward or reverse link for only one user at a given time. The DCH carries logical channels for traffic DTCH and control information DCCH. The transport channels available for FDD and TDD modes are summarized in Table 8.24. Parameters for the channel encoding for error correction are given in Table 8.25. The mapping of transport channels onto the physical channels are summarized in Table 8.26.

Table 8.24 *The transport channels available for FDD and TDD modes*

Common transport channels	Applications
Broadcast Channel (BCH)	A high power forward link channel carries the content of logic channel BCCH such as random access codes, access slot information and information about neighbouring cells. UE decodes BCH to register to the network
Paging Channel (PCH)	A forward link channel broadcasted over the entire cell and carries paging information used when network initiates a connection with certain UE
Random Access Channel (RACH) (Mandatory)	A reverse link channel used by the mobile terminal for initial access and contains limited size data in the TDD mode
Common Packet Channel (CPCH) (Mandatory)	A reverse link channel associated with a dedicated channel on the forward link and used in the FDD mode only
Forward Access Channel (FACH)	Forward link common channel carries cell control information to a mobile terminal and transmitted over the entire cell or part of the cell. When RCN receives a random access request from UE, its response is delivered through FACH. A number of FACH exist in a given cell.
Downlink Shared Channel (DSCH) (Optional)	A forward link channel shared by several users and carries user information logical channel DTCH and dedicated control logical channel DCCH, and is transmitted over the entire cell or part of the cell. DSCH is more efficient than DCH.
Reverse link Shared Channel (RSCH)	A reverse link shared channel carries traffic data and dedicated control information used in TDD mode only
High-Speed Downlink Shared Channel (HS-DSCH)	A forward link channel shared by several terminals and associated with forward link physical channel DPCH

Table 8.25 *Parameters for the channel encoding for error correction*

Transport channel	Encoder	Encoder rate
PCH BCH RACH	Convolutional encoding	1/2
DSCH, CPCH, DCH, FACH	Turbo encoding No coding	1/3, 1/2 1/3

Table 8.26 *Mapping of transport channels onto the physical channels*

Transport channels	Physical channels
Broadcast Channel (BCH)	Primary common control physical channel (P-CCPCH)
Paging Channel (PCH)	Secondary common control physical channel (S-CCPCH), Synchronization channel (SCH)
Forward Access Channel (FACH)	Secondary common control physical channel (S-CCPCH)
Random Access Channel (RACH)	Physical random access channel (PRACH)
Dedicated Channel (DCH)	Dedicated physical data channel (DPDCH), Dedicated Physical control channel (DPCCH), High-speed data physical control channel (HS-DPCCH)
Downlink Shared Channel (DSCH)	Physical downlink shared channel (PDSCH)
Common Packet Channel (CPCH)	Physical common packet channel (PCPCH) Synchronization channel (SCH) Paging indicator channel (PICH) Acquisition indicator channel (AICH) CPCH status indicator channel (CSICH)

8.8.4.3 Physical channels

The physical channels comprise of ten common physical channels which are available to all subscribers. Of these channels, only two can be used in the reverse link and the other eight allocated to the forward link. The physical channels operating in the reverse link and forward link are summarized in Tables 8.27 and 8.28, respectively.

Table 8.27 *Reverse link physical channels*

		Applications
Reverse link common physical channels	Physical Random Access Channel (PRACH)	PRACH carries transport channel RACH with low rate user data and uses slotted ALOHA scheme with fast acquisition. PRACH has preambles that are sent prior to data transmission. Once the preamble has been detected and acknowledged with the acquisition indicator channel (AICH) the data frame is transmitted.
	Physical Common Packet Channel (PCPCH)	PCPCH carries CPCH and uses DSMA-CD scheme with fast acquisition indicator.
Reverse link dedicated physical channels	Dedicated Physical Data Channels (DPDCH)	DPDCH is used between the network and one user and carries dedicated user information in the reverse link.
	Dedicated Physical Control Channels (DPCCH)	DPCCH transfers control information to network such as pilot, and data rate information, transport format combination indicator (TFCI), feedback information (FBI) and transport power control command (TPC).

Table 8.28 *Forward link physical channels*

		Applications
Forward link common physical channels	Page Indicator Channel (PICH)	PICH carries page indicators to indicate the presence of a page message on the PCH.
	Acquisition Indicator Channel (AICH)	AICH carries signatures for the random access procedure.
	Physical Downlink Shared Channel (PDSCH)	PDSCH is a forward link shared channel and carries the DSCH.
	Synchronization Channel (SCH)	There are two sub-channels for SCH; the primary and the secondary SCH. SCH is transmitted during the first 256 chips of each time slot and is mainly used for cell search.
	Secondary Common Control Physical Channel (S-CCPCH)	S-CCPCH carries PCH and FACH transport channels and a cell must always contain at least one S-CCPCH. Its bit rate is fixed and relatively low. One configuration of S-CCPCH is to multiplex the PCH information together with the FACH information.
	Primary Common Control Physical Channel (P-CCPCH)	P-CCPCH carries the BCH transport channel in the forward link and is available to all UEs within the cell. It uses a fixed spreading code so that terminals are able to demodulate its content. It has a bit rate of 30 kb/s and a spreading factor of 256 and is transmitted with high power to reach all UEs.
Forward link dedicated physical channels	Common Pilot Channel (CPICH)	There are two types for CPICH: the primary CPICH and the secondary CPICH. CPICH transmits a pre-defined sequence at fixed rate of 30 kb/s.
	Downlink Physical Channel (DPCH) – Downlink DPDCH	DPDCH is used between the network and one user and carries dedicated user information in the forward link direction. Its size is variable and may carry several calls/connections.
	Downlink Physical Channel (DPCH) – Downlink DPCCH	DPCCH transfers control information to the UE such as power control and data rate information.

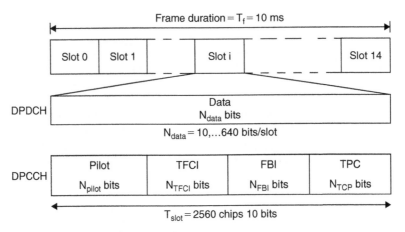

Figure 8.94   *Frame structure for DPDCH/DPCCH.*

8.9   Physical channels spreading and frame structures (3GPP TS25.215, 2002)

8.9.1   Reverse link dedicated physical data/control channels

The frame structure of the reverse link *Dedicated Physical Data Channel* (DPDCH) and *Dedicated Physical Control Channel* (DPCCH) has a length of 2560 chips in 10 ms duration and contains 15 time slots. Thus, 15 bursts are continuously transmitted through the reverse link without a guard time. The frame structure is shown in Figure 8.94.

There can be several DPDCH to carry coded and interleaved user data at $N_{data} = 10 \times 2^k$ bits per slot for $k = 0, \ldots, 6$ and the *spreading factor* (SF) $S_F = \frac{256}{2^k} = 256 \ldots \ldots \ldots 4$. Thus, the maximum and the minimum spreading factors are 256 and 4, respectively, and the maximum transmission speed is 960 kb/s corresponding to minimum spreading factor of 4. The spreading of a DPDCH and DPCCH is shown in Figure 8.95. The DPDCH information is carried over I-branch using spreading code C_d and weight factor β_d. The DPCCH is carried over the Q-branch using a spreading code C_c and weight factor β_c.

A discontinuous transmission is used in the reverse link, such as with a speech service where no information bits are transmitted during the silent period and only information required for the link maintenance, such as low command rate power control bits, are transmitted. The DPDCH and DPCCH are transmitted in parallel to avoid an audible interference in the voice frequency band as shown in Figure 8.96.

The parallel transmission of DPDCH and DPCCH increases the peak to an average power ratio. However, this ratio has to be kept low to provide a better power efficiency for the mobile terminal. This is achieved by the complex scrambling after the spreading modulation. The weight factor $\beta_{d/c}$ defines the power share between DPDCH and DPCCH. The

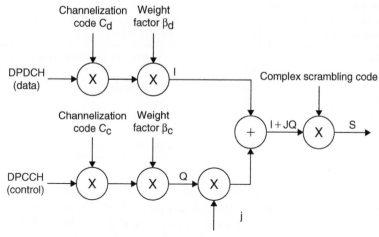

Figure 8.95 *Spreading of DPDCH and DPCCH channels.*

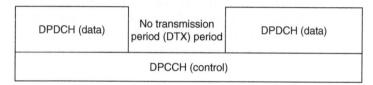

Figure 8.96 *Parallel transmission of DPDCH/DPCCH.*

power difference between DPDCH and DPCCH is specified by 4 bits, i.e. 16 possible combinations with 15 combinations of power differences covering the dynamic range between 0.0 to -23.5 dB and one when no data (i.e. no DPDCH) is transmitted.

There is exactly one DPCCH per connection for signalling information transmission between the physical layers of the terminal and Node B with a spreading factor of 256. The DPCCH slot carries exactly 10 bits corresponding to the bits associated with pilot, *Transport Format Combination Indicator* (TFCI), *Feedback Information* (FBI) and *Transmission Power Control* (TPC). For example, at a given DPCCH transmission, the pilot can be represented by 5 bits, the TFCI by 2 bits, the TPC by 2 bits, and the FBI by 1 bit.

The pilot bits are used for channel assessment, TFCI bits are used to indicate transport formats on the transmitted DPDCH to help the physical layer of the receiver to reconstruct the transport blocks of the individual channels from the data streams of the physical channels. The FBI bits are used for signalling with soft handoff and the TPC bits carried a power control command to increase, maintain or reduce transmitted power. The gross transmission rates of the DPDCH/DPCCH are given in Table 8.29.

A maximum of six parallel DPDCH can be implemented simultaneously to increase the data rate above 960 kb/s using *Orthogonal Variable Spreading Factor* (OVSF). The generation

Table 8.29   *Gross transmission rates of the DPDCH/DPCCH*

Spreading factor	Bits/slot	Bit rate kb/s
256	10	15 (DPCCH and DPDCH Min. rate)
128	20	30
64	40	60 (Typical)
32	80	120
16	160	240
8	320	480
4	640	960 (Max.)

of the OVSF is presented in the next section. The theoretical maximum gross rate is 6×960 kb/s $= 5760$ kb/s but the standard defines a gross rate of 1920 kb/s. A block diagram of combining the physical channels on the reverse link (DPDCH/DPCCH) is shown in Figure 8.97. All channels which are combined in the reverse link are scrambled with the same scrambling code and the channelization codes are orthogonal codes. As seen in Figure 8.97, all data channels have equal power weights which are different from the power weight of the control channel. The power weights β_d and β_c can be used to define different *Quality of Service* (QoS) requirements for different channels. The channels with higher QoS requirements can be transmitted with a relatively higher power level.

8.9.2   Orthogonal variable spreading factor and scrambling codes

The spreading process consists of two stages: spreading and scrambling, and both stages use different codes with different characteristics as shown in Figure 8.98. The spreading uses orthogonal codes and is known as channelization. UTRAN uses OVSF codes created through the use of a code tree and each node of the tree has exactly two branches. Each branch represents double length code and branches at each node have equal length and thus the same level of spreading factor. Consequently, a code with a spreading factor N can be created from a code with $\frac{N}{2}$ nodes. Considering the finite number of OVSF codes, they have to be re-used in more than a single cell which means that the same code can be allocated to other users in adjacent cells. The OVSF code tree is shown in Figure 8.99.

Scrambling is carried out after the spreading, and used to identify Node B to the terminal on the forward link and to identify the terminal to Node B on the reverse link. A scrambling code can be either a short code spanning one symbol duration or a long code spanning several symbols.

A complex-valued scrambling process is used in the reverse link DPDCH/DPCCH channels using either long or short codes. The long scrambling codes are chosen from a set of Gold sequences of 38,400 chips generated using two binary m-sequences generated using polynomials of degree 25. The first m-sequence $x_n(i)$ is generated using polynomial $x^{25} + x^3 + 1$ and the second m-sequence $y_n(i)$ is generated using polynomial $x^{25} + x^3 + x^2 + x + 1$.

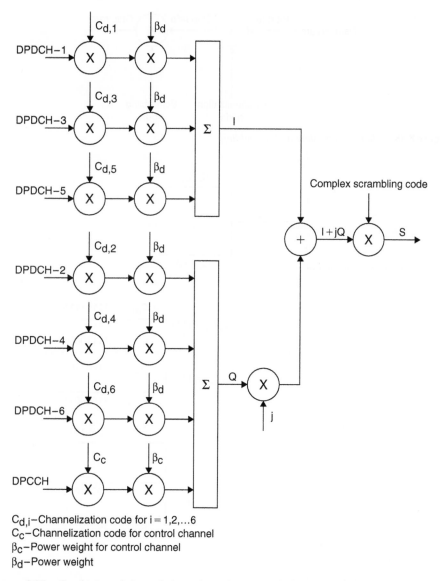

$C_{d,i}$—Channelization code for i = 1,2,...6
C_c—Channelization code for control channel
β_c—Power weight for control channel
β_d—Power weight

Figure 8.97 *Combining of physical channels on the reverse link.*

The algorithm for the generation of the long and short scrambling codes is given in 3 GPP TS 25.213 (2002).

8.9.3 Reverse link physical common packet channels

The *Physical Common Packet Channel* (PCPCH) carries the Common Packet Channel (CPCH). The CPCH uses the *Digital Sense Multiple Access–Collision Detection* (DSMA-CD) scheme for transmission. The structure of the CPCH access transmission is shown

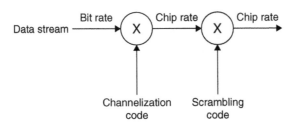

Figure 8.98 *Channel spreading and scrambling.*

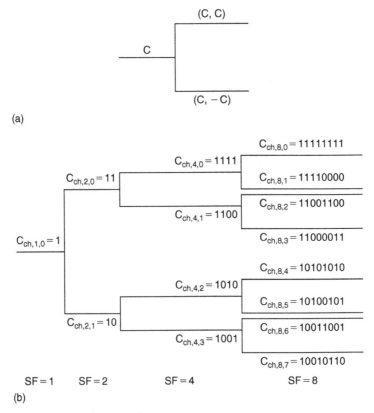

Figure 8.99 *Generation of OVSF codes.*

in Figure 8.100 and consists of one or several access preambles of length 4096 chips $P_0, P_1, \ldots, P_{j-1}$, one collision detection preamble of length 4096 chips $\mathbf{P}_j$, DPCCH power control preamble (PC–P) of either 0 slot or 8 slots in length, and a message of variable length N × 10 ms.

Each CPCH message consists of N × 10 ms frames and each frame splits into 15 slots. The length of each slot is 2560 chips and each slot consists of a data part that carries higher layer information and a control part that carries layer 1 control information. The data and

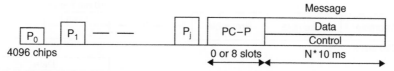

Figure 8.100 *Structure of the CPCH random access transmission.*

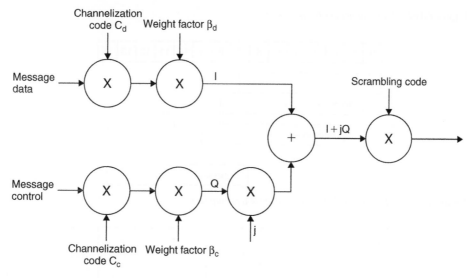

Figure 8.101 *Spreading and scrambling of PCPCH message.*

control parts are transmitted in parallel. The CPCH spreading and scrambling is similar to the DPDCH/DPCCH transmission shown in Figure 8.97. The spreading of the PCPCH message is shown in Figure 8.101.

8.9.4 Physical random access channel

The *Physical Random Access Channel* (PRACH) carries the random access channel and uses a slotted ALOHA technique for transmission. The UE transmits packets at the start of the slot time only and, in the case of re-transmission, it waits for a random number of time slots and then resends. The structure of the random access transmission is shown in Figure 8.102. The transmission consists of one or several preambles. The length of each preamble is 4096 chips. Each preamble is constructed from a 16 chip signature repeated 256 times, i.e. $256 \times 16 = 4096$ chips. The message length is 10 ms or 20 ms.

There are 15 access slots per two frames. The duration of each slot is 5120 chips. The random access slots for slotted ALOHA scheme is shown in Figure 8.103.

The 10 ms message of RACH consists of data part of 10×2^k bits where $k = 0, 1, 2, 3$ corresponding to a spreading factor 256, 128, 64, and 32. The control part of the message

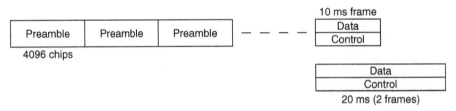

Figure 8.102   *Random access transmission.*

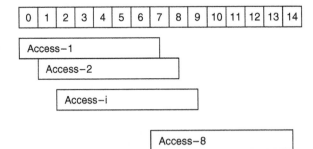

Figure 8.103   *Slotted ALOHA random access transmission.*

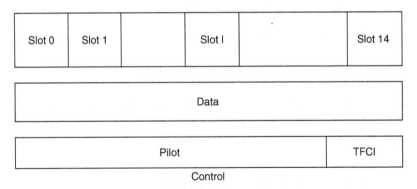

Figure 8.104   *Structure of the random access message.*

consists of 8 known pilot bits used for channel estimation for coherent detection and 2 TFCI bits. The spreading factor of the message control part is 256. Structure of the random access message is shown in Figure 8.104. The spreading of the PRACH message part is shown in Figure 8.105 which is similar to the DPDCH/DPCCH channel in Figure 8.95.

8.10   Forward link physical channels (3GPP TS25.211, 2002)

8.10.1   *Dedicated forward link physical channels*

The *Dedicated forward link Physical Channel* (DPCH) carries data generated at link layer (layer 2) of the DPDCH time multiplexed with control information of the DPCCH generated

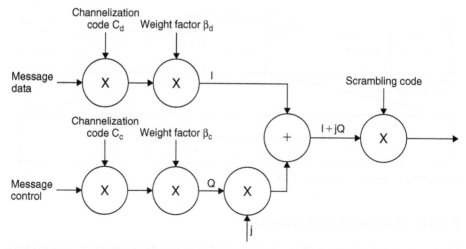

Figure 8.105 *Spreading of PRACH message.*

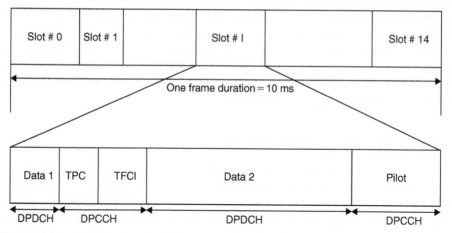

Figure 8.106 *Forward link DPCH frame structure.*

at physical layer (layer 1). The forward DPCH frame is shown in Figure 8.106. Each frame is 10 ms length and made up of 15 slots. Each slot has a length of 2560 chips. The data size is variable given by 10×2^k bits where $k = 0, \ldots, 7$ so the minimum size of the data is 10 bits and the maximum is 1280 bits. The spreading factor is $\frac{512}{2^k}$ giving a minimum spreading factor of 4 and a maximum of 512.

The exact number of bits of different fields (Pilot, Data 1, Data 2, TFCI and TPC) is tabulated in the standards and these numbers depend on bit rate and services being carried by the channel.

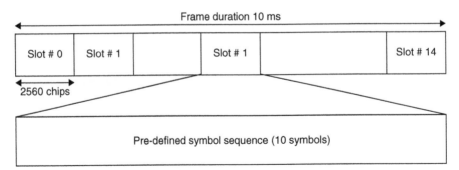

Figure 8.107   *Frame structure for CPICH.*

8.10.2   Common pilot channel

The function of the *Common Pilot Channel* (CPICH) is to aid the mobile terminal in channel estimation. The CPICH is an unmodulated code channel scrambled with the cell specific primary scrambling code. UTRAN has two types of CPICH, namely the primary and the secondary. The primary type uses the primary scrambling code with a pre-defined channelization code sequence. Each cell/sector has only one primary and secondary CPICH broadcast over the entire cell/sector. The secondary channel uses an arbitrarily channelization code of spreading factor 256. The main use of the primary CPICH is in the measurements needed for the handoff and cell selection. The typical application of the secondary CPICH would be with narrow antenna beam in a hot spots operation. The frame structure of the CPICH is shown in Figure 8.107.

8.10.3   Synchronization channel

The *Synchronization Channel* (SCH) is not visible above the physical layer and has a frame consisting of 15 slots. The SCH is used for cell search and consists of two subchannels, primary and secondary, used by the mobile terminal to identify the scrambling code in use by the SCH to enable the terminal to synchronize to the cell. The primary SCH is the same in every cell in the network and contains unmodulated code sequence of 256 chips constructed from 16 shorter (16-chip) sequences. The primary 256-chip sequence is first detected using a matched filtering scheme.

The secondary SCH codes vary from cell to cell and are constructed from 64 different codes which are used to identify the cell and no scrambling is used on top of the 64 codes used for the secondary SCH. Each scrambling sequence is associated with 8 primary scrambling codes giving 512 primary scrambling codes for use in each cell. The SCH is time multiplexed with the primary *Common Control Physical Channel* (CCPCH). The primary and secondary SCH are sent in parallel.

The cell search is initiated when the terminal correlates the primary SCH 256-chip code with the contents of each slot. The peaks correspond to the boundaries of the slots and

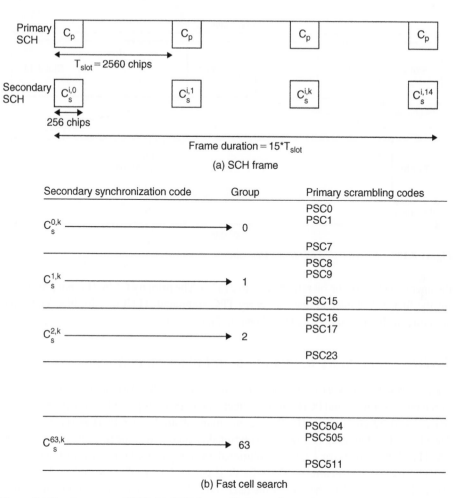

Figure 8.108 *Structure of SCH. (a) SCH frame; (b) Fast cell search.*

hence this step achieves slot synchronization. The terminal then correlates the 64 secondary SCH codes with the content of the 15 slots and the largest peak defines the desired slot and hence the secondary SCH code in use. From the knowledge of the secondary SCH code, the terminal identifies the primary scrambling code group. All the 8 primary scrambling codes in the group have to be tested for the particular slot position to identify the scrambling code in use. The structure of the SCH and the cell search are shown in Figure 8.108.

8.10.4 Primary common control physical channel

The *Primary Common Control Physical Channel* (P-CCPCH) carries the transport channel BCH and has to be demodulated by all terminals. Consequently, all terminals need to know the channel coding and spreading code of 256 chips. The channel uses a $\frac{1}{3}$ rate convolution encoder and transmits at a fixed rate of 30 kb/s. Since the primary CCPCH alternates with

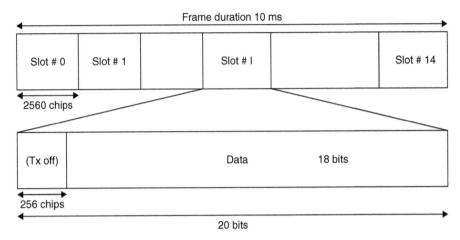

Figure 8.109 *Primary CCPCH frame structure.*

the unencoded SCH, the bit rate reduces to 27 kb/s. The primary CCPCH is not transmitted for the first 256 chips and does not carry TPC command, TFCI or pilot bits. The frame structure of the primary CCPCH is shown in Figure 8.109.

8.10.5 Secondary common control physical channel

The *Secondary Common Control Physical Channel* (S-CCPCH) carries two different transport channels: FACH and PCH. The two subchannels are multiplexed on a single S-CCPCH or can use different S-CCPCHs. Thus, a minimum of single S-CCPCH exists in each cell. The channel coding used is $\frac{1}{2}$ rate convolutional encoder when the S-CCPCH carries both FACH and PCH. Turbo or $\frac{1}{3}$ rate convolutional encoding is used for data transmission with FACH. The frame structure for the S-CCPCH carries TFCI, pilot as well as data bits as shown in Figure 8.110. The size of the data varies from 20 bits minimum to a maximum of 1288 bits and generally is given by 20×2^k bits where $k = 0, \ldots, 6$ depending on the spreading factor in use by the cell.

8.10.6 Physical downlink shared channel

The *Physical Downlink Shared Channel* (PDSCH) carries the transport DSCH and is shared between the active users and their data is multiplexed. The base station indicates to the UE when it has to decode the PDSCH through the TFCI field. The frame structure for the PDSCH is shown in Figure 8.111. The PDSCH carries layer 1 control information. The SF varies between 4 and 256 in 7 values. The bits per slot (Data) is given in Table 8.30.

8.10.7 Paging indicator channel

The *Transport Paging Channel* (PCH) operates with the paging indicator channel to provide the mobile terminal with the sleep mode when not in use. The paging indicator channel uses

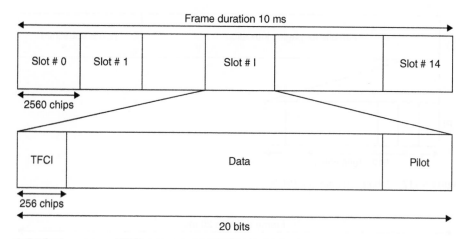

Figure 8.110 *Frame structure for the S-CCPCH.*

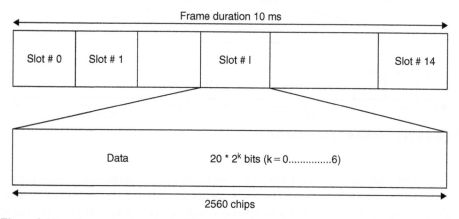

Figure 8.111 *Frame structure for the PDSCH.*

Table 8.30 *PDSCH bit rates and bit/slot*

Spreading factor	Channel bit rate kb/s	Bits/slot
256	30	20
128	60	40
64	120	80
32	240	160
16	480	320
8	960	640
4	1920	1280

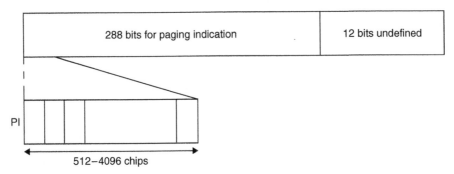

Figure 8.112   *Frame structure for PICH.*

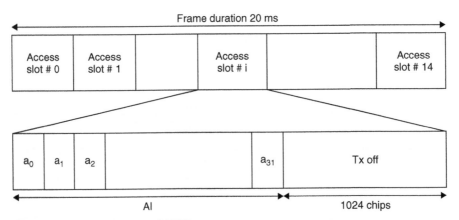

Figure 8.113   *Frame structure of AICH.*

256 chip channelization codes. Each paging indicator channel frame transmits N paging indicator (N = 18, 36, 72 or 144). The frame carries 288 bits and 12 bits idle. The paging indicator channel is broadcasted with relatively high power at a low bit rate. The frame structure of PICH is shown in Figure 8.112.

8.10.8   Acquisition indicator channel

When a mobile terminal wants to access PRACH, it selects an available access slot and one of the 16 preamble sequences. The preamble sequence is transmitted at a lower power and the mobile terminal awaits an acknowledgement over the *Acquisition Indicator Channel* (AICH). If the terminal does not receive an acknowledgement from the base station, or if it receives a negative acknowledgement, it selects a new access slot and a new preamble and transmits the preamble sequence with a relatively higher power. The structure of AICH is shown in Figure 8.113. Each access slot is made up of two parts: an *Acquisition Indicator* (AI) part containing 32 real value symbols a_i and a part without transmission of duration 1024 chips.

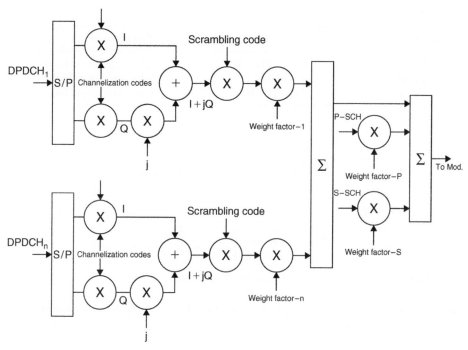

Figure 8.114 *Combining the physical channels on the forward link.*

The forward physical channels are combined together with SCH. Figure 8.114 shows only the DPDCH. Other channels are processed in the same configuration as a DPDCH. All channels, except the SCH, are scrambled with the same scrambling code. The combined spread and scrambled signal is modulated on a QPSK carrier for transmission.

8.11 Rate matching (3GPP TS25.212, 2002)

Transmissions on the physical channels in both reverse and forward directions are structured in frames of fixed length, implying a fixed bit rate for transmission. The information on the transport channels generally occurs in a varying bit rate. Consequently, there is a need to match the two rates for efficient use of the spectrum.

Rate matching in the reverse link channels is achieved by either repeating or puncturing the data. Puncturing is deleting bits before transmission according to a pre-defined scheme. Puncturing is always carried out after channel encoding and, in a sense, it removes some of the redundancies created by the encoder. However, puncturing may weaken the encoding process. On the other hand, it may be required to increase the number of symbols to completely fill in the reverse channel frame through repeating some reverse frame symbols. Clearly, repetition of symbols is preferred since it does not degrade the system channel coding performance.

Table 8.31   *Packet transmission parameters*

Reverse link packet	Forward link packet	Packet properties
CPCH/PCPCH	DSCH/PDSCH	Infrequent N × 10 ms frames
RACH/PRACH	FACH/SCCPCH	Infrequent 10 ms frames
DCH/DPCH	DCH/DPCH	Frequent multiple frames

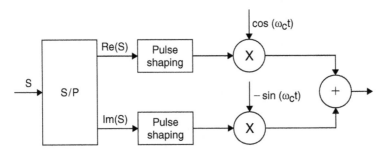

Figure 8.115   *Reverse link and forward link QPSK modulation.*

Similarly, rate matching is required in the forward link and usually the network interrupts the transmission if the rates are not matched. This is called *Discontinuous Transmission* (DTX).

8.12   Packet transmission summary

Table 8.31 itemizes the packet transmissions and their properties.

8.13   Physical channels carrier modulation

The forward physical channels P-CCPCH, S-CCPCH, CPICH, AICH, PICH, PDSCH, HS-SCCH, and DPCH are scrambled, combined with the SCH, and transmitted over a QPSK carrier using *Root Raised Cosine* (RRC) filter for pulse shaping. The reverse dedicated channels DPDCH and DPCCH also use QPSK modulation as shown in Figure 8.115.

8.14   Service multiplexing on the reverse link physical channels

Transport channels are unidirectional and are used between the MAC sub-layer and physical layer. The physical layer operates on 10 ms frames which have to be filled with data from the MAC on the transport channels using rate matching if necessary. The data is generated by the MAC every 10 ms or multiples of 10 ms and sent to the physical layer. The CRC of length 8, 12, 16, and 24 bits is added so that it can be used by the receiver for error checking of the transport blocks. The transport block is segmented, if required, to fit the channel coding used. A channel coding uses rate $\frac{1}{3}$ or $\frac{1}{2}$ convolutional encoder or rate

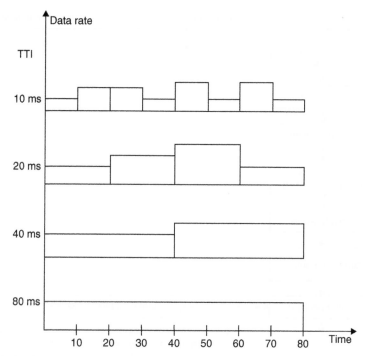

Figure 8.116 *Transmission Time Intervals (TTI).*

$\frac{1}{3}$ Turbo encoder or perhaps channel coding is not used depending on the quality of the channel.

After channel coding, the data is divided into equal blocks when transmitted over frames of length more than 10 ms by padding the necessary bits to produce equal sized blocks per frame.

The *Transmission Time Interval* (TTI) is the inter-arrival time from the MAC to the physical layer which is a multiple of 10 ms, i.e. 10, 20, 40, 80 ms. The TTI informs how often the MAC sends data to the physical layer. The amount of data sent is determined by the block sets size as shown in Figure 8.116.

The first interleaver is related to the TTI. The start positions of the TTI are from different transport channels, which are multiplexed together for a single connection, and are time aligned as shown in Figure 8.117. The data frame is then segmented into 10 ms frames followed by rate matching to match the number of bits to be transmitted to the number available on the frame. Different transport channels are multiplexed in the same frames. A second interleaver known as intra-frame interleaving performs a 10 ms frame block interleaving on individual physical channels and the data is mapped onto the physical channels. The multiplexing of the reverse physical channels is shown in Figure 8.118.

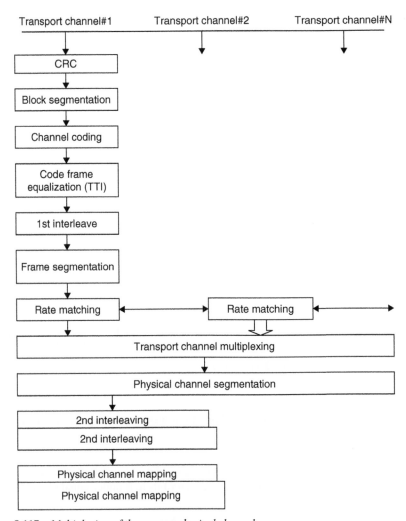

Figure 8.117 *Multiplexing of the reverse physical channels.*

We now consider the mapping of a single transport channel carrying data to the corresponding physical channel. Let the size of the transport block be 132 bits, the CRC be 12 bits and $\frac{1}{2}$ convolutional encoding is in use. Furthermore, assume the TTI to be 20 ms. The steps leading to transport channel multiplexing are shown in Figure 8.119.

8.15   Forward link multiplexing

Multiplexing in the forward link is similar to the multiplexing of different services on the reverse link, with some functions that are performed differently. The interleaving is performed in two parts, inter-frame and intra-frame and rate matching are done in a similar way as in reverse link. However, there are differences in the order in which rate

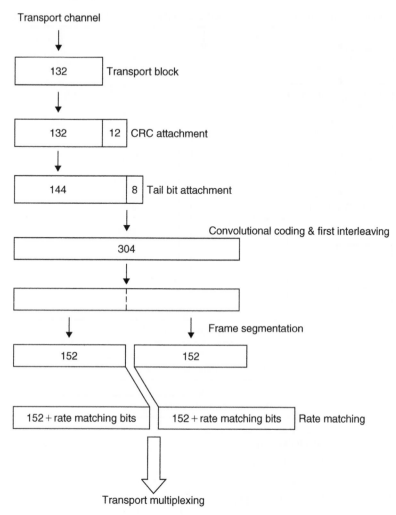

Figure 8.118 *Mapping transport blocks to multiplexed transport channels.*

matching and frame segmentation are performed, as shown in Figure 8.119. The DTx are not transmitted but inserted to inform the transmitter to turn off at these bit positions.

8.16 Power control in UTRAN FDD (3GPP TS25.214, 2002)

The two basic power control methods used in CDMA systems are presented in detail in Chapter 6. In this section, we present the power control adopted in WCDMA standards. Two power control schemes are used in the UTRAN FDD in both reverse and forward links: fast *closed-loop power control* and *outer loop*. The closed-loop, also called 'inner loop', is used for the DPDCH/DPCCH. Another scheme, the *open-loop power control*, is used in the physical random access channel. The power control in the forward link shared

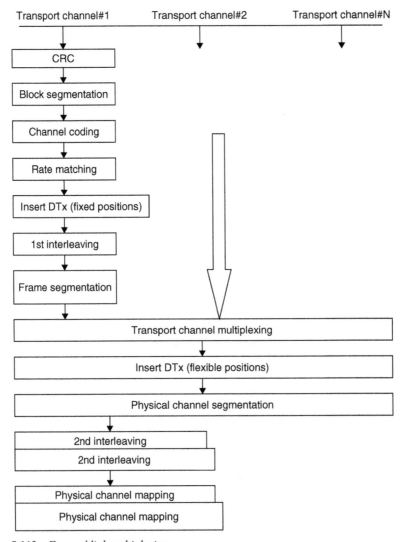

Figure 8.119 *Forward link multiplexing.*

channel is based on the frame error rate rather than specified SIR in a method called *slow closed power control*, which is similar to the fast loop but reacts more slowly to changes in the transmission environment.

8.16.1 Reverse link power control

The reverse link power control is shown diagrammatically in Figure 8.120. The SIR of the received RL is measured by the base station and the RL signal is decoded to adjust the target SIR accordingly during the next transmission from the mobile terminal to the base station. The new SIR generates power control command bits which are multiplexed

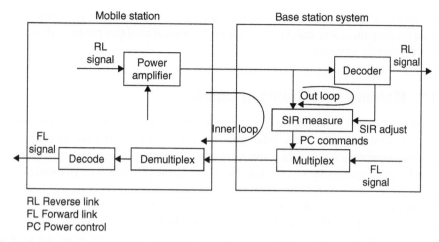

Figure 8.120 *Reverse link power control.*

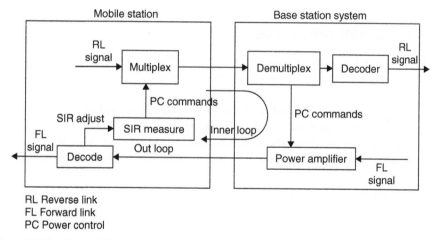

Figure 8.121 *Forward link power control.*

with the FL signal to increase/decrease the mobile transmit power. The mobile terminal demultiplexes the PC commands and uses them to adjust the power amplifier of the mobile terminal to the new transmit power to give the required SIR at the base station.

8.16.2 Forward link power control

The forward link power control is depicted in Figure 8.121. The received RL signal is demultiplexed to extract the power control command bits to adjust the level of the power amplifier at the base station. The SIR of the received FL signal is measured at the mobile terminal and the FL signal is decoded to adjust the target SIR accordingly. The new SIR generates power control commands which are multiplexed with the transmitted RL signal and used to increase/decrease the base station transmit power. The control procedure is repeated to keep the power under control within the defined level that produces the required SIR at the mobile terminal.

8.17   Mobility procedures in UMTS

The user mobility is managed by two procedures which enables the network to provide services to a roaming user. These procedures determine the exact location of a user using *location management* and ensure seamless wireless connection when the user moves from one cell area to another by a procedure known as *Handoff* (HO). The HO switches the user's wireless connection from one base station to another and is commonly associated with connection changes between circuit-switched (i.e. GSM) domains. However, such a procedure between WCDMA domains is not really a HO at all but an intersystem change. We first explain the various HO possibilities followed by deliberation on the location management methods.

8.17.1   Handoff procedures

The WCDMA systems interoperate with circuit-switched as well as with packet-switched networks such as GSM, GPRS, and consequently employ three different types of HOs.

(a) *Hard HO* between GSM and UTRAN TDD domains. A hard HO provides a hard connection switch to the new cell and occurs from frame to frame.
(b) *Soft HO* is employed mainly in packet-switched domains such as WCDMA systems when a terminal communicates simultaneously with more than one base station (Node B).
     In the soft HO, there is no fixed switch over point. Instead a soft connection is transferred from one base station to another as the mobile terminal moves into the new cell indicating a complete termination to the old station and the soft HO is completed.
     On the forward link, the data is split by the RNC to a number of base stations and the received data is combined by the mobile terminal. On the reverse link, the participating base stations forward the received data to the serving RNC to combine and transfer to the CN. The soft HO is also called *macro-diversity* and offers resistance to interference from shadowing, a phenomenon when the wireless connection is cut off by an obstacle such as a high building between the mobile terminal and the serving base station, since there is a possibility that another base station is able to keep the connection alive.
(c) *Softer HO* is a special version of the soft HO in UTRAN FDD applied for intra-cell to change the user's connection from one sector to another in the same coverage area of the serving base station.

The RNC plays a pivotal role in soft HO providing the connection between base stations involved in the soft HO procedure. The *Serving RNC* (S-RNC) combines the received signals and sends them to the *Mobile Switching Centre* (MSC). The RNC, in this case, is responsible for the *Radio Resource Control* (RRC) between the terminal and the UTRAN system. As the roaming user moves to a new *Location Area* (LA), the terminal keeps the connection with at least one of the base stations and initiates a new connection with new base stations belonging to a different RNC, called the *drift* RNC, but the combining of the received signals is carried out by the S-RNC. If the user continues moving to leave all cells controlled by the S-RNC, a relocation process takes place in UTRAN to change the status of the new RNC as the S-RNC, so that all traffic between the CN and the terminal is transferred through the new S-RNC.

8.17.2 Location management procedures

The main function of the location management is to determine the exact location area of the terminal so that the network is able to transfer the incoming calls and data packets to the desired mobile terminal. The mobile terminal detects its location area from the information broadcasted by the base station. When it moves into a new LA, it registers with the network and the new location of the terminal is stored in a database in the CN so that incoming calls can be routed to the correct LA.

When the network wants to make a connection with the terminal, it pages the terminal in all cells in the LA after consulting the location database. In the packet-switched networks, such as GPRS and WCDMA, the coverage area is divided into LAs. Furthermore, a new concept named the *Routing Area* (RA) and the *UTRAN Registration Area* (URA) are defined. The URA consists of one or more cells. The *Serving GPRS Support Node* (SGSN) executes the RA update, which can trace the cell where the terminal is located accurately to relay the incoming packets.

8.18 Evolution of the WCDMA standard

Packet data services, such as web surfing and file download, were previously provided in the first release of the WCDMA standard. Although this is a significant improvement compared to 2G standards, where such services have no or limited support, WCDMA is being continuously evolved to provide a high data rate service. This is called *WCDMA evolved*. The evolution is carried out in two steps: launch of the HSDPA in Release 5, and the introduction of the high-speed uplink packet access in Release 6. In the following sections we will use down-link and uplink to denote the forward link and reverse link, respectively, in accordance with the standard.

8.19 High-speed downlink packet access (Holma and Toskala, 2005; Kaaranen et al., 2005; 3GPP TR25.858, 2002)

High-speed data transmission on the forward link is a necessity for services such as video, on demand, and Internet browsing. Both Release '99 and Release 4 proposed different channels which can be used for forward link packet transmission. The *Dedicated Channel* (DCH) and the *Downlink Shared Channel* (DSCH) are used in Release 4 to handle data traffic, while a low-rate data can also be handled by the *Forward Access Channel* (FACH). However, the maximum downlink data rates provided by these channels are too low for services which include multimedia transmission.

HSDPA is introduced by 3GPP in Release 5 to support services requiring a high data rate in the down link and is defined for both FDD and TDD systems. In this section, the HSDPA is described for FDD systems though the principles are similar in TDD systems.

HSDPA is a complex protocol requiring a significant upgrade to Node Bs to cope with the increase in data throughput. HSDPA increases the data rate in the direction to a terminal within a cell to a theoretical maximum of 14.4 Mb/s with an improved quality, more reliability, and robust services. Consequently, instead of the high-speed data access being limited to a few users in a cell using DCHs as offered in Releases '99, the HSDPA delivers data at a sevenfold rate obtained with the DCHs to more users simultaneously sharing one higher bandwidth channel called the *High-Speed DSCH* (HS-DSCH).

The maximum achievable bandwidth with HSDPA is significantly affected by cell size. The maximum data rate tends to fall away for users at the edge of the cell. For macro cells with a diverse range of users, the peak aggregate data rate will be in the range of 1–1.5 Mb/s. The peak data rate increases to more than 6 Mb/s for the users within a micro cell. Furthermore, a pico cell could get a load data rate of 8 Mb/s. HSDPA achieves the high data rate using higher-level modulation such as 16-QAM with an adaptive coding scheme based on turbo coding with coding rate varying on a per user basis between $\frac{1}{4}$ and $\frac{3}{4}$. Under poor reception conditions, the modulation can revert back to QPSK from the proposed 16-QAM.

Furthermore, the high-speed data transmission offered by HSDPA requires fast control of the Automatic Repeat Request (ARQ) and therefore is moved from the serving RNC to the serving Node B, reducing the response time of the network and providing fast scheduling. However, the fast scheduling increases the system complexity significantly. The TTI for HSDPA is just 2 ms compared with a typical time of 10 ms and a maximum of up to 80 ms in 3GPP previous releases.

When a link error occurs, the data packet can be retransmitted quickly at the request of the mobile terminal and a new HARQ is developed to provide efficient re-transmission of dropped or corrupted packets causing low latency that is essential to HARQ procedures. In addition, the fast re-transmission is implemented with chase combining or incremental redundancy of the erroneous packets at the mobile terminal to correct packets received in error.

The need to support a different *Quality of Service* (QoS) for individual users further complicates the functionality of the scheduler as it cannot deny bandwidth to users with high QoS requirements just because of their poor reception area at a given instant in time.

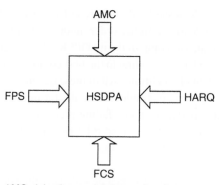

AMC–Adaptive modulation and coding
HARQ–Hybrid automatic repeat request
FCS–Fast cell selection
FPS–Fast packet scheduling

Figure 8.122 *HSDPA concept.*

The fast scheduling per user modulation and coding adaptation procedures require dedicated fast processing and sizable buffering resources almost on a per user or per function basis. For example, one processor may collate information for a processor which just runs the scheduling algorithm allowing scheduling decisions with low latencies. This line of thinking suggests the possibility of parallel processing architecture for HSDPA.

8.19.0.1 HSDPA concept

The HSDPA concept is a combination of several techniques, but the key idea behind the high data throughput is similar to methods already in use in the GSM and the EDGE standards: link adaptation using *Adaptive Modulation and Coding* (AMC) and *Fast Packet Scheduling* (FPS) for transmission/re-transmission techniques using HARQ. The capabilities of HSDPA are enhanced even further using *Fast Cell Selection* (FCS) and *Multiple Input Multiple Output* (MIMO) techniques in Release 7. The main components of the HSDPA concept are shown in Figure 8.122.

The use of AMC in WCDMA represents a significant evolution which replaces the variable SF and the fast power control such that an appropriate coding and modulation combination is selected to improve the signal to interference ratio and thus increase the user's throughput. In fact, the use of HARQ and multicode assignment removes the need for the use of the variable SF.

The aim of the AMC procedure is to select the most appropriate modulation and coding schemes which physical layers can use for transmission to compensate for variations of the radio environments and improve the *Quality of Service* (QoS). The AMC is facilitated by the channel measurements carried out by the mobile terminal using a *Channel Quality Indicator* (CQI) and re-transmission algorithms.

The modulation specified in Release 4 is the QPSK and Release 5 allows use of either QPSK or 16-QAM modulation. Clearly, a higher order modulation, like 16-QAM, provides the possibility of higher data rates as compared with QPSK. HSDPA combines the modulation with the channel encoding process, leading to the best combination of multicode, data rate and modulation for the channel conditions. All of these benefits of the AMC are dependent on a reliable channel measurement conducted by the terminal which may contain some errors when the channel experiences a fast fading process. Consequently, the HSDPA is supported with CQI which uses the received pilot power of the Common Pilot Channel (CPICH) to ensure error-free AMC operation.

In WCDMA system, the terminal communicates with more than one Node B and constructs a list of an active Node B set. When a given Node B begins to indicate degradation, it will be removed from the set. The *Fast Cell Selection* (FCS) allows the UE to scan through the list of the active Node B set and select the cell that indicates the best characteristics to transmit the data packet. This protocol allows the system to achieve a higher data rate since the UE is able to transmit on favourable channel conditions.

8.19.1   HSDPA channels

Three new channels are defined in the physical layer of the HSDPA: the *High-Speed Downlink Shared Channel* (HS-DSCH), the *High-Speed Shared Control Channel* (HS-SCCH) (both of these channels operate over the forward link). The *High-Speed Dedicated Physical Control Channel* (HS-DPCCH) is defined to operate in the reverse link.

As noted previously, the HS-DSCH carries user data in the forward direction with a peak data rate reaching 14.4 Mb/s and has a TTI of 2 ms to ensure a shorter round trip delay between the mobile terminal and Node B compared to the 10, 20, 40 or 80 ms supported in Release '99. It uses a fixed SF of 16 bits and a maximum multicode transmission as well as code multiplexing of different users. The maximum number of codes allocated is 15 but it depends on the capability of individual mobile terminals to receive 5, 10, or 15 codes. The maximum number of channelization codes for SF 16 bits is 16 but as there is a need for at least one control code for use i.e. HS-SCCH, the maximum useable number of channelization codes is 15. In UTRAN systems, several such codes are allocated for a single user for a given instant of time.

The forward link HS-SCCH carries the necessary signalling information from the network to the terminal so that it can be used for demodulation of the data carried by the HS-DSCH. For example, the information indicates which terminal gets data on the next HS-DSCH transmission, the HARQ related information and the terminal's identity. If, for example, the terminal identity is identical to the identity in the HS-SCCH, then the terminal knows that it receives data on the next HS-DSCH frame. Terminals check the information on the HS-SCCHs to find out which code to use for de-spreading the received signal to decode the desired data. The number of HS-SCCHs used by UTRAN corresponds to the number of active users multiplexed using the HS-DSCH, and a maximum of four can be handled by a given terminal in a given time. If there is no HS-DSCH transmission then there will be no need to transmit HS-SCCH.

The HS-SCCH block is made up of two parts covering three-slot durations and each part carries terminal-specific information to allow the terminal to decide whether the control channel is actually intended for it. Both parts of the HS-SCCH use a $\frac{1}{2}$ rate convolutional coding but both parts are encoded separately from each other because the information in the first part of the block is needed immediately by the terminal.

The first part of the HS-SCCH block has a duration one slot and it contains information necessary for the demodulation of HS-DSCH such as the type of codes needed to de-spread, type of modulation used (QPSK or 16-QAM), the ARQ process index (index to which the data belongs), and transmission/re-transmission information. The second part contains the CRC which is used to check the validity of the received HS-DSCH data frame.

Once the data has been proceeded by the high-speed medium access control (MAC-hs), the terminal sends *Positive/Negative Acknowledgement* (ACK/NAK) back to the network on the reverse link control channel HS-DPCCH. Thus, HS-DPCCH carries both ACK/NAK and the channel quality feedback information measured by the terminal to be used by the Node B scheduler to decide which terminal to transmit to in the next TTI and at what data speed. The frame structure of the HS-DPCCH is divided into two parts as shown in Figure 8.123. The first part carries the ACK/NAK to reflect the result of the CRC check, and the second part carries the *Channel Quality Indicator* (CQI) to indicate the transport block size, modulation type (QPSK or 16-QAM), and the number of codes that can be correctly received in the forward link transmission. The CQI information is conveyed in

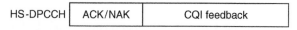

Figure 8.123 *HS-DPCCH frame structure.*

Table 8.32 *Summary of HSDPA channels*

		Functionalities
Forward link channels	High-Speed Physical Downlink Shared Channel (HS-PDSCH)	A data bearer associated with DCH, up to 15 can be used at one time, peak data rate 14.4 Mb/s, spreading factor SF = 16
	High-Speed Shared Control Channel (HS-SCCH)	A signalling channel carries HARQ information, up to 4 logical channels can be used at one time, SF = 128
Reverse link channel	High-Speed Dedicated Physical Control Channel (HS-DPCCH)	A control signalling for packet flow, SF = 256

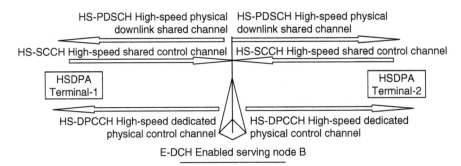

Figure 8.124   *HSDPA physical channels.*

Table 8.33   *HSDPA data rate categories*

Category	Max. number of HS-DSCH codes	Min. TTI ms	Modulation	Peak rate in Mb/s
1	5	3	QPSK/16QAM	1.2
2	5	3	QPSK/16QAM	1.2
3	5	2	QPSK/16QAM	1.8
4	5	2	QPSK/16QAM	1.8
5	5	1	QPSK/16QAM	3.6
6	5	1	QPSK/16QAM	3.6
7	10	1	QPSK/16QAM	7.2
8	10	1	QPSK/16QAM	7.2
9	15	1	QPSK/16QAM	10.2
10	**15**	**1**	**QPSK/16QAM**	**14.4**
11	**5**	**2**	**QPSK only**	**0.9**
12	5	1	QPSK only	1.8

5 bits and converted to suitable scheduler information in the Node B. Table 8.32 presents a summary of HSDPA channels. The signalling between the terminals and the serving Node B on the new physical channels for HSDPA is shown in Figure 8.124.

The 3GPP has defined twelve HSDPA categories, each peak data rate capability as given in Table 8.33.

8.19.2   HSDPA protocol architecture

The transport channels in Release '99 are terminated at the RNC. Consequently, the re-transmission functionality is located in the RLC layer in the serving RNC. Such a high level of transmission is too slow for high-speed data required in HSDPA protocol. Consequently, the RNC in Release '99 provides the scheduling, coding parameters and re-transmission to the terminals. Since these parameters need to be determined for instantaneous channel conditions, the delay incurred by the RNC processing, in addition to the

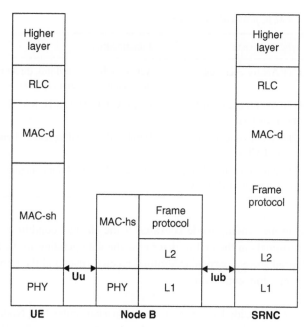

Figure 8.125 *HSDPA protocol architecture.*

fact that individual RNC has to service multiple Node Bs, makes a strong case for modifying Node B to perform these functions not RNC. Furthermore, the new Node B has to perform back compatibility with Release '99 systems with non HSDPA capabilities.

In Release 5, the *Medium Access Control* (MAC-hs) is installed just above the physical layer in Node B. The key functionality of MAC-hs is to handle the HARQ functionality and scheduling as well as transmission priority handling and thus enabling faster re-transmission. The outcome of MAC-hs functionalities is a shorter delay when packet re-transmission is carried out. On the other hand, the scheduler decisions made in the Node B enables the total cell capacity to be allocated to a single user for a very short time if the channel conditions are favourable. HSDPA protocol architecture is shown in Figure 8.125.

The MAC sublayer protocols are in the UE and the UTRAN are given in Table 8.34.

8.19.3   HSDPA algorithm

The HARQ is an acknowledged re-transmission used in HSDPA protocol. The HARQ entity in the terminal checks the correctness of the received packet by Calculating the Checksum (CRC) and comparing the result with the checksum within the received packet. If the two checksums are identical, the terminal sends ACK to the serving Node B. Otherwise, it sends NAK to ask the sending Node B to re-transmit the erroneous packet. At the same time, the terminal has stored the erroneous packet. The *Fast Packet Scheduling* (FPS) carried out

Table 8.34 *HSDPA MAC protocols sublayer*

MAC type	Where it exists	Functionality
MAC-b	UTRAN for each cell	Active in the forward link direct ion to handle the broadcast channel (BCH)
MAC-c/sh	UTRAN for each cell & one in each UE	Deals with PCH, FACH, RACH DSCH
MAC-d	UTRAN has one for each UE for DCHs	Handles dedicated logical and transport channels
MAC-sh	UTRAN for each cell and UE that support HSDPA	Deals with HSDPA functionality.

by the scheduler in the sending Node B evaluates the channel conditions received from different active terminals in the cell to determine the data pending in the buffer of each UE, time has elapsed since a particular terminal received data and the terminals that had requested re-transmission, and so on.

Once the scheduler has decided to serve a terminal in a particular TTI, Node B determines the number of codes, modulation (QPSK or 16-QAM), and type of coding to be used. Node B starts to transmit the HS-SCCH two slots ahead of transmitting the HS-DSCH, to inform the terminal of the necessary parameters to help demodulating the received packet. The terminal monitors the HS-SCCH and decodes part 1 of the received HS-SCCH block and starts to decode the rest of the block. The terminal successfully decodes the HS-DSCH data depending on the results of the CRC check, and the terminal sends the ACK/NAK indicator to Node B.

In case the re-transmitted packet is received in error, the previous packet and the current packet will be combined in an attempt to recover from data errors. The same coding is used in each re-transmission and eventually the packet may recover from the errors in a process called *chase combining*. If the maximum number of re-transmissions is reached and the frame has not recovered from the errors, the erroneous packet recovery is left to the higher layers. The re-transmitted packet can be coded using additional redundant coding information to improve the chance of correcting the packet by combining with previous re-transmitted packet. The scheme is called *Incremental Redundancy* (IR).

8.20 High-speed uplink packet access (3GPP TS25.896, 2004)

The first evolution of the WCDMA, the HSDPA, presented in Section 8.19, offers a theoretical maximum data rate of 14.4 Mb/s on the forward link. However, such evolved WCDMA provision combined the theoretical maximum data rate of 384 kb/s on the reverse link improves the UTRAN system performance compared to the performance of the WCDMA's first release. The evolved system's relative slower data speed on the reverse link inevitably slows the data flow on the forward link particularly when the TCP acknowledgements chock the reverse link causing the forward data flow to slow down. Consequently, what is needed to ensure efficient high-speed wireless system is a high-speed access scheme for the uplink to accompany the data flow on the down link. This section describes the enhanced uplink of UTRA FDD defined in 3GPP Release 6, which improves the performance of the uplink in terms of mobile user experience by reducing delay and increasing system capacity, network coverage area and user throughput. It is worth noting, however, that while service provisions for high-speed users are supported, the optimum performance is only achieved at low to medium speeds. Furthermore, the enhanced uplink achieves significant improvements in system performance when it is compatible with terminals from previous releases such as '99, 4 and 5.

The enhanced down link and uplink can be configured and run simultaneously at a maximum HSDPA and HSUPA rate of 14.4 Mb/s and 5.76 Mb/s, respectively. High data rate on the uplink enables mobile users to send high resolution pictures and video clips while simultaneously communicating by voice. The high-speed data with quality of service offered by the HSUPA technology enables commercial use such as on-line gaming, real time multimedia sharing, *Voice-Over IP* (VoIP), and uploading long files, etc.

8.20.1 UTRAN architecture with HSUPA protocol (3GPP TS25.309, 2006)

The uplink enhancement has distinctive impacts on the UTRAN architecture as demonstrated in Figure 8.126. Therefore, from Figure 8.126(b), the new architecture incorporated a new (enhanced) entity in the MAC sublayer called MAC-e in Node B and shared enhanced MAC entity called MAC-es in the Radio Network Control RNC. Generally, each RNC controls one or more Node Bs and each Node B serves a number of terminals. RNC handles the scheduling and re-transmission on the uplink in the WCDMA first release. Fast

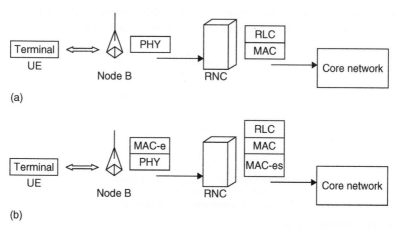

Figure 8.126 *3G Uplink. (a) First release and enhanced Release 6; (b) Uplink protocols.*

scheduling and rapid HARQ ACK/NAK necessitate their location within Node B MAC-e to ensure low delays and fast resource allocation as demanded by HSDPA algorithm. MAC-e has the required functionality to carry out the required soft combining. Furthermore, MAC-es in the RNC supports macro-diversity combining and detects duplicate sequences through activating the re-ordering of the delivered sequences. Additionally, the RLC and MAC entities in WCDMA first release within the RNC provide functions such as mobility, ciphering and backup to the HARQ entity in Node B when the latter fails due to a limited re-transmission attempts.

8.20.2   The HSUPA concept

HSUPA technology adopts some of the techniques used for HSDPA that can suitably be applied for uplink environment. However, the AMC technique is not used in HSUPA since the power capabilities of the terminals are limited and individual terminals may not have enough power for high order modulation to raise transmission data rate. Instead, data rates are raised using multicode transmission with BPSK or QPSK.

HSUPA, however, supports several new features which are not available in previous releases, such as *fast scheduling; fast hybrid ARQ with soft combining* and *short TTI*. The main components of the HSUPA concept are shown in Figure 8.127. Similar schemes have been applied to the HSDPA. However, there are key differences in the motivation for the use of such schemes, and their impacts on the uplink and the down link (Parkvall et al., 2005). The shared resources in the down link are *centralized* by Node B and consist of the transmission power and channelization codes. On the other hand, the shared resource in the uplink is the multiple access interference at Node B which depends on the (decentralized) terminal transmitted power. This difference has implications on the packet transmission scheduling. Fast power control is exercised by the uplink not only in order to limit the interference at Node B but also to handle the Near–Far problem and achieve compatibility with terminals not using the enhancements. Consequently, fast power control is the key to fast link adaptation of the uplink. Compared to the enhancement of the down link, the link (rate) adaptation on the HSDPA requires constant transmission power.

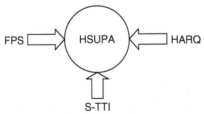

FPS ⇒ HSUPA ⇐ HARQ

S-TTI

HARQ–Hybrid automatic repeat request
FPS–Fast packet scheduling
S-TTI–Short transmission time interval

Figure 8.127   *HSUPA concept.*

8.20.3 HSUPA algorithm

The new *Enhanced Dedicated Channel* (E-DCH) supports the new features in the enhanced uplink as mentioned earlier. A transport block of data is transmitted during each TTI but data rates may change in successive TTIs and, consequently, the corresponding data block size may also change. E-DCH supports a TTI of 2 ms in addition to the 10 ms which is in line with TTI in HSDPA. Clearly, a shorter TTI implies a shorter delay caused by waiting to start data transmission. Also, a shorter processing time is required at both transmitter and receiver due to a smaller amount of data. In addition, a shorter TTI means a shorter end-to-end round trip delay.

An extra set of channelization codes is used in the E-DCH in addition to the codes used in the first release to ensure full backward compatibility with previous release. These new codes are invisible to non E-DCH capable Node B. Each E-DCH is carrying multiple flows of different priorities which are multiplexed into a single transport block per TTI. Each flow has an associated HARQ profile indicating the *required delay and bit rate*. Each HARQ profile is associated with a power offset required to boost the E-DCH transmission power to provide the delay-date rate requirement for the flow. Furthermore, baseband E-DCH data processing is carried out independently of other channels.

The HARQ with soft combining used in uplink is similar to that used for HSDPA. For each data block received by Node B in the uplink, the Node B sends a single bit on the down link to the terminal to indicate a successful decoding (ACK) or erroneous decoding (NAK). When the data block is received by a number of Node Bs, as in soft handoff, a single ACK from a single Node B indicates (to the terminal) a successful decoding.

Incremental Redundancy (IR) is used for soft combining at Node B. The received data is first decoded and stored. If the decoding fails, an ARQ is sent by the serving Node B for re-transmission and the information is transmitted with incremental parity bits. The decoded re-transmission is combined with previously transmitted block to yield a lower bit rate and eventually facilitates the correct decoding. HARQ with soft combining provides protection against interference and improves link efficiency. To explain how this can be achieved, consider the transmission of $'x'$ *Mb/s* using $'y'$ *mW* transmission power to target certain *BER*. It has been established that transmitting given information at data rate nx *Mb/s* using transmission power y *mW* achieves correct decoding with less than n retransmission using IR with soft combining. Such an approach results in $\frac{E_b}{N_0}$ gain, which can be used to improve the service coverage and/or increase the network capacity.

The HARQ in both HSDPA and HSUPA consists of multiple *stop and wait* processes operating in parallel. The HARQ process in any given TTI is identified by the frame number that is known to the terminal and Node B. Consequently, the re-transmission timing is also known to the Node B and can be used to improve data decoding. In addition, knowledge of the re-transmission time is essential for fast scheduling the transmission of other users.

For each re-transmission, the *Re-transmission Sequence Number* (RSN) is incremented starting with 0 for the initial transmission attempt. Node B clears its soft buffer whenever the RSN is set to 0. As a consequence of the HARQ *stop and wait* operation, data blocks may arrive at Node B out of sequence. However, RLC protocol in RNC is unable to process out of sequence data, and so the re-ordering is carried out in the new MAC sublayer MAC-es, below the RLC in the RNC, using the sequence numbers inserted by the MAC-es header by the terminal. Thus, the MAC-es in the RNC handles the duplicates removal.

As we pointed out above, the common resource shared among terminals in the uplink is received power at Node B to limit interference. The amount of power a terminal is transmitting depends on the data rate used. Clearly, the higher the data rate, the larger the required transmission power. On the other hand, the scheduler function is to authorize when and at what rate a terminal may transmit. Generally, the scheduler allocates large transmission power to terminals requiring high data rates and ensures manageable interference level. Fast scheduling enables a larger number of high rate terminals to access the network transmitting in parallel due to the significantly smaller transmit terminal power compared to shared resource of HSDPA due to much larger Node B transmit power. Clearly, different environments and traffic types require different scheduling strategy.

The scheduling algorithm is based on *scheduling grants* sent by Node B scheduler to the terminal as a reply to the terminal *scheduling request*. Two types of grants can be used: *absolute grants* and *relative grants*. The scheduling grants are used to set an upper limit on the data rate the terminal can use. The absolute grants are used to set an absolute upper limit of the terminal power which can be used for transmission; this basically determines the maximum data rate. Relative grants are used to update the resource allocation for a terminal when its transmission environment allows. A relative grant can take one of three values: 'up', 'down' or 'hold', indicating to the terminal to increase, decrease or not change the power level relative to the currently used level. Thus, the relative grants resemble the operation of a power control scheme.

The serving Node B controls the scheduling operation in the cell and it is the only one authorized to issue absolute grants. Other non-serving Node Bs influence the terminal data rates through issuing relative grants to reduce the data rate through values 'down', or keep the rate unchanged using the value 'hold'. A 'down' received from a non-serving Node B indicates that the cell is overloaded and the terminal must therefore reduce its current data rate. However, the relative grant from the serving Node B prevails over a 'hold' received from non-serving Node Bs. Clearly a relative grant from a non-serving Node B serves as an 'overload indicator'.

A simple scheduler may operate without relative grants. The absolute grant is received by all terminals in the cell to fix the maximum transmission power and therefore the maximum data rate. Terminals will gradually increase their transmission power toward the maximum. However, such a scheduling scheme is not effective in controlling the interference.

Terminals place their scheduling request on the uplink indicating to Node B their *buffer status, traffic priority and power resources*. This information is sent on the E-DCH and is used by the serving Node B in the scheduling decision. Prioritized flows have guaranteed data rates by the Node B scheduler. Furthermore, terminals are allowed to transmit data up to a certain rate in non-schedule transmission without requesting for a grant from the scheduler.

8.20.4   HSUPA physical channels

The new physical channels for HSUPA compromise the *E-DCH Dedicated Physical Data Channel* (E-DPDCH) which carries user data at rate up to 5.76 Mb/s. The data transmitted over the E-DPDCH is encoded and spread using SF 256, 128, 64, 32, 16, 8, 4 and 2 and modulated on a BPSK or QPSK carrier. A rate $\frac{1}{3}$ turbo channel coding is used for channel error correction. The E-DPDCH is time aligned with the uplink DPCCH.

Signalling on the uplink is carried over the uplink Dedicated Physical Control Channel (*E-DPCCH*) to transport the scheduling requests transmitted by the terminals carrying the transport block size, the re-transmission sequence number, and data rate required to decode the E-DCH contents. This information is transmitted in parallel to the data transmission on the E-DCH. The E-DCH dedicated physical control channel carries control information such as re-transmission sequence number (2 bits RSN) and 7 bits TFCI. The control information is spread using SF 256. The channel coding used for E-DPCCH is the Reed-Muller code.

Signalling on the down link consists of absolute grants, relative grants, and HARQ ACK/NAKs. The absolute grants are transmitted on the share channel and the intended user identity is attached to each grant. The downlink *E-DCH Absolute Grant Channel* (E-AGCH) is offset by 5120 chips relative to P-CCPCH and carries the absolute grants for the uplink E-DCH scheduling in 6 bits. The information on E-AGCH is spread using an SF of 256 and modulated on QPSK carrier and a channel coding of rate $\frac{1}{3}$ convolutional codes is used. The downlink E-DCH Hybrid ARQ Indicator Channel (E-HICH) carries the ARQ ACK/NAK indicator on QPSK carrier using a spreading factor of 128. The acknowledgements are transmitted with signature sequences.

The downlink E-DCH Relative Grant Channel (E-RGCH) carries relative grants for uplink E-DCH scheduling on QPSK carrier using a SF of 128 and relative grants are transmitted with signature sequences. The transmission of the relative grants and the ACK/NAKs spread over the whole TTI. Multiple terminals share a channelization code and implement user-specific orthogonal signature sequences. The HSUPA physical channels on the uplink and down link are illustrated in Figure 8.128.

8.20.5   E-DCH spreading and modulation (3GPP TS25.213, 2006)

The uplink E-DCH is an enhanced dedicated channel committed for each user in the uplink. This is different from the HSDPA where channels are shared between users sharing the

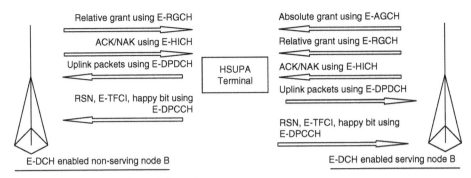

Figure 8.128   *HSUPA physical channels.*

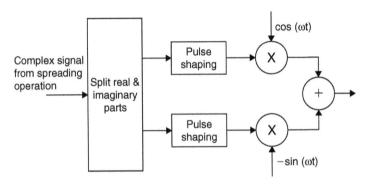

Figure 8.129   *Uplink carrier modulation.*

down link. The uplink modulation is shown in Figure 8.129. The transmit pulse shaping filter is a Root-Raised Cosine (RRC) with roll-off $\alpha = 0.22$ in the frequency domain (3GPP TS25.101, 2006).

As we explained in the previous sections, spreading in the physical channels consists of two operations: the channelization which transforms every data symbol into a number of chips using the SF thereby increasing the signal bandwidth; and the scrambling where a complex scramble code is applied to the complex spread signal.

The spreading operation for the E-DPDCH and E-DPCCH is illustrated in Figure 8.130 which is carried out at a fixed chip rate of 3.84 Mc/s by the channelization codes C_{ec} and $C_{ed,k}$. The channelization operation is followed by weighing of E-DCCH and E-DPDCH signals using gain factors β_{ec} and $\beta_{ed,k}$. E-DPCCH carries real signals and, therefore, is mapped to I-branch.

In the FDD UTRAN system, the terminal can simultaneously configure the following maximum number of uplink dedicated data channels: six DPDCH or one DPDCH plus

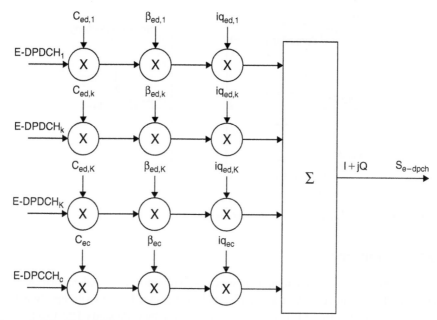

Figure 8.130 *Uplink E-DCH spreading system.*

Table 8.35 *Maximum E-DPDCH channels in uplink*

Number of E-DCH code transmitted	E-DPDCH$_k$	iq$_{ed,k}$
4	E-DPDCH$_1$	1
	E-DPDCH$_2$	J
	E-DPDCH$_3$	1
	E-DPDCH$_4$	J
2	E-DPDCH$_1$	j or 1
	E-DPDCH$_2$	1 or j

two E-DPDCH or four E-DPDCH. Each dedicated data channel requires one dedicated control channel (i.e. HS-DPCCH or E-DPCCH) as shown in Table 8.35.

The values for iq$_{ed,k}$ depend on the maximum E-DPDCH channels as given in Table 8.35.

The Spreading for uplink dedicated channels is shown in Figure 8.131. The scrambling code is used to align the radio frames, i.e. the first scrambling chip corresponds to the beginning of the frame.

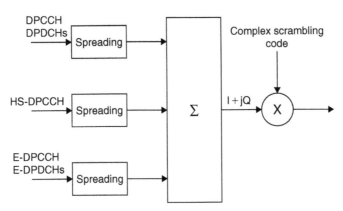

Figure 8.131 *Spreading for uplink dedicated channels.*

8.21 Summary

This chapter introduced the contemporary wireless communication standards that are based on the CDMA technology. These standards can be categorized into two groups. Standards are issued by the 3G Project Partnership which supports the wideband CDMA technology for the 3G systems that supersede the GSM systems. The other category was issued by the 3GPP2 as are an evolution of the IS-95A standard led by Qualcom in North America in the 1990s.

The chapter started with a consideration of the IS-95A in Sections 8.2–8.6. In Section 8.2 we introduced the IS-95A Standard in detail and presented the forward channels and the reverse channels in Sections 8.3–8.4. The mobility issues in the IS-95 system were considered in Section 8.5 considering the procedures for the location update and the handoff.

The evolution of the IS-95A standard to the IS-95B was considered in Section 8.6, explaining the issues that support the high-speed data and the trade-off between the capacity and the coverage in IS-95B system.

A comprehensive presentation of the cdma2000 standards was given in Section 8.7. The physical layer channels were presented in Sections 8.7.1–8.7.14 to cover the forward and reverse channels operating in the physical layer together with the processing carried out in each channel in the forward and reverse links.

The evolution of the cdma2000 was considered in Section 8.7.15 which included revisions 0, A, B, C and D. These revisions provide high-speed data on both the forward and reverse traffic channels and integrate the voice and data to provide high-speed applications such as video streaming, interactive gaming, VoIP and video telephony.

The first release of WCDMA standard were presented in Sections 8.8–8.16. The system architecture was considered in Section 8.8 and the physical layer covered in Sections

8.9–8.16 including the forward and reverse channels, the rate matching, and the carrier modulation. The mobility procedures in the UMTS were presented in Section 8.17.

The evolution of the WCDMA standard was given in Sections 8.18–8.20. The high-speed transmission over the down link (HSDPA) was considered in Section 8.19. The concept supporting HSDPA was explained in Section 8.19.0.1, followed by a description of the HSDPA physical channels and the signalling between Node B and HSDPA enabled terminals given in Sections 8.19.2. Details of the HSDPA algorithm and its protocol architecture were given in Section 8.19.3.

High-speed transmission over the uplink (reverse link) was presented in Section 8.20. The architecture of the HSUPA was given in Section 8.20.1 while HSUPA concept and its algorithm were presented in Sections 8.20.2 and 8.20.3. The physical channels used in HSUPA and the signalling over these channels between Node B and the HSUPA enabled terminals were considered in Section 8.20.4. Finally, spreading and modulation of information used in HSUPA was given in Section 8.20.5.

8.22 List of Standards

EIA/TIA Interim Standard 95 (IS-95): 'Mobile station–base station compatibility standard for dual-mode wideband spread spectrum cellular system', Telecommunication Industry Association, July 1993 (amended as IS-95A in May 1995).

EIA/TIA Interim Standard 95B (IS-95B): 'Mobile station–base station compatibility standard for dual-mode wideband spread spectrum systems', 1997.

EIA/TIA Interim standard 2000 (IS-2000): 'cdma2000 Standard for spread spectrum systems', July 1999.

8.23 3G Partnership Project Specifications

3GPP TS 25.215, V 3.11.0, Physical layer – measurements (FDD) (Release 5), March 2002.

33GPP TS 25.212, V 5.0.0, 'Multiplexing and channel coding' (Release 5), March 2002.

3GPP TR 25.858 V 5.0.0, 'High speed downlink packet access: Physical layer aspects' (Release 5), March 2002.

3 GPP TS 25.213 V 3.8.0, 'Spreading and modulation' (Release 1999), June 2002.

3GPP TS 25.215, V 4.1.0, 'UTRAN overall description' (Release 1999), June 2002.

3GPP TS 25.301, V 5.0.0, Radio interface architecture (Release 1999), June 2002.

3GPP TS 25.214, V 3.11.0, 'Physical layer procedure' (FDD), (Release 1999), Sept 2002.

GPP TS 25.211, V 3.12.0, Physical channels and mapping of transport channels onto physical channels (FDD), (Release 1999), Sept 2002.

3GPP2 C.S0024, 'cdma2000 High rate packet data air interface specification', Release 0, Version 4.0, 25th Oct., 2002.

3GPP2 C.S0002-D, 'Physical Layer Standard for cdma2000 Spread Spectrum Systems', Revision D, Version 1.0, February 13, 2004.

3GPP TS 25.896, V 5.0.0, Feasibility study for enhanced uplink for UTRAN FDD (Release 6), March 2004.

3GPP2 C.S0024-A, 'cdma2000 High rate packet data air interface specification', Revision A, Version 1.0, March 2004.

3GPP2 C.S0002-C, 'Physical Layer Standard for cdma2000 Spread Spectrum Systems', Revision C, Version 2.0, July 23, 2004.

3GPP TS 25.309 V 6.1.0 'FDD enhanced uplink overall description' (Release 6), March 2006.

3GPP TS 25.213 V 6.5.0 'Spreading and modulation (FDD)' (Release 6), March 2006.

3GPP2 C.S0024-B, 'cdma2000 High rate packet data air interface specification', Revision B, Version 1.0, May 2006.

3GPP TS 25.101 V 7.4.0 'User equipment (UE) radio transmission and reception (FDD)' (Release 7), June 2006.

References

Bhushan, N., Lott, C., Black, P., Attar, R., Jou, Y.-C., Ghosh, D. and Au, J. (2006) cdma2000 1XEV-DO Revision A: A physical layer and MASC layer overview, *IEEE Communication Magazine*, pp. 75–87, Feb.

Bi, Q. (2005) *A forward link performance study of the 1xEV – DO Rev. 0 system using field measurements and simulations*, Bell laboratories, Lucent Technologies.

C.-L., I. and Nanda, S. (1996) Load and interference based demand assignment for wireless networks, *Proc. IEEE GlobeCom*.

Derryberry, R.T., Hsu, A. and Tamminen, W. *Overview of cdma2000 Revision D*, Nokia research centre. Nokia Research Centre. Taxes USA: www.cdg.org/resources/white_papers/files/Overview_of_cdma 2000_Revision_D.pdf

Ejzak, R.P., Knisely, D.N., Kumar, S., Laha, S. and Nanda, S. (1997) BALI: A solution for high-speed CDMA data, *Bell Labs Technical Journal*, 2(3), 124–151.

Holma, H. and Toskala, A. (2005) *WCDMA for UMTS*, John Wiley & Sons ISBN 0-470-87096-6 (CB), Chapter 11.

Kaaranen, H., Ahtiainen, A., Laitnen, L., Naghian, S. and Niemi, V. (2005) *UMTS Network Architecture, Mobility and Services*, John Wiley & Sons. ISBN 0-470-01103-3, Chapter 4.

Kinsely, D.N., Kumar, S., Laha, S. and Nanda, S. (1998) Evolution of wireless data services: IS-95 to cdma2000, *IEEE Communication Magazine*, 140–149, Oct.

Kumar, S. and Nanda, S. (1999) High data rate packet communications for cellular networks using CDMA: Algorithms and performance, *IEEE Journal on Selected Areas in Communications*, 17(3), 472–492.

Lee, J.S. and Miller, L.E. (1998) *CDMA Systems Engineering Handbook*, Artech House, ISBN 0-89006-990-5, pp. 379, 380.

Parkvall, S., Peisa, J., Torsner, J., Sagfors, M. and Malm, P. (2005) WMCDA enhanced uplink: Principles and basic operation, *Proceedings of the IEEE Vehicular Technology Conference VTC 2005*, Spring, Vol. 3, pp. 1411–1415, 30 May–1 June.

Soong, A.C.K., Oh, S.-J., Damnjanovic, A.D. and Yoon, Y.C. (2003) Forward high-speed wireless packet data service in IS-2000 – 1×EV-DV, *IEEE Communication Magazine*, pp. 170–177, Aug.

Yavus, M., Diaz, S., Kapoor, R., Grob, M., Black, P., Tokgoz, Y. and Lott, C. (2006) VoIP over cdma2000 1xEV-DO Rev. A, *IEEE Communications Magazine*, pp. 88–95, Feb.

Appendix 8.A

Hadamard Matrix 64 × 64 bits

1.
2.
3.
4.
5.
6.
7.
8.
9.
10.
11.
12.
13.
14.
15.
16.
17.
18.
19.
20.
21.
22.
23.
24.
25.
26.
27.
28.
29.
30.

31.	0011110011000011	1100001100111100	0011110011000011	1100001100111100
32.	0110100110010110	1001011001101001	0110100110010110	1001011001101001
33.	0000000000000000	0000000000000000	1111111111111111	1111111111111111
34.	0101010101010101	0101010101010101	1010101010101010	1010101010101010
35.	0011001100110011	0011001100110011	1100110011001100	1100110011001100
36.	0110011001100110	0110011001100110	1001100110011001	1001100110011001
37.	0000111100001111	0000111100001111	1111000011110000	1111000011110000
38.	0101101001011010	0101101001011010	1010010110100101	1010010110100101
39.	0011110000111100	0011110000111100	1100001111000011	1100001111000011
40.	0110100101101001	0110100101101001	1001011010010110	1001011010010110
41.	0000000011111111	0000000011111111	1111111100000000	1111111100000000
42.	0101010110101010	0101010110101010	1010101001010101	1010101001010101
43.	0011001111001100	0011001111001100	1100110000110011	1100110000110011
44.	0110011010011001	0110011010011001	1001100101100110	1001100101100110
45.	0000111111110000	0000111111110000	1111000000001111	1111000000001111
46.	0101101010100101	0101101010100101	1010010101011010	1010010101011010
47.	0011110011000011	0011110011000011	1100001100111100	1100001100111100
48.	0110100110010110	0110100110010110	1001011001101001	1001011001101001
49.	0000000000000000	1111111111111111	1111111111111111	0000000000000000
50.	0101010101010101	1010101010101010	1010101010101010	0101010101010101
51.	0011001100110011	1100110011001100	1100110011001100	0011001100110011
52.	0110011001100110	1001100110011001	1001100110011001	0110011001100110
53.	0000111100001111	1111000011110000	1111000011110000	0000111100001111
54.	0101101001011010	1010010110100101	1010010110100101	0101101001011010
55.	0011110000111100	1100001111000011	1100001111000011	0011110000111100
56.	0110100101101001	1001011010010110	1001011010010110	0110100101101001
57.	0000000011111111	1111111100000000	1111111100000000	0000000011111111
58.	0101010110101010	1010101001010101	1010101001010101	0101010110101010
59.	0011001111001100	1100110000110011	1100110000110011	0011001111001100
60.	0110011010011001	1001100101100110	1001100101100110	0110011010011001
61.	0000111111110000	1111000000001111	1111000000001111	0000111111110000
62.	0101101010100101	1010010101011010	1010010101011010	0101101010100101
63.	0011110011000011	1100001100111100	1100001100111100	0011110011000011
64.	0110100110010110	1001011001101001	1001011001101001	0110100110010110

Appendix 8.B

Reverse traffic channel/access channel input symbols at rate 9600 bps

1	33	65	97	129	161	193	225	257	289	321	353	385	417	449	481	513	545
2	34	66	98	130	162	194	226	258	290	322	354	386	418	450	482	514	546
3	35	67	99	131	163	195	227	259	291	323	355	387	419	451	483	515	547
4	36	68	100	132	164	196	228	260	292	324	356	388	420	452	484	516	548
5	37	69	101	133	165	197	229	261	293	325	357	389	421	453	485	517	549
6	38	70	102	134	166	198	230	262	294	326	358	390	422	454	486	518	550
7	39	71	103	135	167	199	231	263	295	327	359	391	423	455	487	519	551
8	40	72	104	136	168	200	232	264	296	328	360	392	424	456	488	520	552
9	41	73	105	137	169	201	233	265	297	329	361	393	425	457	489	521	553
10	42	74	106	138	170	202	234	266	298	330	362	394	426	458	490	522	554
11	43	75	107	139	171	203	235	267	299	331	363	395	427	459	491	523	555
12	44	76	108	140	172	204	236	268	300	332	364	396	428	460	492	524	556
13	45	77	109	141	173	205	237	269	301	333	365	397	429	461	493	525	557
14	46	78	110	142	174	206	238	270	302	334	366	398	430	462	494	526	558
15	47	79	111	143	175	207	239	271	303	335	367	399	431	463	495	527	559
16	48	80	112	144	176	208	240	272	304	336	368	400	432	464	496	528	560
17	49	81	113	145	177	209	241	273	305	337	369	401	433	465	497	529	561
18	50	82	114	146	178	210	242	274	306	338	370	402	434	466	498	530	562
19	51	83	115	147	179	211	243	275	307	339	371	403	435	467	499	531	563
20	52	84	116	148	180	212	244	276	308	340	372	404	436	468	500	532	564
21	53	85	117	149	181	213	245	277	309	341	373	405	437	469	501	533	565
22	54	86	118	150	182	214	246	278	310	342	374	406	438	470	502	534	566
23	55	87	119	151	183	215	247	279	311	343	375	407	439	471	503	535	567
24	56	88	120	152	184	216	248	280	312	344	376	408	440	472	504	536	568
25	57	89	121	153	185	217	249	281	313	345	377	409	441	473	505	537	569
26	58	90	122	154	186	218	250	282	314	346	378	410	442	474	506	538	570
27	59	91	123	155	187	219	251	283	315	347	379	411	443	475	507	539	571
28	60	92	124	156	188	220	252	284	316	348	380	412	444	476	508	540	572
29	61	93	125	157	189	221	253	285	317	349	381	413	445	477	509	541	573
30	62	94	126	158	190	222	254	286	318	350	382	414	446	478	510	542	574
31	63	95	127	159	191	223	255	287	319	351	383	415	447	479	511	543	575
32	64	96	128	160	192	224	256	288	320	352	384	416	448	480	512	544	576

Reverse traffic channel/access channel input symbols at rate 4800 bps

1	17	33	49	65	81	97	113	129	145	161	177	193	209	225	241	257	273
1	17	33	49	65	81	97	113	129	145	161	177	193	209	225	241	257	273
2	18	34	50	66	82	98	114	130	146	162	178	194	210	226	242	258	274
2	18	34	50	66	82	98	114	130	146	162	178	194	210	226	242	258	274
3	19	35	51	67	83	99	115	131	147	163	179	195	211	227	243	259	275
3	19	35	51	67	83	99	115	131	147	163	179	195	211	227	243	259	275
4	20	36	52	68	84	100	116	132	148	164	180	196	212	228	244	260	276
4	20	36	52	68	84	100	116	132	148	164	180	196	212	228	244	260	276
5	21	37	53	69	85	101	117	133	149	165	181	197	213	229	245	261	277
5	21	37	53	69	85	101	117	133	149	165	181	197	213	229	245	261	277

6	22	38	54	70	86	102	118	134	150	166	182	198	214	230	246	262	278
6	22	38	54	70	86	102	118	134	150	166	182	198	214	230	246	262	278
7	23	39	55	71	87	103	119	135	151	167	183	199	215	231	247	263	279
7	23	39	55	71	87	103	119	135	151	167	183	199	215	231	247	263	279
8	24	40	56	72	88	104	120	136	152	168	184	200	216	232	248	264	280
8	24	40	56	72	88	104	120	136	152	168	184	200	216	232	248	264	280
9	25	41	57	73	89	105	121	137	153	169	185	201	217	233	249	265	281
9	25	41	57	73	89	105	121	137	153	169	185	201	217	233	249	265	281
10	26	42	58	74	90	106	122	138	154	170	186	202	218	234	250	266	282
10	26	42	58	74	90	106	122	138	154	170	186	202	218	234	250	266	282
11	27	43	59	75	91	107	123	139	155	171	187	203	219	235	251	267	283
11	27	43	59	75	91	107	123	139	155	171	187	203	219	235	251	267	283
12	28	44	60	76	92	108	124	140	156	172	188	204	220	236	252	268	284
12	28	44	60	76	92	108	124	140	156	172	188	204	220	236	252	268	284
13	29	45	61	77	93	109	125	141	157	173	189	205	221	237	253	269	285
13	29	45	61	77	93	109	125	141	157	173	189	205	221	237	253	269	285
14	30	46	62	78	94	110	126	142	158	174	190	206	222	238	254	270	286
14	30	46	62	78	94	110	126	142	158	174	190	206	222	238	254	270	286
15	31	47	63	79	95	111	127	143	159	175	191	207	223	239	255	271	287
15	31	47	63	79	95	111	127	143	159	175	191	207	223	239	255	271	287
16	32	48	64	80	96	112	128	144	160	176	192	208	224	240	256	272	288
16	32	48	64	80	96	112	128	144	160	176	192	208	224	240	256	272	288

Reverse traffic channel input symbols at rate 2400 bps

1	9	17	25	33	41	49	57	65	73	81	89	97	105	113	121	129	137
1	9	17	25	33	41	49	57	65	73	81	89	97	105	113	121	129	137
1	9	17	25	33	41	49	57	65	73	81	89	97	105	113	121	129	137
1	9	17	25	33	41	49	57	65	73	81	89	97	105	113	121	129	137
2	10	18	26	34	42	50	58	66	74	82	90	98	106	114	122	130	138
2	10	18	26	34	42	50	58	66	74	82	90	98	106	114	122	130	138
2	10	18	26	34	42	50	58	66	74	82	90	98	106	114	122	130	138
2	10	18	26	34	42	50	58	66	74	82	90	98	106	114	122	130	138
3	11	19	27	35	43	51	59	67	75	83	91	99	107	115	123	131	139
3	11	19	27	35	43	51	59	67	75	83	91	99	107	115	123	131	139
3	11	19	27	35	43	51	59	67	75	83	91	99	107	115	123	131	139
3	11	19	27	35	43	51	59	67	75	83	91	99	107	115	123	131	139
4	12	20	28	36	44	52	60	68	76	84	92	100	108	116	124	132	140
4	12	20	28	36	44	52	60	68	76	84	92	100	108	116	124	132	140
4	12	20	28	36	44	52	60	68	76	84	92	100	108	116	124	132	140
4	12	20	28	36	44	52	60	68	76	84	92	100	108	116	124	132	140
5	13	21	29	37	45	53	61	69	77	85	93	101	109	117	125	133	141
5	13	21	29	37	45	53	61	69	77	85	93	101	109	117	125	133	141
5	13	21	29	37	45	53	61	69	77	85	93	101	109	117	125	133	141
5	13	21	29	37	45	53	61	69	77	85	93	101	109	117	125	133	141
6	14	22	30	38	46	54	62	70	78	86	94	102	110	118	126	134	142
6	14	22	30	38	46	54	62	70	78	86	94	102	110	118	126	134	142
6	14	22	30	38	46	54	62	70	78	86	94	102	110	118	126	134	142
6	14	22	30	38	46	54	62	70	78	86	94	102	110	118	126	134	142

7	15	23	31	39	47	55	63	71	79	87	95	103	111	119	127	135	143
7	15	23	31	39	47	55	63	71	79	87	95	103	111	119	127	135	143
7	15	23	31	39	47	55	63	71	79	87	95	103	111	119	127	135	143
7	15	23	31	39	47	55	63	71	79	87	95	103	111	119	127	135	143
8	16	24	32	40	48	56	64	72	80	88	96	104	112	120	128	136	144
8	16	24	32	40	48	56	64	72	80	88	96	104	112	120	128	136	144
8	16	24	32	40	48	56	64	72	80	88	96	104	112	120	128	136	144
8	16	24	32	40	48	56	64	72	80	88	96	104	112	120	128	136	144

Reverse traffic channel input symbols at rate 1200 bps

1	5	9	13	17	21	25	29	33	37	41	45	49	53	57	61	65	69
1	5	9	13	17	21	25	29	33	37	41	45	49	53	57	61	65	69
1	5	9	13	17	21	25	29	33	37	41	45	49	53	57	61	65	69
1	5	9	13	17	21	25	29	33	37	41	45	49	53	57	61	65	69
1	5	9	13	17	21	25	29	33	37	41	45	49	53	57	61	65	69
1	5	9	13	17	21	25	29	33	37	41	45	49	53	57	61	65	69
1	5	9	13	17	21	25	29	33	37	41	45	49	53	57	61	65	69
1	5	9	13	17	21	25	29	33	37	41	45	49	53	57	61	65	69
2	6	10	14	18	22	26	30	34	38	42	46	50	54	58	62	66	70
2	6	10	14	18	22	26	30	34	38	42	46	50	54	58	62	66	70
2	6	10	14	18	22	26	30	34	38	42	46	50	54	58	62	66	70
2	6	10	14	18	22	26	30	34	38	42	46	50	54	58	62	66	70
2	6	10	14	18	22	26	30	34	38	42	46	50	54	58	62	66	70
2	6	10	14	18	22	26	30	34	38	42	46	50	54	58	62	66	70
2	6	10	14	18	22	26	30	34	38	42	46	50	54	58	62	66	70
2	6	10	14	18	22	26	30	34	38	42	46	50	54	58	62	66	70
3	7	11	15	19	23	27	31	35	39	43	47	51	55	59	63	67	71
3	7	11	15	19	23	27	31	35	39	43	47	51	55	59	63	67	71
3	7	11	15	19	23	27	31	35	39	43	47	51	55	59	63	67	71
3	7	11	15	19	23	27	31	35	39	43	47	51	55	59	63	67	71
3	7	11	15	19	23	27	31	35	39	43	47	51	55	59	63	67	71
3	7	11	15	19	23	27	31	35	39	43	47	51	55	59	63	67	71
3	7	11	15	19	23	27	31	35	39	43	47	51	55	59	63	67	71
3	7	11	15	19	23	27	31	35	39	43	47	51	55	59	63	67	71
4	8	12	16	20	24	28	32	36	40	44	48	52	56	60	64	68	72
4	8	12	16	20	24	28	32	36	40	44	48	52	56	60	64	68	72
4	8	12	16	20	24	28	32	36	40	44	48	52	56	60	64	68	72
4	8	12	16	20	24	28	32	36	40	44	48	52	56	60	64	68	72
4	8	12	16	20	24	28	32	36	40	44	48	52	56	60	64	68	72
4	8	12	16	20	24	28	32	36	40	44	48	52	56	60	64	68	72
4	8	12	16	20	24	28	32	36	40	44	48	52	56	60	64	68	72
4	8	12	16	20	24	28	32	36	40	44	48	52	56	60	64	68	72

Index

Printed and bound by CPI Group (UK) Ltd, Croydon, CR0 4YY

03/10/2024

01040335-0006